ADVANCED STRUCTURAL MATERIALS

PROPERTIES, DESIGN OPTIMIZATION, AND APPLICATIONS

MATERIALS ENGINEERING

1. Modern Ceramic Engineering: Properties, Processing, and Use in Design. Second Edition, Revised and Expanded, *David W. Richerson*
2. Introduction to Engineering Materials: Behavior, Properties, and Selection, *G. T. Murray*
3. Rapidly Solidified Alloys: Processes • Structures • Applications, *edited by Howard H. Liebermann*
4. Fiber and Whisker Reinforced Ceramics for Structural Applications, *David Belitskus*
5. Thermal Analysis of Ceramics, *Robert F. Speyer*
6. Friction and Wear of Ceramics, *edited by Said Jahanmir*
7. Mechanical Properties of Metallic Composites, *edited by Shojiro Ochiai*
8. Chemical Processing of Ceramics, *edited by Burtrand I. Lee and Edward J. A. Pope*
9. Handbook of Advanced Materials Testing, *edited by Nicholas P. Cheremisinoff and Paul N. Cheremisinoff*
10. Ceramic Processing and Sintering, *M. N. Rahaman*
11. Composites Engineering Handbook, *edited by P. K. Mallick*
12. Porosity of Ceramics, *Roy W. Rice*
13. Intermetallic and Ceramic Coatings, *edited by Narendra B. Dahotre and T. S. Sudarshan*
14. Adhesion Promotion Techniques: Technological Applications, *edited by K. L. Mittal and A. Pizzi*
15. Impurities in Engineering Materials: Impact, Reliability, and Control, *edited by Clyde L. Briant*
16. Ferroelectric Devices, *Kenji Uchino*
17. Mechanical Properties of Ceramics and Composites: Grain and Particle Effects, *Roy W. Rice*
18. Solid Lubrication Fundamentals and Applications, *Kazuhisa Miyoshi*
19. Modeling for Casting and Solidification Processing, *edited by Kuang-O (Oscar) Yu*
20. Ceramic Fabrication Technology, *Roy W. Rice*
21. Coatings of Polymers and Plastics, *edited by Rose A. Ryntz and Philip V. Yaneff*
22. MicroMechatronics, *edited by Kenji Uchino and Jayne Giniewicz*
23. Ceramic Processing and Sintering, Second Edition, *edited by M. N. Rahaman*
24. Handbook of Metallurgical Process Design, *edited by George Totten*
25. Ceramic Materials for Electronics, Third Edition, *Relva Buchanan*
26. Physical Metallurgy, *William F. Hosford*
27. Carbon Fibers and Their Composites, *Peter Morgan*
28. Chemical Processing of Ceramics: Second Edition, *Burtrand Lee and Sridhar Komarneni*
29. Modern Ceramic Engineering, Third Edition, *David Richerson*
30. Engineering Design with Polymers and Composites, *James C. Gerdeen, Harold W. Lord and Ronald A. L. Rorrer*
31. Self-Organization During Friction: Advanced Surface-Engineered Materials and Systems Design, *edited by German Fox-Rabinovich, and George E. Totten*
32. Advanced Structural Materials: Properties, Design Optimization, and Applications, *edited by Winston O. Soboyejo with T. S. Srivatsan*

ADVANCED STRUCTURAL MATERIALS

PROPERTIES, DESIGN OPTIMIZATION, AND APPLICATIONS

EDITED BY

WINSTON O. SOBOYEJO
WITH T. S. SRIVATSAN

CRC Press
Taylor & Francis Group
Boca Raton London New York

CRC Press is an imprint of the
Taylor & Francis Group, an **informa** business

CRC Press
Taylor & Francis Group
6000 Broken Sound Parkway NW, Suite 300
Boca Raton, FL 33487-2742

First issued in paperback 2019

ISBN-13: 978-1-57444-634-0 (hbk)
ISBN-13: 978-0-367-38960-4 (pbk)

Library of Congress Cataloging-in-Publication Data

Advanced structural materials : properties, design optimization, and applications / edited by Wolé Soboyejo.
    p. cm.
   ISBN 1-57444-634-7
   1. Metals. 2. Materials. I. Soboyejo, W. O.

TA459.A38 2006
620.1'1--dc22   2006044676

Preface

In recent years, the concept of advanced structural materials has changed from advanced composites and intermetallics to microelectromechanical systems (MEMS), cellular materials, biomaterials, shape memory alloys, amorphous alloys, and nanostructured materials. Many of the intermetallic and composite systems that appeared promising just a decade ago are no longer considered by many to be serious candidates for near-term applications in the next decade. Within this context, a number of existing structural materials, such as titanium and cobalt alloys, have also embraced advanced applications in ways that enhance their status as advanced structural alloys/systems.

This book is written for both the non-specialist and the specialist. It is written for those who want to develop an understanding of the breadth and depth of current advanced structural materials. Although each chapter has been written by an expert in the field, no prior knowledge of a given material system is assumed. Each chapter, therefore, presents the fundamental concepts (structure and properties of materials) and the applications of advanced structural materials.

Due to the huge nature of the field, we have been forced to define advanced structural materials as cutting-edge systems that are currently in structural use, or future systems that appear to have the promise for near-term structural applications. Hence, we do not include chapters on advanced intermetallics and ceramics, which remain as long-term contenders for future large-scale structural applications. Nevertheless, we hope that the rich array of selected topics will provide readers with useful insights into the structure, properties, and applications of some of the systems that are currently considered advanced structural materials.

The book is divided into four sections. In section I, a broad introduction to advanced structural materials is presented. This is followed by section II, in which materials at the frontiers of emerging applications are presented. These include some aspects of biomaterials, MEMS, amorphous materials, and nanotechnology. In section III, existing advanced structural alloys are described before focusing on high temperature structural materials in section IV.

We are grateful to authors for taking time out of their busy schedules to prepare their chapters. We are also grateful to Ms. Betty Adam of Princeton University for her tireless efforts in coordinating the correspondence with the authors, and synthesizing their inputs into a coherent document. It is hard to imagine how this book could have been completed without her skilled help.

This book was initiated by Dawn Wechsler and Janet Sachs of Marcel Dekker. Since their initial efforts, we have been guided by Shelley Kronzek of Taylor & Francis Books, CRC Press. We would like to thank her for her vision and her patience. We hope that this book will be useful to senior undergraduate and graduate students, practicing materials scientists and engineers, researchers, and those who simply want to learn more about advanced structural materials.

Much of my current understanding of advanced structural materials has been nurtured by program managers who have supported my research over the past two decades. I would, therefore, like to thank Charles Whitesett (McDonnell Douglas), Dick Lederich (McDonnell Douglas), Shankar Sastry (McDonnell Douglas/Washington University), Oscar Dillon (NSF), Dan Davis (NSF), Jorn-Larsen Basse (NSF), George Yoder (ONR), Julie Christodolou (ONR), Chuck Ward (AFOSR), Bruce MacDonald (NSF), Majia Kukla (NSF), Ulrich Strom (NSF), Tom Rieker (NSF), Joe Akkara (NSF), Tom Weber, Lance Haworth, Adriaan Graaf, and Carmen Huber (NSF) for their support of my efforts. I would also like to thank NSF (DMR Grant No. 0231418) for providing the financial support used to coordinate the preparation of this book.

Finally, I would like to thank my dear wife, Morenike, and my children, Rotimi, Deji, and Wolé for allowing me the time to work on yet another book project. I hope that the time spent on this project will help to enrich the lives of others, just as it has enriched mine.

Wolé Soboyejo
Princeton, NJ

Author

 **Wolé Soboyejo** was educated in England. He received his bachelor's degree in mechanical engineering from King's College, London University in 1985. He then went on to Cambridge University, where he received his Ph.D. degree in materials science and metallurgy in 1988. Between 1988 and 1992, he was a research scientist at the McDonnell Douglas Research Laboratories. He then worked briefly as a principal research engineer at the Edison Welding Institute (EWI) before joining the faculty in the Department of Materials Science and Engineering at The Ohio State University. Between 1997 and 1998, he was a Visiting Martin Luther King Professor in the Department of Materials Science and Engineering and the Department of Mechanical Engineering at MIT. In 1999, Professor Soboyejo moved to Princeton University and was appointed full professor in the Department of Mechanical and Aerospace Engineering and the Princeton Institute of Science and Technology of Materials (PRISM). He is the director of the Undergraduate Materials Program at PRISM. He is the director of the US/Africa Materials Institute (USAMI) and the chair of the African Scientific Committee of the Nelson Mandela Institution. Professor Soboyejo is the recipient of two National Young Investigator Awards (NSF and ONR) and the Bradley Stoughton Award for Excellence in the Teaching of Materials Science. He is a Fellow of ASME and the Materials Society of Nigeria. He has published more than 350 papers and one textbook on the mechanical properties of engineered materials.

Contributors

Seyed M. Allameh
Department of Physics and Geology
Northern Kentucky University
Highland Heights, Kentucky

Tiffany Biles
NASA Glenn Research Center
Cleveland, Ohio

Yifang Cao
Department of Mechanical and
    Aerospace Engineering
Princeton Institute for the Science and
    Technology of Materials (PRISM)
Princeton University
Princeton, New Jersey

Kwai S. Chan
Southwest Research Institute
Mechanical and Materials
    Engineering Division
San Antonio, Texas

M. Freels
Department of Materials Science
    and Engineering
The University of Tennessee
Knoxville, Tennessee

L. Jiang
Corporate Research and Development Center
General Electric Company
Schenectady, New York

K. S. Kumar
Department of Materials Science
Brown University
Providence, Rhode Island

D. L. Klarstrom
Haynes International, Inc.
Kokomo, Indiana

John Lewandowski
Department of Materials Science
    and Engineering
Case Western University
Cleveland, Ohio

Peter K. Liaw
Department of Material Sciences
    and Engineering
The University of Tennessee
Knoxville, Tennessee

Jun Lou
Department of Mechanical Engineering
    and Materials Science
Rice University
Houston, Texas

Fred McBagonluri
Siemens Hearing
Piscataway, New Jersey

Ronald Noebe
NASA Glenn Research Center
Cleveland, Ohio

Santo A. Padula II
NASA Glenn Research Center
Cleveland, Ohio

J. H. Perepezko
Department of Materials Science
    and Engineering
University of Wisconsin-Madison
Madison, Wisconsin

R. Sakidja
Department of Materials Science
    and Engineering
University of Wisconsin-Madison
Madison, Wisconsin

Gary J. Shiflet
Department of Materials Science
    and Engineering
University of Virginia
Charlottesville, Virginia

W. O. Soboyejo
Department of Mechanical and
    Aerospace Engineering
Princeton Institute for the Science and
    Technology of Materials (PRISM)
Princeton University
Princeton, New Jersey

T. S. Srivatsan
Department of Mechanical
  Engineering
University of Akron
Akron, Ohio

Satish Vasudevan
Department of Mechanical
  Engineering
University of Akron
Akron, Ohio

Jikou Zhou
Materials Science and Technology Division
Lawrence Livermore
  National Laboratory
Livermore, California

Aiwu Zhu
Department of Materials Science
  and Engineering
University of Virginia
Charlottesville, Virginia

Table of Contents

SECTION 1: Introduction

Chapter 1 Introduction to Advanced Materials 1

W. O. Soboyejo

SECTION 2: Novel Materials

Chapter 2 Small Scale Contact and Adhesion in
Nano- and Bio-Systems 15

Yifang Cao and W. O. Soboyejo

Chapter 3 Mechanical Characterization of Thin Film
Materials for MEMS Devices 35

Jun Lou

Chapter 4 Silicon-Based Microelectromechanical
Systems (Si-MEMS) 65

Seyed M. Allameh

Chapter 5 Porous Metallic Materials. 103

Jikou Zhou

SECTION 3: Advance of Structural Materials

Chapter 6 A Thermodynamic Overview of Glass Formation
Abilities: Application to Al-Based Alloys 125

Aiwu Zhu and Gary J. Shiflet

Chapter 7 NiTi-Based High-Temperature Shape-Memory
Alloys: Properties, Prospects, and Potential Applications 145

Ronald Noebe, Tiffany Biles, and Santo A. Padula II

Chapter 8 Cobalt Alloys and Composites 187

M. Freels, Peter K. Liaw, L. Jiang, and D. L. Klarstrom

Chapter 9 The Science, Technology, and Applications of
Aluminum and Aluminum Alloys. 225

T. S. Srivatsan and Satish Vasudevan

Chapter 10 Metal Matrix Composites: Types, Reinforcement,
Processing, Properties and Applications. 275

T. S. Srivatsan and John Lewandowski

Chapter 11   Titanium Alloys: Structure, Properties, and Applications 359

Fred McBagonluri and W. O. Soboyejo

SECTION 4:   High Temperature Materials

Chapter 12   Niobium Alloys and Composites. 401

Kwai S. Chan

Chapter 13   Mo-Si-B Alloys for Ultrahigh
Temperature Applications. 437

J. H. Perepezko, R. Sakidja, and K. S. Kumar

Chapter 14   Nickel-Base Alloys 475

W. O. Soboyejo

Index 493

1 Introduction to Advanced Materials

W. O. Soboyejo
Department of Mechanical and Aerospace Engineering,
Princeton Institute for the Science and Technology
of Materials (PRISM), Princeton University

CONTENTS

1.1   Introduction 1
1.2   Applications of Advanced Materials 2
    1.2.1   Materials in Aeroengines...2
    1.2.2   Materials for the National Aerospace Plane 4
    1.2.3   Materials in Sporting Goods 8
    1.2.4   Materials for Human Prosthetic Devices...8
    1.2.5   Materials for Automotive Applications 8
1.3   Engineering of Balanced Properties 11
1.4   Summary...12
References 13

1.1   INTRODUCTION

Advanced structural metallic materials have had a considerable impact on the development of a wide range of strategic technologies. However, only a few specialists are aware of the basic scientific concepts that have guided the design of new alloys, intermetallics, and metal matrix composites. These concepts are described in this book in an effort to make such knowledge widely accessible to engineers and scientists who have a strong tendency to become rather focused on a few particular systems. We also hope that the book will serve as a useful overview to the public at large who have faith in these advanced materials without knowing much about the properties that make them suitable for structural applications.

We hope that this book will provide a simple picture of the basic concepts that guide the development of new alloys and composite materials. For this reason, excessive detail has been avoided in the individual sections in an effort to retain clarity in the presentation of key ideas. However, the more inquiring reader is provided with an extensive list of references at the end of each chapter.

Finally, it is important to emphasize that this book is not intended to be a comprehensive overview of advanced metallic materials. No authors could possibly provide a complete review of all the relevant material available in the literature. Instead, we hope that this text will provide a useful picture of the basic ideas used to control the fracture properties of engineered structural metallic materials. We also hope that our limited choice of examples will serve to illustrate the critical role that advanced structural metallic materials will play in the aerospace, biomedical, automotive, sporting goods, and other industries in the twenty-first century. An overview of advanced metallic materials is presented in this chapter.

1.2 APPLICATIONS OF ADVANCED MATERIALS

Since the Stone Age, and perhaps before, the ability to process materials has had significant economic and political implications. The shaping of stones into weapons of war often determined success on the battlefield, just as advanced structural materials and superior computer technology can influence the outcome of modern day warfare. Man's ability to process materials with enhanced properties has therefore provided economic and political advantages to the groups with materials that are relatively advanced compared to those of their counterparts. This was particularly apparent in ancient Japanese society where the Samurai's ability to process steels with complex laminated microstructures guaranteed victory in the battlefield [1]. Similar examples of superior sword-making abilities abound in the bible, where the Hittites and other groups used their superior materials to their advantages in wide variety of military adventures.

In more recent times, advanced materials have played a key role in the arms race between the countries of the old Eastern and Western blocs. In fact, it is easy to argue that the arms race was the primary driving force behind the development of new materials between 1945 and 1991 when the Berlin Wall came down. Other factors that contributed to the development of new materials include various international space programs, transportation and materials processing industries, emerging biotechnology industries, and the rapid growth of the microelectronics and computer industries during the last quarter of the twentieth century.

The historical development of some important materials systems is summarized in Figure 1.1. Advanced materials have therefore played a pivotal role in providing us with many of the products that have become part of our daily lives. However, most people are unaware of the central role advanced materials play in their lives in spite of their ubiquitous nature [2]. For this reason, a few examples of the existing and emerging applications of advanced materials are presented in the remainder of this chapter as follows:

1. Materials in aeorengines
2. Materials for the National Aerospace Plane
3. Materials in sporting goods
4. Materials for human prosthetic devices
5. Materials for automotive applications

A wide range of other examples will also be presented in the individual chapters throughout this book. Nevertheless, we hope that the few examples presented in this section will provide a useful introduction to the advanced materials systems that will be described in detail in subsequent chapters.

1.2.1 MATERIALS IN AEROENGINES

Our ability to power modern airplanes depends largely on the thrusts generated by aeroengines. Since speed and safety are critical in both commercial and military aircrafts, the selection of materials for aeroengine applications requires a careful balance of performance and risk [3]. Material cost is also an important factor, especially in commercial aeroengines where plane crashes can attract global attention, particularly in cases where human lives are lost.

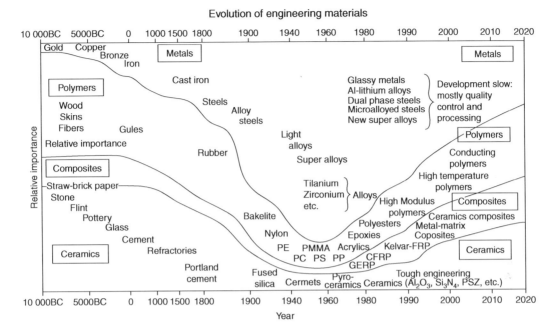

FIGURE 1.1 Evolution of the historical evolution of the relative importance of the different classes of materials.

Materials in modern day aeroengines are, therefore, introduced only after extensive materials development and testing have been performed over a period of ∼15 years. This has led to unprecedented levels of safety and performance in most of the aeroengines that are currently in service.

A schematic of a modern day aeroengine is presented in Figure 1.2. This shows the three main sections of the engine, i.e., the fan, compressor, and turbine. Air is sucked in by the fan, which is normally fabricated from a titanium alloy (such as Ti–6Al–4V) or a polymer matrix composite. The air is then passed through multiple stages of the compressor, which consists primarily of disks and blades that are also fabricated primarily from titanium alloys such as Ti–6Al–4V. The temperatures in the fan and aeroengine sections of the aeroengine are relatively low (close to room temperature). However, higher temperatures (∼400°C–500°C) may occur in the high-pressure stages. Beyond the compressor, the air is ignited in a combustor, before expanding through multiple stages of the turbine and the nozzles at the back end of the engine. Nickel and cobalt-based superalloys are often used in the combustor and turbine sections of the aeroengine due to high temperatures experienced in these systems (from ∼500°C to 650°C in the disks to ∼1200°C at the tips of the blades in the turbines).

The compositions and microstructures of the superalloys are also tailored to provide the required combinations of strength, creep, fatigue, and oxidation resistance required for service in aeroengines that revolve at angular speeds as high as 20,000 revs/minute. Beyond the turbine section of the aeroengine, air is expanded through nozzles that are typically fabricated from coated niobium-based alloys (Figure 1.3).

The above materials systems have been largely optimized by empirical processing and alloy design schemes. However, many of the systems have inherent limits in properties that may not be overcome by alloying and heat treatment. For example, the densities and moduli of metallic materials may not be altered significantly by alloying. Hence, new materials are often needed to exceed the intrinsic limits of the existing structural metallic systems in modern aeroengines [3–6]. The emerging systems that have been identified as potential candidates to replace existing aeroengine materials are listed in Table 1.1, along with existing aeroengine materials. Note that these

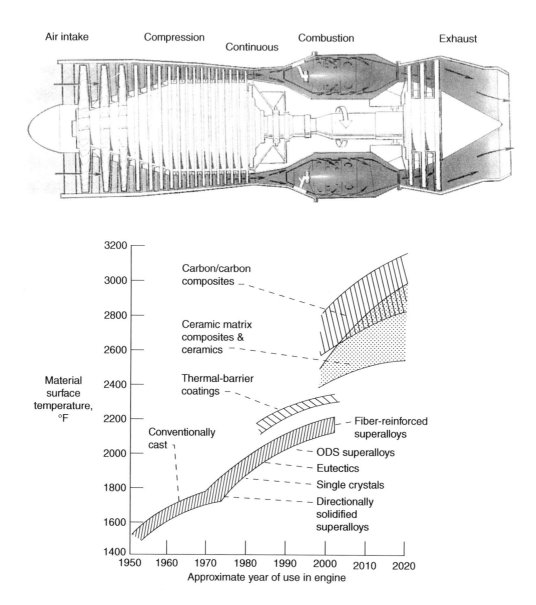

FIGURE 1.2 The role of high temperature materials in the historical evolution of the turbine blade surface temperature in aeroengines.

generally have lower densities than the existing aeroengine materials. However, their fatigue and fracture properties are generally not as attractive as those of the existing materials at room temperature. The current text will therefore devote a significant number of sections to the toughening and strengthening strategies that are being used to guide the design of damage-tolerant aeroengine materials. As in most applications, the trick is to design new materials with the required balance of properties. This theme is one that will recur throughout this book.

1.2.2 MATERIALS FOR THE NATIONAL AEROSPACE PLANE

The National Aerospace Plane (NASP) vehicle was conceived in the 1980s in the U.S. It was proposed as a Mach-10 hypersonic vehicle that could travel from Tokyo to New York in two hours [9]. As with

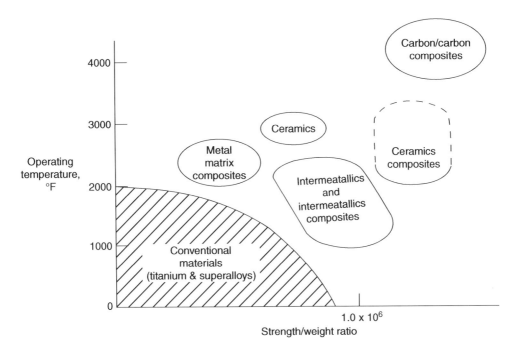

FIGURE 1.3 The ranges of operating temperature and strength/weight ratio for different classes of materials.

most aerospace concepts, the main limitation to this hydrogen-powered and hydrogen-cooled vehicle was the availability of materials with the required combinations of mechanical properties and oxidation resistance. This should be immediately apparent to the reader after reviewing Figure 1.4, in which a schematic of the NASP vehicle is presented. Note that the temperature at the nose of the X-33 derivative of the plane is as high as 1800°C. Also, the temperatures in the trailing edges of the wing are between

TABLE 1.1
Densities of Existing and Emerging Aeroengine Materials

Section(s) of Aeroengine	Existing Alloy System(s)	Density of Existing Alloy System(s) (g/cm^3)	Emerging Alloy Systems(s)	Density of Emerging Alloy System(s) (g/cm^3)
Fan hubs and disks	Ti–6Al–4V	4.5	Polymer matrix composites	2
Compressors	α and/or β titanium alloys, e.g. Ti–6Al–4V	4.5	Orthorhombic Ti alloys/composites	4.5–5.0
Combustors	Ni-base superalloys	8–10	Ceramic matrix composite	2
Turbines	Ni–Co-base superalloys	8–10	Titanium aluminide	4.0–4.5
			Niobium aluminide	5.1–5.5
			Niobium silicide	6
Nozzles	Nb alloys	10–12	Niobium aluminide	5.1–5.5
			Niobium silicide	6

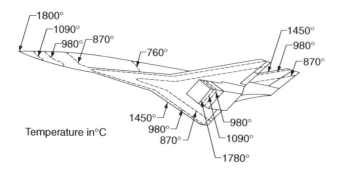

FIGURE 1.4 Temperature variations on the skin of the X-33 NASP vehicle.

~ 1000°C and 1450°C. Such temperatures clearly exceed the "useful" temperature limits of existing superalloys and refractory Nb alloys. A significant materials development effort [7] was therefore initiated to produce the required high temperature materials for the NASP vehicle in a record time of about 10 years (note that the materials development is generally ~ 15 years). The initial effort was not successful in spite of several billions of dollars of research expenditures. Nevertheless, the X-33 vehicle did fly successfully in 2004, after adopting a more realistic approach to materials and structures. In any case, the NASP program did stimulate extensive materials efforts to develop new high temperature materials with the required balance of mechanical properties. Some of the more promising systems considered for use in the NASP vehicle are listed in Table 1.2, along with their inherent temperature limits.

It is important to note that the NASP program did have some very important beneficial effects on other materials development efforts. For example, the gamma-based titanium aluminide systems have been selected for use as shroud seals and nozzle liner files in the engineering of the engine of the YF-22 airplane (Figure 1.5). The frame of this futuristic military jet is produced by the Lockheed–Martin Corporation, while the aeroengine is supplied by Pratt & Whitney. Other potential aerospace applications for γ-TiAl alloys include low-pressure turbine blades and small structural

TABLE 1.2
Potential Materials Systems for Hypersonic Vehicles

Temperature Range (°C)	Materials System	Main Limitation(s)
400–600	α/β Ti alloys	Oxidation and creep resistance
600–650	Orthorhombic Ti alloys	Oxidation resistance and fabricability
650–760	Gamma-based titanium Aluminides	Damage tolerance and creep Resistance
	Niobium aluminide	Oxidation resistance
650–1200	Nickel aluminides	Damage tolerance and creep
	Molybdenum disilicides	Resistance
		Damage tolerance
1200–1400	Molybdenum disilicides	Damage tolerance and creep
	Niobium silicides	Resistance
		Damage tolerance
1400–1800	Ceramic matrix composites	Damage tolerance
1800–2000	Carbon–carbon composites	Oxidation resistance

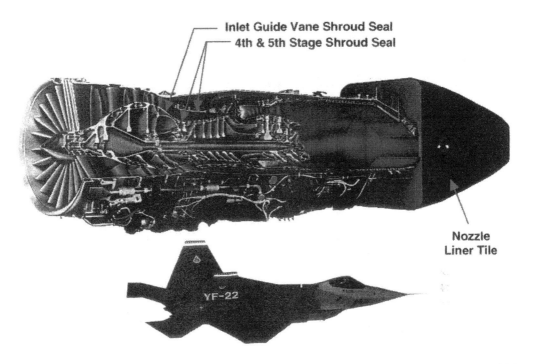

FIGURE 1.5 Cross-section of aeroengine the YF-22 airplane - applications of gamma titanium aluminides indicated by arrows.

elements (Figure 1.5). It is perhaps the first test-bed for the use of titanium aluminide in an aeroengine application.

The Toyota Motor Company of Japan, General Motors, and Ford Motor Company in the U.S. are also exploring possible applications of gamma titanium aluminides in valve and turbo-charged

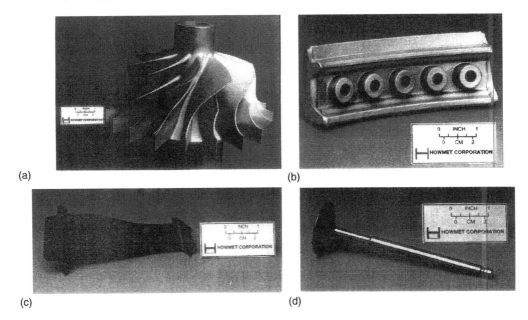

FIGURE 1.6 Applications of gamma titanium aluminides: (a) turbo-charger, (b), (c) turbine blade, and (d) automotive valve (Courtesy of The Howmet Corporation).

applications. These companies are interest in gamma-based titanium aluminides because improvements in performance can be engineered by the use of lightweight (lighter than the highly alloyed steels that are currently used) titanium aluminide alloys. The use of TiAl valves (Figure 1.6) could also result in reductions in fuel consumption of ~1–2 miles per gallon and higher engineering operating pressure/temperatures that can be used to control engine emissions to meet standards imposed by the Environmental Protection Agency (EPA) in the U.S. For this reason, the manufacturers of diesel engines (for large trucks) are currently developing gamma-based titanium aluminide valves for the next generation of diesel truck engines.

1.2.3 MATERIALS IN SPORTING GOODS

Advanced materials have played an increasing role in sporting applications during the past 30 years. One sport where the influence has been particularly important has been the ancient Scottish game of golf. State-of-the-art golf clubs are now fabricated from titanium alloys and newly developed amorphous alloy (Figure 1.7) that absorb minimal amounts of energy upon contact with golf balls [8]. The current state-of-the-art amorphous glass alloy is Al–62Zr–10Ni–10Cu–3.5Be (weight %). This alloy has a strength of ~1900 MPa and a modulus close to that of titanium. It is used in the fabrication of golf clubs that cost up to a few thousand dollars each! However, they are extremely popular among professional or avid golfers due to their high coefficients of restitution. Similar advantages are also beneficial to the design of baseball bats where very thin sections of precipitation strengthened aluminum alloys are used. However, these baseball bats are almost too effective, i.e., the sudden bounce of the ball from the baseball bat is sudden, and an unanticipated bounce can be dangerous. In fact, there have been recorded cases of players developing detached retinas after being struck by fast recoiling baseballs traveling at unprecedented velocities after being hit by these strengthened aluminum baseball bats. Other sports in which advanced materials/composites play a strong role include tennis, boat racing, and canoeing.

1.2.4 MATERIALS FOR HUMAN PROSTHETIC DEVICES

Few metallic materials are biocompatible with the human body, i.e., most metallic materials that can be used as implants are rejected by the human body [9]. However, titanium and cobalt are two metals that are compatible with the human body. These two metals are therefore used widely in a range of human prosthetic devices (human implants). In many cases, pure titanium and pure cobalt are preferred because most of the possible alloying elements such as vanadium that have been shown to have undesirable toxic side effects are not present. However, the pure metals may not have sufficient balance of strength and other mechanical properties for prosthetic applications in the human body. Titanium and cobalt-based alloys (mixtures with other elements) have, therefore, been developed for applications in which higher levels of strength and fatigue resistance are required [9,10].

Titanium alloys are particularly attractive for hip implants because of their exceptional combinations of corrosion resistance, moderate density (~ 4.5 g/cm^3), yield strength (500–800 MPa), and fracture toughness (~ 40–100 MPa m$^{1/2}$). However, titanium alloys have limited wear resistance [10]. Cobalt-based alloys are, therefore, preferred in applications where wear resistance is critical. These include knee implants and interfacial layers between polyethylene caps and titanium hip implants that fit into socket joints. Stainless steels are also used in some corrosion-resistant devices, while carbon–carbon materials tend to be favored in some heart-valve applications that undergo several million fatigue cycles (heartbeats) per year!

1.2.5 MATERIALS FOR AUTOMOTIVE APPLICATIONS

Advanced materials offer some unique opportunities in the design of lightweight vehicles with improved fuel consumption and performance. However, low cost is also a very strong factor in the selection and application of materials in the automotive industry [11]. Unlike aerospace materials,

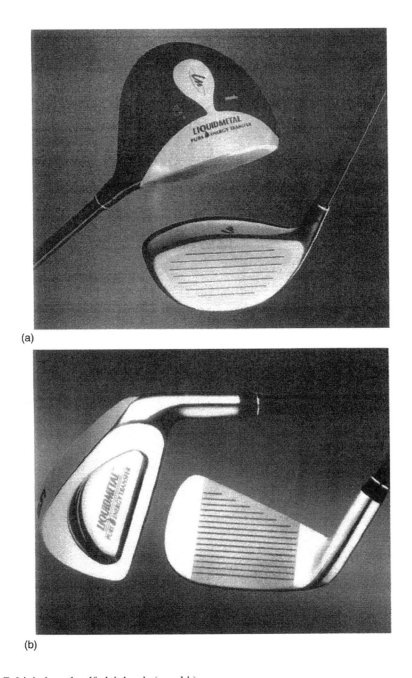

(a)

(b)

FIGURE 1.7 Liqiudmetal golf club heads (a and b).

cost is perhaps the biggest factor in the selection of new materials for most automotive applications. This is particularly true for high volume compact cars and family automobiles. However, luxury vehicles and high performance sports cars/racing cars have provided a valuable test-bed for the introduction of advanced materials into the next generation of automobiles.

One example of an advanced materials system that has been introduced into Toyota motor vehicles is a new class of in situ titanium matrix composites reinforced with TiB whiskers [11–14]. The whiskers in these structures are produced in situ reactions that occur during processing via powder [11,12] or ingot [13,14] metallurgy routes. They are formed by the

20 µm

FIGURE 1.8 Microstructure of In situ titanium matrix composite reinforced with TiB whiskers (Ti-8Al-1V-1Mo+TiB whiskers).

reaction with the titanium that is present in titanium alloy mixtures. A micrograph of an in situ titanium matrix composite is presented in Figure 1.8. This shows aligned TiB whiskers in a matrix of Ti–8Al–1V–1Mo. Note that the TiB whiskers were aligned by the extrusion process. Otherwise, the whiskers would be randomly oriented.

The in situ titanium matrix composites have higher strength, wear resistance, and stiffness than the unreinforced titanium matrices. They also have improved elevated-temperature strength and creep resistance and moderate density ($\sim$ 4.1 g/cm^3). Furthermore, they are relatively easy to process via powder metallurgy, casting, and wrought processing techniques [11–14] at costs that are comparable to those of unreinforced titanium alloys. For these and other reasons, the Toyota Motor Company of Japan has explored the possible application of in situ titanium matrix composites in a wide range of automobile applications. These include applications in connecting rods, gears, and valves (Figure 1.9). The TiB-reinforced values were tried successfully in racing cards in 1997 and are currently being introduced into the next generation of Toyota sports cars and luxury vehicles such as the Lexus.

As discussed earlier in Section 1.2.2, titanium aluminide intermetallics based on TiAl are also being considered for valve and turbocharger applications (Figure 1.6) in the next generation of automotive vehicles [15]. These materials offer a temperature advantage (up to $\sim$ 750°C–800°C) over titanium alloys which are generally limited to applications below $\sim$ 500°C–600°C [12]. They also have moderate density ($\sim$ 4.1–4.5 g/cm^3) and stiffness ($\sim$ 150 GPa). However, titanium aluminides are relatively brittle (fracture toughness $\sim$ 20–35 MPa m$^{1/2}$ and ductility $\sim$ 1%–2%) compared to the valve steels that they may replace. There are also some concerns about their cost (compared with valve steels) and wear resistance [15]. Nevertheless, most major automobile manufacturers are currently exploring possible applications of titanium aluminides in intake and exhaust valves. The Howmet Company in Whitehall, MI, is also developing low-cost methods for the fabrication of titanium aluminide valves. In particular, there is considerable interest in the application of titanium aluminides in the diesel engines of large trailer trucks. It is envisaged that such applications could ultimately lead to significant reductions in NO$_x$ emissions that are required by the EPA. Materials producers estimate that future sales from such valves could amount to hundreds of millions of dollars over the next few years. An example of an investment cast titanium aluminide valve produced by Howmet is shown in Figure 1.6. This valve can be produced at costs that are close to those of highly alloyed steels used in current vehicles. However, further development work is needed to optimize the production of wear resistant, low cost titanium aluminide valves [15].

FIGURE 1.9 A range of automotive components fabricated from in situ titanium matrix composites-gears, connecting rods, and inlet/exhaust.

Other projected applications of advanced metallic materials in automotive vehicles include the use of aluminum matrix composites in connecting rods, engine blocks, cylinder liners, and brake calipers [5]. However, existing applications of advanced materials in automotive vehicles are still largely restricted to polymer matrix composites used primarily in high-performance vehicles. There are also some limited applications of particle- or whisker-reinforced particle- or whisker-reinforced aluminum matrix composites in some engine applications [5].

1.3 ENGINEERING OF BALANCED PROPERTIES

It should be clear from the above discussion that a balance of properties is required for the safe application of advanced materials in structural and nonstructural applications. In the case of high temperature alloys/composites for transportation systems, a combination of strength and light-weight (lower density) is often a basic requirement that must be satisfied before the more advanced properties (such as fracture toughness and fatigue resistance) are fully considered.

Some of the most promising metallic materials systems (those with the basic combinations of strength and density) are shown in Figure 1.10. Note that gamma-based titanium aluminide inter-metallics have excellent combinations of low density (~ 4.5 g/cm^{-3}) and strength for intermediate-temperature applications up to $\sim 760°C$. However, as discussed earlier, gamma-based titanium aluminides are limited by their room-temperature damage tolerance (fracture toughness and fatigue crack growth resistance). Titanium alloys also have comparable levels of strength and density to gamma-based titanium aluminides. However, they are generally limited to applications at temperatures below 600°C. This is due largely to their limited oxidation and strength at temperatures above this limit.

In recent years, Nb–Al–Ti-based intermetallics have been developed for potential structural applications [16,17]. These B2 + orthorhombic intermetallics have combinations of strength and fracture toughness that are comparable to those of steel sand and other structural metallic materials.

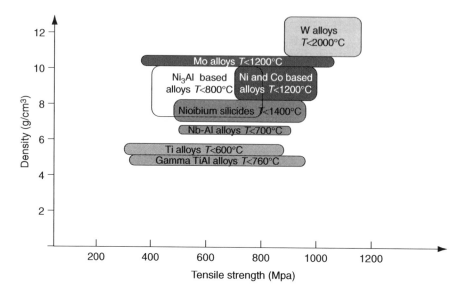

FIGURE 1.10 The ranges of density and tensile strength for different types of high temperature materials.

They also have good oxidation resistance below $\sim 700°C$–$750°C$. In the temperature range between 800 and 1200°C, nickel and cobalt alloys are currently used in a wide range of engines. However, damage tolerant nickel aluminide (Ni_3Al-based) intermetallics with boron and hafnium additions [18,19] have emerged in recent years as promising materials for applications at the lower end of this temperature regime. Unfortunately, however, near commercial nickel aluminide intermetallics are just as heavy (density between 8 and 10 g/cm^3) as the nickel base and cobalt base alloys that they are slated to replace. Nevertheless, there has been considerable interest in their potential applications in land-based engines [19].

In the temperature range between 1000°C and 1400°C, a number of very high temperature materials systems are being considered for potential applications in the next generation of engines. These include in situ composites of niobium silicides and niobium solid solution alloys [20,21], molybdenum disilicide intermetallics [22,23], molybdenum and its alloys [24], and a new class of Mo–Si–B alloys [25]. The latter have excellent oxidation resistance and phase stability at temperatures up to 1400°C. However, very little is known about their mechanical properties.

For applications at the very highest temperature (up to 2000°C), refractory alloys are being considered as alternatives to ceramic matrix composites and carbon–carbon composites. However, the refractory alloys and carbon–carbon composites have very limited oxidation resistance at the required service temperatures. There is, therefore, a need for further work to develop high temperature coatings and alloy systems with the potential for high temperature structural applications in this temperature regime. Further details on the above alloy systems will be presented later in subsequent chapters of this book.

1.4 SUMMARY

A brief introduction to advanced materials and their applications has been presented in this chapter. The information is presented at a basic level and is clearly not intended to be comprehensive. Instead, the intent is to provide a framework for the introduction of advanced metallic materials, and some of the fundamental concepts that will be presented in greater detail in subsequent chapters. Following a brief historical review of the economic and military/political implications of "advanced" materials, selected examples were presented to illustrate the current and future applications of advanced materials in the aerospace, biomedical, sporting goods, and automotive industries. Critical issues in the engineering of balanced properties in high temperature metallic materials were then highlighted in the final section of the chapter.

REFERENCES

1. Bain, E. C., Nippon-to, an introduction to Old Swords of Japan, *Journal of the Iron and Steel Institute*, 200, 265–282, 1962.
2. Matthews, F. L. and Rawlins, R. D., *Composite Materials: Engineering and Science*, 1st ed., Chapman and Hall, New York, 1994.
3. Schwartz, M. M., *Emerging Engineering Materials: Design, Processes and Applications*, Technomic Publishing Company, Lancaster, PA, 1996.
4. Taya, M. and Arsenault, R. J., *Metal Matrix Composites: Thermomechanical Behavior*, Pergamon Press, New York, 1989.
5. Suresh, S., Mortensen, A., and Needleman, A., *Fundamentals of Metal Matrix Composites*, Butterworth-Heinemann, Boston, 1993.
6. Clyne, T. W. and Withers, P. J., *An Introduction to Metal Matrix Composites*, Cambridge University Press, Cambridge, 1993.
7. Strauss, B. and Hulewicz, J., X-33 advanced metallic thermal protection system, *Advanced Materials and Processes*, 51, 55–58, 1977.
8. Larsen, D., Alloy process hits hole in one, *Machine Design News*, 42, October 1998.
9. Semlitsch, M. F., Weber, H., Steicher, R., and Schon, R., Joint replacement components made of hot-forged and surface-treated Ti–6Al–7Nb alloy, *Biomaterials*, 13, 781–788, 1992.
10. Dowson, D., Friction and wear of medical implants and prosthetic devices. *ASM Handbook*, ASM International, OH, pp. 656–664, 1992.
11. Saito, T., *Development of Low Cost Titanium Matrix Composite*, TMS, Warrendale, PA, 1995.
12. Soboyejo, W. O., Lederich, R. J., and Sastry, S. M. L., Mechanical behavior of damage tolerant TiB reinforced in-situ titanium matrix composites, *Acta Metallurgica et Materialia*, 42, 2031–2047, 1997.
13. Dubey, S., Lederich, R. J., and Soboyejo, W. O., Fatigue and fracture of damage tolerant ingot metallurgy in-situ titanium matrix composites, *Metallurgical and Material Transactions*, 26A, 2031–2047, 1997.
14. Lederich, R. J., Soboyejo, W. O., and Srivatsan, T.S., Preparing damage tolerant in-situ titanium matrix composites, *Journal of Metals*, 68–71, 1994.
15. Kim, Y. W., Ordered intermetallic alloys, mechanical behavior of damage tolerant TiB reinforced in-situ titanium matrix composites, Part III, *Journal of Metals*, 46, 30–40, 1994.
16. Hou, D. H., Shyue, J., Yang, S. S., and Fraser, H. L., In *Alloy Modeling and Design*, Stocks, G. M. and Turchi, P. Z. A., Eds., TMS, Warrendale, PA, 291–298, 1994.
17. Ye, F., Mercer, C., Soboyejo, W. O., *Metallurgical and Materials Transactions*, 2361–2374, 1998.
18. Liu, C. T. and White, C. L., Design of ductile polycrystalline Ni3Al alloys, In *Proceedings of the Symposium on High Temperature Ordered Intermetallic Alloys*, Koch, C. C., Liu, C. T., and Stoloff, N. S., Eds., Vol. 39, Materials Research Society, Warrendale, PA, pp. 365–380, 1985.
19. Sikka, V. K., Nickel aluminides, *New Advanced Alloys, Materials and Manufacturing Processes*, 4, 1–24, 1989.
20. Subramanian, P. R., Nediratta, M. G., and Dimiduk, M., *Journal of Metals*, 33–38, 1996.
21. Bewlay, B. P., Lewandowski, J. J., and Jackson, M. R., *Journal of Metals*, 44–67, 1997.
22. Vasudevan, A. K. and Petrovic, J. J., A comparative overview of molybdenum disilicide composites, *Materials Science and Engineering*, A155, 1–17, 1992.
23. Soboyejo, W. O., Venkateswara Rao, K. T., Sastry, S. M. L., and Ritchie, R. O., Strength, fracture, and fatigue behavior of advanced high-temperature intermetallics reinforced with ductile phases, *Metallurgical Transactions*, 24A, 585–600, 1993.
24. Shields, J. A., Molybdenum and its alloys, *Advanced Materials and Processes*, 28–36, 1992.
25. Brezic, D., *Method for Enhancing the Oxidation Resistance of a Molybdenum Alloy*, U.S. Patent No. 5,595,616, 1997.

2 Small Scale Contact and Adhesion in Nano- and Bio-Systems

Yifang Cao and W. O. Soboyejo
Department of Mechanical and Aerospace Engineering,
Princeton Institute for the Science and Technology
of Materials (PRISM), Princeton University

CONTENTS

2.1   Introduction..15
2.2   Overview of Nanoindentation Methods...16
2.3   Fundamentals of Contact Mechanics 19
2.4   Fundamentals of Adhesion and Biological Adhesion 21
2.5   Stamping Methods for Fabricating Organic Electronic Devices.......................26
2.6   Methods for Studying Cell Mechanical Properties and Prestresses.....................28
References 30

2.1   INTRODUCTION

In recent years, there has been significant interest in the material properties and behaviors of nano- and bio-systems relevant to microelectronics and biomedical applications. Soft materials are beginning to play an increasingly important role in this field, especially in the bio-microelectromechanical systems (bioMEMS) area. Some disadvantages of silicon, such as high cost and stiffness, have begun to limit its applications in bioMEMS. Therefore, although people have been successful in making silicon based valves and microfluidic bioMEMS devices, alternative fabrication techniques using nontraditional soft materials, including hydrogels [1], plastics, and elastomers [2,3], are becoming more popular. Besides, compact, lightweight, and flexible implantable bioMEMS are more desirable for decreasing discomfort in patients. Ideally, those devices should be almost unnoticeable to the patient, much the same as our nerves are integrated into our tissues. Figure 2.1a and b show bioMEMS fluidic devices, made of silicon and polydimethylsiloxane (PDMS), respectively.

In an effort to realize biocompatible and mechanically flexible bioMEMS, we have to pursue a good understanding of both nanotechnology and biotechnology. On the nano side, a rigorous understanding of the manufacturing processes for organic electronic devices could be necessary for the fabrication of flexible bioMEMS, which will ultimately based on flexible organic electronics technology, instead of rigid Si technology. On the bio side, a fundamental knowledge of the mechanisms and structures with which cells interact with materials, specifically the cytoskeleton and the cellular adhesion apparatus, will be helpful for designing biomedical implant devices.

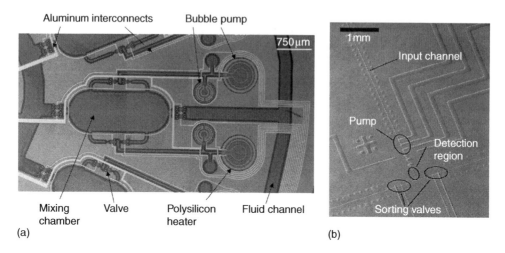

FIGURE 2.1 BioMEMS fluidics made of silicon and PDMS: (a) A silicon-based fluidic platform. (Adapted from Evans, J., Liepmann, D., and Pisano, A. P., *Proceedings of the 10th IEEE Workshop on Micro Electro Mechanical Systems (MEMS 97)* (Nagoya, Japan: Jan. 1997), 96–101. With permission.), (b) An integrated PDMS Microfluidic device. (Reproduced from Quake, S. R. and Scherer, A., *Science*, 290, 1536–1540, 2000. With permission).

Both the nano and bio aspects, and in particular, the small-scale contact and adhesion theories and phenomena, will be reviewed in this chapter. The chapter is divided into six sections. After the introduction in Section 2.1, nanoindentation methods for characterizing material properties at small scales will be reviewed in Section 2.2. This is followed by Section 2.3, in which a brief description of basic contact mechanics models will be presented. In Section 2.4, the fundamentals of adhesion and biological adhesion will then be reviewed. Subsequently, in Section 2.5, adhesion and contact-induced phenomena are discussed for the fabrication of organic/flexible electronics structures using stamping techniques involving cold welding or van der Waals bonding. Finally, in Section 2.6, various methods for studying biological cell property and adhesive interaction with biomaterials are summarized.

2.2 OVERVIEW OF NANOINDENTATION METHODS

A detailed knowledge of the mechanical properties and adhesion of the soft and hard material components used to construct the material systems is crucial for a well-rounded understanding and quantitative analysis of the whole nano- and bio-systems. New experimental techniques that provide novel insights into the mechanical properties and adhesion of materials need to be developed and utilized. The nanoindentation method based on depth sensing indentation (DSI) load–displacement data is commonly used as a tool to explore the material properties. Typical load displacement curves as well as experimental setups are shown in Figure 2.2.

The most comprehensive method for determining the hardness and modulus from indentation load displacement data was developed initially by Doerner and Nix [4] and later refined by Oliver and Pharr [5,6]. In the theory, the Meyer's definition of hardness, H was adopted [5]. This gives

$$H = \frac{P_{max}}{A} \tag{2.1}$$

where P_{max} is the maximum load and A is the projected contact area. The recorded load–displacement data were used to relate the stiffness S from the slope of the initial unloading curve, to the reduced elastic modulus, E_r

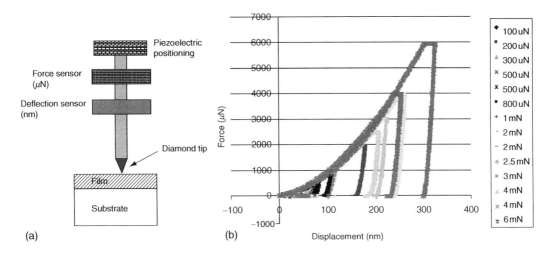

FIGURE 2.2 (a) A schematic showing typical nanoindentation experiments on film/substrate systems, (b) The load-penetration curves obtained from the nanoindentation experiments performed to different loads on the 500 nm Au film on a Si substrate.

$$E_r = \frac{\sqrt{\pi}}{2} \frac{S}{\sqrt{A}} \tag{2.2}$$

where S is the contact stiffness corresponding to the slope of the load–penetration curve at the beginning of the unloading, and E_r is expressed in terms of the elastic moduli (E) and Poisson's ratios (ν) of the indenter (i) and the indented material (im):

$$\frac{1}{E_r} = \frac{1 - \nu_i^2}{E_i} + \frac{1 - \nu_{im}^2}{E_{im}} \tag{2.3}$$

In the case of film/substrate systems, the indentation response is a combined elastic–plastic response from the film and the substrate. A solution can be found from Saha and Nix [7] to decouple the combined indentation response from the film and the substrate, and extract the intrinsic mechanical properties of the film. Their modified solution assumes that a flat punch is situated at the tip of the indenter. This yields

$$\frac{1}{E_r} = \frac{1 - \nu_i^2}{E_i} + \frac{1 - \nu_f^2}{E_f} \left(1 - e^{\frac{\alpha(t-h)}{\sqrt{A}}} \right) + \frac{1 - \nu_s^2}{E_s} \left(e^{\frac{\alpha(t-h)}{\sqrt{A}}} \right) \tag{2.4}$$

where t is the film thickness, E_f and ν_f are the modulus and Poisson's ratio of the film, E_s and ν_s are the modulus and Poisson's ratio of the substrate, α is a numerically determined scaling parameter that is a function of indentation size [7,8] and can be determined from expressions in Reference [7].

For a Berkovich tip, the projected contact area A has a calibrated relationship with the contact depth h_c. This is given by [5]

$$A(h_c) = 24.5 h_c^2 + C_1 h_c^1 + C_2 h_c^{1/2} + \cdots + C_8 h_c^{1/128} \tag{2.5}$$

where C_1 through C_8 are constants. However, after the tip rounding effects [9] and indentation pile-up [7] are taken into account, the corrected contact area is given by

$$A(h_c) = 24.5(h_c + \xi + h_p)^2 + C_1(h_c + \xi + h_p)^1 + C_2(h_c + \xi + h_p)^{1/2} + \cdots$$
$$+ C_8(h_c + \xi + h_p)^{1/128} \tag{2.6}$$

where ξ is the calculated blunting distance and h_p is the measured pileup height. The coefficients C_1 through C_8 can be empirically calibrated for the Berkovich indenter.

For contact-induced deformation of metallic materials in the micro- and nano-scale regimes, the hardness increases, as the indentation size decreases [9–11]. This so-called indentation size effect has been explained recently using the mechanism-based strain gradient (MSG) plasticity theory, established by Nix and Gao [10]. It gives the following characteristic form of the indentation depth dependence of hardness:

$$\frac{H}{H_0} = \sqrt{1 + \frac{h^*}{h}} \tag{2.7}$$

where h is the depth of indentation, H_0 is the plateau hardness in the limit, where there are no strain gradient effects, and h^* is a characteristic length scale that depends on the shape of indenter, the shear modulus, and H_0. This model was used to derive the following law for strain gradient plasticity:

$$\left(\frac{\sigma}{\sigma_0}\right)^2 = 1 + \hat{l}\chi \tag{2.8}$$

where σ is the effective flow stress in the presence of a gradient σ_0 is the flow stress in the absence of a gradient, χ is the effective strain gradient, and $\hat{l}$ is a characteristic material length scale. $\hat{l}$ can be expressed in terms of the Burger's vector b and shear modulus μ [9]. This gives

$$\hat{l} = \frac{b}{2}\left(\frac{\mu}{\sigma_0}\right)^2 \tag{2.9}$$

Hence, the length scale $\hat{l}$ can be thought of as a formalism that enables the strain contributions to plasticity from strain gradients to be modeled within a continuum framework. For the case of pure FCC metals, this length scale positively increases with the mean spacing between statistically stored dislocations (SSD) L_s through the following expression [10]:

$$\hat{l} = \frac{4}{3}\frac{L_s^2}{b} \tag{2.10}$$

For soft materials such as polymers and small molecular weight materials, it is well known that there are some peculiar concerns in an effort to achieve correct measurements using nanoindentation techniques. For example, the creep behavior of polymeric materials while unloading can alter the slope of an unloading curve (leading to negative S in extreme cases) and thus result in a failure of power law fits that are based on the Oliver–Pharr method [5,6]. One way to correctly obtain S in such cases is to use a linear approximation to the high load portion of the unloading curve [12].

In addition to the viscoelasticity issues of polymers, the induced penetration into the soft materials at a preset load for initial contact is no longer insignificant relative to maximum indentation depth [13,14]. Hence, it is important to determine the initial contact so that we can transform the translated displacement into actual separation or indentation depth [14,15]. Various contact mechanics models have been used to study the initial contact regime in the indentation experiment [14,16,17]. A nanoindentation method for the determination of the initial contact, adhesion characteristics, and the elastic moduli of soft materials has been reported in Reference [18]. In Section 2.3, a brief description of basic contact mechanics models will be presented.

2.3 FUNDAMENTALS OF CONTACT MECHANICS

Various continuum models have been used to study the effects of contact forces. These include Hertzian contact mechanics [19]; the Johnson–Kendall–Roberts (JKR) model [20]; the Derjaguin–Muller–Toporov (DMT) model [21]; and most recently, the Maugis–Dugdale (MD) model [22], in which the JKR/DMT transition is obtained by the incorporation of Dugdale zone into the region of contact.

The Hertzian contact model [19] describes the elastic resistance of the sample surface to a sphere that is being pushed into it, assuming there are no surface forces. It gives the relationship between the applied force F and indentation depth δ through the following two equations:

$$F = \frac{Ka^3}{R} \tag{2.11}$$

$$\delta = \frac{a^2}{R} \tag{2.12}$$

where R is the radius of curvature of the tip, a is the radius of the contact area, K is the reduced elastic modulus, having a relationship with E_{tip} and E, which are the tip and sample Young's moduli, and ν_{tip} and ν are the Poisson's ratios of the tip and sample, respectively. The relationship between K and these parameters is given by

$$\frac{1}{K} = \frac{3}{4} \left(\left(\frac{1 - \nu_{\text{tip}}^2}{E_{\text{tip}}} \right) + \left(\frac{1 - \nu^2}{E} \right) \right) \tag{2.13}$$

The JKR model [20] applies to the case of short-range surface forces (i.e., low elastic modulus, high adhesion, and large tip radius systems). It gives the relationship between the applied force F and indentation depth δ through the following two equations:

$$F = \frac{Ka^3}{R} - \sqrt{6\pi a^3 W_{12}} \tag{2.14}$$

$$\delta = \frac{a^2}{R} - \frac{2}{3} \sqrt{\frac{6\pi a W_{12}}{K}} \tag{2.15}$$

where W_{12} is the work of adhesion, and the other terms have their usual meaning. The adhesive force on separation F_{ad} in the JKR model is given by

$$F_{\text{ad}} = \frac{3}{2} \pi W_{12} R \tag{2.16}$$

The DMT model [21] applies to the case of long-range surface forces (i.e., high elastic modulus, low adhesion, and small tip radius systems). It gives the relationship between the applied force F and indentation depth δ through Equation 2.12 and

$$F = \frac{Ka^3}{R} - 2\pi RW_{12} \qquad (2.17)$$

the adhesive force F_{ad} in the DMT model can be expressed as

$$F_{ad} = 2\pi W_{12}R \qquad (2.18)$$

A nondimensional physical parameter referred as Tabor's parameter [23] μ can be used to bridge the two limits (JKR model and DMT model), which is defined as

$$\mu = \left(\frac{16RW_{12}^2}{9K^2Z_0^3}\right)^{1/3} \qquad (2.19)$$

where Z_0 is the equilibrium separation of the surfaces. Furthermore, μ can be shown to be the ratio of the elastic deformation just before the surfaces separate to the equilibrium separation Z_0. The JKR model applies to cases with large μ, while the DMT model is more appropriate for small μ.

For the intermediate regime between the JKR model and DMT model, the Maugis–Dugdale model [22] provides an analytical solution. It shows that the surface interaction is characterized by nondimensional transition parameter λ, which is similar to μ ($\lambda = 1.16\mu$ [23]) and defined as

$$\lambda = 2\sigma_0\left(\frac{R}{\pi W_{12}K^2}\right)^{1/3} \qquad (2.20)$$

where σ_0 is the constant adhesive stress in the Dugdale model used by Maugis [22]. This acts over a range of distance until the separation of the two surfaces is reached. An adhesion map of normalized load $\bar{P}$ vs. $\lambda = 1.16\mu$ is presented in Figure 2.3. This is taken from Reference [23].

When $\lambda < 0.1$, the DMT model applies, and when $\lambda > 5$, the JKR model applies. In intermediate cases, the following three equations are needed to relate F and δ [24].

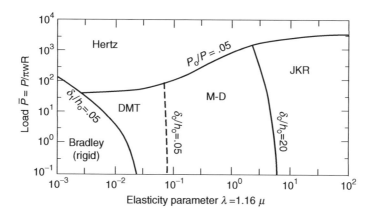

FIGURE 2.3 The adhesion map for the contact of elastic spheres. (Reproduced from Johnson, K. L. and Greenwood, J. A., *J. Colloid Interface Sci.*, 192, 326, 1997. With permission.)

$$\frac{\lambda \bar{a}^2}{2} \left[(m^2 - 2)\tan^{-1}\sqrt{m^2 - 1} + \sqrt{m^2 - 1}] + \frac{4\lambda^2 \bar{a}}{3}[(m^2 - 1)\tan^{-1}\sqrt{m^2 - 1} - m + 1 \right] = 1 \quad (2.21)$$

$$\bar{F} = \bar{a}^3 - \lambda \bar{a}^2 \left[m^2 \tan^{-1}\sqrt{m^2 - 1} + \sqrt{m^2 - 1} \right] \quad (2.22)$$

$$\bar{\delta} = \bar{a}^2 - \frac{4\lambda \bar{a}}{3}\sqrt{m^2 - 1} \quad (2.23)$$

where the parameter m is the ratio between the contact radius a and an outer radius c, at which the gap between the surfaces reaches the limit where the adhesive stress no longer acts; $\bar{F}$, $\bar{a}$, and $\bar{\delta}$ are dimensionless parameters defined as

$$\bar{a} = a \left(\frac{K}{\pi W_{12} R^2} \right)^{1/3} \quad (2.24)$$

$$\bar{F} = \frac{F}{\pi W_{12} R} \quad (2.25)$$

$$\bar{\delta} = \delta \left(\frac{K^2}{\pi^2 W_{12}^2 R} \right)^{1/3} \quad (2.26)$$

Since cumbersome numerical iterations are needed to obtain solutions to these equations, recently, a generalized equation has been developed by Pietrement and Troyon [24]. This approximates the Maugis–Dugdale model within an accuracy of 1% or better. It gives the elastic indentation depth and load via the following expression:

$$\delta = \frac{a_0^2}{R} \left[\left(\frac{\alpha + \sqrt{1 + F/F_{ad}}}{1 + \alpha} \right)^{4/3} - S_{(\alpha)} \left(\frac{\alpha + \sqrt{1 + F/F_{ad}}}{1 + \alpha} \right)^{2\beta_{(\alpha)}/3} \right] \quad (2.27)$$

where $0 < \alpha < 1$, is a dimensionless parameter corresponding to the transition parameter λ; $S_{(\alpha)}$ and $\beta_{(\alpha)}$ depend only on α; a_0 is the contact radius at zero load; $\bar{a}_0$ and $\bar{F}_{ad}$ are the dimensionless forms of a_0 and F_{ad}, defined through Equation 2.24 and Equation 2.25 respectively, depending only on α. The conversion table between α and λ, and the expressions for $S_{(\alpha)}$, $\beta_{(\alpha)}$, $\bar{a}_0$, and $\bar{F}_{ad}$ are all given in Reference [24].

When $\alpha < 0.07$, the DMT model applies, and when $\alpha > 0.98$, the JKR model applies. For the intermediate cases, α can be estimated using Equation 2.27, which can be used to fit the experimental data relating F to δ.

2.4 FUNDAMENTALS OF ADHESION AND BIOLOGICAL ADHESION

One of the goals of this chapter is to investigate small-scale contact and adhesion phenomena in small structures relevant to the fabrication of organic electronic devices (by stamping process) and cell/surface interactions with biomaterials that are needed to realize biocompatible and mechanically flexible bioMEMS. A brief description of adhesion mechanism theories, as well as biological adhesion at interfaces will be given in this section.

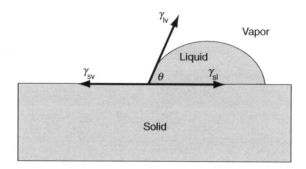

FIGURE 2.4 The balance of surface tensions on a contact angle measurement experiment.

Wetting equilibria is commonly used to assess adhesion phenomena [25]. The wetting of a coating onto the substrate is the basic contributory factor in improving adhesion. Wetting can be quantitatively analyzed using the configuration shown in Figure 2.4, where a liquid drop rests on a solid surface. The commonly known Young's equation can be used to characterize the relation between those surface tensions and the equilibrium contact angle θ_e [25]. It can be written as follows:

$$\gamma_{sl} = \gamma_{sv} - \gamma_{lv} \cos \theta_e \tag{2.28}$$

where γ_{sl} is the interfacial surface free energy between the solid and the liquid; γ_{lv} is the surface tension of the liquid; γ_{sv} is the surface free energy of the solid substrate resulting from the adsorption of vapor from the liquid, which may be considerably lower in value than the surface free energy of solid in vacuo γ_s.

The thermodynamic work of adhesion W_a is often required to separate a unit area of a solid and a liquid phase. It can be related to the surface and interfacial free energies by the Dupre equation [25], which is given by

$$W_a = \gamma_s + \gamma_{lv} - \gamma_{sl} \tag{2.29}$$

It is also well recognized that the formation of intimate molecular contact is a necessary condition for developing strong adhesion. There are several types of forces between molecules that could contribute to the adhesion at the intimate contact surfaces [26]. Table 2.1 summarizes the typical intermolecular forces. These include ionic bond, covalent bond, hydrogen bond, ion–dipole interaction, ion–induced dipole interaction, dipole–dipole interaction, dipole–induced dipole interaction, and induced dipole–induced dipole interaction.

Several different mechanisms also contribute to the adhesion [25]. These include mechanical interlocking, diffusion, electrostatic double layer, adsorption, and chemical bonding [27]. The mechanical interlocking theory postulates that adhesion is established as a consequence of material penetration into surface irregularities of another material, therefore, resulting in "interlocking." The interdiffusion or diffusion theory is usually associated with intermolecular entanglements across the interface, after materials have interdiffused into one another. In the electrostatic adhesion mechanism, it is postulated that two dissimilar materials coming in contact will result in a charge transfer, therefore, leading to the formation of an electrical double layer. The attractive electrostatic forces inherent in the double layer contribute to adhesion. The most widely applicable adhesion theory is the adsorption theory. It proposes that, given sufficiently intimate contact achieved at the interface, the materials will adhere due to the interatomic and intermolecular forces that are established between the atoms and molecules in the two surfaces. In chemical

TABLE 2.1

Typical Intermolecular Forces r is the Distance between Interacting Atoms or Molecules

Types of Intermolecular Forces	Molecules Involved	Interaction Range	Typical Energy
Ion–ion	Two ions	Short range	Proportional to $1/r$
Covalent	C–C bond	Short range	Complicated
Hydrogen bond	A–H and B where A, B = F, O, N	Short range	Complicated, roughly proportional to $1/r^2$
Ion–dipole	An ion and a polar molecule	Variable, long	Proportional to $1/r^2$ or $1/r^4$
Ion–induced dipole	An ion and a non-polar molecules	Variable, long	Proportional to $1/r^4$
Dipole–dipole	Stationary polar molecules	Variable, long	Proportional to $1/r^3$ or $1/r^6$
Dipole–induced dipole	Polar and non-polar molecules	Variable, long	Proportional to $1/r^6$
Induced dipole–induced dipole	Non-polarmolecules	Variable, long	Proportional to $1/r^6$

Source: Adapted from Israelachvili, J. N., *Intermolecular and Surface Forces*, Academic Press, San Diego, 1992; Ge, J., *Interfacial Adhesion in Metal/polymer Systems for Electronic*, PhD Thesis, Helsinki University of Technology, Espoo, Finland, 2003.

bonding theory, the adhering materials undergo chemical reactions with each other to form chemical bonds at the interface, therefore, the adhesion due to this tends to be very strong.

Biological adhesion between cells and other cells or biomaterial surfaces is currently attracting lots of attention due to its significant effects in cellular physiological functions [28–31] and artificial device development [32,33]. Hence, the knowledge of adhesion is obviously helpful for understanding biological adhesion.

Wetting theory [25] can be utilized to characterize the kinetics of cell adhesion and spreading behavior by treating a cell as a drop of highly viscous fluid, with apparent viscosity η bounded by a cortex bearing a tension τ, since the spreading of a cell is very similar to a liquid drop wetting a surface [34].

Figure 2.5a and b show the schematics of cell contact and spreading to a biomaterial substrate. Initially, the cell contacts the substrate with a round shape (Figure 2.5a). Typical images of human osteosarcoma (HOS) cells in a rounded shape before spreading and a spherical cap shape after spreading are presented in Figure 2.5c and d. Note that these experimental observations are in close agreement with the schematics (Figure 2.5a and b) proposed earlier for the characterization of the kinetics of cell spreading [34]. Cell spreading/wetting usually starts after this in an effort to achieve the macroscopic equilibra between the cortical tension τ and the work of adhesion with the substrate W_a, as shown in Figure 2.5b. The commonly known Young's equation can be used to characterize the relation between τ, W_a, and the equilibrium contact angle θ_e [25,34,35] (given that $\tau = r_{ev}$). These can be expressed as

$$W_a = \tau(1 + \cos\theta_e) \tag{2.30}$$

Geometrical consideration in Figure 2.5b of a spreading cell on a substrate leads to the following relation between the cell-substrate contact radius R and the contact angle θ [34,36]:

$$R = \left[\frac{4\sin\theta(1+\cos\theta)}{(1-\cos\theta)(2+\cos\theta)}\right]^{1/3} R_0 \tag{2.31}$$

A Poiseuille flow approximation can give the contact radius increasing rate dR/dt [34], which can be written as

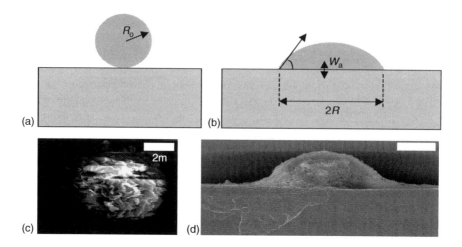

FIGURE 2.5 Schematics of cell contact and spreading to a biomaterial substrate. Initially, cell contacts the substrate with a round shape in (a) cell spreading/wetting then starts, (b) in an effort to achieve the macroscopic equilibra between the cortical tension τ and the work of adhesion with the substrate W_a, (c) and (d) show typical images of human osteosarcoma (HOS) cells in a rounded shape before spreading and a spherical cap shape after spreading, respectively.

$$\frac{dR}{dt} = \frac{\tau\theta(\cos\theta_e - \cos\theta)}{3\eta\ln\alpha} \tag{2.32}$$

where α is the ratio of the equilibrium contact radius to the molecular size of an adhesive protein [34]. Typically, α is calculated to be 10^4, since the order of the equilibrium contact radius is 10 μm and the molecular size of an adhesive protein is about 1 nm [34].

The derivative of θ with respect to t can then be obtained from Equation 2.31 and Equation 2.32. This is given by

$$\frac{d\theta}{dt} = \frac{\tau\theta(\cos\theta - \cos\theta_e)(2 + \cos\theta)^2 \left[\cos^2(\theta/2)\cot(\theta/2)/(2 + \cos\theta)\right]^{2/3}}{3\eta R_0 \ln\alpha \cot^2(\theta/2)} \tag{2.33}$$

Equation 2.33 can be numerically integrated to obtain the relationship between the contact angle θ and the culture time t. Then after Equation 2.31 is incorporated, the contact radius R can be calculated with respect to t.

It is also very important to note that the feature distinguishing cell adhesion from other adhesion phenomena is the specific binding between cell surface receptors and counteradhesion molecules (ligands) on cells or substrates [29]. Figure 2.6a shows the regions of cell non-specific attraction from membrane surfaces and cell specific attraction from adhesion receptors in a plot of interaction energy vs. cell separation. The receptor-ligand recognition is usually thought as a lock-and-key mechanism. Hundreds of receptors and the associated complementary ligands are involved in cell adhesion. Most receptors are cell surface proteins or glycoprotein macromolecules. Typically, they are embedded in the cell membrane surrounding the cell, and have sections that are extracellular (typically 2–50 nm in length), transmembraneous (typically 6–8 nm in length), and intracelluar. There are several families of cell adhesion receptors, including (a) integrins, (b) immunoglobulin adhesion molecules, (c) selectins, and (d) cadherins. A schematic of these molecules is presented in Figure 2.6b. Note that there is occasional cross-reactivity between receptor and ligands, although receptors bind ligands with high affinity and selectivity.

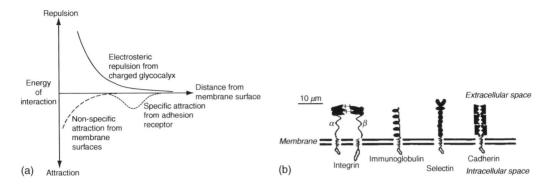

FIGURE 2.6 (a) Energies of interaction in cell–cell recognition include nonspecific attraction, electrosteric repulsion from the charged glycocalyx, and specific attraction from intersurface bonding, (b) Schematic diagram of four of the major classes of adhesion receptors. (Reproduced from Hammer, D. A., *Annu. Rev. Mater. Sci.*, 26, 651–691, 1991. With permission.)

Those individual adhesion molecules in cells often enable the formation of focal adhesion sites, specialized and discrete regions of the plasma membrane through which cells adhere to the substrate surfaces by large adhesion protein complexes. Those focal adhesion sites are closed junctions with 10–20 nm gap between a cell membrane and substrate surface [32], consisting of a complex network of extracellular matrix, membrane, and cytoplasmic components.

Typical cellular adhesion apparatus and cytoskeletal structures are shown in Figure 2.7. It is shown that several actin-binding proteins, such as vinculin, tallin, and paxillin, play roles in mediating the connection between the actin microfilaments in cytoskeleton and the membrane receptor proteins, such as integrins [32,33,37]. Basically, the small intracellular domain of integrins binds to the actin-binding proteins, which in turn bind to actin. Therefore, membrane receptors can relay information about the extracellular environment (ECM or substrate) to inside the cell

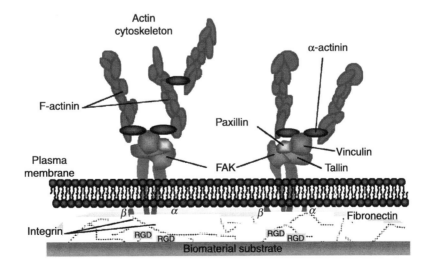

FIGURE 2.7 Schematics of cellular adhesion apparatus and cytoskeletal structures during cell contact and adhesion to a biomaterial substrate. (Reproduced from Mwenifumbo, S., *Cell/Surface Interactions and Adhesion on Biomedical and Biomems Surfaces: Effects of Laser Micro-Texturing and Titanium Coatings*, Princeton University Thesis, Princeton, NJ, 2004. With permission.)

(nucleus, cytoplasm, and cytoskeleton) via signaling pathways. This in turn can affect cell adhesion, spreading, or locomotion, and consequently regulate cell growth, differentiation, and gene expression [33,37].

Cell adhesion is also very dynamic, since the adhesion receptor tendency to bind ligands can be altered by changes in temperature, pH value, or ionic strength [29]. The change of cell adhesion can also be due to the mechanical stimuli that often lead to changes in the cytoskeleton that may include actin microfilament depolymerization and stress fiber alignment [38,39].

2.5 STAMPING METHODS FOR FABRICATING ORGANIC ELECTRONIC DEVICES

Some interesting interfacial contact and adhesion phenomena in soft materials are associated with stamping processes involved in the fabrication of organic electronic devices. These techniques have been introduced as a potentially low cost option for organic electronic device fabrication [40–43]. A rigorous understanding of these manufacturing processes for organic electronic devices could be necessary for the fabrication of flexible bioMEMS, which will be ultimately based on flexible organic electronics technology, instead of rigid Si technology.

Rogers et al. [44] have successfully demonstrated "paperlike" electronic displays made from organic thin film transistors (OTFTs) using a stamping method (micro-contact printing) for the patterning of source/drain electrodes. High-resolution patterning of organic polymer light-emitting pixels was also reported by using imprinting lithography techniques [45]. However, those techniques were limited to the substrate patterning before the deposition of an organic layer [46]. The involvements of either the wet etchant in micro-contact printing [44], or relatively high temperature in imprinting lithography [45], can easily result in the degradation of the organic layer, if the substrate patterning is done after the organic layer deposition.

Cold-welding techniques reported by Kim et al. [40–43], as shown in Figure 2.8, provided a good opportunity for the realization of post-deposition electrode patterning (i.e., the patterning of a metal film after an organic layer is deposited). This method is based on the transfer of a thin film of metal between substrate and stamp, using a cold-welding process. It results in the formation of metallic bonds at room temperature. The bond is formed between two metallic surfaces subjected to pressure.

Initially, subtractive cold welding was studied [40] (Figure 2.8a). In this technique, submicrometer patterning of the metal electrodes was achieved through selective lift-off of the metal cathode layer, by pressing a pre-patterned, metal-coated silicon stamp on the unpatterned organic light-emitting device layers to form strong cold-welds, and then separating the stamp from the substrate to remove the cathode metal in the regions contacted by the stamp. Here, the metal layer subtraction relied on both the cold-welding process and fracture of the cathode metal coating on the top of organic layers. Typically, a very high pressure (300 MPa) was required for a successful patterning.

In the additive cold-welding technique [41] (Figure 2.8b), the metal film on the stamp was transferred to the substrate due to the strong cold welds and relatively weaker adhesion reduction layer, upon separation of the stamp from the substrate. This additive process differed from the subtractive process in that it relied only on the cold-welding process and the delamination occurred on a stamp rather than on a substrate. Therefore, the required pressure (150 MPa) was significantly reduced compared to that for the subtractive method (300 MPa). However, high pressure was still the remaining problem of the cold-welding techniques. For example, during the source and drain patterning of pentacene OTFTs, the pentacene film was occasionally wrinkled as a result of the high shear stress at the pentacene–insulator interface [46].

By using flexible PDMS stamps, Kim et al. [42] found that cold-welding of gold layers can be used to fabricate organic light emitting devices (OLEDs) at pressures $\sim$1000 times lower than

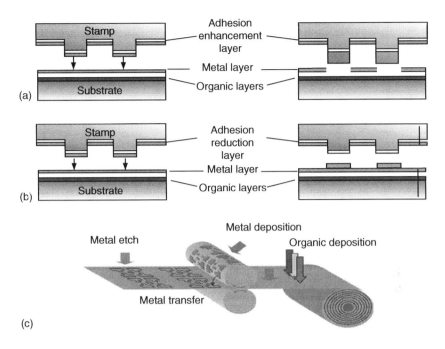

FIGURE 2.8 Direct micro/nanopatterning of an organic electronic device by cold-welding technique: (a) Subtractive method, (b) Additive method, and (c) Conceptual diagram of continuous and very-low-cost roll-to-roll manufacture of organic electronic devices. (Adapted from Forrest, S. R., *Nature*, 428, 911–918, 2004. With permission.)

those required using rigid Si stamps. Other researchers [47–49] have also found significant pressure differences between those required by the bonding of either two metal coatings, or one metal coating and a substrate using flexible vs. rigid stamps. Both theoretical and computational models have been used to explore the physics underlying cold-welding and stamping processes [40,50]. These have provided some initial insights into the contact-induced deformation and stress distributions in the films. Prior experimental and modeling efforts have demonstrated that soft materials on soft stamps can be transferred to substrates more easily than those on hard stamps due to the conformation of soft stamps to intrinsically rough substrate surfaces [2,50,51]. However, in cold-welding [42], the layer to be transferred was a stiff gold layer (with a corresponding Young's modulus of 106 GPa [52]). A soft PDMS stamp did not result in sufficient deformation of the gold and compensation for the intrinsic substrate surface roughness. In this case, the cold-welding pressure differences between stiff and compliant stamps might be explained partly by the differences in the deformation of the stamps around sub-micron and microscale dust particles that could be present. Therefore, the study of effects of dust particles and layer properties on organic electronic devices fabricated by stamping is needed [53].

It has also been shown that organic materials can be patterned using a similar method, by directly transferring an organic thin film from a stamp to a substrate, employing van der Waals adhesion in organic–organic interfaces [54]. This process extended the range of application of patterning via stamping to devices, where the active organic materials must be locally deposited on the substrate. However, on the step edges of a substrate with pre-existing patterns, some organic materials fractured, and therefore, still remained on the stamp, instead of being transferred to the substrate, due to the fragility of organic materials. Typically, many different organic thin film transfer modes exist depending on different film thicknesses, mechanical properties, and pattern geometries [55–57]. Since the patterning of organic materials on a substrate with pre-existing patterns and morphologies is crucial to the integration of various devices and interconnects [46], there is a need to investigate the mechanisms of patterning in an effort to provide new insights and guidance for the real device fabrication processes. Furthermore, computational, analytical, and experimental studies of the van der Waals bonding-assisted stamping process on patterned substrates have not been conducted [43].

2.6 METHODS FOR STUDYING CELL MECHANICAL PROPERTIES AND PRESTRESSES

As stated previously, a complete understanding of cell biophysics such as cell mechanical properties and cell adhesion and tractions is yet to emerge. There are various methods for studying the cell properties and cell interactions with biomaterials utilizing small scale contact and adhesion.

Micropipette suction [58–61] can be used to study the mechanical behavior of living cells by aspirating the surface of a cell into a small glass tube and then monitoring the displacement of the cell membrane extension via optical microscopy. A schematic of this experimental technique is shown in Figure 2.9a. Applied aspiration pressure can range from 0.1 to 1000 Pa, with a resolution of 0.1 Pa. The membrane displacement resolution can reach 25 nm by using light microscopy [61]. The applied stress state is relatively complex. Continuum approximations have been often used to extract the mechanical properties of the cell deformed by aspiration. The viscoelastic properties of various cells (articular chondrocytes and intervertebral disc cells) have been explored using this method [58,59].

Compression between glass microplates [36,62–64] has been used to investigate the response of the biological cells to mechanical forces under an optical microscope. A schematic of this experimental technique is shown in Figure 2.9b. Measurements of the uniaxial force applied to the cell and the resulting deformation were easy to determine using this technique. The main advantage of

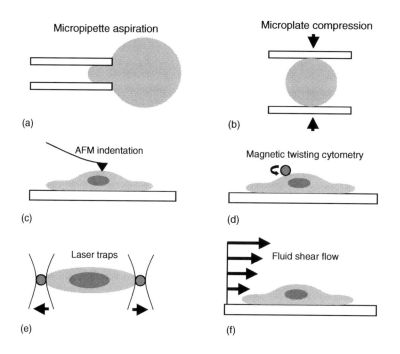

FIGURE 2.9 Schematics of various experimental techniques used for studying the cell mechanical properties: (a) micropipette aspiration, (b) Microplate compression, (c) Atomic force microscopy indentation, (d) Magnetic bead twisting cytometry, (e) Optical tweezers/laser traps, and (f) Fluid shear assay.

microplates is their simple geometry and surface properties [63]. A stiffening of fibroblasts was observed during spreading using this method. This is possibly correlated with the structural organization of the cytoskeleton during compressive deformation [36].

Commercially available atomic force microscopes (AFM) [65–69] can also be used to measure the mechanical properties of cells as well as to image cell surfaces (Figure 2.9c). An AFM typically produces estimates of cell stiffness based on the tip deflection data obtained as the tip is pushed into the cell surface. Also, an AFM can be used to characterize cell heterogeneity by indenting different regions of a cell. However, some atomic force microscopy studies of cell deformation have often suggested higher values of cell mechanical properties, than those obtained from other techniques [70]. Such results are probably due to substrate effects on the measured properties of cells [64,65].

A wide range of other techniques has also been used to measure the mechanical properties of cells [61,71–75]. These include magnetic bead cytometry [71–73], laser tweezers (also called optical tweezers, optical traps, or laser traps) [61,74,75], and force spectroscopy [61,76,77]. Schematics of these two experimental techniques are shown in Figure 2.9d and e, respectively. These techniques have also been used to explore the responses of biological cells to mechanical stimuli, i.e., mechano-transduction. In the case of laser tweezers and magnetic cytometry, control forces can be applied to cell surfaces by attaching particles to the cells with the use of an adhesive ligand or antibody to a specific receptor. Forces are generated on the particle through either magnetic fields [71–73] or laser trapping [74,75]. These two methods can be viewed as complementary to each other, since a fixed displacement can be applied, by using optical trapping and a fixed force can be applied by using magnetic trapping [70]. However, it tends to be difficult to determine the contact area between the bead and cell membrane. This results in some uncertainties in the local stresses applied on the cell [70]. Furthermore, the application of magnetic or laser fields puts the biological cells to be studied in a non-natural environment.

Shear assay experiments (Figure 2.9f) using a parallel-plate flow chamber have also been used to study the in vitro behavior of osteoblast-like cells subjected to shear flow [78–80]. The experimental shear flow setup is very similar to the natural environment of biological cells. For example, one of the major mechanical stimuli that bone cells (osteoblasts and osteocytes) respond to is the deformation-generated fluid shear stress exerted by the insterstitial fluid flow through the lacunar/canalicular spaces [78,79]. Though a great deal of research has used the shear assay technique to explore fluid flow shear stress-induced deformation [80,81], biochemical changes [82], and cell proliferation and differentiation [78,83], little is devoted to how this technique is used to jointly explore the cell adhesion and viscoelasticity, which tends to be very important to understanding the intrinsic features of cell response to mechanical stimuli [84].

The preexisting mechanical stresses in biological cells also play important roles in both physiological and pathological processes, such as cell growth [28], locomotion [85], remodeling of extracelluar matrix [86], even vertebrate brain formation [87], and cancer development [88]. Those so-called cell prestresses/tractions are commonly believed to be generated by the internal contraction capability of the cell cytoskeletal structures [89,90]. Tensegrity cell models have been developed to describe cytoskeletal prestresses by considering elastic and contractile actin microfilaments as tension elements and microtubules as compression-resistant elements [90,91]. In recent years, significant efforts have been made to understand cytoskeletal prestresses and its relationship with cell morphology, motility [85,92], and elasticity [77] using different experimental methods such as the wrinkling of silicone substrata [85], microfabricated structures [92–94], and a flexible gel substrate with embedded microbeads [76,77].

For example, as in the References [92–94], micropatterned substrates fabricated using lithographic techniques can be used for pointwise control of cell adhesion and traction measurement through the displacement of the pattern features. As is typical in the approach of microfabricated post array detectors (mPAD), Tan et al.[92] successfully manipulated and measured mechanical interactions between cells (Bovine pulmonary artery smooth muscle cells, 3T3 mouse fibroblasts, and Bovine pulmonary artery endothelial cells) and their underlying substrates by using micro-fabricated arrays of elastomeric, microneedle like PDMS posts.

Another typical example is the embedded particle tracking method [61,76,77], in which micro-scale beads were embedded within a flexible polymeric substrate, and traction forces exerted by adherent cells were measured at many points of cell-surface contact by analyzing bead displacements.

Although traction stresses generated by fibroblasts were first explored using thin silicone rubber substrata [85], where wrinkles appeared as a result of compression and stretching. However, no known theoretical method currently exists for cell traction estimation. This is due to the inherently nonlinear and chaotic process [95], that gives rise to wrinkling of silicone substrata [85].

These methods described above, such as wrinkling of silicone substrata [85], microfabricated structures [92–94], and a flexible gel substrate with embedded microbeads [76,77], provide direct quantitative or semiquantitative characterization of cell tractions mainly from the cell top view. The cell tractions and associated deformation observed from the side view will be complementary to the prior work. Recently, deformation/bending induced by the cell tractions on the side of the ridges of microgrooved PDMS surfaces has been reported [96]. This issue needs to be taken into account in the design of implantable bioMEMS, since significant substrate deformation could affect the performance of implantable bioMEMS structures.

REFERENCES

1. Beebe, D., Moore J., Bauer, J., Yu, Q., Liu, R., Devadoss, C., and Jo, B., Functional hydrogel structures for autonomous flow control inside microfluidic channels, *Nature*, 404, 588–590, 2000.
2. Xia, Y. and Whitesides, G. M., Soft lithography, *Angew. Chem. Int. Ed.*, 37, 551–575, 1998.
3. Quake, S. R. and Scherer, A., From micro- to nanofabrication with soft materials, *Science*, 290, 1536–1540, 2000.
4. Doerner, M. F. and Nix, W. D., A method for interpreting the data from depth-sensing indentation instruments, *J. Mater. Res.*, 1, 601–609, 1986.
5. Oliver, W. C. and Pharr, G. M., An improved technique for determining hardness and elastic modulus using load and displacement sensing indentation experiments, *J. Mater. Res.*, 7, 1564, 1992.
6. Pharr, G. M., Oliver, W. C., and Brotzen F., On the generality of the relationship among contact stiffness, contact area, and elastic modulus during indentation, *J. Mater. Res.*, 7, 613, 1992.
7. Saha, R. and Nix, W. D., Effects of the substrate on the determination of thin film mechanical properties by nanoindentation, *Acta Mater.*, 50, no. 1, 23–28, 2002.
8. King, R. B., Elastic analysis of some punch problems for a layered medium, *Int. J. Solids Struct.*, 23, 1657–1664, 1987.
9. Lou, J., Shrotriya, P., Buchheit, T., Yang, D., and Soboyejo, W. O., Nanoindentation study of plasticity length scale effects in LIGA Ni microelectromechanical systems structures, *J. Mater. Res.*, 18, 719–728, 2003.
10. Nix, W. D. and Gao, H., Indentation size effects in crystalline materials: a law for strain gradient plasticity, *J. Mech. Phys. Solids*, 46, 411–425, 1998.
11. Ma, Q. and Clarke, D. R., Size-dependent hardness of silver single-crystals, *J. Mater. Res.*, 10, 853–863, 1995.
12. Strojny, A., Xia, X., Tsou, A., and Gerberich, W. W., Techniques and considerations for nanoindentation measurements of polymer thin film constitutive properties, *J. Adhes. Sci. Technol.*, 12, 1299, 1998.
13. VanLandingham, M. R., Villarrubia, J. S., Guthrie, W. F., and Meyers, G. F., Nanoindentation of polymers: an overview, *Macromol. Symp.*, 167, 15, 2001.
14. Sun, Y., Akhremitchev, B., and Walker, G. C., Using the adhesive interaction between atomic force microscopy tips and polymer surfaces to measure the elastic modulus of compliant samples, *Langmuir*, 20, 837, 2004.
15. Gillies, G., Prestidge, C. A., and Attard, P., Determination of the separation in colloid probe atomic force microscopy of deformable bodies, *Langmuir*, 17, 7955, 2001.

16. Briscoe, B. J., Fiori, L., and Pelillo, E., Nano-indentation of polymeric surfaces, *J. Phys. D: Appl. Phys.*, 31, 2395, 1998.

17. Briscoe, B. J. and Sebastian, K. S., The elastoplastic response of poly(methyl methacrylate) to indentation, *Proc. R. Soc. A*, 452, 439, 1996.

18. Cao, Y., Yang, D. and Soboyejo, W. O., Nanoindentation method for determining the initial contact and adhesion characteristics of soft polydimethylsiloxane, *J. Mater. Res.*, 20, no. 8, 2004–2011, 2005.

19. Hertz, H., ber die berhrung fester elastischer korper, *J. Reine Angew. Math.*, 92, 156, 1882.

20. Johnson, K. L., Kendall, K., and Robertz, A. D., Surface energy and the contact of elastic solids, *Proc. R. Soc. London A*, 324, 301, 1971.

21. Derjaguin, B. V., Muller, V. M., and Toporov, Y. P., Effect of contact deformations on the adhesion of particles, *J. Colloid Interface Sci.*, 53, 314, 1975.

22. Maugis, D., Adhesion of spheres-the JKR-DMT transition using a dugdale model, *J. Colloid Interface Sci.*, 150, 243, 1992.

23. Johnson, K. L. and Greenwood, J. A., An adhesion map for the contact of elastic spheres, *J. Colloid Interface Sci.*, 192, 326, 1997.

24. Pietrement, O. and Troyon, M., General equations describing elastic indentation depth and normal contact stiffness versus load, *J. Colloid Interface Sci.*, 226, 166, 2000.

25. Kinloch, A. J., *Adhesion and Adhesives: Science and Technology*, London, Chapman and Hall, 1987.

26. Israelachvili, J. N., *Intermolecular and Surface Forces*. San Diego: Academic Press, 1992.

27. Kinloch, A., Review: the science of adhesion - part 1. surface and interfacial aspects, *J. Mater. Sci.*, 15, 2141, 1980.

28. Chen, C. S., Mrksich, M., Huang, S., Whitesides, G. M., and Ingber, D. E., Geometric control of cell life and death, *Science*, 276, 1425–1428, 1997.

29. Hammer, D. A., Biological adhesion at interfaces, *Annu. Rev. Mater. Sci.*, 26, 651–691, 1996.

30. Dubin-Thaler, B. J., Giannone, G., Dobereiner, H.-G., and Sheetz, M. P., Nanometer analysis of cell spreading on matrix-coated surfaces reveals two distinct cell states and steps, *Biophys. J.*, 86, 1794–1806, 2004.

31. Raucher, D. and Sheetz, M., Cell spreading and lamellipodial extension rate is regulated by membrane tension, *J. Cell Biol.*, 148, 127–136, 2000.

32. Walboomers, X. F. and Jansen, J. A., Cell and tissue behavior on micro-grooved surfaces, *Odontology*, 89, 2–11, 2001.

33. Anselme, K., Osteoblast adhesion on biomaterials, *Biomaterials*, 21, 667–681, 2000.

34. Frisch, T. and Thoumine, O., Predicting the kinetics of cell spreading, *J. Biomech.*, 35, 1137–1141, 2002.

35. Evans, E., Physical actions in biological adhesion, in *Lipowsky, R., Sackmann, E. (Eds), Handbook of biological Physics Series, vol 1b. Structure and Dynamics of Membranes. Elsevier, Amsterdam.*, 723–1754.

36. Thoumine, O., Cardoso, O., and J.-J. Meister, Changes in the mechanical properties of fibroblasts during spreading: a micromanipulation study, *J. Eur. Biophys.*, 28, 222–234, 1999.

37. Mwenifumbo, S., *Cell/Surface Interactions and Adhesion on Biomedical and Biomems Surfaces: Effects of Laser Micro-Texturing and Titanium Coatings*. Princeton, NJ: Princeton University, 2004.

38. Decave, E., Rieu, D., Dalous, J., Fache, S., Brechet, Y., Fourcade, B., Satre, M., and Bruckert, F., Shear flow-induced motility of dictyostelium discoideum cells on solid substrate, *J. Cell Sci.*, 116, no. 21, 4331, 2003.

39. Givelekoglu-Scholey, G., Orr, A. W., Novak, I., Meister, J. J., Schwartz, M. A., and Mogilner, A., Model of coupled transient changes of rac, rho, adhesions and stress fibers alignment in endothelial cells responding to shear stress, *J. Theor. Biol.*, 232, 569, 2005.

40. Kim, C., Burrows, P. E., and Forrest, S. R., Micropatterning of organic electronic devices by cold-welding, *Science*, 288, 831–833, 2000.

41. Kim, C., Shtein, M., and Forrest, S. R., Nanolithography based on patterned metal transfer and its application to organic electronic devices, *Appl. Phys. Lett.*, 80, 4051–4053, 2002.

42. Kim, C. and Forrest, S. R., Fabrication of organic light-emitting devices by low-pressure cold welding, *Adv. Mater.*, 15, 541–545, 2003.

43. Kim, C., Cao, Y., Soboyejo, W. O., and Forrest, S. R., Patterning of active organic materials by direct transfer for organic electronic devices, *J. Appl. Phys.*, 97, 113 512–1, 2005.

44. Rogers, J. A., Bao, Z., Baldwin, K., Dodabalapur, A., Crone, B., Raju, V. R., Kuck, V., Katz, H., Amundson, K., Ewing, J., and Drzaic, P., Paper-like electronic displays: Large-area rubber-stamped plastic sheets of electronics and microencapsulated electrophoretic inks, *Proc. Natl. Sci. U.S.A.*, 98, 4835, 2001.

45. Cheng, X., Hong, Y. T., Kanicki, J., and Guo, L. J., High-resolution organic polymer light-emitting pixels fabricated by imprinting technique, *J. Vac. Sci. Technol. B*, 20, 2877, 2002.

46. Kim, C., *Patterning of organic electronic devices*. Princeton, NJ: Princeton University Ph.D Thesis, 2005.

47. Ferguson, G., Chaudhury, M. K., Sigal, G. B., and Whitesides, G. M., Contact adhesion of thin gold-films on elastomeric supports — cold welding under ambient conditions, *Science*, 253, 776–778, 1991.

48. Jacobs, H. O. and Whitesides, G. M., Submicrometer patterning of charge in thin-film electrets, *Science*, 291, no. 5509, 1763–1766, 2001.

49. Loo, Y., Willett, R. L., Baldwin, K. W., and Rogers, J. A., Additive, nanoscale patterning of metal films with a stamp and a surface chemistry mediated transfer process: Applications in plastic electronics, *Appl. Phys. Lett.*, 81, 562–564, 2002.

50. Michel, B., Bernard, A., Bietsch, A., Delamarche, E., Geissler, M., Juncker, D., Kind, H., Renault, J.-P., Rothuizen, H., Schmid, H., Schmidt-Winkel, P., Stutz, R., and Wolf, H., Printing meets lithography: Soft approaches to high-resolution printing, *IBM J. Res. & Dev.*, 45, 697–719, 2001.

51. Hui, C. Y., Jagota, A., Lin, Y. Y., and Kramer, E. J., Constraints on microcontact printing imposed by stamp deformation, *Langmuir*, 18, 1394–1407, 2002.

52. Cao, Y., Allameh, S., Nankivil, D., Sethiaraj, S., Otiti, T., and Soboyejo, W., Nanoindentation measurements of the mechanical properties of polycrystalline au and ag thin films on silicon substrates: effects of grain size and film thickness, *Mat. Sci. and Eng. A*, 427, 232–240, 2006.

53. Cao, Y., Kim, C., Forrest, S. R., and Soboyejo, W. O., Effects of dust particles and layer properties on organic electronic devices fabricated by stamping, *J. Appl. Phys.*, 98, 033 713-1, 2005.

54. Kim, C., Cao, Y., Soboyejo, W. O., and Forrest, S. R., Fabrication of organic light-emitting devices by direct transfer of active organic materials using organic-organic adhesion, In *Proceedings of The 17th Annual Meeting of the IEEE Laser & Electro-Optics Society (LEOS)*, 1, 336, 2004.

55. Wang, Z., Yuan, J., Zhang, J., Xing, R., Yan, D., and Han, Y., Metal transfer printing and its application in organic field-effect transistor fabrication, *Adv. Mater.*, 15, 1009, 2003.

56. Bao, L.-R., Cheng, X., Huang, X., Guo, L., Pang, S., and Yee, A., Nanoimprinting over topography and multilayer three-dimensional printing, *J. Vac. Sci. Technol. B*, 20, no. 6, 2881–2886, 2002.

57. Tan, L., Kong, Y., Bao, L. R., Huang, X., Guo, L., Pang, S. W., and Yee, A., Imprinting polymer film on patterned substrates, *J. Vac. Sci. Technol. B*, 21, no. 6, 2742–2748, 2003.

58. Guilak, F., H. Ting-Beall, Baer, A., Trickey, W., Erickson, G., and Setton, L., Viscoelastic properties of intervertebral disc cells - identification of two biomechanically distinct cell populations, *SPINE*, 24, 2475, 1999.

59. Guilak, F., Tedrow, J., and Burgkart, R., Viscoelastic properties of the cell nucleus, *Biochem. Biophys. Res. Commun.*, 269, 781, 2000.

60. Hochmuth, R. M., Micropipette aspiration of living cells, *J. Biomech.*, 33, 15–22, 2000.

61. Van-Vliet, K., Bao, G., and Suresh, S., The biomechanics toolbox: experimental approaches for living cells and biomolecules, *Acta Mater.*, 51, 5881–5905, 2003.

62. Thoumine, O. and Ott, A., Time scale dependent viscoelastic and contractile regimes in fibroblasts probed by microplate manipulation, *J. Cell Sci.*, 110, 2109–2116, 1997.

63. Thoumine, O., Ott, A., Cardoso, O., and Meister, J. J., Microplates: a new tool for manipulation and mechanical perturbation of individual cells, *J.Biochem. Biophys. Methods.*, 39, 47–62, 1999.

64. Caille, N., Thoumine, O., Tardy, Y., and Meister, J., Contribution of the nucleus to the mechanical properties of endothelial cells, *J. Biomech.*, 35, 177, 2002.

65. Shiga, H., Yamane, Y., Ito, E., Abe, K., Kawabata, K., and Haga, H., Mechanical properties of membrane surface of cultured astrocyte revealed by atomic force microscopy, *J. Appl. Phys.*, 39, 3711, 2000.

66. Charras, G. T. and Horton, M. A., Determination of cellular strains by combined atomic force microscopy and finite element modeling, *Biophys. J.*, 83, 858, 2002.

67. Haga, H., Sasaki, S., Kawabata, K., Ito, E., Ushiki, T., and Sambongi, T., Elasticity mapping of living fibroblasts by afm and immunofluorescence observation of the cytoskeleton, *Ultramicroscopy*, 82, 253, 2000.

68. Sokolov, I., AFM study shows old cells lose their elasticity, *APS(American Physical Society) News*, page 1, May 2004.

69. Wu, H., Kuhn, T., and Moy, V., Mechanical properties of l929 cells measured by atomic force microscopy: Effects of anticytoskeletal drugs and membrane crosslinking, *Scanning*, 20, 389, 1998.

70. Huang, H., Kamm, R. D., and Lee, R. T., Cell mechanics and mechanotransduction: pathways, probes, and physiology, *Am J Physiol Cell Physiol*, 287, C1–C11, 2004.

71. Bausch, A., Ziemann, F., Boulbitch, A., Jacobson, K., and Sackmann, E., Local measurements of viscoelastic parameters of adherent cell surfaces by magnetic bead microrheometry, *Biophys. J.*, 75, 2038, 1998.

72. Bausch, A., Moller, W., and Sackmann, E., Measurement of local viscoelasticity and forces in living cells by magnetic tweezers, *Biophys. J.*, 76, 573, 1999.

73. Lo, C.-M. and Ferrier, J., Electrically measuring viscoelastic parameters of adherent cell layers under controlled magnetic forces, *Eur Biophys J*, 28, 112, 1999.

74. Dao, M., Lim, C. T., and Suresh, S., Mechanics of the human red blood cell deformed by optical tweezers, *J. Mech. Phys. Solids.*, 51, 2259–2280, 2003.

75. Dai, J. and Sheetz, M. P., Mechanical properties of neuronal growth cone membranes studied by tether formation with laser optical tweezers, *Biophys J*, 68, 988–996, 1995.

76. Dembo, M. and Wang, Y., Stresses at the cell-to-substrate interface during locomotion of fibroblasts, *Biophys J*, 76, 2307–2316, 1999.

77. Wang, N., Tolic-Norrelykke, I.-M., Chen, J., Mijailovich, S., Butler, J., Fredberg, J., and Stamenovic, D., Cell prestress. i. stiffness and prestress are closely associated in adherent contractile cells, *Am. J. Physiol. Cell Phys.*, 282, c606–c616, 2002.

78. Kapur, S., Baylink, D. J., and Lau, K. H. W., Fluid flow shear stress stimulates human osteoblast proliferation and differentiation through multiple interacting and competing signal transduction pathways, *Bone*, 32, 241, 2003.

79. Sikavitsas, V. I., Temeno, J. S., and Mikos, A. G., Biomaterials and bone mechanotransduction, *Biomaterials*, 22, 2581, 2001.

80. McGarry, J. G., Klein-Nulend, J., Mullender, M. G., and Prendergast, P. J., A comparison of strain and fluid shear stress in stimulating bone cell responses - a computational and experimental study, *Faseb J.*, 18, 1, 2004.

81. Suwanarusk, R., Cooke, B., Dondorp, A., Silamut, K., Sattabongkot, J., White, N., and Udomsang-petch, R., The deformability of red blood cells parasitized by plasmodium falciparum and p-vivax, *J. Infect. Dis.*, 189, 190, 2004.

82. Nauman, E. A., Satcher, R. L., Keaveny, T. M., Halloran, B. P., and Bikle, D. D., Osteoblasts respond to pulsatile fluid flow with short-term increase in PGE2 but no change in mineralization, *J. Appl. Physiol.*, 90, 1849, 2001.

83. Liegibel, U., Sommer, U., Bundschuh, B., Schweizer, B., Hischer, U., Lieder, A., Nawroth, P., and Kasperk, C., Fluid shear of low magnitude increases growth and expression of TGF beta 1 and adhesion molecules in human bone cells in vitro, *Exp. Clin. Endocrinol. Diabetes*, 112, 356, 2004.

84. Cao, Y., Bly, R., Moore, W., Gao, Z., Cuitino, A. M., and Soboyejo, W., Investigation of the viscoelasticity of human osteosarcoma cells using a shear assay method, *J. Mater. Res.*, 21, 1922–1930, 2006.

85. Harris, A. K., Wild, P., and Stopak, D., Silicone rubber substrata: a new wrinkle in the study of cell locomotion, *Science*, 208, 177–9, 1980.

86. Barocas, V. H. and Tranquillo, R. T., An anisotropic biphasic theory of tissue-equivalent mechanics: the interplay among cell traction, fibrillar network deformation, fibril alignment, and cell contact guidance, *Biomech. J.. Eng.*, 119, 137–145, 1997.

87. VanEssen, D., A tension-based theory of morphogenesis and compact wiring in the central nervous system, *Nature*, 385, 313–318, 1997.

88. Huang, S. and Ingber, D. E., Cell tension, matrix mechanics, and cancer development, *Cancel Cell*, 175–176, Sept. 2005.

89. Chicurel, M. E., Chen, C. S., and Ingber, D. E., Cellular control lies in the balance of forces, *Curr. Opin. Cell Biol.*, 10, 232–239, 1998.

90. Ingber, D. E., Tensegrity : the architectural basis of cellular mechanotransduction, *Annu Rev Physiol.*, 59, 575–579, 1997.

91. Wang, N., Naruse, K., Stamenovic, D., Fredberg, J. J., Mijailovich, S. M., I. M. Tolic-Norrelykke, Polte, T., Mannix, R., and Ingber, D. E., Mechanical behavior in living cells consistent with the tensegrity model, *Proc. Natl. Acad. Sci. USA*, 98, 7765–7770, 2001.

92. Tan, J., Tien, J., Pirone, D., Gray, D., Bhadriraju, K., and Chen, C., Cells lying on a bed of micro-needles: an approach to isolate mechanical force, *Proc. Natl. Acad. Sci. USA*, 100, 1484–9, 2003.

93. Galbraith, C. G. and Sheetz, M. P., A micromachined device provides a new bend on fibroblast traction forces, *Proc. of the Natl. Acad. Sci. USA*, 94, 9114–9118, 1997.

94. Parker, K. K., Brock, A. L., C. Brangwynne, Mannix, R. J., Wang, N., Ostuni, E., Geisse, N. A., Adams, J. C., Whitesides, G. M., and Ingber, D. E., Directional control of lamellipodia extension by constraining cell shape and orienting cell tractional forces, *FASEB*, 16, 1195–1204, 2002.

95. Munevar, S., Wang, Y., and Dembo, M., Traction force microscopy of migrating normal and h-ras transformed 3t3 fibroblasts, *Biophys. J.*, 80, 1744–1757, 2001.

96. Cao, Y., Chen, J., and Soboyejo, W., Investigation of the spreading and adhesion of human osteo-sarcoma cells on smooth and microgrooved polydimethylsilaxone surfaces, 2006, to be submitted.

97. Ge, J., Ph.D Thesis: *Interfacial adhesion in metal/polymer systems for electronics.* Espoo, Finland: Department of Electrical and Communications Engineering, Helsinki University of Technology, 2003.

98. Forrest, S. R., The path to ubiquitous and low-cost organic electronic appliances on plastic, *Nature*, 428, 911–918, 2004.

99. Evans, J., Liepmann, D., and Pisano, A. P., Planar laminar mixer, In *Proc. 10th IEEE Workshop on Micro Electro Mechanical Systems (MEMS 97), Nagoya, Japan*, 96–101, 1997.

3 Mechanical Characterization of Thin Film Materials for MEMS Devices

Jun Lou
Department of Mechanical Engineering and Materials Science,
Rice University

CONTENTS

3.1 Introduction..36
3.2 MEMS Basics and Challenges..37
    3.2.1 MEMS Fabrication Methods...37
    3.2.2 Selected MEMS Applications..38
    3.2.3 Introduction to LIGA Technologies and Applications......................39
    3.2.4 Materials and Reliability Challenges in MEMS...............................41
3.3 Overview of Mechanical Testing Methods at Microscale and Below.........42
    3.3.1 Uniaxial Mechanical Testing...43
    3.3.2 Membrane Testing...44
    3.3.3 Microbeam Testing..46
    3.3.4 Indentation Testing..48
3.4 Size-Dependent Plasticity and the Fracture and Fatigue Behavior
    of LIGA Ni Thin Film Materials 51
    3.4.1 Size-Dependent Plasticity...51
        3.4.1.1 The Concept of Strain Gradient Plasticity.......................52
        3.4.1.2 Plasticity Size Effects in LIGA Ni Thin Film Materials........53
    3.4.2 Fracture of LIGA Ni Thin Film Materials.......................................55
    3.4.3 Fatigue Behavior of LIGA Ni Thin Film Materials.........................56
3.5 Concluding Remarks 61
Acknowledgments..61
References..61

Abstract

Microelectromechanical systems (MEMS) combine mechanical and electronic functions in devices at very small scales and have become increasingly important for future technological developments. Advances in materials science and technology, including the understanding of material properties at small scales, have played key roles in the maturation of this exciting technology, and will remain to be an important and integral part of MEMS technology in the years to come. In this chapter, a brief overview of the basics of MEMS technology and some critical materials and reliability issues will first be given. Then, experimental testing methods used to extract the mechanical properties of MEMS thin film materials will be discussed. Finally, discussed are some of the theoretical developments and experimental findings from recent studies on the mechanical properties of LIGA (a German acronym for lithography, electroplating, and molding) Ni MEMS thin film materials at small scales comparable to MEMS devices.

3.1 INTRODUCTION

In his milestone talk, "There's Plenty of Room at the Bottom," given to the American Physical Society at the California Institute of Technology in 1959 [24], Nobel laureate Richard Feynman revealed an exciting picture of an entirely new world "at the bottom" for generations of people to explore. Feynman predicted the emergence of the micromachines that have now begun to flourish and that play an increasingly important role in many areas of science and technology today. Four decades later, scientists and engineers from many different disciplines have created machines (which usually mean "large, slow, and expensive") and integrated circuit devices (which usually mean "small, fast, and cheap") that have combined beautifully with each other, creating the term microelectromechanical systems (MEMS). Although MEMS-related research began as early as the 1950s, it was not until the 1990s that MEMS made its way from being a laboratory curiosity to becoming commercialized/developed in many industries, including the automobile, medical, electronics, communication, and defense industries [52].

Generally speaking, MEMS consist of both electronic and mechanical components. They can sense changes in the environment through sensor components, process information obtained in a control unit, and make a response through an actuation component if necessary. Furthermore, the dimensions of MEMS devices range from a few micrometers to a few millimeters, with key components sized in the micrometer range. Although all the sensing, processing, and actuating happen at the micron scale, MEMS structures are capable of generating macroscale effects [52]. Most importantly, MEMS is not just about the miniaturization of existing mechanical components and devices. It is about completely rethinking microsystem technology, i.e., the integration of suitable materials, technologies, and IC batch fabrication processes [62].

Despite being considered as one of the most promising technologies of the twenty-first century (some may even compare it to the success of the revolutionary integrated circuit microchips of the twentieth century), there are still many challenges and technical obstacles that must be addressed and overcome before MEMS can reach its true potential. Some of these challenges include materials and reliability issues, packaging and testing issues, and mass production and cost issues [2,62]. Among the materials-related issues are problems such as plasticity at the micron scale, fracture and fatigue at the micron scale, and tribological and adhesion problems. These problems have attracted considerable attention from the research community, simply because they are the fundamental issues that hinder the development of MEMS technologies, which ultimately impact everyday lives. Furthermore, at the micron scale, the physical and mechanical properties (Young's modulus, plastic behavior, fracture and fatigue, etc.) of materials deviate largely from those at the bulk scale [3,79]. There is therefore a need to study the physical and mechanical properties of MEMS structures. Moreover, a majority of current MEMS devices are made of silicon due to the convenience of utilizing the existing integrated circuit (IC) fabrication process, but silicon is known as not being a very robust engineering material. This promotes the use of "conventional" stronger metallic materials especially in the load-bearing mechanical components in MEMS devices. Similarly, new techniques like LIGA have been developed to accommodate these "nonconventional" MEMS developments, which are being considered for a number of emerging applications. This provides the motivation to study plasticity, fracture, and fatigue behavior in metallic thin films such as LIGA Ni thin film materials, which will be reviewed at the end of this chapter.

Therefore, this chapter will introduce a brief overview of MEMS fabrication methods with examples of applications, followed by a specific emphasis on LIGA technology and LIGA Ni MEMS applications. This is followed by a brief review of some critical MEMS materials and reliability issues, which have motivated recent interest within the materials science and mechanics communities in the small-scale mechanical behavior of materials used for MEMS applications. Then experimental testing methods to extract the mechanical properties of thin film materials at the micron scale and below will be discussed in some detail. Finally, the theoretical developments and experimental findings from recent research on size-dependent plasticity, and the fracture and fatigue behavior of LIGA Ni thin film materials at small scales comparable to MEMS devices will be discussed.

3.2 MEMS BASICS AND CHALLENGES

3.2.1 MEMS FABRICATION METHODS

There are three general classifications of MEMS fabrication methods: surface micromachining, bulk micromachining, and molding/LIGA processing [52]. We will focus on surface and bulk micromachining first, and then the LIGA process will be discussed in detail in the next section.

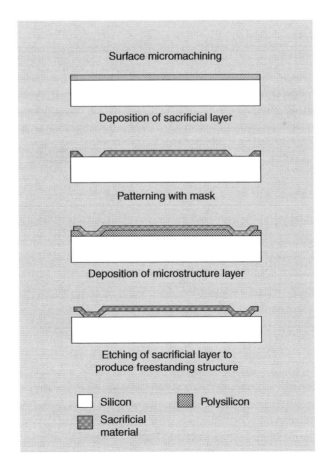

FIGURE 3.1 Schematic illustration of steps involved in surface micromachining. (From Madou, M., *Fundamentals of Microfabrication: The Science of Miniaturization*, 2nd ed., CRC Press, Boca Raton, FL, 2002. With permission.)

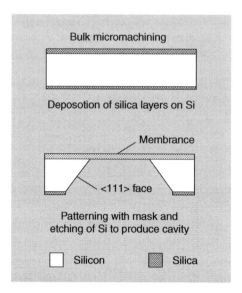

FIGURE 3.2 Demonstration of steps used in bulk micromachining. (From Madou, M., *Fundamentals of Microfabrication: The Science of Miniaturization*, 2nd ed., CRC Press, Boca Raton, FL, 2002.)

Surface micromachining involves an additive process on a substrate (typically a silicon wafer): The sacrificial and structural layers of thin films are deposited and patterned on the substrate to create a freestanding structure. These steps are illustrated schematically in Figure 3.1 [52] with an example of a simple freestanding polysilicon bridge. In contrast, bulk micromachining is a subtractive method and involves the removal of part of the bulk substrate, and is achieved by wet anisotropic etching or a dry etching method such as reactive ion etching (RIE), which creates large pits, grooves, and channels [62]. Figure 3.2 shows a simple demonstration of the steps used in bulk micromachining. Both methods produce two-dimensional (2D) structures [40,52], although more layers of thin film can be added to create much more complex multilayer MEMS structures [68].

3.2.2 SELECTED MEMS APPLICATIONS

In today's high technology market, a high volume of MEMS devices can be found in a range of applications [52]. The automobile industry has been the biggest user of MEMS technology, as MEMS-based accelerometers are used for airbag deployment in vehicles. More than 60 million MEMS accelerometers have been sold and put to use over the past ten years. The first commercialized air bag sensor manufactured by Analog Devices (1990) is shown in Figure 3.3a [52].

Another successful example of MEMS is the ink jet print heads invented by Hewlett Packard (HP) which make use of the thermal (caused by an electric driven resistor) expansion of ink vapor [7]. Disposable pressure sensors (shown in Figure 3.3b) have also been widely used to monitor blood pressure in hospitals [62]. These disposable pressure sensors have been the biggest MEMS medical application to date: The replacement of the large, burdensome, and expensive external sensors with the smaller and more economical sensors has completely changed industry views about the future of MEMS technology in the medical field. The digital micromirror device (DMD) (Figure 3.3c) manufactured by Texas Instruments contains more than a million tiny pixel-mirrors that can rotate $\pm 10°$ and more than 1000 times per second [36]. This MEMS technique is widely used for displays of PC projectors and high definition TV (HDTV).

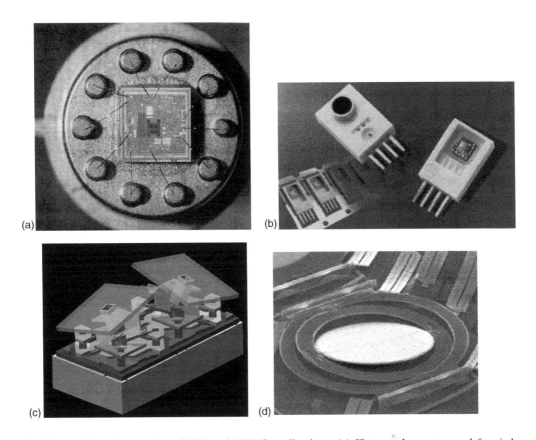

FIGURE 3.3 Selected examples of Si-based MEMS applications. (a) First accelerometer used for air bag deployment (Analog Devices), (b) Disposable blood pressure sensor (Motorola), (c) Schematic of digital micromirror devices (DMD) (Texas Instruments), and (d) MEMS optical switch consists of microscopic mirrors (Agere Systems).

MEMS are also actively pursued as optical switches that can be used to control and switch optical signals directly. The advantages of the MEMS-based optical switch are obvious: It saves the trouble of converting the optical signals to electrical signals when transferring data and converting the electrical signals back to optical signals when receiving. Agere Systems (previously part of Lucent Technologies) has an inspiring example of such devices and is shown in Figure 3.3d. Other potential MEMS applications including Bio-MEMS and RF MEMS are also emerging quickly in recent years [62].

3.2.3 Introduction to LIGA Technologies and Applications

LIGA, an acronym for the German words for Lithographic, Galvanoformung, Abformun (lithography, electroplating, and molding), is a technique used to produce MEMS from a wide range of materials including metals, ceramics, polymers, and plastics. The LIGA process was first developed at the Forschungszentrum Karlsruhe (FZK) in Karlsruhe, Germany in the mid-1980s. Since then, there has been a global effort in processing research, technology maturation, and materials research on LIGA technology [35,37,52]. A simple flow chart of the LIGA process is shown in Figure 3.4. This is followed by a schematic illustration of the process in Figure 3.5. Basically, the LIGA technique employs x-ray synchrotron radiation to expose polymethylmethacrylate (PMMA)

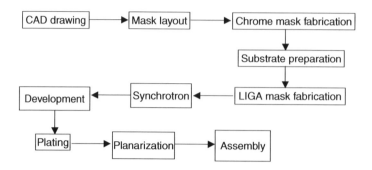

FIGURE 3.4 Flow chart of LIGA process.

x-ray photoresists. After chemically resolving the exposed area, metals (or other materials) can then be electrodeposited into the resist mold. Finally, a freestanding metal structure can be obtained after removal of the resist mold [52]. This method is particularly attractive in comparison to the material limitations of the bulk or surface micromachining methods (mainly Si-based materials). Also, given that Si is not the most robust engineering material available, exploring other methods that can be used to fabricate metallic or plastic MEMS structures is important.

LIGA parts typically have 1–10 μm minimum lateral feature sizes, and thicknesses between hundreds of microns and a few millimeters. This is especially advantageous for micromechanical systems, such as micromotors and micropumps, because the high aspect ratios allow high torques to be produced. Also, for microfluid devices, the high aspect ratios allow larger amounts of fluid to flow through the channels, which is often important to the measurement.

As stated above, MEMS devices produced by the LIGA technique have unique advantages in terms of load-bearing capabilities of mechanical components, as well as larger processing volumes in microfluidic channels. So far, this area of discussion has been focused on LIGA Ni for two reasons. First, there is a strong knowledge base on the electroplating of Ni [12,20,52], which in recent years has been applied to the plating of Ni [12,20,52], Ni–Fe alloys [20,52], and Ni–Co alloys [52] in MEMS structures. Second, LIGA Ni MEMS structures have been shown to have

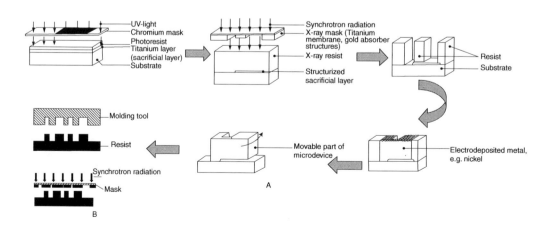

FIGURE 3.5 Schematic illustration of the LIGA process. (From Madou, M., *Fundamentals of Microfabrication: The Science of Miniaturization*, 2nd ed., CRC Press, Boca Raton, FL, 2002.)

FIGURE 3.6 LIGA Ni microturbine rotor with integrated optical fiber. (From Madou, M., *Fundamentals of Microfabrication: The Science of Miniaturization*, 2nd ed., CRC Press, Boca Raton, FL, 2002.)

attractive combinations of tensile properties (~ 500 MPa), ductility ($\sim 8\%$), Young's modulus (~ 180 GPa) [9,12,20,70,84], and fatigue resistance [17].

Commercial or nearly commercial LIGA Ni MEMS applications are emerging in Germany and the United States. Figure 3.6 shows a LIGA nickel microturbine rotor integrated with an optical fiber fabricated at the Forschungszentrum Karlsruhe. It can be used as the central part of a volumetric flow sensor, as the rotation of the rotor is caused by gas flow and can be recorded by the optical fiber to determine the flow speed [52]. LIGA Ni has also been pursued as an acceleration sensor with a capacitive readout (Figure 3.7). The advantages of the LIGA Ni accelerometer over the Si-based accelerometer are due to two facts. LIGA Ni accelerometer's ability to increase the density of seismic mass and Young's modulus of bending beams can increase the detected G-range. Also, the very high aspect ratios attainable with the LIGA technique result in improved signals and sensitivities [13]. For the same reason, electromagnetic micromotors made out of LIGA Ni produce much larger torques than surface micromachined Si motors [23]. Other potential LIGA Ni MEMS applications include inductive position sensors, electrical and optical interconnections, microfiltration membranes, microactuators, microvalves, and Fresnel lenses [52].

In summary, LIGA technology, especially LIGA Ni MEMS devices are playing increasingly important roles in the MEMS community. In order for this exciting class of devices to achieve their tremendous potential, the mechanical behaviors of LIGA Ni MEMS thin film materials need to be carefully studied at size scales comparable to those of LIGA Ni MEMS structures and devices.

3.2.4 MATERIALS AND RELIABILITY CHALLENGES IN MEMS

It has long been recognized that the greatest challenge to the success of MEMS technology is in improving its reliability [2,56]. The central part of this challenge is closely associated with the materials issues. Hence, it is not simply a case of obtaining the basic mechanical properties of materials being used in the device at comparable size scales to those of MEMS components (although this alone is not a trivial endeavor), but also to try gain an understanding of the materials' responses that lead to plasticity, fracture and fatigue. Identifying possible failure modes to provide valuable guidance to MEMS designers is also very important because MEMS is a new technology with poorly understood failure mechanisms. Furthermore, many emerging applications of MEMS are used in critical systems where the cost of failure is very high. For example, lives could be

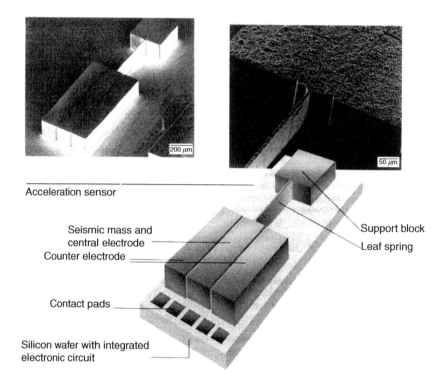

FIGURE 3.7 LIGA Ni acceleration sensor with a capacitive readout. (From Madou, M., *Fundamentals of Microfabrication: The Science of Miniaturization*, 2nd ed., CRC Press, Boca Raton, FL, 2002.)

endangered if air bag deployment accelerometers fail. Also, by identifying and eliminating just one failure mode of a MEMS device can save months of development time and hundreds of thousands of dollars of unnecessary redesign, fabrication and test expenses. Hence, developing an under-standing of the materials that could lead to improvements in the reliability of MEMS structures is of real importance.

3.3 OVERVIEW OF MECHANICAL TESTING METHODS AT MICROSCALE AND BELOW

As stated in the introduction, mechanical properties of thin films used as MEMS materials are very different from those of bulk materials [3,79]. There is therefore a strong need to develop new testing methods to extract mechanical properties at comparable size scales [10,46]. Over the years, efforts in this area have been focused on developing reliable, convenient, and accurate methods that can guide the design and reliability assessment of real MEMS devices, as well as provide insights for development of theories and mechanics models. Although the initial priority was for silicon-based materials, most of these developments have been subsequently extended to metallic systems as well. The following is a brief review of some of the testing techniques that have been utilized or have emerged over the past 10 to 15 years. The focus of this discussion will be on the testing methods of the freestanding thin films that are most commonly used in MEMS devices. Details of the experimental setup and procedures used for the character-ization of LIGA Ni thin film materials will also be introduced in the sections corresponding to the specific technique employed.

3.3.1 UNIAXIAL MECHANICAL TESTING

The uniaxial tensile testing method is the most straightforward technique available, at least in terms of the testing procedures. The basic setup of these so-called microtensile experiments is not much different from their large-scale counterparts. However, as both sample size and size of testing facilities decrease, issues like sample handling and alignment, as well as strain measurement, become more critical. An overview of uniaxial tensile testing of freestanding thin films, testing devices development, and some early results is given by Brotzen in a 1994 review paper. Since then, efforts have been devoted to improvements in sample handling, alignment, and strain measurement. Read et al. [65] proposed the concept of releasing tensile samples by etching away the substrate with a standard microfabrication process. Later, Sharpe et al. [69,71] proposed the use of engraved grips and an SiC fiber guide idea to improve sample alignment. Sample loading modules that can be detached from the testing machine and put under a microscope for more precise manipulation and better sample alignment have also been used in several groups [48].

The greatest advancement in testing has been achieved in strain measurements. Early efforts by Koskinen et al. [45] used only crosshead motion to estimate strain. This was improved by a variety of methods, such as strain gauges [33], eddy-current sensors [65], speckle interferometry [44,64], optical diffraction [38,66], and laser interferometry [19,69]. A method combining direct video measurements of local displacements (marker made by focus ion beam (FIB) etching process) and image analysis was also used [48].

The microtensile testing system assembled for the measurements of LIGA Ni thin film materials is shown schematically in Figure 3.8a. A photograph of the actual system used is presented in Figure 3.8b. The system is based on a design originally created by Sharpe and co-workers (1996) [69]. It consisted of a Velmex (Bloomfield, NY) unislide drive with a gear ratio of 1500:1 and provided the strain rate of $\sim 1.2 \times 10^{-4} \ \text{sec}^{-1}$ used in the tensile tests. The loads were measured with a microload cell obtained from Entran (Fairfield, NJ) with load range of 67 N with a precision of 0.1 N. The loads were recorded with a Labview (National Instruments, Austin, TX) software package. The load cell was attached to a frictionless Nelson (Milford, NH) air bearing, on which the mobile platform was mounted. The entire detachable loading module, including the stationary and mobile platforms, were machined simultaneously to produce a self-aligned integrated stage for the mounting and testing of the microsamples.

To facilitate the in situ observation of the displacement, the specimen was illuminated by a light source. An image acquisition system, coupled with an optical microscope and a digital camera, was used to capture the images of gauge displacement between two FIB markers prescribed onto the specimen. Continuous recording of the images of the two markers proceeded at a frequency of 5 Hz using IMAQ Vision in a Labview environment. The markers were located by block matching in terms of their coordinates in the series of images obtained during the test. The measured coordinates were used to calculate the displacements and strains. The LIGA Ni tensile specimens were loaded to failure in tension at a constant strain rate of $0.00012 \ \text{sec}^{-1}$. The data from the load cell and image analysis were synchronized and captured using a commercial data acquisition system.

In addition to the monotonic uniaxial mechanical testing techniques discussed above, cyclic load was also applied in the uniaxial setup to study the fatigue behavior of thin film materials in tension–tension mode [1,19]. The microfatigue testing system (Figure 3.9a and b) was developed based on the previously discussed microtensile testing system, with a piezoelectric actuator obtained from Polytech (Auburn, MA) to apply cyclic loading to specimens. This was attached in a series to a gripping block mounted on a vibration isolation table. Please note that other loading mechanisms, such as a voice-coil actuator were also used to apply cyclic loading in the micro-fatigue testing [18,19].

LIGA Ni thin film samples in dog-bone shape, similar to those used in the tensile tests, were mounted in the gripping blocks. The samples were connected to the load cell at one end and to

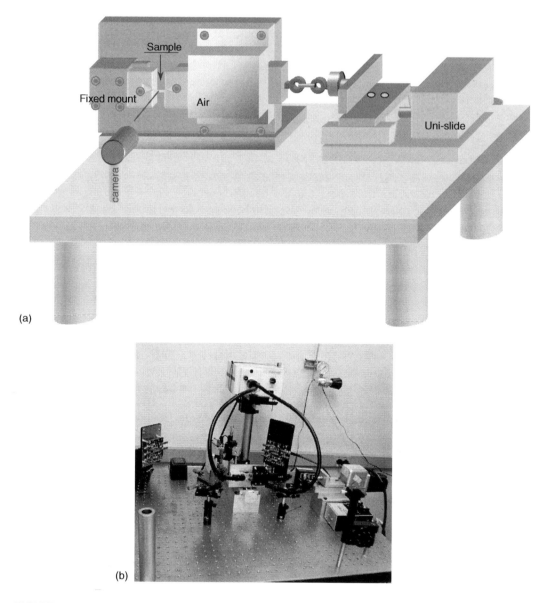

(a)

(b)

FIGURE 3.8 Microtensile testing system. (a) Schematic and (b) Photograph of actual system.

a rotational/translational stage at the other end. Alignment was performed with an optical microscope to ensure pure axial loading. The fatigue tests were performed under load control at a stress ratio, $R = \sigma_{min}/\sigma_{max}$, of 0.1. A cyclic frequency of 10 Hz was used, and the tests were continued until specimen failure occurred. By using this technique, the stress-life curve can be obtained for different freestanding thin film samples.

3.3.2 MEMBRANE TESTING

Another method developed and refined is the so called "bulge test" [5]. In this setup, the thin membrane sample is deflected with pressure on one side and the deflection at the center is measured.

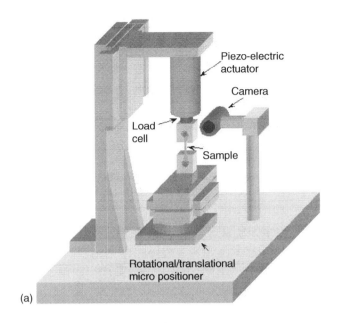

FIGURE 3.9 Microfatigue testing system. (a) Schematic and (b) Photograph of actual system.

A Schematic diagram of a bulge test is shown in Figure 3.10. Beams first developed this method in 1959 [5]. It was later refined by Nix and co-workers [34,80]. The stress and strain in the membrane are related to the pressure (P) and the bulge height (H), assuming plane strain conditions in the central portion of the membrane. This is given for a rectangular membrane (with an aspect ratio of more than two) by Vlassak and Nix [34,80]:

$$\sigma = \frac{Pa^2}{2Ht} \quad \text{and} \quad \varepsilon = \varepsilon_0 + \frac{2H^2}{3a^2} \tag{3.1}$$

where $2a$ is the width, and t is the thickness of the membrane. Thus, basic mechanical properties, such as modulus and yield strength can be obtained from the stress–strain curve. Recent work by Xiang and Vlassak [80,83] also extends this technique to study cyclic deformation behavior of thin film samples

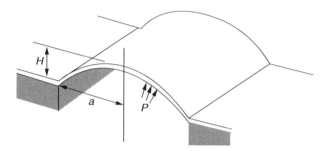

FIGURE 3.10 Schematic illustration of bulge test. (From Kraft, O. and Volkert, C. A., *Adv. Eng. Mater.*, 3, 99, 2001.)

under tension-compression mode. By microfabricating a bilayer structure consisting of targeted metallic thin film and a much stiffer Si_3N_4 film, they were able to introduce tensile stress in metal thin film during elastic loading of the composite structure and compression stress during the unloading procedure.

Since the freestanding thin membrane's edges are securely fixed, defects on the sample edges due to the fabrication process are negligible with respect to property measurement, and the application of biaxial stress to the thin film samples is a simple method. Due to the large surface-to-volume ratio of thin film samples, their insensitivity to edge defects in the bulge test is especially advantageous compared with the uniaxial tensile testing technique [10].

However, the samples used in a bulge test generally involve time-consuming microfabrication processes (which also complicate the experiments) and the thickness ranges of membranes that can be tested are also limited. Moreover, while attempting to dissolve the supporting substrate of metallic thin films on silicon substrates, pinholes can be developed in the films during the ablation process, creating more uncertainties in the bulge testing procedures [10].

3.3.3   MICROBEAM TESTING

This class of testing methods uses elastic beam theory to extract mechanical properties. For static beam tests (or microbeam deflection tests), the mechanical properties of a film can be investigated by measuring the deflection of a cantilever beam when a load is applied [82], as shown schematically in Figure 3.11. This method can be used for both freestanding and substrate-supported thin films. Florando et al. [28] improved this technique by modifying the beam geometry from

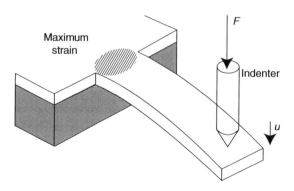

FIGURE 3.11 Schematic illustration of microbeam deflection test. (From Kraft, O. and Volkert, C. A., *Adv. Eng. Mater.*, 3, 99, 2001.)

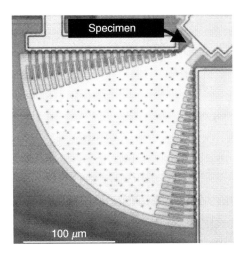

FIGURE 3.12 Real image of resonant microbeam actuated by capacitive comb drive.

rectangular to triangular to ensure that the bending moment per unit width is constant throughout the beam, resulting in uniform strain along the beam. Improving the technique made it possible to compare the beam deflection tests results to those obtained from the other techniques described above.

For resonant beam tests, the thin film samples being tested are attached to an excitation framework—usually a capacitive comb actuator [11,41,42], as shown in Figure 3.12. The advantage of this method is that exciting the test structure over a wide range of frequencies is easy. Detection of the resonant frequency is also easy. This is especially useful for the accelerated fatigue tests. However, the fabrication of the sample involves extensive microfabrication processes. Also, the release of the samples is often challenging due to stiction problems. Furthermore, the air-damping effect should be considered in non-vacuum environments.

Microbeam bending is the third type of microbeam testing method explored over the last few years. Stölken and Evans [76] pioneered the use of the microbend technique as a method to study the length scale effect. This method was then adapted and modified by Lou et al. [47,50] and then by Shrotriya et al. [72] to explore these phenomena further. The advantage of this method is that obtaining the applied bending moment from the beam theory is very straightforward. The experiments are also relatively easy to perform since there are no grips or load cells involved. This can effectively reduce the uncertainty of the measurement error associated with loading. The disadvantage is that relating the measured moment and surface strain directly to the stress–strain curve is difficult.

The experimental procedure for microbend experiments is illustrated schematically in Figure 3.13. As-deposited LIGA Ni beams of different thicknesses prepared by TEM sample preparation methods were bent around a series of tungsten fibers of different diameters. The final curvature of the deformed beams was measured after the elastic springback that occurred upon the release of the moment applied to the beams. Since the unloading is strictly elastic and the Young's modulus of LIGA Ni was known, the change in curvature due to elastic springback was used to determine the applied bending moment, M. The moment curvature relationship of the elastic springback is illustrated in Figure 3.13d.

An image acquisition system was utilized to capture the images of specimens during the microbend test to determine the in situ curvature of the beams. Figure 3.14 presents the images of the initially undeformed beam of 50 μm thickness and the final deformed shape after elastic

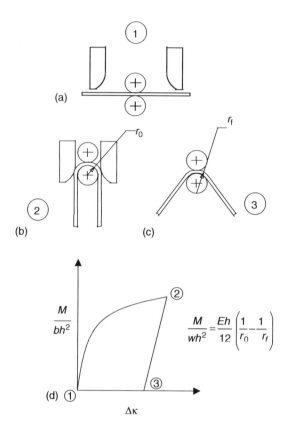

FIGURE 3.13 Schematic of the microbend test developed by Stölken and Evans (1998). (a) Initial configuration, (b) Bent configuration, (c) Final configuration due to elastic springback upon release, and (d) Extraction of bending moment from curvature measurements. (From Stölken, J. S. and Evans, A. G., *Acta Mater.*, 46, 5109–5115, 1998.)

springback. The images of the deformed beam were analyzed using image correlation and curve fitting techniques to determine the final curvature.

3.3.4 INDENTATION TESTING

Indentation tests have been widely used due to their simplicity—very little sample preparation is required. Traditionally, hardness is defined as the ability of a material to resist penetration by an indenter. However, the value of hardness is affected by many factors such as yield strength, strain hardening, surface roughness, and material pile-up and sink-in around the indentation making the interpretation of the results rather complicated. Furthermore, since the tested thin film sample is always supported by a substrate, it is the rule of thumb that the indentation depth should not exceed 10% of the sample thickness to avoid the effect of the substrate. Recently, with the advancement of depth-sensing indentation techniques (the indentation depth can be monitored continuously as the load is applied) that are commercially available, additional information such as Young's modulus, can be extracted from the load-displacement curve.

A typical load-displacement curve is shown in Figure 3.15. Following Oliver and Pharr's approach [60], the measured load-displacement curves were analyzed according to the equation

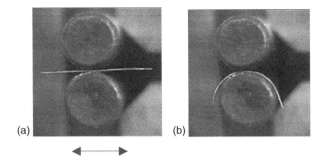

(a)   (b)

FIGURE 3.14 Images of a 50 μm thick LIGA Ni specimen undergoing microbend test. (a) Initial configuration and (b) Final configuration after elastic springback.

$$S = \frac{dP}{dh} = \frac{2}{\sqrt{\pi}} E_r \sqrt{A} \tag{3.2}$$

where S is the experimentally measured stiffness of the upper portion of the unloading data, A is the projected area of the elastic contact, and E_r is the reduced modulus, which is defined to effectively account for non-rigid indenters on the load-displacement behavior as follows:

$$\frac{1}{E_r} = \frac{(1-\nu^2)}{E} + \frac{(1-\nu_i^2)}{E_i} \tag{3.3}$$

E and ν are Young's modulus and Poisson's ratio for the specimen, and E_i and ν_i are the same parameters for the indenter. Rearranging Equation gives

$$E_r = \frac{\sqrt{\pi}}{2} \frac{S}{\sqrt{A}} \tag{3.4}$$

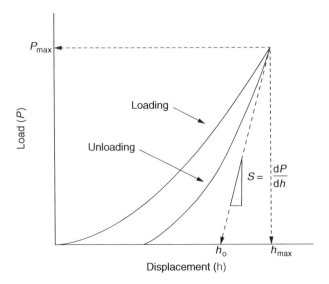

FIGURE 3.15 Typical load-displacement curve (After Oliver and Pharr, 1992).

So if the contact area at the peak load is known, the reduced modulus can be calculated from the measured stiffness. To determine the area functions for each tip, a series of indents at various contact depths were performed on a standard fused silica specimen and the contact area calculated using Figure 3.2 with necessary modifications as will be discussed later. A plot of the computed area as a function of contact depth is plotted and a fitting procedure is employed to fit contact area vs. contact depth curve to obtain a polynomial of the form given by the following:

$$A = C_0 h_c^2 + C_1 h_c + C_2 h_c^{1/2} + C_3 h_c^{1/4} + \cdots \qquad (3.5)$$

where A is the computed contact area, h_c is the contact depth that can be computed according to the following equation:

$$h_c = h_{max} - \varepsilon \frac{P_{max}}{dP/dh} \qquad (3.6)$$

for which the notations are given in Figure 3.15. The value of ε depends on the geometry of the indenter [10,60]: $\varepsilon = 1$ for flat punch, $\varepsilon = 0.75$ for a paraboloid of revolution, and $\varepsilon = 0.75$ for a conical indenter.

The hardness of the material is then defined in the usual way [78] as the mean pressure exerted by the indenter at maximum load. This gives

$$H = \frac{P_{max}}{A} \qquad (3.7)$$

where P_{max} is the maximum load applied during the indentation process and A is, again the projected contact area. Note that the above approach is the classical Oliver–Pharr approach for depth-sensing indentation experiments. However, some modifications are needed to account for the effects of tip rounding and Poisson's ratio [14,49,53], as well as material pile-up effect on the estimation of contact depth, and thus, contact area [8,15,49,55]. Other modifications involve the use of the so-called β and γ functions and were also discussed in the literature [16,21,30,43,49].

The nanoindentation experiments for LIGA Ni MEMS thin film materials (Lou et al. 2003b) [49] were performed using a TriboScope (TriboScope is a registered trademark of Hysitron, Inc., Minneapolis, MN) nanomechanical testing system. The force and displacement resolution of the Triboscope system are 1 nN and 0.0002 nm, respectively. The noise floor for the force is 100 nN, and the noise floor for the displacement is 0.2 nm. A sharp North Star indenter (a three-sided cube-corner tip) and a Berkovich indenter (a three-sided pyramidal tip) were used in the hardness measurements. The loading rate was kept at 1000 μN/sec, and a holding period of 15 sec was applied. Peak load ranges between 50 and 11,000 μN were applied in an effort to study the effects of impression size from a few microns to tens of nanometers.

Contact mode atomic force microscopy scans were also obtained before and after the indentation tests. Such contact mode AFM images were used for two purposes: first, to ensure that the indentations were only performed on spots with relatively low roughness. This minimized the possible effects of RMS on hardness measurement [29]. RMS, the root-mean-square height of the surface features, is one of the measurements for surface roughness. Indents were only introduced into regions with RMS <6 nm; and second, to estimate the effects from material piling up around indents on hardness measurements [49]. Also, a load-displacement curve was recorded during each test. Finally, from known area functions of the calibrated tips, hardness and Young's modulus could be calculated using a modified Oliver–Pharr approach as discussed before [49].

In addition, some special cautions were taken in the experiments to ensure that the results provided a true representation of property measurements. First, to account for load frame compliance, the contact compliance was obtained by subtracting the machine compliance from the total measured compliance of the testing system. Also, the tests were done in a quiet room with the temperature maintained at a constant 20°C. An enclosure was also placed around the nanoindentation system to minimize the possible effects of airflow currents. Furthermore, relatively short holding periods were used to reduce the possible effects of thermal drift on the measured nanoindentation hardness data.

3.4　SIZE-DEPENDENT PLASTICITY AND THE FRACTURE AND FATIGUE BEHAVIOR OF LIGA NI THIN FILM MATERIALS

After establishment of the most commonly used experimental setup to study the microscale mechanical behavior of materials, we are now ready to further explore how those techniques can be utilized. Experimental procedures have already been discussed, and the details of the material processing and characterization of microstructures and microtextures have been published elsewhere [47–50,72]. Thus, in this section, we will focus on a review of recent theoretical developments, as well as some experimental findings for *size-dependent plasticity and the fracture and fatigue* behavior of LIGA Ni thin film materials. Please note that the thickness of these thin film samples varies from tens of microns to 100s of microns. This range of thicknesses was chosen to account for the needs of performing mechanical testing on specimens with dimensions that roughly scale with the size of typical LIGA components [20,71].

3.4.1　SIZE-DEPENDENT PLASTICITY

When a material is loaded beyond its elastic limit, it deforms plastically. The term "plasticity" refers to the permanent shape change of a deformed material, i.e., the deformation is not reversible. Plastic deformation can be accomplished by the diffusion of atoms, movement of dislocations, or homogeneous shear (twinning, martensitic transformations). The stress needed to cause irreversible deformation is defined as the flow stress. The theory of plasticity has been studied extensively over the last century and there is a significant body of literature on deformation mechanisms in bulk solids [31,58,73]. However, as the dimensions of new technologies such as MEMS devices decrease, the understanding is more limited. For example, it is not clear how materials plastically deform at the micron- and nanoscales. This has stimulated strong interest in the materials science and mechanics communities to develop relevant theories and experiments concerning the size-dependent plasticity of materials [3,39,59]. Moreover, a very careful study of plasticity at small scales is also required for the design of a large number of LIGA Ni MEMS or other metallic-based MEMS applications that involve components such as flexural parts that could be subject to plastic deformation in service.

Plastic deformation of metals or crystalline solids is often accomplished by the generation, accumulation, and interaction of dislocations. However, depending on the materials, microstructures, and testing conditions (including temperatures), the plasticity of solids may involve a range of mechanisms. These were well summarized in the literature for the different types of solids such as metals, ceramics, and polymers [73].

The classical theory of plasticity has been developed to fully characterize bulk crystal plasticity [32,58]. It describes the mechanics of plastic deformation in most engineering bulk solids and is based on an experimentally determined constitutive relation between stress and strain. Mathematical expressions have been developed for polycrystalline aggregates under simple loading conditions [32]. These are generally phenomenological in nature. However, the empirical flow rules of the crystal plasticity theory have been verified for their general applicability to a wide range of engineering materials. They have also been adopted by engineers for decades [73] and widely used to relate the plastic behavior of material in uniaxial tensile tests to the behavior of the same material under multiaxial stress states. Please note that these rules apply largely to isotropic materials. It is also assumed that the stress at one point is only a function of the strain at that point.

In the following section, a review of the recently developed strain gradient plasticity (SGP) theory relevant to size-dependent plasticity at small scales, as well as the reported length scale effects in plasticity will be given. The connections of small-scale plasticity to dislocation theories will also be discussed. Then, findings from recent experimental work on LIGA Ni thin film materials will be briefly reviewed.

3.4.1.1 The Concept of Strain Gradient Plasticity

Existence of a plasticity length scale in metals has been known for several decades. Since the pioneering work of Ashby [4], the existence of the plasticity length scale, i.e., size effects or length scale effects, has been attributed to the contributions of the geometrically necessary dislocations associated with strain gradients. The geometrically necessary dislocations are needed to maintain compatibility in the presence of strain gradients. They contribute a dislocation density, ρ_G, in addition to the statistically stored dislocations of density, ρ_s.

During plastic deformation, the density of statistically stored dislocations increases due to a wide range of processes that lead to the production of new dislocations. However, the density of geometrically necessary dislocations is directly proportional to the plastic strain gradient, i.e., $\partial \dot{\varepsilon}^P / \partial x \propto \dot{\rho}_G b$, and the geometrically necessary dislocations accumulate at a rate that is consistent with $\partial \dot{\varepsilon}^P / \partial x \propto \dot{\rho}_G b$. Also, if L is the average distance traveled by a newly generated dislocation, then the rate of accumulation of strain due to statistically stored dislocations scales is $\dot{\varepsilon}^P = \dot{\rho}_s bL$, while the rate of strain accumulation of geometrically necessary dislocations scales is $\dot{\varepsilon}^P \propto (\partial \dot{\varepsilon}^P / \partial x)L$. Depending on the relative magnitudes of these two strain increments, plastic strain accumulation may be dominated either by statistically stored dislocations, or by geometrically necessary dislocations.

In bulk metals and their alloys, the contributions from statistically stored dislocations to strain accumulations are much greater than those due to geometrically necessary dislocations. However, at the microscale, the increments in the two dislocations densities may be comparable, i.e., $\rho_s \sim \rho_G$. The length, l, is the length scale at which plastic strain gradients, $\partial \varepsilon^p / \partial x$, significantly influence the strain increments through the Taylor relationship, $\tau = \alpha \mu b \sqrt{\rho_s + \rho_G}$, where α depends on the crystal type, μ is the shear modulus, and b is the Burgers vector.

The length scale, l, has been derived by Nix and Gao [59] to be $\sim b(\mu/\sigma_0)^2$, where σ_0 is a reference stress in the absence of strain gradient. This length scale, l, has also been determined experimentally. The early experiments for determining length scale parameter were performed by Fleck et al. [25] They performed torsional experiments on copper wires and showed that the thin wires ($\sim 15\ \mu m$) required significantly higher torques than thicker wires for the same strain. Fleck and Hutchinson [25–27] have also developed phenomenological theories of strain gradient plasticity, which are amenable to finite element implementation.

Subsequent efforts to explain the length scale effects associated with indentation experiments [6,51,61], fracture [81], and microbend experiments [76] resulted in the introduction of two length scale parameters: one associated with the rotational gradients, l_R, and the other associated with stretch gradients, l_s [6,25]. The basic equations of the Fleck–Hutchinson deformation theory (as extensions of conventional J_2 theory), definitions of length scale parameters, and the mechanism-based strain gradient plasticity (MSG) approach [22,59,75], discussed here, are based on detailed consideration of the contributions of geometrically necessary and statistically stored dislocations (GNDs and SSDs) to the overall plasticity. Given that these topics are well documented in the literature, further discussion will be omitted here due to the scope of the current work. Instead, we will briefly summarize the experimental results for LIGA Ni thin film materials in the following section.

3.4.1.2 Plasticity Size Effects in LIGA Ni Thin Film Materials

Following the approach described above for nanoindentation experiments, the hardness data obtained by North Star cube corner tip were analyzed. This data exhibited a strong size dependence on the as-received LIGA Ni film (200 μm thick). As suggested by Nix and Gao [59], the square of the hardness can be plotted against the reciprocal of the depth of indentation. The linear fit gives estimates of H_0 and h^* as 1.64 GPa and 1362.3 nm, respectively. Where H_0 is the hardness in the absence of strain gradient and h^* is a fitting parameter related to Burgers vector, indenter geometry, and material property. Also, the data can be displayed as a plot of $(H/H_0)^2$ vs. $1/h$, as shown in Figure 3.16. A very good linear relationship is shown clearly in this figure, indicating that the mechanism-based strain gradient plasticity (MSG) model can be extended to a nanoscale regime for very fine grained polycrystalline material in the presence of a sharp enough indenter tip, i.e., significant strain gradients. The microstructural length scale parameter, $\hat{l}$, can be computed based on the MSG theory [59]. Using the measured value of H_0, and taking the Tabor factor as 3, b as 0.25 nm for the f.c.c. nickel structure, and the shear modulus as 73 GPa of LIGA Ni films, we obtain $\hat{l} = 2.2$ μm. The extracted microstructural length scale parameter, $\hat{l}$, is actually related to the material length scale parameter, l, defined by Fleck and Hutchinson [26]. The relationship is given by the following Equation [59]:

$$l \approx \hat{l} \left(\frac{\sigma_0}{\sigma_{\text{ref}}} \right)^2$$ (3.8)

where σ_{ref} is taken to be a measure of the yield stress and σ_0 is the flow stress in the absence of a strain gradient effect, i.e., the contributions from statistically stored dislocations. So, taking a yield stress of 277 MPa obtained from larger samples of LIGA Ni MEMS material [12] as σ_0 is not unreasonable. Using the yield stress of 437 MPa for LIGA Ni, the material length scale parameter can be calculated from Equation 3.8 to be $l \sim 0.9$ μm.

For microbend experiments, the measured changes in beam curvatures, $\Delta\kappa$, are converted into the bending moment, M, by using the following expression:

$$\frac{M}{wh^2} = \frac{\bar{E}h\Delta\kappa}{12}$$ (3.9)

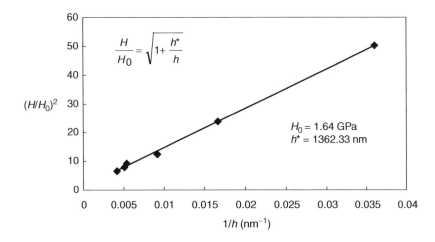

FIGURE 3.16 Depth dependence of the hardness of the as-plated LIGA Ni from sulfamate bath.

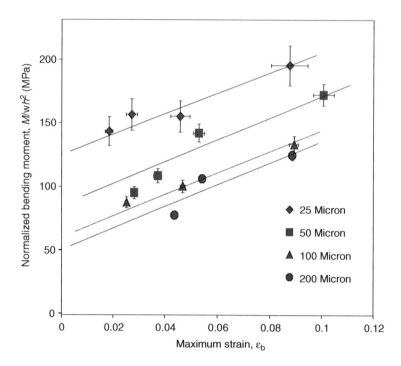

FIGURE 3.17 Plots of bending moment against surface strains for all film thickness.

where w is the width of the beam, h is the beam thickness, and $\bar{E}$ is the plane strain, Young's modulus for LIGA Ni ($\bar{E} = 189$ GPa). The results are plotted as a function of surface strain, ε_b, for four different beam thicknesses (25, 50, 100, and 200 μm) in Figure 3.17. The regression fits for different thicknesses are also shown. The normalized moments needed to induce similar strains are significantly larger for thinner beams ($\sim 25\ \mu$m) than the thicker beams ($\sim 200\ \mu$m). In the absence of a size-dependent plastic response, the normalized bending moment curves corresponding to different thicknesses should be coincident. Accordingly, the differences in material responses are attributed to length scale dependence [76].

The bending moments obtained from the microbend test are utilized to determine the composite length scale parameter, l_c, following the strain gradient plasticity analysis developed by Stölken and Evans [76]. This gives

$$l_c = l_R \sqrt{1 + \frac{8}{5}\left(\frac{l_s}{l_R}\right)^2} \qquad (3.10)$$

where l_c is the composite length scale parameter, l_R is the rotational gradient length scale parameter, and l_s is the stretch gradient length scale parameter. From the framework of Fleck and Hutchinson [26], the strain energy density in the presence of strain gradients can be defined as [72]

$$w = \frac{\varepsilon}{2}(E_P\varepsilon + 2\Sigma_0) \qquad (3.11)$$

where E_P (hardening coefficient) and Σ_0 (effective yield strength) are the material constants obtained to be: $\Sigma_0 = 400 \pm 20$ MPa and $E_P = 1.03 \pm 0.09$ GPa. Then the energy density was integrated [72,76] throughout the thickness to obtain the total energy of the beam, and differentiated with respect to the neutral axis curvature in order to express the bending moment, M, as a function of the surface strain, ε_b,

$$\frac{4M}{\sum_0 wh^2} = \left(\frac{2}{\sqrt{3}}\right)\left(\sqrt{2\beta^2+1} + 2\beta^2\ln\left(\frac{\sqrt{2\beta^2+1}+1}{\sqrt{2\beta^2+1}-1}\right)\right) + \left(\frac{\varepsilon_b E_P}{\sum_0}\right)\left(\frac{8}{9} + \frac{16}{3}\beta^2\right) \quad (3.12)$$

where $\beta = l_c/h$. The nonlinear data fitting was performed for the bending moment, M, to all of the bending data, with l_c as the only unknown using the built-in functions of the symbolic computation software Mathematica (Wolfram, Illinois). The length scale parameter was thus determined to be $l_c = 5.6 \pm 0.2$ μm.

Now, we may extract the rotational gradient length scale parameter as defined before from the obtained composite length scale parameter, l_c, using Equation 3.10. From the nanoindentation experiments, we just calculated the material length scale parameter to be $l \sim 0.9$ μm. This length scale parameter corresponds essentially to the stretch gradient parameter since the stretch gradient plays a dominant role in the nanoindentation tests. The computed parameter is on the same order of what Begley and Hutchinson [6] obtained from earlier studies on W, Cu, and Ag single crystal, as well as Cu polycrystal. These gave an estimate of l on the order of $0.25 \sim 1$ μm. From the composite length scale parameter, l_c of 5.6 μm, the rotational gradient length scale parameter can be estimated to be ~ 5.5 μm. The measured values of l_R and l_s can now be incorporated into the extended J_2 plasticity theory to obtain the new constitutive law. This is given by

$$\varepsilon = \left(\left(\frac{2}{3}\varepsilon'_{ij}\varepsilon'_{ij}\right) + (0.9^2\eta'_{ijk}\eta'_{ijk}) + \left(\frac{2}{3}5.5^2\chi_{ij}\chi_{ij}\right)\right)^{1/2} \quad (3.13)$$

where ε'_{ij} is the deviatoric part of the strain, η'_{ijk} is the deviatoric part of the strain gradient, and χ_{ij} is the deformation curvature.

Incorporation of above measured plasticity length scale parameters into the strain gradient plasticity framework of [26], therefore, provided a framework for continuum modeling from the micron to the submicron scales. Within this regime, the extended J_2 framework may be incorporated into analytical or numerical models for the determination of: local stress–strain distributions, cracktip fields, and failure criteria for thin film materials and structures used for MEMS applications.

3.4.2 FRACTURE OF LIGA Ni THIN FILM MATERIALS

It is well known that electroplating processes can often be tailored to optimize the properties required for the best performance of the nickel electrodeposits. Several factors such as solution conditions (including composition, temperature, pH level, organic additives, etc.), impurities, current densities, and agitations can greatly affect the mechanical properties of Ni electrodeposits [67]. Thus, the diverse and ample data for mechanical properties measurements on electrodeposited Ni that exist in the literature are due to this complexity.

The microtensile test results obtained for the 200 μm thick LIGA Ni sample are plotted in terms of true stress and true strain in Figure 3.18a. This shows that the plastic deformation starts at very low strain levels (~ 0.01). However, because of compliance problems with the machine frame, obtaining accurate measurements of the elastic deformation of the sample is very difficult. A stiffer load frame/load train design is therefore needed to obtain measurements of elastic deformation and the Young's modulus. In this study, a Young's modulus of 189 GPa is obtained by a linear fit of limited data in an elastic regime from a 50 μm sample test and is used as a material constant in our estimation for a 0.2% offset yield stress in other cases. Nevertheless, the ultimate tensile strength measured for the 200 μm sample was 547 MPa. The maximum load applied at the onset of sample failure was ~ 16 N. In the case of the 100 μm sample, the stress–strain behavior was similar to that

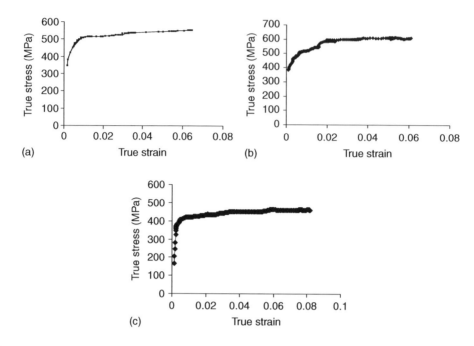

FIGURE 3.18 Microtensile results of the LIGA Ni sample plated from sulfamate baths with different thicknesses. (a) 200 μm, (b) 100 μm, and (c) 50 μm.

of the 200 μm thick sample (Figure 3.18a and b). The ultimate tensile strength of the 100 μm structure was 587 MPa and the maximum load at the onset of sample failure was $\sim$12.7 N. Similarly, as shown in Figure 3.18c, early evidence of plastic deformation is apparent in the 50 μm sample. The ultimate tensile strength in this case was 497 MPa and the maximum load at the onset of sample failure was $\sim$3.9 N. There is therefore no evidence of a thickness effect in LIGA Ni MEMS structures with thicknesses between 50 and 200 μm. Mazza et al. [54] reported a yield stress of $\sim$405 MPa, an ultimate strength of $\sim$782 MPa, and a Young's Modulus of $\sim$202 MPa. The difference in numbers is probably because the material they were testing has a different microstructure and is obviously stronger. Nevertheless, the values of tensile properties are in fairly good agreement with the work of Sharpe et al. [70,84] and Christenson et al. [20] for LIGA nickel films with similar columnar microstructures.

In all the thin film samples, there was clear evidence of necking and final shear fracture at an angle of $\sim$45° to the loading direction (Figure 3.19a–c). Rough transgranular fracture modes were observed in all cases, along with evidence of secondary cracks (Figure 3.19a–c). In the case of the 50 μm thick structure, a sharp edge is formed, following the onset of necking and shear fracture (Figure 3.19c). However, similar sharp edges were not observed in the thicker samples. Fracture in 100 and 200 μm thick foils occurs by a combination of faceted fracture and shear fracture. In the case of the 50 μm foil, the formation of a sharp edge is presumably a result of the combination of necking and cracking along shear planes that are at an angle of $\sim$45° to the loading axis.

3.4.3 FATIGUE BEHAVIOR OF LIGA NI THIN FILM MATERIALS

The word "fatigue" is from the Latin expression "fatigare" which means, "to tire" [77]. It was believed that a material "tires" over time when subjected to repeated loading and finally fails. Fatigue occurs as a result of repeated stresses or strains which lead to crack initiation/propagation prior to catastrophic failure. In most cases, materials fail at stress well below the static yield or

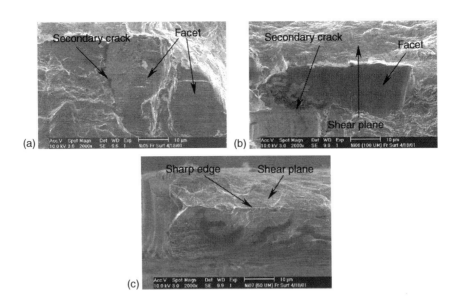

FIGURE 3.19 Typical fracture modes obtained from structures with different thicknesses. (a) 200 μm, (b) 100 μm, and (c) 50 μm.

fracture stress. It has been estimated that approximately 80%–90% of all industrial failures can be attributed to some forms of fatigue, resulting in many cases in severe financial loss or even loss of human lives [77]. Moreover, as MEMS devices start to push toward higher mechanical power levels and many of them may be subject to very high numbers of fatigue cycles during their service time due to the inherent high operating frequencies, fatigue also become a major concern in the small scales [74].

Mohr et al. [57] first studied the fatigue curve (S–N curve) for micromechanical elements in their acceleration sensor made of LIGA nickel using electromagnetic activation of cantilever deflection tests. For 10 μm thick and 200 μm long cantilevers, they found that the long-term fatigue behavior is comparable to that known from macroscopic nickel materials. Later, Hemker and co-workers [19] conducted fatigue tests for dog-bone shape LIGA Ni samples with thicknesses of ∼250 μm using a microfatigue tester driven by a voice-coil actuator. They employed a sinusoidal tension–tension cyclic loading condition with an R ratio of 0.1 and a frequency of 200 Hz in their fatigue tests. They suggested that the fatigue limit of LIGA Ni is approximately 195 MPa and the endurance ratio is 0.35, which is comparable to commercial annealed nickel. From a scanning electron microscopy (SEM) fracture surface study, they also proposed that the fatigue crack started at the corner of the microsamples and then propagated across the sample in a transgranular mode.

More recently, Boyce et al. [9] used a fixed-free cantilever beam subjected to fully-reversed bending to study the fatigue behavior of LIGA Ni. The width, thickness, and length of the microbeams were ∼26, ∼250, and ∼750 μm, respectively. From their fatigue-life study, the endurance limit was determined to be 0.37. A SEM inspection of the fatigued surface showed evidence of localized extrusions and intrusions associated with persistent slip bands (PSBs). Moreover, a thick (up to 400 nm) oxide–NiO layer was found on the surface of the PSBs and thought to be the source of crack initiation. This thick oxide layer and the associated oxygen-suppression mechanism decreased the fatigue life of LIGA Ni by more than an order of magnitude compared with tests done in low-oxygen environments. The thickening of the oxide is attributed to the disruption of oxide by motions of PSB extrusions and intrusions.

Stress-fatigue life data obtained for the 70 and 270 μm thick LIGA Ni specimens are compared with prior results by Cho et al. [18,19] and Mohr et al. [57] in Figure 3.20. Also included in

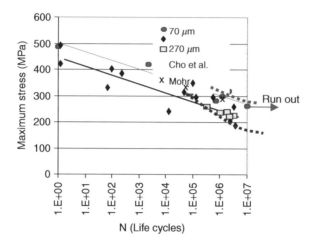

FIGURE 3.20 S–N curves for 70 and 270 μm thick LIGA Ni microsamples. Results of Cho et al. (2002, 2003) (for 250 μm thick samples). Mohr et al. (1992) are also included along with S–N curves for bulk nickel.

Figure 3.20 are data for bulk Ni. The thinner (70 μm thick) samples exhibited longer fatigue lives and increased endurance limits ($\sim$260 MPa) than the thicker samples (endurance limits $\sim$195 MPa). The high cycle fatigue results of Cho et al. [18,19] merge with the high cycle fatigue data obtained from the thicker (270 μm) samples. However, the data from the thicker samples were also close to those reported for bulk Ni in the annealed state [57].

The results of fatigue tests performed on the thinner ($\sim$70 μm thick) specimens were comparable to those reported for bulk Ni in the hardened condition examined by Mohr et al. [57] The results suggest that the 70 μm thick structures have better fatigue resistance than the 270 μm thick specimens. However, the increased fatigue strength is associated largely with the increased tensile strength of the thinner samples. This is consistent with prior work by Cho et al. [18,19], who have shown that the fatigue resistance of LIGA Ni MEMS thin films scales with tensile strength. Furthermore, as in classical face-centered cubic materials in bulk solids [73,77], the endurance limits corresponded to $\sim$35% of the tensile strengths.

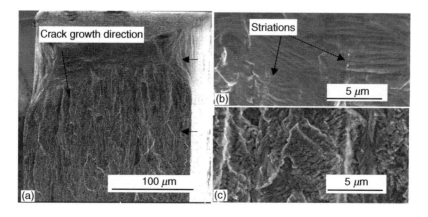

FIGURE 3.21 Fracture surface of 270 μm thick LIGA sample tested under fatigue at low stresses (high life cycles). (a) Overall view, (b) High magnification of fatigue crack propagation regime, and (c) High magnification of fast fracture regime.

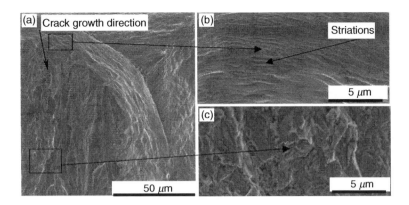

FIGURE 3.22 Fracture surface of 270 μm thick LIGA sample tested under fatigue at intermediate stress levels (intermediate life cycles). (a) Overall view, (b) Fatigue crack propagation regime, and (c) High magnification of fast fracture regime.

The fatigue fracture modes and the deformed gauge sections are presented in Figure 3.21 through Figure 3.25 for the 70 and 270 μm thick specimens. In the case of the 270 μm specimens tested at low stresses, corner fatigue cracks were observed initiating from the surfaces of the specimens. These propagated inwards at an angle of $\sim 45°$ (stage I fatigue cracks) via a flat fracture mode (Figure 3.21a). A transition to a rougher fracture mode, consisting of fatigue striations was then observed after the cracks reached lengths of ~ 20–30 μm (Figure 3.21b). Finally, final failure occurred by a ductile dimpled fracture mode (Figure 3.21c). Similar fatigue fracture modes were observed at different stress levels (Figure 3.22). However, the extent of stable fatigue crack growth was less at higher stresses. At very high stress levels (very low fatigue life) as shown in Figure 3.23, the fracture mode resembles the tensile fracture mode observed in monotonic loading condition, presumably due to overload type of failure (Figure 3.23).

In the case of the thinner 70 μm thick specimens, a significant amount of plastic deformation and necking occurred prior to final fracture (Figure 3.24a). However, the fatigue crack propagation generally occurred on planes that were mainly perpendicular to the loading axis. The fatigue fracture surfaces exhibited clear evidence of transcolumnar fracture (Figure 3.24b). The features observed at lower stresses were more consistent with mechanical fatigue. At intermediate and higher stresses, the fracture surfaces were dominated by evidence of plastic deformation (necking down to chisel-shaped wedges), secondary cracking, and faceted trans-granular crack growth. At the highest stresses, extensive necking was observed with limited evidence of rupture by ductile rupture (Figure 3.25). The fatigue of the thinner samples is therefore dominated by

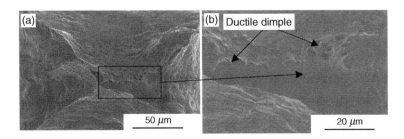

FIGURE 3.23 Fracture surface of 270 μm thick LIGA sample tested under fatigue at high stress levels (low life cycles). (a) Overall view and (b) Large ductile dimples.

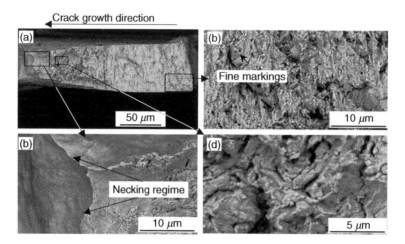

FIGURE 3.24 Fracture surface of 70 μm thick LIGA sample tested under fatigue at low stresses. (a) Overall view, (b) High magnification of fatigue mode fracture area, (c) Tensile mode fracture area and (d) Transition region.

extensive necking and ductile dimpled ruptures at intermediate and higher stresses. In contrast, transcolumnar fatigue fracture modes dominate at lower stresses.

The above results suggest that the stress-life behavior of LIGA Ni MEMS is associated with significant plastic deformation and classical stage I and stage II mechanisms [73,77]. However, the underlying fatigue deformation and crack growth mechanisms depend on the magnitudes of the applied stress ranges and the sample thickness.

In the case of the 70 μm thick structures, more extensive plastic deformation is observed along with predominantly stage II growth by transcolumnar crack growth modes. No evidence of fatigue striations was observed in these structures. However, in the case of the 270 μm thick structures, less extensive (but significant) plastic deformation is observed along with stage I and stage II crack growth. The stage I cracks initiate from the corners and extend inwards until they transform into stage II cracks. Most of the stage II crack growth in the 270 μm thick structures occurs by classical cracktip blunting mechanisms [77].

FIGURE 3.25 Fracture surface of 70 μm thick LIGA sample tested under fatigue at high stresses. (a) Overall view, (b) High magnification of fatigue mode fracture area, (c) Tensile mode fracture area, and (d) Transition region.

3.5 CONCLUDING REMARKS

Materials issues in MEMS technology and particularly the mechanical characterizations of MEMS thin film materials have been discussed in this chapter. After briefly reviewing the status of MEMS technology, the focus has been on the development of experimental techniques for the mechanical testing of thin films and small structures. Recent studies on size-dependent plasticity and the fracture and fatigue behavior of LIGA Ni thin film materials for MEMS applications were discussed in some detail. The purpose was to demonstrate how these experimental developments could be utilized to further our understanding of the mechanical behavior of materials at small scales, and to help resolve design and reliability issues in MEMS technology.

ACKNOWLEDGMENTS

The author wishes to acknowledge Professor W. O. Soboyejo for his encouragement and helpful discussions in the writing of this chapter. Contributions from Dr. S. Allameh and Dr. P. Shrotriya are also gratefully acknowledged.

REFERENCES

1. Allameh, S., Lou, J., Kavishe, F., and Soboyejo, W. O., *Mater. Sci. Eng. A*, 371, 256–266, 2004.
2. Arney, S., *MRS Bull.*, 26, 296–299, 2001.
3. Arzt, E., *Acta Mater.*, 46, 5611–5626, 1998.
4. Ashby, M. F., *Phil. Mag.*, 21, 399, 1970.
5. Beams, J. W., The structure and properties of thin film, *Mechanical Properties of Thin Films of Gold and Silver*, Wiley, New York, p. 183, 1959.
6. Begley, M. R. and Hutchinson, J. W., *J. Mech. Phys. Solids*, 46, 2049–2068, 1998.
7. Bishop, D., Heuer, A., and Williams, D., *MRS Bull.*, 26, 282–283, 2001.
8. Bolshakov, A. and Pharr, G. M., *J. Mater. Res.*, 13, 1049–1058, 1998.
9. Boyce, B., Kotula, P. G., and Michael, J. R., *Acta Mater.*, 52(6), 1609–1619, 2004.
10. Brotzen, F. R., *Int. Mater. Rev.*, 39, 24–45, 1994.
11. Brown, S. B., Povuk, S., and Connally, J. A., *MEMS-93*, Fort Lauderdale, FL, pp. 99–104, 1993.
12. Buchheit, T. E., LaVan, D. A., Michael, J. R., Chrinstenson, T. R., and Leith, S. D., *Metall. Mater. Trans.*, 33A, 539–554, 2002.
13. Burbaum, C., Mohr, J., Bley, P., and Ehrfeld, W., *Sens. Actuators*, A25–27, 559–563, 1991.
14. Cheng, Y.-T. and Cheng, C.-M., *J. Mater. Res.*, 13, 1059–1064, 1998.
15. Cheng, Y.-T. and Cheng, C.-M., *Phil. Mag. Lett.*, 78, 115–120, 1998.
16. Cheng, Y.-T. and Cheng, C.-M., *J. Appl. Phys.*, 84, 1284–1291, 1998.
17. Cho, H. S., Babcock, W. G., Last, H., and Hemker, K. J., *Mater. Res. Soc. Symp. Proc.*, 657, EE5231–EE5236, 2001.
18. Cho, H. S., Hemker, K. J., Lian, K., and Goettert, J., *Technical Digest. MEMS 2002 IEEE International Conference. Fifteenth IEEE International Conference on Micro Electro Mechanical Systems*, Piscataway, NJ, pp. 439–442, 2002.
19. Cho, H. S., Hemker, K. J., Lian, K., Goettert, J., and Dirras, G., *Sens. Actuators, A: Phys.*, 103(1–2), 59–63, 2003.
20. Christensen, T., Buchheit, T., Schmale, D. T., and Bourcier, R. J., *Microelectromechanical structures for materials research*, Brown, S. et al., Eds., *Materials Research Society*, Vol. 518, pp. 185–191, 1999.
21. Dao, M., Chollacoop, N., Van Vliet, K. J., Venkatesh, T. A., and Suresh, S., *Acta Mater.*, 49, 3899, 2001.
22. De Guzman, M. S., Neubauer, G., Flinn, P., and Nix, W. D., *Mater. Res. Symp. Proc.*, 308, 613–618, 1993.
23. Ehrfeld, W., *LIGA at IMM*, Banff, Canada, 1994.

24. Feynman, R., *There is Plenty of Room at Bottom*, American Physical Society Conference, Pasadena, CA, 1959.

25. Fleck, N. A., Muller, G. M., Ashy, M. F., and Hutchinson, J. W., *Acta Met. Mater.*, 42(2), 475–487, 1994.

26. Fleck, N. A. and Hutchinson, J. W., In *Advances in applied Mechanical*, Hutchinson, J. W. and Wu, T. T., Eds., Vol. 33, p. 295–361, 1997.

27. Fleck, N. A. and Hutchinson, J. W., *J. Mech. Phys. Solids*, 49, 2245–2271, 2001.

28. Florando, J., Fujimoto, H., Ma, Q., Kraft, O., Schwaiger, R., and Nix, W. D., *Materials Reliability in Microelectronics IX, Materials Research Society Symposium Proceedings*, Vol. 563, p. 231, 1999.

29. Gerberich, W. W., Yu, W., Kramer, D., Strojny, A., Bahr, D., Lilleodden, E., and Nelson, J., *J. Mater. Res.*, 13, 1–19, 1998.

30. Hay, J. C., Bolshakov, A., and Pharr, G. M., *J. Mater. Res.*, 14, 2296–2305, 1999.

31. Hertzberg, R. W., *Deformation and Fracture Mechanics of Engineering Materials*, 3rd ed., Wiley, New York, 1989.

32. Hill, R., *The Mathematical Theory of Plasticity*, Oxford University Press, Oxford, UK, 1950.

33. Hommel, M., Kraft, O., and Arzt, E., *J. Mater. Res.*, 14, 2374, 1999.

34. Hong, S., Weihs, T. P., Bravman, J. C., and Nix, W. D., *J. Electron. Mater.*, 19, 903–909, 1990.

35. Hormes, J., Gottert, J., Lian, K., Desta, Y., and Jian, L., *Nucl. Instrum. Methods Phys. Res. B*, 199, 332–341, 2003.

36. Hornbeck, L., *Proceedings of SPIE*, 1150, 86, 1990.

37. Hruby, J., *MRS Bull.*, 26, 337–340, 2001.

38. Huang, H., Mechanical properties of freestanding polycrystalline metallic thin films and Multilayers, Ph.D. Thesis, Harvard University, 1998.

39. Hutchinson, J. W., *Int. J. Solids Struct.*, 37, 225, 2000.

40. Jaeger, R. C., *Introduction to Microelectronic Fabrication*, Vol. 5, Addison-Wesley, Reading, MA, 1988.

41. Kahn, H., Stemmer, S., Nandakumar, K., Heuer, A. H., Mullen, R. L., Ballarini, R., and Huff, M. A., *IEEE Micro Electro Mechanical Systems*, pp. 343–348, 1996.

42. Kiesewetter, L., Zhang, J. M., Houdeau, D., and Steckenborn, A., *Sens. Actuators A*, 35, 153–159, 1992.

43. King, R. B., *Int. J. Solids Struct.*, 23, 1657, 1987.

44. Klein, M., Hadrboletz, A., Weiss, B., and Khatibi, G., *Mater. Sci. Eng. A*, 319–321, 924, 2001.

45. Koskinen, J., Steinwall, J. E., Soave, R., and Johnson, H. H., *J. Micromech. Microeng.*, 35, 13–17, 1993.

46. Kraft, O. and Volkert, C. A., *Adv. Eng. Mater.*, 3, 99, 2001.

47. Lou, J., Shrotriya, P., Allameh, S., Yao, N., Buchheit, T., and Soboyejo, W. O., In *Proceedings of Materials Science of MEMS Devices IV*, Ayon, A. A. et al., Eds., Vol. 687, MRS, Warrendale, PA, pp. 41–46, 2001.

48. Lou, J., Allameh, S., Buchheit, T., and Soboyejo, W. O., *J. Mater. Sci.*, 37(14), 3023–3034, 2003.

49. Lou, J., Shrotriya, P., Buchheit, T., Yang, D., and Soboyejo, W. O., *J. Mater. Res.*, 18(3), 719–728, 2003.

50. Lou, J., Shrotriya, P., and Soboyejo, W. O., *ASME J. Eng. Mater. Technol.*, 127(1), 16–22, 2005.

51. Ma, Q. and Clarke, D. R., *J. Mater. Res.*, 10, 853–863, 1995.

52. Madou, M., *Fundamentals of Microfabrication: The Science of Miniaturization*, 2nd ed., CRC Press, Boca Raton, FL, 2002.

53. Malzbender, J., de With, G., and den Toonder, J., *J. Mater. Res.*, 15, 1209–1212, 2000.

54. Mazza, E., Abel, S., and Dual, J., *Microsyst. Technol.*, 2(4), 197–202, 1996.

55. McElhaney, K. W., Vlassak, J. J., and Nix, W. D., *J. Mater. Res.*, 13, 1300, 1998.

56. Miller, W. M., Tanner, D. M., Miller, S. L., and Peterson, K. A., *MEMS Reliability: The Challenge and the Promise*, http://www.mdl.sandia.gov/Micromachine, 1998.

57. Mohr, J. and Strohrmann, M., *J. Micromech. Microeng.*, 2, 193–195, 1992.

58. Nadai, A., *Theory of Flow and Fracture of Solids*, McGraw-Hill Book Co., New York, 1950.

59. Nix, W. D. and Gao, H., *J. Mech. Phys. Solids*, 46, 411–425, 1998.

60. Oliver, W. C. and Pharr, G. M., *J. Mater. Res.*, 7, 1564–1583, 1992.

61. Poole, W. J., Ashby, M. F., and Fleck, N. A., *Scripta Mater.*, 34(4), 559–564, 1996.

62. Prime Faraday Technology Watch, *Introduction to MEMS*, 1999.
63. Read, D. T., Use of electronic speckle pattern interferometry (ESPI) in measurement of Young's Modulus of thin films, *Presented at the ASME Mechanics and Materials Conference*, Johns Hopkins University, Baltimore, MD, 1996.
64. Read, D. T., *Int. J. Fatigue*, 20, 203, 1998.
65. Rudd, J. A., Interface Stress and Mechanical Properties of Multilayered Thin Films, Ph.D. Thesis, Harvard University, 1992.
66. Safranek, W. H., *The Properties of Electrodeposited Metals and Alloys*, American Elsevier, New York, 1986.
67. Sandia National Laboratories, SUMMiT Technologies, http://www.mems.sandia.gov
68. Sharpe, W. N., Jr., Yuan, B., and Vaidyanathan, R., *Proceedings, of the SPIE—The International Society for Optical Engineering, Symposium on Microlithography and Metrology in Machining II*, Vol. 2880, pp. 78–91, 1996.
69. Sharpe, W. N., Jr., LaVan, D. A., and Edwards, R. L., *Transducers 97—1997 International Conference on Solid-State Sensors and Actuators, Digest of Technical Papers*, Vol. 1, pp. 607–610, 1997.
70. Sharpe , W. N. Jr., and Turner, K., *Fatigue 99*, pp. 1837–1844, 1999.
71. Shrotriya, P., Allameh, S. M., Lou, J., Buchheit, T., and Soboyejo, W. O., *Mech. Mater. J.*, 35, 233–243, 2003.
72. Soboyejo, W. O., *Mechanical Properties of Engineering Materials*, Marcel Dekker, Princeton University, NJ, 2003.
73. Spearing, S. M., *Acta Mater.*, 48, 179–196, 2000.
74. Stelmashenko, N. A., Walls, M. G., Brown, L. M., and Milman, Y. V., *Mechanical Properties and Deformation Behavior of Materials Having Ultra-Fine Microstructures*, Nastasi, M., Parkin, D. M., and Gleiter, H., Eds., NATO ASI Series E 233, pp. 602–610, 1993.
75. Stölken, J. S. and Evans, A. G., *Acta Mater.*, 46, 5109–5115, 1998.
76. Suresh, S., *Fatigue of Materials*, 2nd ed., Cambridge University Press, Cambridge, UK, 1998.
77. Tabor, D., *Hardness of Metals*, Oxford University Press, Oxford, 1951.
78. Vinci, R. P. and Baker, S. P., *MRS Bull.*, 27, 12–13, 2002.
79. Vlassak, J. J. and Nix, W. D., *J. Mater. Res.*, 7, 3242, 1992.
80. Wei, Y. and Hutchinson, J. W., *J. Mech. Phys. Solids*, 45, 1253–1273, 1997.
81. Weihs, T. P., Hong, S., Bravman, J. C., and Nix, W. D., *J. Mater. Res.*, 3, 931, 1988.
82. Xiang, Y. and Vlassak, J. J., *Scripta Mater.*, 53, 177–182, 2005.
83. Xie, Z. L., Pan, D., Last, H., and Hemker, K. J., *Materials Science of Microelectromechanical Systems (MEMS) Devices II*, de Boer et al., Eds., *Materials Research Society*, Vol. 605, pp. 197–202, 2000.

4 Silicon-Based Microelectromechanical Systems (Si-MEMS)

Seyed M. Allameh
Department of Physics and Geology, Northern Kentucky University

CONTENTS

4.1	Introduction	66
4.2	Materials	66
	4.2.1 Silicon	66
	4.2.1.1 Physical Properties	66
	4.2.1.2 Mechanical Properties	67
	4.2.1.3 Electrical Properties	68
	4.2.1.4 Thermal Properties	70
	4.2.2 Silicon Oxide, Silicon Carbide, and Silicon Nitride	70
	4.2.3 Other Materials	70
4.3	Microfabrication Processes	71
	4.3.1 Growth of Topical Films	72
	4.3.1.1 Epitaxy	72
	4.3.1.2 Oxidation	72
	4.3.1.3 Sputtering	72
	4.3.1.4 Evaporation	73
	4.3.1.5 Chemical Vapor Deposition	73
	4.3.1.5.1 Deposition of Polysilicon	73
	4.3.1.5.2 Deposition of Silicon Oxides	74
	4.3.1.5.3 Deposition of Silicon Nitride	74
	4.3.1.6 Spin-On method	74
	4.3.2 Lithography	75
	4.3.3 Etching	76
	4.3.3.1 Isotropic Wet Etching of Silicon	76
	4.3.3.2 Anisotropic Wet Etching of Silicon	76
	4.3.3.3 Dry Etching of Silicon	77
	4.3.3.4 Wet Etching of Silicon Oxides	79
4.4	Applications	79
	4.4.1 MEMS Sensors and Actuators	79
	4.4.1.1 Sensing Techniques	79
	4.4.1.2 Actuation Techniques	81
	4.4.1.3 Smart Sensors	84

	4.4.2	Mechanical Sensors	84
		4.4.2.1 Pressure Sensors	84
		4.4.2.2 Force and Torque Sensors	85
		4.4.2.3 Inertial Sensors	86
		4.4.2.4 Flow Sensors	88
	4.4.3	Biochemical Sensors	88
		4.4.3.1 Conductimetric Gas Sensors	89
		4.4.3.2 Potentiometric Devices	90
	4.4.4	Automotive Applications	90
	4.4.5	Photonics Applications	90
	4.4.6	Life Science Applications	91
	4.4.7	RF Applications	92
4.5	Reliability Issues		93
	4.5.1	Mechanical Testing of MEMS	93
	4.5.2	Reliability of Si-MEMS	94
4.6	Summary		95
Acknowledgment			96
References			96

4.1 INTRODUCTION

Microelectromechanical systems (MEMS), also known as microsystems technology (MST), has seen an explosive growth during the past two decades. Miniaturizing technology with benefits such as new functionality, cost reduction, and space saving, has helped MEMS applications span over numerous fields including automotive, aerospace, photonics, telecommunications, life sciences, biochemistry, biology, biomedicine, and drug delivery to name a few. When it comes to sensors and actuators, MEMS is a strong competitor for the conventional manufacturing processes. Whenever a new functionality becomes possible by going small (e.g., biological applications), or when mass production at small scale reduces production costs (e.g., automotive applications), or when space is a major constraint (e.g., aerospace applications), utilizing MEMS becomes an obvious choice. This chapter focuses on Si-based MEMS with the main emphases placed on silicon properties, device fabrication, device applications, and the related mechanical and reliability related issues.

4.2 MATERIALS

4.2.1 SILICON

Silicon is the material of choice for most MEMS devices. This arises mainly from the economic benefits due to the well-established semiconductor manufacturing technology that provides the industrial infrastructure needed for MEMS fabrication. This is in addition to the desirable properties of silicon including electrical, optical, and mechanical, linked to various crystal structures. The well-established micromachining techniques with additive and subtractive processes make the design and mass production of Si-MEMS easy and economical. Si-based MEMS may have other materials that are compatible with silicon. These include silicon oxides, silicon nitrides, silicon carbides, and metals such as Al, W, Cu, and polymers such as polyimide.

4.2.1.1 Physical Properties

In terms of microstructure, silicon used for MEMS can be single crystalline, polycrystalline (polysilicon), or amorphous. The type of application and manufacturing process are the main factors that determine the type of silicon used. The thickness of the material fabricated, either grown by deposition or fabricated by wafer bonding and etching/electrochemical polishing, is usually less

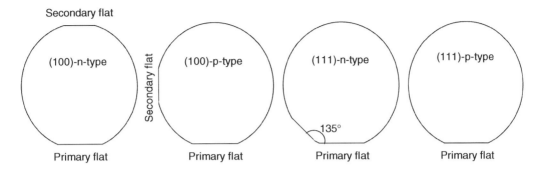

FIGURE 4.1 Crystal orientation, flat orientation and type of dopant for silicon single crystal wafers (Drawing created after Maluf, N., *An Introduction to Microelectromechanical Systems Engineering*, Artech House, Boston, MA, 2004.)

than 5 μm. Single crystal silicon wafers used for MEMS are commonly 100 and 150 mm in diameter with thicknesses in the range of 525–650 μm for single-sided wafers and 100 μm thinner for the double-sided polished wafers. Crystal orientation of the wafers are commonly (100) and (111) with dopant type being n or p. Figure 4.1 shows the primary and secondary flats of silicon single crystal wafers denoting their crystal orientation and the type of doping. The importance of orientation becomes obvious noting that the etching process, using KOH, leaves (100) wafers with etch pits bounded by {111} planes. This property can be utilized to etch holes with walls making an angle of 54.7° with the top surface.

While silicon is compatible with most service conditions for which MEMS devices are considered, its biocompatibility is being currently investigated. Bare silicon surfaces have shown to be less than favorable for amplification of genetic DNA materials. However, for most biological applications such as implants, Si has been shown to be biocompatible with no toxic products being released.

4.2.1.2 Mechanical Properties

Single crystal silicon is hard and brittle with an elastic behavior that continues all the way to fracture at ~7 GPa. Young's modulus of silicon varies with orientation, from ~130 GPa (in [100] direction) to 169 GPa (in [9] direction). Its Poisson's ratio is 0.22. The strength of crystalline silicon can be understood visualizing the weight of a man (~70 kg) being suspended from a silicon fiber as thin as his hair (~100 μm diameter). Table 4.1 presents the mechanical properties of silicon, silicon oxide, silicon nitride, and silicon carbide. For pure single crystal silicon wafers, the mechanical properties are uniform across the whole lot leading to devices

TABLE 4.1
Mechanical Properties of Si and Si Alloys [10]

Material	Young's Modulus (GPa)	Yield/Fracture Strength (GPa)	Poisson's Ratio
Silicon	160	7	0.22
Silicon oxide	73	8.4	0.17
Silicon nitride	323	14	0.25
Silicon carbide	450	21	0.14

that have no venialities arising from the difference in microstructure. Doping, however, may lead to internal stresses, especially at levels higher that $10^{20}/cm^3$.

While single crystal wafers are nearly stress free due to their purity, this is not the case with polysilicon. After deposition, polysilicon must be annealed at high temperatures (e.g., 900°C) to reduce the intrinsic stresses developed during its growth. The tensile/compressive nature of stresses developed in polysilicon during deposition changes with temperature. Careful tailoring of deposition temperatures and film thicknesses in polysilicon multilayer films, grown alternately at lower and higher temperatures, can lead to nearly stress-free films. This is particularly important for polysilicon cantilevers where intrinsic stresses can cause curling of the beams if they are not properly annealed.

Mechanical properties of single crystal silicon at elevated temperatures (up to 700°C) are not significantly different from those of silicon at room temperature. The onset of softening and plastic deformation takes place at 700°C. Polysilicon, on the other hand, starts to show time-dependent stress annealing at temperatures as low as 250°C leading to instabilities that may interfere with its mechanical functions.

4.2.1.3 Electrical Properties

Due to its electrical properties, silicon has had dual use both as a structural material to fabricate MEMS components, and a semiconductor to make electric circuits, P–N junctions and other electronic components necessary to drive and control mechanical components. Some of the more important electrical properties of silicon and its alloys are presented in Table 4.2. Important electrical properties include piezoresistivity, piezoelectricity, and thermoelectricity. Silicon in single crystalline and polycrystalline and even amorphous forms exhibit piezoelectricity characteristics. The effect is much larger for single-crystal silicon, however, it varies with crystallographic orientation, maximized in [110] direction, but vanishing in [100] direction for p-type wafers doped below $10^{18}/cm^3$. Table 4.3 shows the variation piezoresistivity coefficient for p- and n-type silicon single crystals with crystal orientation. Piezoresistivity also varies with Young's modulus, temperature, type, and concentration of dopant. Defined as the change in resistivity due to applied stress, piezoresistivity can be related to the stresses applied parallel and perpendicular to silicon resistor as follows:

$$\frac{\Delta\rho}{\rho} = \pi_{\text{II}}\sigma_{\text{II}} + \pi_{\perp}\sigma \tag{4.1}$$

where π_{II} and $\pi_{\perp}$ denote proportionality constants and σ_{II} and $\sigma_{\perp}$ denote stresses in the two orthogonal directions. According to Equation 4.1, stresses parallel to the p-type resistor in [9] direction will increase resistance while tensile stresses perpendicular to the resistor decreases the

TABLE 4.2
Electrical Properties of Si and Si Alloys [10]

Material	Relative Permittivity (ε_r)	Dielectric Strength (V/cm$\times 10^6$)	Electron Mobility (cm^2/Vs)	Electron Mobility (cm^2/Vs)	Bandgap (eV)
Silicon	11.7	0.3	1500	400	1.12
Silicon oxide	3.9	5–10			8.9
Silicon nitride	4–8	5–10			
Silicon carbide	9.7	4	1000	40	2.3–3.2

TABLE 4.3
Piezoresistivity Coefficients for p- and n-Type Silicon Single Crystal Doped below 10^{18}/cm^3

Material	Direction	$\pi_{\parallel}$ (10^{-11} m^2/N)	$\pi_{\perp}$ (10^{-11} m^2/N)
Silicon p-type	$\langle 100 \rangle$	7	-1
Silicon p-type	$\langle 110 \rangle$	72	-66
Silicon n-type	$\langle 100 \rangle$	-102	53
Silicon n-type	$\langle 110 \rangle$	-31	-18

resistance. The proportionality constants are functions of temperature and change by $-0.3\%/°$C at a doping level of 10^{18}/cm^3. A photosensitivity characteristic of silicon allows fabrication of load/pressure sensing devices out of crystalline and amorphous silicon.

For polycrystalline silicon, piezoresistivity is averaged over all grains and all directions yielding an overall effect that is nearly one-third of that of single crystal silicon. A gauge factor, called K, relates the applied stress to the relative change in resistivity and ranges from -30 to $+40$. While gauge factor drastically decreases as doping concentration exceeds 10^{19}/cm^3, it is less sensitive to temperature changes, making polysilicon advantageous for pressure sensors used in temperature-varying conditions. The temperature coefficient of resistivity (TCR) of polysilicon is 0.04% compared to a TCR of 0.14% for single crystal silicon.

Piezoelectricity effect arises from asymmetry in the primitive unit cell of a crystalline material combined with ionic or semi-ionic bonding of its atoms. Examples of piezoelectric materials are quartz, zinc oxide, lead–titanate–zirconate (PTZ), barium titanates (BaTiO$_3$), and lithium niobate (LiNbO$_3$). Silicon is symmetric in crystal structure (diamond crystal structure, cubic) with covalent bonding and therefore lacks piezoelectric capability. However, piezoelectric materials may be applied to a silicon substrate as thin films by either deposition (e.g., sputtering of ZnO or PZT), sol–gel process (PZT), or spin coating (PVDF). Currently, there is no easy way to deposit quartz on Si substrate.

Thermoelectric properties of polysilicon make it a suitable material for applications in which the Seebeck effect, Peltier effect, and Thomson effect play the main role. Seebeck effect used in thermocouples allows for the measurement of temperature differences, while Peltier effect is utilized in thermoelectric coolers and refrigerators. To measure temperature difference between hot and cold sources, two elements with different Seebeck coefficients of α_1 and α_2 are joined and the joint is placed at the hot source (e.g., furnace). The other ends of the two elements are placed at the cold source (e.g., a water/ice mixture). The temperature difference can be related to the measured electric field generated between the two elements as follows:

$$\Delta V = (\alpha_2 - \alpha_1)(T_{\text{hot}} - T_{\text{cold}}) \tag{4.2}$$

In the Peltier effect, two dissimilar materials are joined together. When an electrical current passes through the couple, a heat flux is generated. This flux cools down one side of the couple relative to the other. While Peltier devices are made of n-type and p-type bismuth telluride elements, it is difficult to implement Peltier effect in thin film applications. Seebeck coefficients for n-type polysilicon with resistivities of 30 and 2600 Ω cm are -100 and -400 μV/K respectively. For p-type polysilicon with a resistivity of 400 Ω cm, the Seebeck coefficient is 270 μV/K. These numbers can be compared with those of Bi (-73.4 μV/K), Ni (-14.8 μV/K), Al (4.2 μV/K), Mg (4.4 μV/K), Ag (7.4 μV/K), and Au (7.8 μV/K) [10].

TABLE 4.4
Physical/Thermal Properties of Si and Si Alloys [10]

Material	Density	Melting/Sublima-tion Point (°C)	Specific Heat (J/g K)	Coefficient of Thermal Expansion (CTE) $(10^{-6}/°C)$	Thermal Conductivity (W/m K)
Silicon	2.4	1415	0.7	2.6	157
Silicon oxide	2.2	1700	1.0	0.55	1.4
Silicon nitride	3.1	1800	0.7	2.8	19
Silicon carbide	3.2	1800	0.8	4.2	500

4.2.1.4 Thermal Properties

Silicon and silicon carbide have comparable coefficients of thermal expansion (CTE). However, silicon dioxide has a CTE nearly one-fourth of that of silicon giving rise to significant thermal stresses at the Si/SiO_2 interface. For silicon nitride, CTE is twice that of silicon leading to milder interfacial stresses upon heating and cooling of Si/Si_3N_4 bilayers. Table 4.4 shows the density, melting point, specific heat, CTE, and thermal conductivity of silicon and its compounds. As seen from Table 4.4, silicon carbide and to a lesser degree silicon are thermally conductive while silicon oxide and to a lesser degree silicon nitride are insulators.

4.2.2 Silicon Oxide, Silicon Carbide, and Silicon Nitride

Native silicon dioxide forms immediately on the bare surfaces of silicon at the instance of exposure to oxygen or an oxidizing agent. The thickening of this layer up to a few tens of nanometers provides the structure with a solid adherent protective layer that effectively prevents further oxidation of silicon. The inert nature of silicon dioxide makes silicon structures very immune to the stress corrosion cracking that affects many other structural materials. Nevertheless, as will be seen later, this will not hold true if the loadbearing structural components made of silicon are exposed to moisture (Table 4.3).

Silicon oxides are widely used in surface micromachining for insulators and sacrificial layers to name two applications. Silicon dioxide (SiO_2) is thermally grown by exposing silicon to oxidizing atmospheres at temperatures above 800°C. Other oxides of silicon are grown by chemical vapor deposition (CVD), sputtering, and spin coating. Silicon oxides become soft at temperatures above 700°C and therefore are of limited use as beams and cantilevers at higher temperatures. Silicon dioxide growth leads to large stresses at the interface of Si/SiO_2.

Silicon nitrides coatings act as barriers to the diffusion of most mobile ions including the sodium and potassium present in biological environments. Young's modulus of Si_3N_4 is larger than that of Si making it an excellent choice for coatings on cutting tools. It also makes silicon nitride coating a suitable mask for alkaline etch processes.

4.2.3 Other Materials

A number of metals are used for MEMS applications including Ag, Al, Au, Cr, Cu, Ir, Ni, Pd, Pt, Ti, W, as well as alloys such as NiCr, TiNi, and TiW. While Al ($\rho = 2.7$ μΩ cm) mostly used for interconnects and for optical reflection of visible and infrared light, Au ($\rho = 2.4$ μΩ cm) is used for high temperature interconnects and optical reflection in infrared radiation. Au and Ag ($\rho = 1.58$ μΩ cm) are used for electrochemistry applications due to their noble nature. Metals such as Ni ($\rho = 6.8$ μΩ cm) do not adhere to Si surfaces. An example of the delamination of Ni

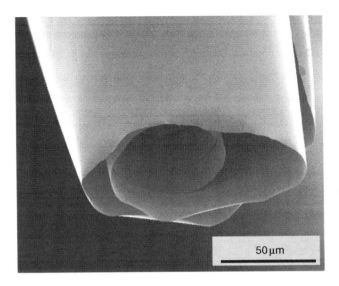

FIGURE 4.2 Delamination and rolling of thick ($\sim$500 nm thick Ni film deposited on Si substrate). (From Allameh, S. M., *Materials and Manufacturing*, 2006. With permission.)

film from silicon substrate is presented in Figure 4.2. For the coating of Au, Ag, and Ni metals, adhesion layers of Cr ($\rho = 12.9$ $\mu\Omega$ cm) or Ti ($\rho = 42$ $\mu\Omega$ cm) or TiW with a thickness of a few nanometers would be used. Cu ($\rho = 1.7$ $\mu\Omega$ cm) makes low resistivity interconnects, however due to its fast diffusion in Si, many deposition chambers are sealed against the use of Cu to prevent possible cross contamination of integrated chips (ICs). Permalloy and Ni are the materials of choice for magnetic transducers. NiCr ($\rho = 200$–500 $\mu\Omega$ cm) is used for laser trimmed resistors. Ti and Ti6Al4V make excellent biocompatible coatings. TiNi is used for shape memory alloy actuation. TiW ($\rho = 75$–200 $\mu\Omega$ cm) is used as near zero TCR and W ($\rho = 5.5$ $\mu\Omega$ cm) is used as high temperature interconnect as well as for thermionic emitter applications. Other materials used for MEMS include SnO_2 for chemoresistance in gas sensors, TaN for negative TCR laser trimmed resistor, SiCr ($\rho = 2000$ $\mu\Omega$ cm) for laser trimmed resistors, and Indium–tin-oxide (ITO) ($\rho = 300$–3000 $\mu\Omega$ cm) as transparent conductive layer for liquid crystal displays (LCD).

4.3 MICROFABRICATION PROCESSES

Si-MEMS fabrication benefits from the industrial infrastructure established by the semiconductor industry. What distinguishes micromachining from conventional bulk machining is mainly the size, implying that at least one-dimension of the product is in micron range. Additionally, micromachining usually comprises parallel fabrication of multiple devices, occasionally thousands per wafer with dozens of wafers being processed simultaneously. Micromachining involved in the manufacture of MEMS can be divided into two main groups: bulk and surface micromachining. Structures are rarely made by self-assembly, self-synthesis, or self-organization of species. Usually some kind of patterning will be needed using lithography and pattern transfer followed by micromachining. The basic processes include deposition of layers (e.g., SiO_2) followed by patterning and etching. Deposition methods include epitaxy, oxidation, sputtering, evaporation, CVD/LCVD/PECVD, spin-on method, and sol–gel. In addition to deposition methods, additive processes include anodic bonding and fusion bonding. Pattering techniques include optical lithography and double-sided lithography. Finally, etching techniques include wet isotropic, wet anisotropic, plasma, reactive ion etching (RIE), and deep reactive ion etching (DRIE). In this section, some of the more common techniques are briefly described. Other methods of fabrication of MEMS such as LIGA, electrodischarge machining (EDM), laser beam machining (LBM), hot embossing, spin coating, used to make metallic, and ceramic and plastic MEMS are not discussed here.

4.3.1 GROWTH OF TOPICAL FILMS

For Si-MEMS, fabrication processes usually start with the growth of a topical film such as an oxide, nitride, metal, insulator, or polymer thin film on the silicon substrate. The growth techniques include epitaxy, oxidation, sputtering, evaporation, chemical vapor deposition, and spin-on method.

4.3.1.1 Epitaxy

In epitaxial growth, the crystal structure of the films grown on silicon is affected by that of the substrate. An orientation relationship (OR) between the two crystals at the two sides of the interface determines the crystallographic planes and directions of the overgrowth. The simplest OR is cube-on-cube in which the crystal orientation of the substrate is mimicked by that of overgrowth. Depending on the magnitude of the mismatch between the lattice periodicities of the two crystals, other orientation relationships may occur. Epitaxial layers of single crystal Si can be grown on silicon wafers with a high degree of purity to thicknesses of 1–20 μm. Single crystal silicon deposition is accomplished by chemical vapor deposition (CVD) using silane (SiH_4), dichlorosilane (SiH_2Cl_2) or silicon tetrachloride ($SiCl_4$). Doping of the growing topical silicon film is performed simultaneously by dissociating of dopant-bearing gases. These include arsine and phosphine (ASH_3 and PH_3, both of which are very toxic) for n-type and diborane (B_2H_6) for p-type silicon. The type and concentration of the dopant used for the overgrowth is usually different from that of the substrate. Called heteroepitaxy, Si films are occasionally grown on other crystalline materials such as sapphire (Al_2O_3), however, lattice mismatch limits the thickness of the epitaxial silicon grown on these substrates.

4.3.1.2 Oxidation

Growth of silicon dioxide can be achieved by providing oxygen to the silicon substrate in the absence of moisture (dry oxidation) or in its presence (wet oxidation) at temperatures between 850°C and 1150°C. Growth of this oxide is usually associated with the development of thermal stresses due to the mismatch between the CTE of Si and SiO_2 as well as the mismatch between their lattice periodicities. Stresses are often compressive in nature and cause bowing of the substrate or curling of silicon cantilevers and beams.

4.3.1.3 Sputtering

A target made of the material to be deposited is bombarded by energetic inert gas ions such as argon at pressures of 0.1–10 Pa. This causes the atomic species of the target material to be blasted away from the surface of the target and deposited onto the surface of substrates such as silicon wafers. The depositing layer will have the same chemical composition as that of the target material making this process particularly suitable for the deposition of compound materials made of multiple elements.

For deposition of conductive materials, the target can be electrically biased with respect to the silicon wafer substrate by the application of a direct current (dc) electrical field. Called "dc-glow discharge," this method further accelerates the inert gas ions between the target and the substrate. In radio frequency (RF) sputtering, the target and the substrate form two parallel surfaces. RF tension, applied to the target, causes excitation of the target material and its subsequent deposition on the silicon wafer substrates.

Ion beam milling comprises the formation of gaseous ions such as nitrogen in remote plasmas and their acceleration in an electric field of a few thousand volts toward the target, which can be either a conductor or an insulator. The *ion*-beam density of the gaseous ions can be intensified by the presence of a magnetic field established in magnetron sputtering. Deposition rates can be as high as 1 μm/min for materials such as Al. If the gaseous ions are allowed to react with the target material, the depositing material will have a chemical composition different from that of the target. This technique, called "reactive ion sputtering" allows for the deposition of materials such as titanium nitride and titanium dioxide by using a Ti target along with reactive ions of nitrogen and oxygen.

Materials, sputter deposited for MEMS, include Al, Ti, Cr, Pt, Pd, W, Al/Si, Ti/W, amorphous Si, glasses, and ceramics such as PZT and ZnO. Step coverage of the deposited films is good, with fine grain size and uniform film thickness. However, stress develops in the sputter deposited films. The type of the stress (e.g., being negative or positive) and its magnitude greatly depend on experimental parameters such as the properties of the depositing material and the substrate (e.g., atomic size, crystal structure, and melting point), deposition rate, and substrate temperature [11,12].

4.3.1.4 Evaporation

Many materials used for MEMS can be deposited by evaporation from their respective bulk source. These include Al, Ti, Cr, Au, Cu, Ni, Mo, Ni/Cr, and even ceramics like Al_2O_3. The evaporation of the source material can be achieved by either resistive heating or electron beam heating. Using a set of deflection coils for the latter case, an electron beam is scanned on the surface of the source material until it heats up and causes evaporation at the desired deposition rates. Evaporation rate scales with the heat input into the material influenced by electron beam intensity, voltage level, dwell time, and the melting point of the source material among other factors. The source material is biased with respect to the electron gun such that electrons are accelerated in an electric field of about 10 kV. The crucibles used to contain molten source materials include alumina, graphite, tungsten, or even copper. Electron beam deposition yields higher quality films compared with resistive heating. Evaporation is carried out in a vacuum, of which the quality affects not only the purity of the depositing film, but also its crystalline structure and orientation [13]. The relative orientation of the substrate with respect to the molecular beam will influence the film coverage, grain size, and occasionally growth direction.

4.3.1.5 Chemical Vapor Deposition

Chemical reactions take place between gaseous reactants on or above the substrates leading to the formation of a new substance, which deposits on the surface of the substrate. Deposition of materials by CVD takes place at relatively higher temperatures (e.g., 300–800°C). Si-MEMS materials deposited by CVD include polysilicon, silicon carbide, silicon nitride, and some low-permittivity dielectric insulators used for electrical interconnects. Deposition atmosphere can be at ambient (APCVD), low pressure (LPCVD), plasma enhanced (PECVD), or high density plasma (HDPCVD). The first two methods are performed at higher temperatures (400–800°C) while the latter two processes occur at ~ 300°C.

4.3.1.5.1 Deposition of Polysilicon

Polysilicon films ranging in thicknesses from nanometers to microns are deposited by chemical vapor deposition using SiH_4 (silane) in a LPCVD reactor. To be crystalline, silicon must be deposited at temperatures above 630°C; otherwise, amorphous silicon forms below 600°C. Deposition rate will depend on temperature, partial pressure, and flow rate of silane, and ranges from 6 to 70 nm for deposition temperatures of 620°C and 700°C. For structures with aspect ratios below 10,

CVD deposition of polysilicon produces reasonably uniform film thickness on the flat surfaces as well as on the sidewalls.

Polysilicon can be doped during deposition by introducing dopant gases such as arsine or phosphine for n-type or diborane for p-type silicon to the deposition chamber. Dopant concentration can be reached as high as $10^{20}/cm^3$ leading to resistivities of $1-10$ mΩ cm. Polysilicon films can develop residual (intrinsic) stresses as high as 500 MPa during deposition. The magnitude of these stresses scale with the thickness of the film being deposited as well as deposition temperature. Annealing polysilicon at 900°C will reduce residual stresses to levels below 50 MPa.

4.3.1.5.2 Deposition of Silicon Oxides

Silicon oxide films are grown on silicon to form a dielectric insulator between metal layers or to make a sacrificial layer that will be dissolved in later stages. Low temperature oxide (LTO) of silicon can be deposited by APCVD, LPCVD or PECVD processes using silane and oxygen at temperatures below 500°C. Phosphosilicate glass (PSG) or borophosphosilicate glass (BPSG) forms when LTO is doped with phosphine or a combination of diborane and phosphine. When annealed at 1000°C, they soften and conform to the surface topography of the substrates. Deposition rate varies from 25 to150 nm/min for temperatures of 400–450°C respectively.

Silicon dioxide is grown by LPCVD from tetraethoxysilane [$Si(OC_2H_4)_4$]. Deposition rates of 5–50 nm/min are typical for temperatures of 650–750°C. Reacting dichlorosilane with N_2O in a LPCVD process can also lead to the deposition of silicon dioxide at 900°C, however, this method is less common than the other methods mentioned above. Silicon oxides formed by CVD are amorphous even after annealing at temperatures of 600–1000°C. Densification of polysilicon takes place in this temperature range. Residual stresses in silicon oxide films, grown by CVD, are usually compressive in nature and range from 100 to 300 MPa.

4.3.1.5.3 Deposition of Silicon Nitride

Silicon nitride (Si_3N_4) is used as masks for selective etching of silicon in basic solutions such as hot KOH. SiN_4 is deposited either by LPCVD from dichlorosilane reacting with ammonia or by atmospheric pressure CVD from silane reacting with ammonia at 700–900°C. Tensile stresses (as high as 1000 MPa) result from the chemical vapor deposition of silicon nitride films. For MEMS applications, the flow of dichlorosilane is increased to enrich Si in the film being deposited at 800–850°C resulting in reduced levels of residual stresses (below 100 MPa). Residual stresses can be controlled better during the PECVD deposition of monostoichiometric silicon nitride from reaction of silane with ammonia or nitrogen at temperatures below 400°C. Reducing the excitation frequency of plasma in a PECVD process, from 13.56 MHz to 50 kHz, reduces residual stresses from 400 to 200 MPa.

4.3.1.6 Spin-On Method

Dielectric insulators, as well as organic polymers, can be dispensed in liquid suspension forms onto the center of substrates rotating at 500–5000 rpm. This process results in uniform films with thicknesses of 0.5–20 μm. For polymers, a subsequent baking allows the evaporation of solvent and increased rigidity of the films. Spin-on glass (SOG) in thick (5–100 mm) and thin (0.1–0.5 mm) layers are used for planarization of uneven surfaces and as insulating material between metal layers. After spin coating, densification of SOG is performed at a temperature of 300–500°C, leading to tensile residual stresses of ~ 200 MPa. Photoresist, polyimides, and epoxy-based SU-8 are spun on silicon substrates. The thickness of some of these materials can reach 200 μm.

4.3.2 Lithography

Similar to ICs, fabrication of Si-MEMS involves transference of patterns. Lithography is used to delineate areas that are going to be added or subtracted. The first step in lithography is preparation of a mask that contains the desired pattern. Several masks are usually needed during various stages of the micromachining process. Masks are generally made of transparent fused quartz or soda lime glass substrates coated with a topical film of chromium or iron oxide. The pattern is engraved into the topical film by e-beam or laser beam writing dictated by an AutoCAD file.

The next step in photolithography is to spin the photoresist on the surface of the substrate to a thickness of 0.5–10 μm followed by baking of the film to remove the solvent. For inorganic resins, exposure of the positive resist layers to light, emitting through the mask, causes chemical changes in the composition of the exposed areas. For the exposure of the resist layer to light, masks can be placed on top of the substrate (contact method) or just held above it at a distance of ∼25 μm (proximity method). Alternatively, the pattern can be projected onto the surface of the resist using optical lenses. Upon developing, the exposed areas of the positive resist will dissolve while the unexposed areas will remain on the substrate being further stabilized in fixing solution. Figure 4.3 shows exposure methods used in contact and non-contact lithography.

Resolution of surface features fabricated by lithography is 1–2 μm for contact method and ∼5 μm for the proximity method (e.g., holding mask above the resist) and about 1 μm for the projection method. These correspond to lithography methods using light wavelengths of ∼400 nm along with stepper optical apertures common in MEMS fabrication. While depth of focus is poor for contact and proximity methods, the focal plane can be adjusted within a few microns in the projection method. For current Si-MEMS, resolution is not a critical issue, however, other issues (such as exposure time for thick resists) arise in lithography of certain MEMS devices.

Thick resists (20–100 μm) can be used as protection against etching deep structures; however, feature resolution degrades due to the limitations on optical focal depth. As a general rule, feature

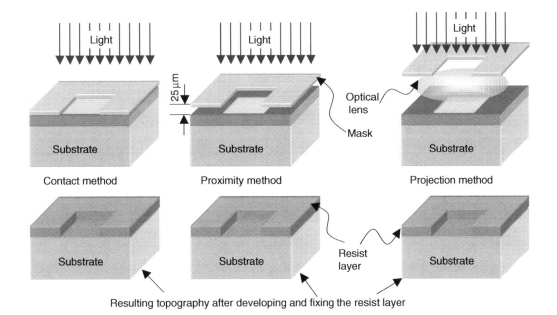

FIGURE 4.3 Three methods of exposure of resist: contact, proximity and projection methods. (Drawing created after Maluf, N., *An Introduction to Microelectromechanical System Engineering*, Artech House, Boston, MA, 2004.)

sizes (like width of beams or gaps) must be larger than one-third of the resist thickness. Furthermore, uniformity of the thickness may not be maintained for topographies typical of MEMS. While thinning of resist takes place at external angles (e.g., convex corners), its thickening occurs at internal angels (concave corners or cavities). To mitigate the non-uniformity of resist on sever topographies, spray or electroplating method can be used. Many Si-MEMS require micromachining on both sides of the silicon wafer. Double-sided polished wafers are used along with techniques that expose the resist on both sides of the wafer. Called double-sided lithography, it ensures alignment of features on the backside (such as cavities) with the features on the front side (such as sensors, detectors, and cantilevers). Resolution of double-sided lithography is typically ~2 μm.

4.3.3 ETCHING

Subtractive micromachining includes a number of techniques that selectively remove material from the surface or bulk of the substrates to accomplish the desired MEMS structure. These techniques include wet and dry etching, focused ion-beam milling (FIB), laser machining, ultrasonic drilling, electrical discharge machining (EDM), and traditional precision machining. Some of the most important subtractive techniques suitable for Si-MEMS are discussed here. Etching is a pattern transfer technique that selectively removes areas not protected by a topical maskant layer by chemical or physical processes. Selective etching may be performed on a metal or oxide layer, which in turn, can be used for further etching of the material beneath it. Bulk machining is achieved by deep etching of the silicon substrate to make Si-MEMS devices. The type of the etchant and process variables for etching are mostly determined based on the material to be removed, type of application, desired etching rate, selectivity of the etch to other materials (already incorporated into the structure to be etched), and isotropy.

4.3.3.1 Isotropic Wet Etching of Silicon

Wet etching is preferred for low cost batch processing. *Iso etch* and *poly etch* solutions are basically combinations of hydrofluoric (HF), nitric acid (HNO_3), and acetic acid ($HCH3COO$) called HNA. Once exposed to HNA, polysilicon is oxidized by nitric acid and subsequently dissolved in HF. Etching rate can be controlled by changing the volume fraction of the three acids in the etching solution. This allows etching rates in the range of 0.1–100 μm/min. In isotropic etching, the material is removed uniformly in all directions with no preferential etching in any particular direction.

4.3.3.2 Anisotropic Wet Etching of Silicon

If directional material removal is desired, anisotropic wet etching can be used. Orientation-dependent etchants (ODEs) will selectively etch certain crystallographic planes faster than others leaving flat surfaces, which may be inclined with respect to the surface of the wafer. ODEs used for anisotropic etching of silicon include NaOH, KOH, CsOH, NH_4OH, $N(CH_3)_4OH$, and ethylene-diamine mixed with pyrochatechol (EDP) in water. Etching process is usually performed at 70–100°C. For KOH, the most popular etchant for Si, etching rate of (100)-type planes is 50 times that of (110)-type planes, which is in turn, twice that of (111)-type planes. When (100)-type wafers are used, the sidewalls of the etched areas create an angle of 54.74° with respect to the top surface of the wafer. Dissolution reaction of silicon is accompanied by the release of four electrons as follows:

$$Si + 2OH^- \rightarrow Si(OH)_2^{++} + 4e^-$$

(4.3)

$$Si(OH)_2^{++} + 4e^- + 4H_2O \rightarrow Si(OH)_2^{--} + 2H_2$$

(4.4)

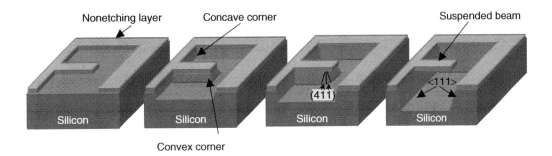

Nonetching layer   Concave corner   Suspended beam

(411)   <111>

Silicon   Silicon   Silicon   Silicon

Convex corner

FIGURE 4.4 Orientation-dependent etching of silicon to form cavity under a suspended cantilever. Note (111)-type planes bounding the pit with (411)-type planes appearing at convex corners. (Drawing created after Maluf, N., *An Introduction to Microelectromechanical System Engineering*, Artech House, Boston, MA, 2004.)

The faster rate of etching of (111)-type planes is believed to be due to its higher charge density providing the electrons needed for the Reaction 4.4 to proceed to the right. This may also explain the much slower rate of the p-type silicon used as an etch-stop material. The etching rate of n-type silicon (0.4–5 μm/min) is nearly 500 times that of a p-type at a dopant concentration of $10^{20}/cm^3$. Silicon nitride is an excellent mask material for patterning silicon by etching with KOH. Fast formation of a protective oxide layer on (111)-type facets has also been proposed for the slow rate of etching of (111)-type planes compared to (110)- and (100)-type planes. Figure 4.4 shows the formation of (111)- and (411)-type planes at the sides of the etch pit made in a single crystal silicon.

4.3.3.3 Dry Etching of Silicon

Dry etching comprises vapor phase, physical, chemical, or physical/chemical material removal. Advantages include cleaner surfaces, less corrosion of metal layers, less undercutting, less broadening of the photoresist features, and more effectiveness in producing submicron features. Plasma-assisted dry etching comes in two configurations: glow discharge and ion-beam milling, as will be discussed below.

Glow Discharge: This technique involves the dissociation of a feed gas such as SF_6 to form energetic species accelerated in high energy electric and magnetic fields. These include neutrals, ions, photons, and electrons. In glow discharge plasma, neutrals such as free radicals outnumber the ions by a ratio in range of $10^4:1$ to $10^6:1$. This becomes important in deciding which species perform the actual etching action.

DC Plasma Etching: The strength of the plasma determines the choice of dry etching technique. It depends on ions and neutrals density, ion energy, electron density, and ion current density. In the typical dc plasma dry etch, the substrate is placed on the cathode (target) and the argon plasma is oriented toward the substrate. Ion etching takes place by the physical removal of the substrate atoms by energetic Ar ions accelerated in an electric field of 200–1000 V. Plasma ion flux can be determined from [14]

$$J_i = qn_i\mu_i E \tag{4.5}$$

where q is the charge, μ_i is the ion mobility, n_i is the ion density, and E is the applied electric field.

RF Plasma Etching: Charge carriers are accelerated in an alternative current electric field established between two electrodes at a radio frequency (typically 13.56 MHz). Electrons eventually gain enough energy to cause the ionization of the gas between the two electrodes leading to the formation of plasma at pressures lower than that necessary for dc plasma (10 vs. 40 mTorr). Unlike dc plasma, with RF plasma, dielectric materials can be etched as well as metals. To etch a substrate with RF plasma, it is placed on the cathode of a parallel plate reactor at low pressure (e.g., 10^{-1} to 10^{-2} Torr) with the anode grounded. Within the RF plasma, electrons, cations, anions, free radicals, and photons are formed. Like dc plasma, the ratio of ions to neutral species is in the range of 10^4:1 to 10^6:1. The energy with which the target is bombarded with anions is related to a self-biased voltage [14]

$$V_{dc} = \frac{KT_e}{2e} \ln \frac{T_e m_e}{T_i m_e} \tag{4.6}$$

where V_{dc} represents the self-biased voltage build up induced to cathode by the plasma, T_e and T_i are the electrons (e.g., 2×10^4 K) and ions (e.g., 500 K) with temperatures defined by v_e/k and v_i/k; v_e and v_i are cathode and anode voltages, k is Boltzmann's constant, and m_e and m_i are the mass of electrons and mass of ions respectively. Using Equation 4.5 for a typical RF power density of 4 kW/m^2, the energy of anions hitting the substrate is about 300 eV.

Ion Etching/Ion Beam Milling: In the absence of a reactive etching process, the removal of material from the substrates can be achieved by physical etching. This comprises the collision of energetic inert ions such as argon with the substrate causing the ejection of surface atoms. Ion etching takes place when the substrate is placed on the cathode plate of an RF reactor where two electrodes are used. Usually noble gases are used due to their heavy ions being effective in momentum transfer. Physical etching rates of different species within the substrate are comparable (e.g., the energy of colliding ions is much greater than that of the chemical bonds associated with various types of atoms within the substrate). This results in a material nonselective etching process.

For ion beam milling, the substrate is not placed on the cathode plate; rather a third electrode is used. As before, a dc or RF plasma provides ions which are directed toward the surface of the substrate causing the removal of the material. The electrons accumulated in the substrate can be extracted by a thermionic cathode (third electrode) rendering the substrate neutral. The discharge voltage for ion beam milling is usually 30 to several keV. For single crystal and polycrystalline silicon, the resolution is better than 10 nm. However, the aspect ratios accomplished by this method will be about unity.

Modification of ion beam milling leads to enhanced milling as briefly described here. In magnetron sputtering, the energy and free path lengths of electrons are increased by the use of magnetic coils leading to enhanced ionization rates. Reactive ion beam etching (RIBE) is achieved by replacing the inert gas species with a reactive gas in the ion beam-milling chamber. In direct ion beam etching, free radicals do most of the etching, reacting with the atoms of the substrate, effecting a chemical/physical removal process.

Deep Reactive Ion Etching (DRIE): Reactive ion etching uses reactive gases that not only impart energy to the substrate through momentum transfer, but also react chemically with the atoms on the surface. Examples of such gases are C_2F_6–Cl_2 for silicon dioxide and O_2–CF_4 for Si. To improve material selectivity and induce anisotropy, C_2F_6–Cl_2, CHF_3–SF_6, and O_2–(50%)-SF_6 are used for Si and H_2–CF_4 is used for SiO_2. Some newer gas etchants include $SiBr_4$, HI, and I_2. For deep trenches and high aspect ratio structures, deep reactive ion etching is used. It uses high-density ($> 10^{11}$/cm^3) plasma source at low pressures (1–20 mTorr). One such high-density source is inductively coupled plasma (ICP) which operates at 13.56 MHz. An RF voltage creates a magnetic field that couples with the ion-producing electrons. Electron density surpasses 10^{12}/cm^3 enhancing etching rates of fluorine-based gases to 6 μm/min for silicon, combined with a selectivity ratio of 150:1 for Si:SiO^2 and a uniformity better than $\pm 5\%$.

Isotropic etching can also be performed without plasma by the use of xenon difluoride (XeF$_2$) at room temperature and 1–4 Torr pressure. Reaction that leads to etching of Si can be written as

$$Si + 2XeF_2 \rightarrow 2Xe + SiF_4 \tag{4.7}$$

If XeF_2 is exposed to water or moisture, HF is formed which can attack SiO_2. Otherwise, this process yields a high etch rate and needs minimal maskant thicknesses. Maskants can be silicon dioxide, silicon nitride, 50 nm of aluminum or a single layer of hard-baked photoresist.

SCREAM Process: Devised by Cornell nanofabrication lab, single crystal reactive etching and metallization provides a single mask micromachining method that allows fabrication of self-aligned high-aspect ratio lateral beams at a temperature lower than 300°C. Additional advantages are its compatibility with IC circuitry and the relatively short processing time (8 h total). The crucial part of the process is the protection of the sidewalls of the beams by a silicon dioxide layer during release of the beams using SiF_6 isotropic etch that undercuts their bottom side. Resonating structures have been made by SCREAM with beams as wide as 0.5–5 μm with an aspect ratio exceeding 10.

4.3.3.4 Wet Etching of Silicon Oxides

Etching of silicon dioxide is performed using buffered oxide etch (BOE), which is a dilute solution of HF buffered with NH_4F to avoid the depletion of fluoride ions. BOE is used to remove underlying sacrificial oxide layers of suspending structures. It removes unwanted silicon dioxide on patterned silicon wafers and removes parasitic oxide in preparation for thermal oxidation. The chemical reaction that leads to the dissolution of SiO_2 is

$$SiO_2 + 6HF \rightarrow H_2SiF_6 + 2H_2O \tag{4.8}$$

BOE can be used for etching of both thermally grown oxides as well as deposited silicon oxides films; however, the etching rate of the latter is much higher.

4.4 APPLICATIONS

MEMS applications can be categorized in terms of their function or the type of industry they are intended for. The type of function can be mechanical, optical, biological, chemical, or acoustic. Industries include automotive, aerospace, electronics, military, and pharmaceutical among many others. Issues involved are the selection of design tools and simulation of the product/process/function, fabrication, packaging, testing for reliability and functionality, marketing, and services.

4.4.1 MEMS Sensors and Actuators

4.4.1.1 Sensing Techniques

Mechanical sensing methods are based on the physical processes affected by motion. These include piezoresistivity, piezoelectricity, variable capacitance, optical, and resonance effects.

Resistivity Techniques: Strain gauges (Figure 4.5) are sensitive to changes in the dimension of a conductive element denoted by gauge factor (GF) relating relative change in resistivity ($\Delta R/R$) to the applied strain (ε):

$$\varepsilon = \frac{\Delta R/R}{GF} \tag{4.9}$$

The change in resistivity of the materials undergoing deformation is contributed to by both geometric change and piezoresistive effect. For materials with a dominant piezoresistive contribution, the gauge factor can be related to the conductivity ($\Delta \rho/\rho$), longitudinal strain (ε_l), and Poisson's ratio (v) by the following equation:

$$GF = \frac{\Delta \rho/\rho}{\varepsilon_l} + (1 + 2v) \tag{4.10}$$

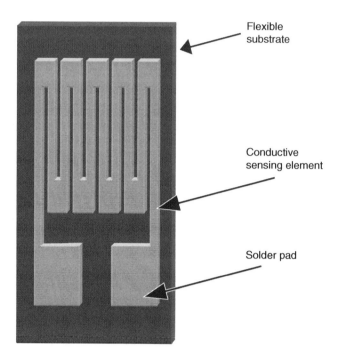

FIGURE 4.5 Strain gauge.

While the gauge factor is 2–5 for metal foils, it ranges from -125 (n-type) to $+200$ (p-type) for single crystal silicon and -30 (n-type) to $+30$ (p-type) for polysilicon. For elastic deformation of materials, the applied strain is related to the resulting stress magnitude (σ) through Young's modulus (E) by $E = \sigma/\varepsilon$. If Young's modulus is known, the reading of the strain gauge can be directly related to the stress in the component undergoing deformation.

Piezoelectricity Techniques: Piezoelectric materials can act as sensors as well as actuators. Below their Curie temperature, piezoelectric materials *deform* under an applied electric field. On the other hand, they *produce electricity* under applied strain. Silicon lacks piezoresistivity; however, piezoelectric materials films are grown on Si-MEMS for actuation or sensing.

Capacitive Techniques: These are based on the relative motion of electrodes. Capacitance sensors are sensitive to temperature change with a nonlinear relationship between the sense signal and the extent of motion. However, precise sensing can be achieved by the integration of a temperature sensor with electronics on the same chip. A capacitive sense signal can show changes in the gap between the two capacitance plates, a change in the area, or a change of a dielectric material between the two plates (Figure 4.6).

Optical Techniques: These methods utilize sensing of the changes in the following electromagnetic wave properties: intensity, phase, wavelength, spatial position, frequency, and polarization. Detection in intensity modulation is performed by photodiodes and phototransistors directly. An LED or laser can be used for light source in intensity modulation sensors. Change in the phase of a wave is detected by mixing it with a second wave leading to interferometry phenomena that is detected by the photodetectors. Modulation of a source spectrum upon interaction with the microsensors forms the basis for wavelength-based sensing. This technique is independent of the intensity variation of the light source. It can measure multiple parameters using wavelengths of interest within the light spectrum. Spatial position sensing is achieved by the motion of the microsensors. Called triangulation, this technique is also independent of light source intensity variation. Laser Doppler velocimetry is based on the frequency shift of an incident beam when it is reflected

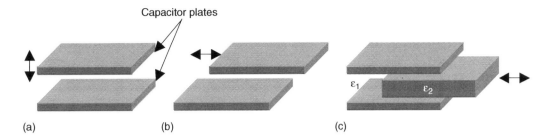

FIGURE 4.6 Capacitive motion sensors: (a) sensing gaps, (b) sensing capacitance area change, (c) sensing change in dielectric property.

from a moving target. The velocity of the target can be calculated based on this Doppler frequency shift detected as intensity variation due to interferometry of the reflected beam with a reference beam.

Resonating Techniques: Resonant structures are made of single crystal materials with specific resonant frequencies that are very stable (e.g., quartz tuning fork is used as a time base in watches). The shift in the resonant frequency of the device by alteration of mass, stiffness, or shape of the resonator can be accurately sensed. Compared with the accuracy of piezoresistive and capacitive sensors, resonant structures have 5–10 times higher accuracies (0.01–0.1%). Mechanical sensors based on resonating techniques are force, pressure, and acceleration sensors that use the resonating structure as a strain gauge. In density and level sensors, a liquid or gas surrounds the resonating structure, effectively lowering its resonant frequency. If molecules of gases (e.g., in gas sensors) are deposited on the resonating structure, its mass will change and the resonant frequency will reduce. Rate and thickness sensors used for monitoring film deposition (PVD, CVD, etc.) are good examples of resonating sensors.

4.4.1.2 Actuation Techniques

Actuation techniques for mechanical sensors include electrostatic, piezoelectric, thermal, and magnetic.

Electrostatic Actuation: A schematic of an electrostatic actuator is presented in Figure 4.7. An applied voltage (V) creates an attraction force (F) to reduce the gap (g) between two attracting plates of area (A) with opposite charges as follows:

$$F = \frac{\varepsilon_0 \varepsilon_r A V^2}{2g^2} \tag{4.11}$$

where ε_0 and ε_r are the permittivity of free space and the relative permittivity of the medium between the two plates. According to Equation 4.11, the relationship between force and voltage

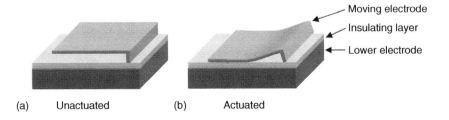

FIGURE 4.7 Electrostatic actuation, (a) unactuated, (b) actuated.

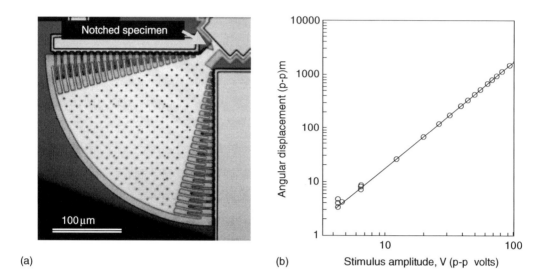

(a)   (b)   Stimulus amplitude, V (p-p volts)

FIGURE 4.8 (a) Silicon MEMS resonating structure, (b) Dependence of angular displacement on the applied voltage. (From Allameh, S. M. et al., *Journal of Microelectromechanical Systems*, 12(3), 313–324, 2003. With permission.)

is nonlinear. For a resonating structure, actuation is achieved by supplying an alternate current to two interdigitating comb structures shown in Figure 4.8a. Capacitively actuated, they are known as comb drives. Instead of moving the gap between the fingers, motion occurs laterally and changes the effective capacitive area between the fingers. The extension of motion is again proportional to the voltage squared. This relationship is seen in the linear plot of Figure 4.8b obtained for the resonant structure of Figure 4.8a. The out of plane motion must be minimized by proper design. One set of combs are rigidly attached to the substrate, allowing the other set to move in and out relative to the first one. In addition to linear motion, rotary motion can also be generated by electrostatic actuation. Rotation is induced in capacitive plates surrounding a rotor by energizing them with voltages of correct phases. In scratch drive actuators (SDA), the buckling of the moving top electrode due to electrostatic actuation results in a small motion of the end of the top plate parallel to the two plates which can be used for stepwise linear motion.

Piezoelectric Actuation: Silicon MEMS cantilevers can be actuated by depositing a piezo-electric material on their top surface. This is due to the silicon's lack of piezoelectric effect. Micropumps use piezoelectric-type actuation that deflects thin membranes. This, in turn, displaces the volume of the micropump, leading to a pumping action through inlet and outlet ports shown schematically in Figure 4.9. Zinc oxide (ZnO) piezoelectric films have been used to actuate micropump diaphragms [15].

Thermal Actuation: Thermal actuators utilize the difference between the CTE of two adjoining layers causing the bilayer to bend up or down depending on whether the material with higher CTE is on top or at the bottom. Called bimorphs, these act similar to the bimetal strips used in thermostat switches. The bimorphs can be fabricated by depositing Al on silicon supports. Another type of thermal actuation is by the thermal expansion of a fluid sealed in a cavity behind a membrane. Upon heating, the fluid will expand causing the membrane to move outward generating a relatively large force. Si-based thermal actuators are made by the bulk micromachining of the cavity in silicon sealed by the anodic bonding of a Pyrex wafer, which contains the resistive element. For industrial use, these actuators are coupled with microvalves for applications such as medical instrumentation, gas mixtures, and process control equipments [2].

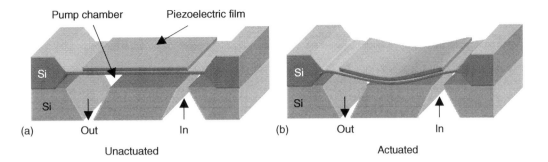

FIGURE 4.9 Si-Based micropump, (a) unactuated, (b) actuated after. (Drawing created after Beeby, S. et al., *MEMS Mechanical Sensors*, Artech House, Inc., Norwood, MA, 2004.)

Thermal actuation is also possible using shape memory alloys (SMA). Upon heating, the crystal structure of SMA undergoes a phase transformation to take a predetermined shape. Upon cooling, it returns to its original shape. As an example of this class of materials, Nitol (TiNi) is sputter deposited as thin films producing a strain up to 5% upon actuation. The transition temperature can be adjusted (from $-100°C$ to $+100°C$) by controlling impurity concentration. A thermal actuator used in positioning optical fiber is shown in Figure 4.10.

Magnetic Actuation: The use of permanent magnets is not compatible with current MEMS fabrication processes. Electromagnetic actuation is therefore performed by passing a current through a coil that generates a magnetic field. A force is generated perpendicular to the current direction and to the magnetic field. Permalloy can be deposited with a thickness of 7 μm on Si cantilevers and placed in a magnetic field. Such a beam will deflect with angular displacements exceeding 90° [16]. Magnetostrictive actuation generates strains upon magnetization much greater than those achieved by piezoelectric materials. Terfenol-D (Tb-Dy-Fe) produces strains as large as 0.2%.

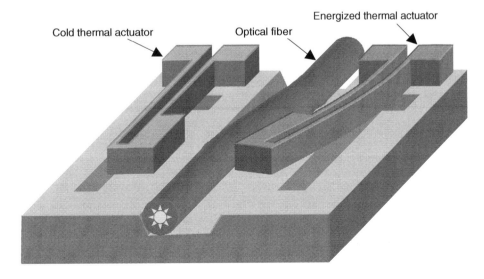

FIGURE 4.10 Optical microdevices: Micromachine 1×2 optical fiber swtich with moving. (Drawing created after Field, L. A., *Sensors and Actuators: A (Physical) International Solid-State Sensors and Actuators Conference-TRANSDUCERS '95*, 25–29 June 1995, A53(1–3), 311–316, 1996.)

4.4.1.3 Smart Sensors

Many sensors come with some electronics packaged with them. Called smart or intelligent sensors, they do elementary signal processing tasks such as filtering or may even perform functions that are more complicated. The integrated modules that combine MEMS sensors with onboard electronics can be treated as a black box by the host system, isolating it from the sensors' complexities. Further, they may house additional sensors such as local temperature monitors. Housing amplifiers (e.g., charge amplifiers for capacitive drives) will produce better signal to noise ratios. The reason is that high gains can be achieved next to the sensors and away from the front-end analog circuitry. In addition to signal filtering and amplification, other tasks such as analog and digital processing, analog/digital data conversion, data communication, along with components such as memory, and control processor may be housed within the sensors.

4.4.2 Mechanical Sensors

These include pressure, force, torque, inertial, and flow sensors.

4.4.2.1 Pressure Sensors

With a wide range of applications from automotive manifold and tire pressure to complex hydraulic systems, microphones and blood pressure measurement, Si-MEMS pressure sensors constitute a major portion of the commercial MEMS market. Sensing techniques range from piezoresistivity-based methods to more advanced resonating techniques. Direct readings of pressure can be used to infer other properties such as the flow rate in a pipe, volume of a liquid in a tank, and altitude. MEMS pressure sensors are mostly based on diaphragms that deflect with an increase in the pressure of the fluid (liquid or gas). The magnitude of such deflection is nonlinear with the applied pressure. Nevertheless, linearity of pressure sensors is defined as the closeness with which the sensing data for various pressures fits a straight line. Other issues involved in pressure sensors are the sensitivity of the signal to applied pressure, long-term drift of the signal over a period of time, and temperature effects. The pressure sensing function of MEMS sensors is analogous to mass-spring systems. Therefore, their dynamic response depends on the stiffness, mass, and degree of damping. Sensing frequency should not be close to the resonant frequency of such mass-spring system [2].

Types of Pressure Sensors: Most Si-based pressure sensors operate by the deflection of a diaphragm; however, there are other types that utilize piezoresistive, capacitive and resonating sensing techniques. The most common method for fabrication of diaphragm based pressure sensors is anisotropic wet silicon etching using (100)-type silicon single crystal wafers. Etching of the backside of the diaphragm must be stopped by a boron doped etch stop. This allows precise control over the thickness of the diaphragm. Equations have been developed to relate the deflection of the diagrams to the applied pressure. Dimensions of the diaphragm as well as its shape and the magnitude of the deflection affect this relationship. Analytical solutions to such relationships are also available for membranes (very thin diaphragms).

Due to its piezoresistivity, silicon is an obvious choice of material for pressure sensors. Diffused or implanted resistors are placed along the edges of the silicon diaphragm to measure strain in each of x and y directions. This type of pressure sensor has been commercially in use for applications such as the absolute pressure of manifolds used in automobiles. To improve their function, rigid centers can be added to the diaphragms (e.g., bossed diaphragms). This enhances the linearity and sensitivity of pressure sensors. Addition of a temperature sensor to the pressure-sensing element will enable correction of the temperature effect on the measured pressure. Use of SOI has greatly improved the fabrication technique for the diaphragms. The silicon dioxide layer acts as an etch stop layer allowing a more precise control of diaphragm thickness. Surface-machined pressure sensors allow reduction in the size of the sensors resulting in a reduction in the fabrication cost and better compatibility with IC technology. This method allows fabrication of the diaphragm from polysilicon, silicon nitride or other materials. In addition, shapes that are more complex can be adopted for pressure sensing structures.

Capacitive Pressure Sensors use a parallel plate gap sensing technique to measure the deflection of the diaphragms. A reference plate is mounted beneath the diaphragm. An applied pressure causes the diaphragm to deflect, moving toward the reference plate, reducing the gap between the two plates. The reduction in the gap is monitored and related to the applied pressure. Silicon vacuum sensors with a sensitivity of 1 pF/mbar have been developed by NASA using a circular diaphragm. Fused quartz has been used to make capacitive pressure sensors. The main drawback of capacitive sensors is the nonlinearity of the sensor output that will require complex electronic circuitry. Capacitive pressure sensors have a commercial application in measuring blood pressure.

Resonant Pressure Sensors use a resonating structure to measure the deflection of the diaphragm. The fabrication of the resonating structure, its actuation mechanism, and the circuitry associated with sensing its displacement is difficult; nevertheless, they are more accurate than other pressure sensors. The change in the deflection of the diaphragm causes a change in the frequency of the resonating structure. Both resonating structure and diaphragms are made from silicon using a boron etch stop method. Differential pressure sensors using dual resonating structures have been developed to reduce temperature effects on the pressure readings, and therefore, offer more accuracy (e.g., sensing pressures less than 1 mTorr). A variety of techniques are used to fabricate these structures including Si fusion bonding, surface-micromachining combine with bulk etches, entirely surface machining, and SOI wafer technology.

Microphones: MEMS-Based microphones are pressure sensors specifically developed to convert acoustic waves into electrical signal. Membranes for microphones must be of sufficient sensitivity but insensitive to static pressure with desired dynamic range. They must also have the capability to be packaged properly for applications such as hearing aids. Types of MEMS-based microphones include capacitive, piezostrictive, piezoelectric and modifications of each of them. More recent, but less known techniques of pressure sensing include optical techniques, surface acoustic wave technique, MOS transresistors, inductive coupling, and force balance sensing.

4.4.2.2 Force and Torque Sensors

Loadcells have home applications in scales as well as industrial applications in retailing, automotive, and aerospace. One way to measure a load is to apply it to a spring and measure the spring's extension by resistive, capacitive, or resonant sensors. The load (P) can be related to this extension (x) by the spring stiffness (k) through Hook's law as $P = kx$. Due to the high power required by metallic resistive strain gauges and their smaller sensitivity (response is in millivolts), there is a move toward other types of strain measurement techniques such as SAW, optical, and magnetoelastic technologies [2].

Si-Based Devices: Silicon strain sensors with a gauge factor of ± 130 are highly sensitive arising from silicon's piezoresistive capability that allows strain measurements in the nanostrain range (10 nanostrain). To mitigate their temperature sensitivity, silicon strain gauges are used in pairs with opposite gauge factors. Mounting one pair in x direction and one pair in y direction, strain gauges form a bridge, which produces a signal proportional to the applied strain. With piezoresistivity characteristics, amorphous silicon films can be grown on a load-carrying member to monitor the loads applied to it. Silicon-based bulk machined micro-torque sensors have been developed [17] for the use of the watch-manufacturing industry. In differential force mode, they can measure forces as small as -200 to $200 \mu N$ with resolutions better than $0.5 \mu N$. For loadcells used to measure loads less than 150 kg, a bending silicon beam can be used; however, for higher loads silicon should be used only in compressive loading mode.

Resonating Structures: Si-Based resonating loadcells have been developed using a tuning fork consisting of a capacitively driven comb or a parallel plate drive. Application of load to a pair of these loadcells will cause a change in the resonant frequency of their tuning fork, which is registered. The differential operation of the loadcells will allow measurement of load. For non-contact torque measurements, SAW-based sensors have been fabricated. These are used in pairs, oriented perpendicular to each other. Measurement of the difference between the changes in the phase or resonant frequency of the pair allows the determination of the applied torque with minimal effects of temperature or shaft bending on the output signal.

Optical, Capacitive, and Magnetic Devices: Optical loadcells, immune to electromagnetic interferences, are commercially produced to monitor the torque in the electric power-assisted steering systems (EPAS) of automobiles. An optical system measures the torque angle of a known mechanical link leading to the calculation of the torque. Optical force measurement is implemented in atomic and scanning force microscopes (AFM, SFM) where a silicon cantilever flexes when coming in contact with the surface being scanned. The magnitude of the deflection is registered by optical measurement using a laser beam reflected from the backside of an AFM tip. Si-based capacitive force sensors with variable capacitance gaps provide for the non-contact measurement of load. In double stack configuration, a middle plate between two other plates provides two capacitive signals from which the difference is utilized to calculate the load. With a gap of 10 μm and capacitance of 1 pF, loads in the range of 0.01–10 N can be measured accurately [18]. Capacitive force measurement has applications such as nanoindentation for studying the mechanical behavior of materials on nano-scale.

Tactile Sensors: To minimize invasive endoscopic surgery, Si-based tactile sensors have been developed and incorporated in the grips of a surgical grasper measuring loads of 0.1–2 N with the piezoelectric technique [19,20]. Usually, silicon substrates are coated with polymer based tactile sensor films (polyimides and polyvinylidene fluoride have been used). Micro-Newton tactile sensors have been made using piezoresistive silicon films capable of resolutions as high as 0.02 mV/μN.

4.4.2.3 Inertial Sensors

Single and multi-axis accelerometers and gyroscopes have been made of micromachined silicon for the measurement of angular and linear acceleration. Application of these sensitive devices has been extended from aerospace and military to automotive for the monitoring and control of components such as seat belts, air bags, active suspension, and traction control.

Other applications include smart ammunitions and inertial guidance in military and monitoring patients with Parkinson's diseases in medicine. Developmental work is in progress for the use of inertial sensors in the image stabilization of camcorders, virtual reality head-mounted displays, GPS backup systems, vibration and shock monitoring during shipping of packages, and pointing devices for computer control. Except for vibration sensors and smart ammunitions operating at frequencies up to 100 kHz, most other inertial sensors have frequencies in the range of 0–1 kHz. Resolutions of better than 1 μG for accelerometers and 10^{-4} °/s for gyroscopes have been accomplished.

Most micromachined accelerometers [21] operate based on a proof mass suspended from a spring in parallel with a dashpot. Acceleration will move the proof mass (m) for which the displacement (x) can be measured. The magnitude of the acceleration (a) can be calculated from [2]

$$\frac{x(s)}{a(s)} = \frac{1}{s^2 + \frac{b}{m} + s\frac{k}{m}} \tag{4.12}$$

where b is the damping coefficient of the dashpot, k is the spring stiffness, and s is the Laplace operator. Damping of the sensing element is crucial for maximum bandwidth. Position measurement mechanisms used for the calculation of displacement (x) include the following types: capacitive, piezoresistive, piezoelectric, optical, and tunneling current.

Multiaxial Accelerometers: A single proof mass accelerometer, capable of moving in three perpendicular directions has been made by surface micromachining. Acceleration in two directions parallel to the device plane is sensed by the in-plane motion of interdigitating combs. The third component of acceleration is monitored capacitively by the change in the air gap of the capacitor comprising of an electrode under the proof mass and the central portion of the back of the proof mass itself [22].

Commercial Accelerometers: Analog devices makes commercially available ADXL accelerometers that are made by surface micromachining with electronics incorporated on the same chip. The ADXL-50 used for air bags has a dynamic range of ± 50 G with a bandwidth of 6 kHz and a sensitivity of 19 mV/G. Their more recent accelerometer, ADXL311, is very inexpensive ($2.50 in quantities over 10,000) and contains dual-axis sensors that operate from a single 3 V power supply. Motorola's MMA121P, with a dynamic range of ± 40 G, is also made by surface micromachining and operates in the temperature range of $-40°C$ to 85°C. Other companies that make micromachined accelerometers include Applied MEMS, Colibrys, Bosch, Endevco, Honeywell, MEMSIC, Kionix, Kistler, SensoNor, and STMicroelectronics [2].

Micromachined Gyroscopes: The operation of Si-based gyroscopes is mainly based on secondary oscillations induced by Coriolis forces caused by the primary oscillation of the same or a different resonating structure oscillating nearby. Angular acceleration is calculated from the amplitude of this secondary oscillation. Coriolis force (F) and acceleration ($\vec{\alpha}$) can be formulated as

$$\vec{F} = m\vec{\alpha}$$
$$\vec{\alpha} = 2\vec{\Omega} \times \vec{v}_r$$

(4.13)

where $\vec{\Omega}_r$ is angular acceleration and $\vec{v}_r$ is radial velocity.

Clark and Howe made the first single-axis micromachined gyroscopes form 2-μm thick polysilicon driven by comb drive actuators with large drive amplitudes [23]. A schematic of a silicon-based gyroscope devised by Greiff et al. is presented in Figure 4.11. There are two built-in torsional supports for the two gimbals, the first of which is electrostatically forced to oscillate at 3 kHz. In the presence of a rotation about the vertical axis, the inner gimbal will gain energy and will also

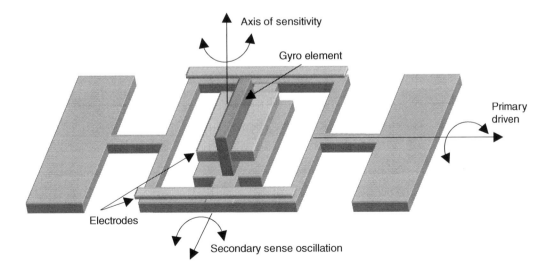

FIGURE 4.11 Gyroscope using two gimbal structure. (Drawing created after Grieff, P. et al., *1991 International Conference on Solid-State Sensors and Actuators,* Jun 24–28, 1991, San Francisco, CA, IEEE, Piscataway, NJ, 1991.)

oscillate by 3 kHz. While the frequency of oscillation of the inner gimbal is induced by the outer gimbal, the amplitude of oscillation is variable. The amplitude of the oscillation of the inner gimbal will be proportional to the rate of energy received from the rotation and will be a measure of the angular acceleration about vertical axis. Dual-axis gyroscopes have also been developed by surface micromachining of 7-μm thick polysilicon with resolutions as high as 0.1 °/s [24].

Commercial Gyroscopes: Si-Based micromachined gyroscopes are commercially available from Silicon Sensing Systems use magnetic actuation and detection along with a ring type sensing element. Analog Devices offers a series of angular-rate sensors (ADXRS family) that are surface micromachined using a BIMOS process used for making air bag accelerometers for automotive use. As an example, ADXRS150, a single chip yaw rate gyro comes with signal conditioning, a temperature sensor output, a range of ± 150 °/s and 2000 g powered shock survivability. Samsung and SensoNor are also in the process of making microgyroscopes [2].

4.4.2.4 Flow Sensors

Si-MEMS devices have numerous applications in fluidics, including flow sensors. Flow sensors are usually an integral part of a microfluidic system that may contain micropumps, microvalves, and micromixers. The main application of flow sensors is in the automotive industry, which is continuously increasing the number of sensors used in engines (30 sensors are expected to be used in engines by 2010). Electronic fuel injection system uses flow sensors to determine the mass flow rate of air entering the engine. Other applications include pneumatics, bioanalysis, metrology (wind velocity), civil engineering (wind forces), medical technology (respiration, blood flow, and surgical tools) indoor climate control (ventilation and air conditioning), and home applications (vacuum cleaners and air dryers) [2].

Flow sensing mechanisms are based on various physical, thermal, and chemical phenomena. Thermal flow sensors constitute the majority of flow sensors and have application in auto air intake systems. They are divided into three groups: anemometers, calorimetric flow sensors, and time-of-flight sensors. Thermal variation is sensed by electro-resistivity techniques. Silicon, silicon carbide, germanium, gold, platinum, and Ni–ZrO$_2$ cement films have been used for sensing temperature variation in flow sensors. A number of thermistors and thermocouples (Al/polysilicon, Pt/polysilicon, and Au/polysilicon) have also been investigated. Thermal flow sensors are produced by companies like Bosch GmbH (Germany) for the automotive industry and by HL Planartechnik GmbH (Germany), as well as by the Fraunhofer Institute for Silicon Technology (Germany) among many others.

There are other types of thermal flow sensors including pressure difference flow sensors, force transfer flow sensors (based on drag force, lift force, and Coriolis force), and static turbine flow meters. Nonthermal flow sensors include electrohyrodynamic, electrochemical flow sensors, as well as sensors based on Faraday's Principle, or based on periodic flapping motion. Use of optics in characterization of flow has lead to developments of Si-MEMS with applications in flow measurement. This technique allows fluid velocity measurement, particle detection and counting, and multiple phase flow detection.

4.4.3 Biochemical Sensors

Microsensors used for biochemical applications consist of functionalized (sensitive) surfaces coupled with sensing transducers that produce electrical signals when species to be analyzed (analyte) react with the functionalized surfaces. Figure 4.12 shows the main components of a biochemical sensor. The type of the reaction depends on the shape, charge, and nature of the analyte molecule. This reaction can be reversible (e.g., adsorption and desorption of an organic vapor for a polymer material) or irreversible (e.g., catalytic reactions where analyte molecules are

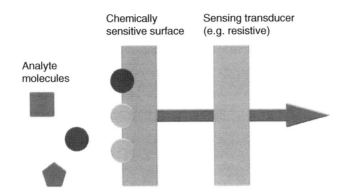

FIGURE 4.12 Biochemical sensors main components. (Drawing created after Gardner, J. W. and Bartlett, P. N., Bok review: Electronic noses, *Nature*, 402(6760), 351, 1999.)

consumed). Selectivity of the material tends to be a function of the shape of the molecule with the lock-key type mechanism to be the most selective.

Biochemical sensors have been classified based on the type of sensing function performed by their sensitive layer [3] (Figure 4.13). The type of signal produced by the sensitive layer can be conductimetric (change in electrical resistance), potentiometric (change in work function), calorimetric (change in heat of reaction) or mechanical.

4.4.3.1 Conductimetric Gas Sensors

For these sensors, the reaction of the target gas molecules with the sensitive surfaces causes a change in the electrical resistance of a material (e.g., SnO_2) that will be transduced into an electrical signal. A commercial example of this type of microsensor is the Taguchi-type oxide gas sensor made by Figaro Engineering Co. (Japan) [3]. While the Taguchi sensor is sensitive, it has to operate at high temperatures (usually a heating coil is used) and it is not very selective.

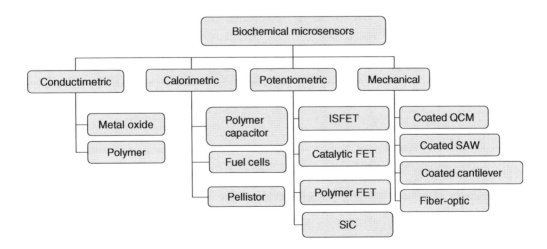

FIGURE 4.13 Classification of biochemical microsensors. (Chart drawn after Gardner, J. W., Varadan, V. K., and Awadel Karim, O. O., *MicroSensors MEMS and Smart Materials*, John Wiley and Sons, Ltd, West Sussex, England, 2001.)

4.4.3.2 Potentiometric Devices

Based on metal–insulator semiconductor structures, potentiometric sensors work on the basis of a change in electric potential of the devices when target gas species are detected. Silicon micro-machined metal oxide semiconductor fields effect transistor, MOSFET-based, gas microsensors have been made with a silicon microplatform and a thermal plug. A temperature sensor may be incorporated into the device. The gate of the MOSFET transistor can be made of Pt, Pd, and Ir, and gases sensed include ammonia, ethanol, and hydrogen sulfide. When a current passes through the transistor, hydrogen present in the atmosphere dissociates and causes a shift in the threshold voltage monitored by a characteristic voltage response. These devices have shown a shift in the threshold voltage equal to 220 mV when exposed to a pulse of 20 ppm ammonia. Commercial MOSFET gas sensors are produced by Nordic Sensors (Sweden) for the detection of odors and gases.

4.4.4 AUTOMOTIVE APPLICATIONS

Since mid 1970s, MEMS devices have been contributing to the safety and performance of auto-mobiles. With 55 million passenger vehicles sold [25] worldwide (2002), the automotive market remains a lucrative market for MEMS devices. The U.S. car sales alone exceed 16 million (2004) which increases every year. The MEMS market of just under $1 billion in 2002 is predicted to grow to nearly 1.5 billion in 2007 [25]. Other predictions are even higher [14]. Application of MEMS and microelectronics include air bags, antilock brakes, active/dynamic suspension, alarms, antennas, atmospheric pressure, climate control, comfort programs, crash severity, electronic fuel injection, fuel tank vapor, GPS, headlights, heads-up displays, keyless entry, keyless ignition, night vision, occupant detection/weight, oil pressure monitoring, rollover detection, tire pressure monitoring, and electronic stability system. MEMS devices used in automotives include inertial sensors, pressure sensors, biometric, humidity and infrared sensors, and optical MEMS.

4.4.5 PHOTONICS APPLICATIONS

One main commercial application of Si-MEMS in photonics is fiber optics-based communications where microdevices are used for tunable lasers, wavelength lockers, optical attenuators, and optical switches. Si-MEMS are also used in the manufacture of various components of electronic displays such as digital light processing (DLP). The commercially available Si-MEMS based infrared-ray-imager from Honeywell, Inc. (Minneapolis, Minnesota) is rivaling the cooled cameras made from II–VI materials. The microdevice used for each image pixel senses infrared radiation by a tempera-ture sensitive resistive element, which gets hot when radiated by infrared rays. Made of metal-oxide-coated silicon nitride, this element is thermally isolated form the substrate by suspension over a reflective metal layer. Suitable metal oxides for heat sensing are vanadium oxides and lanthanum manganese oxides with a fraction of La ions replaced with Ca, Sr, Ba, or Pb.

Producing sharp images, digital micromirror displays developed by Texas Instruments (Dallas, TX) have been commercially produced for the last ten years. The micromirrors that produce image pixels are suspended from torsional hinges and can be actuated electrostatically by energizing electrodes beneath them. In the on-state, each micromirror is tilted by $+10°$, reflecting an incident beam through a lens positioned in front of the micromirror. In the off-state, micromirror is tilted $-10°$, directing the incident beam away from the lens. Pixel elements can be bright, dark, or can be of various levels of gray. The dwell time of the micromirrors, in the on-state, determines the level of gray corresponding to the image pixel elements. Two-dimensional micro-mirror arrays of 800×600 and 1280×1024 are commonly used for DMD-based display projectors. Al metallization is used to provide electrodes, hinges, and reflectivity for the Si-based micromir-rors. While the mirrors can withstand shock levels of 1500 G, their failure is usually caused by the creep of metal hinges leading to residual tilt in the unactuated state. Color production is by the use of a color wheel placed in the path of incident light before it reaches the DLP chip. The timing of the actuation of each pixel will determine if the reflected light is red, blue, green, or a combination of them. Alternatively, three DLP chips can be used, each dedicated to one primary color.

A simpler design adopted by Silicon Light Machines (Sunnyvale, CA) has allowed the reduction of the two-dimensional arrays of micromirrors to simply one by making them out of reflecting ribbons. Suspended from their end supports they are actuated electrostatically acting as grating light valves. In an actuated state, the ribbons are pulled down toward the substrate by a one half-wavelength distance. This effectively changes the phase of the reflected beam by 180°. The use of an aperture in front of the ribbons allows the passage of first order reflections through a lens but blocks the higher-order reflections. For each pixel, six ribbons are used, two for each primary color (red, blue, and green). The ribbons are actuated alternatively. One advantage of this technology, called grating light valve (GLV) is the speed of pixel actuation, which is 1000 times faster than that of DMD (about 20 ns). This allows the optical scanning of one single row of ribbons to produce the whole two-dimensional image. Other advantages include a simpler design and fewer mirrors, which reduces the manufacturing cost and increases the reliability of these devices.

Fiber optics applications of MEMS include numerous devices with design variations implemented by the manufacturers, tuned with the particular applications for which they are developed. Some of the most recent work has been done in the areas of magnetically actuated MEMS optical switches [26–28], Lorentz force microactuators [27,29], vertical Si micromirrors [30], 1×2 [8] and 2×2 [31] optical fiber switches, and coupling platforms for optical fiber switching in free space [32]. Examples are presented here with the name of one company in each category: external cavity tunable lasers (Iolon, Inc., San Jose, CA); distributed feedback (DFB) lasers (Santure Corp., Fremont, CA); wavelength lockers (Digital Optics Corp., Charlotte, NC); digital optical switches (JDS Uniphase, San Diego, CA); beam steering micromirrors for photonic switches and cross connects (Integrated Micromachines, Inc., Irwindale, CA); and achromatic variable optical attenuators (Lightconnect, Inc., Newark, CA). An example of a micromachined 1×2 optical fiber switch [8] is presented in Figure 4.10.

4.4.6 LIFE SCIENCE APPLICATIONS

Since most biological analyses involve fluids, the area of Si-MEMS based microfluidics has found many life-science applications. Some of the biochemical sensors and pressure transducers, discussed earlier, have biological applications. A holistic approach to bioanalysis has lead to the concept of the micro-total analysis system (μ-TAS) in which all stages of biochemical analysis are performed using microdevices. Microfluidics has applications not only in bio-MEMS, but also in chemical analysis, drug synthesis, and drug delivery. The latter incorporates Si-based microsensors, micropumps, microvalves, and some other micromachined devices. One example of the biomedical application of MEMS is shown in the schematic of a pill camera presented in Figure 4.14. The wireless tiny camera installed in the capsule allows endoscopy of the lining and middle part of gastrointestinal tract [33–35]. The system includes light emitting diodes, a CMOS image sensor, batteries, and an application specific integrated circuit (ASIC) transmitter along with an antenna. Additional devices such as a micromachined biopsy tool [36] and a remotely propelled magnetic actuator [37,38] may be added to the capsule type endoscope. A newer design of such microcapsules has been proposed using biologically inspired microfibrillar adhesives [39].

Si-MEMS micropumps produce various types of flow including differential pressure driven flow, electrophoreses flow, or electroosmotic flow. Flow rates are much smaller for microfluidics compared to macroscale (e.g., 1 cm/s for a channel diameter of 30 μm and a Reynolds's number of

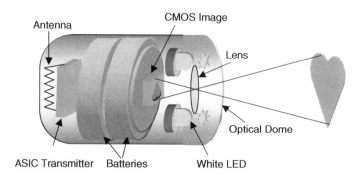

FIGURE 4.14 Schematic of wireless capsule endoscope. (Drawing created after Malauf, N., *An Introduction to Microelectromechanical System Engineering*, Artech House, Boston, MA, 2004.)

0.03). This makes mixing possible mainly by diffusion, which is advantageous for systems like Cell LabChip (Agilent Inc., Palo Alto, CA) where cells are tagged and detected using fluorescent dyes.

DNA analysis is an important implementation of Si-MEMS in biological sciences. A polymerase chain reaction (PCR) on a chip utilizes microdevices to do the amplification process where many identical copies of DNA are made from one single DNA strand. Advantages of microscale PCR include faster thermal cycling, use of smaller quantities of reagents, and the ability to integrate with detection microdevices such as electrophoretic separation or TaqMan tagging. Electrophoretic-on-a-chip is another application of Si-MEMS in biochemical analysis. For DNA sequencing, amplified DNA strands are labeled with specific fluorescent or radioactive tags and later detected by the analysis of tagged fragments. When using microdevices for electrophoresis, smaller lengths of DNA fragments (~ 100 µm) are needed. This in turn requires smaller voltages to drive the DNA fragments through shorter channels. The process will also be faster due to the shorter distances and smaller volumes involved. MEMS-based electrophoretic devices are available from Caliper Life Sciences Inc. (Hopkinton, MA). Other applications of Si-MEMS in biological sciences include DNA hybridization arrays, microelectrode arrays for DNA addressing, and cell culture.

More recent applications of MEMS in the biomedical field include micromachined silicon antennas for RF telemetry used for implantable bio-MEMS sensors [40]. The use of silicon-based microfluidic to amplify bacterial cell numbers by orders of magnitude along with their detection purely by electrical means has been recently reported [41]. A MEMS-based differential mobility spectrometer has been demonstrated for the simultaneous detection of multiple chemical and biological agents with potential application in counter terrorism operations [42].

4.4.7   RF APPLICATIONS

During the past two years, research in the area of RF application of silicon MEMS has been focused on the fabrication of silicon spirals for monolithic MEMS LC circuits [43,44]; surface micromachined SiC vertical resonators [45]; novel structures for low-loss, low-cost, high-yield RF-MEMS switches [46]; use of SiDex process for fabrication of integrated RF MEMS inductors [47]; digital type single crystal RF MEMS variable capacitors [48]; reconfigurable antennas with RF-MEMS switches [49,50]; high-Q integrated RF-Si-MEMS [51]; piezoelectric-driven RF MEMS switches [52]; single crystal RF-MEMS using SiOG process [4,53]; fabrication of RF MEMS with SiOG method [53]; and RF-MEMS switches for space applications [53]. The details of the techniques used to fabricate RF-MEMS components, namely inductors, capacitors, and switches on silicon substrates are recently reviewed [55].

4.5 RELIABILITY ISSUES

Unlike silicon structures used for electronic integrated circuits, Si-MEMS devices have numerous moving parts exceeding hundreds of thousands in optical applications such as DLP projectors. Many of these carry substantial loads that generate significant levels of stress in the presence of stress risers such as sharp corners. Cyclic deformation, both thermal and mechanical will complicate the problem, especially in the presence of moisture or an aggressive environment. Failure of MEMS devices due to mechanical failure will briefly be discussed here. Some preventive measures to mitigate the chances of failure of silicon components will also be presented.

4.5.1 MECHANICAL TESTING OF MEMS

Mechanical tests have been performed to establish not only the behavior of mechanical components at microscale, mimicking service life conditions, but also to devise and verify preventive measures to enhance the reliability of the devices, predict their service life, and help improve the next generation silicon MEMS. While Young's modulus is size independent, the mechanical behavior of bulk silicon differs from that of thin films made of polysilicon. This is in addition to the orientation dependence of most properties for single crystal silicon. Etch holes [56], sharp corners, and other stress risers will degrade the mechanical properties of Si-MEMS. On-the-chip test structures are examples of microscale samples loaded by electrostatic actuation. A picture of such a test sample, a resonating structure is shown in Figure 4.8. This structure has been extensively used for the study of fatigue in silicon MEMS [7,57–63].

Fatigue of Silicon MEMS: Si-MEMS Structures have been studied by various groups of researchers [7,58,60,62,64–79]. Bulk silicon is known to resist fatigue failure. However, in thin films, such as silicon structures used for accelerometers in cars, failure has been reported at stresses much below monotonic loading fracture strength [72,79–83]. In the presence of moisture, this effect is more pronounced [80]. Actuation of the resonating polysilicon test structure at room temperature, in atmospheres with various levels of moisture, has shown that fatigue strength increases as relative humidity decreases. Further, low cycle tensile fatigue testing of on-the-chip single crystal silicon structures, cyclically stretched to strains over 3.5% with a frequency of 10 Hz, has shown a reduction in the probability of failure when the applied strain is lower than the mean fracture strain [64]. Surface topography evolution of resonating structures shown in Figure 4.8 has shown an increase in the in the amplitude of surface perturbations at loci of high stress levels (e.g., at the notch root) [7,60]. Deepening of the grooves has been observed during the in situ AFM imaging of the resonating structures. The rate of deepening of the grooves were reported to be proportional to the magnitude of stress [60]. Similar resonant structures made from single crystal silicon show cyclic failure stresses nearly half of the fracture strength of silicon.

Several mechanisms have been reported for the fatigue failure of silicon-MEMS. Most experimental evidence point to some kind of oxide formation or damage accumulation during the crack ignition stage of fatigue [61]. Surface roughening during crack initiation observed by AFM imaging, may indicate thickening of silicon oxide layer up to 90 nm before fatigue failure [7,60]. In addition to the surface oxide formation and dissolution mechanisms for fatigue, called reaction layer mechanism and cyclic stress-assisted dissolution models respectively, other mechanisms such as mechanically induced subcritical crack growth model have been suggested [84]. The latter mechanism, however, does not account for the surface modifications observed for silicon resonant structures [7,68].

Residual Stresses: Formation of tensile and compressive residual stresses under different experimental conditions during the film deposition was discussed. It was mentioned that it is possible to tailor near-stress free films by combining such tensile and compressive layers. In fact multilayered polysilicon films can be designed such that they will have a prescribed curvature [12].

For example, LPCVD polysilicon films develop tensile and compressive stresses depending on the deposition temperature. Tensile layers deposited at 570°C can be combined with compressive layers deposited at 615°C. Depending on the desired curvature, the thickness of each layer can be determined [11,12]. Annealing of films will relieve most of the internal stresses developed during implantation of various species to levels of 50–100 MPa.

Effect of Environment: While detrimental for fatigue of Si-MEMS including cyclic damage accumulation of actuated structures [61,75,80], presence of moisture on the surfaces of mating parts may help reduce friction during actuation [85]. Temperature does not appear to affect the performance of Si-MEMS. Tests on Sandia microengines, conducted at a temperature range of −55°C to 200°C, show no detrimental effect on the performance of these microengines [86]. While actuated MEMS are susceptible to shock exceeding certain limits, unactuated structures better withstand shock pulses. Unactuated Si-based microengines exposed to shock loads of 0.5–40 kg with durations of 0.2–1 ms remain structurally sound. However, this is not the case for actuated components; shorts have been reported for actuators exposed to shock loads as small as 4 kg [87]. It is postulated that debris from the edges will cause such shorts. Unactuated actuators exposed to vibration may suffer by adhesion of rubbing parts [88].

Creep of Si-MEMS: Silicon has a low brittle-to-ductile-transition (BDT) temperature of 550°C. Silicon is the material of choice for many Si-MEMS applications. However, problems arise for devices such as microengines which are intended to work at temperatures in the range of 600–750°C [89,90]. Single crystal silicon components incorporated in microengines, microrocket engines, or even the microturbo pumps may suffer from creep deformation at their service temperatures [91]. Deformed cantilevers tested at a temperature in the range of 600–850°C, in 4-point bend configuration under 50–200 MPa, show slip bands in <111> directions [91]. Arrhenius-type dependence of creep rate on temperature is observed for these samples with an activation energy of 224 kJ/mol. Creep deformation is found to be localized having implications for silicon members used for MEMS with high temperature applications [91]. Creep tests performed on silicon whiskers exposed to stress levels of 10–500 MPa and temperatures of 300–100 K, suggest that heterogeneous nucleation of dislocations at high stress points associated with thermal fluctuations are responsible for the observed creep behavior [92]. Creep bending of boron-doped silicon shows negative creep attributable to redistribution of vacancies and impurities in the transverse field of mechanical stresses [93].

4.5.2 RELIABILITY OF Si-MEMS

Silicon reliability has been linked with a number of parameters including fatigue [94], shock [95], stiction [96], wear [97], vibration [95], frequency [98], and environment [99]. Most reliability studies involve development of a number of tools such as those needed for the electrical actuation of surface machined actuators, in situ visual inspection of components during their normal operation, test data acquisition, and performance characteristics extraction [100]. Silicon MEMS that have been studied for reliability issues include microengines [101], microactuators [102], aerospace MEMS [103–105], microrelays [106], telecommunications MEMS [107–109], resonating structures [80], and structural MEMS [110]. Other issues such as packaging [111,112], interfacial properties [113], and size effect [114] have also been addressed in terms of their effects on the reliability of MEMS devices.

For reliability studies, usually a large number of MEMS devices need to be examined while operating in parallel. This is needed to ensure statistical validity of the reliability data. The data can be then used to model the behavior and service life of MEMS devices. These models will then be able to predict the life of the devices operating under different service conditions [86,96,97,100,115]. Test methodologies developed for reliability design have been discussed for resonating structures made of silicon along with solutions to reliability issues [108]. Often simulations such as Monte Carlo are used to assess the stochastic nature of local processes such

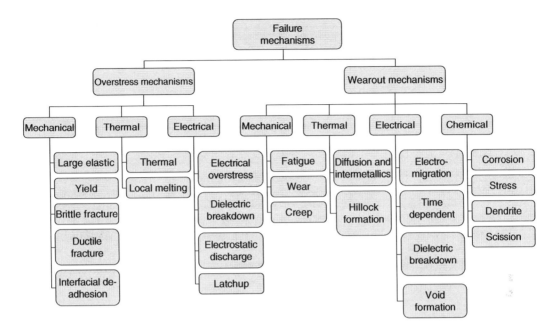

FIGURE 4.15 Failure mechanism. (Chart drawn after Pecht, M., DasGupta, A., and Evans, J. W., *Quality Conformance and Qualification of Microelectronic Packages and Interconnects*, Wiley, New York, 1994. With Permission.)

as via fatigue to study reliability of complex MEMS modules [116]. Early studies of MEMS reliability focused on the root cause of failure employing an "upstream problem solving" technique [116].

Failure mechanism assessment, as the main focus of reliability studies, comprises a series of tasks that may be categorized in the hierarchical structure [4] of Figure 4.15. Not all failure mechanisms may be operative at the same time. However, failure may originate from a physical process effected by thermal, mechanical electrical, or chemical stresses [117]. Solder joints that connect electrical components including actuators, fail by fatigue, and creep. In aggressive environments, oxidation and fatigue may combine to cause stress corrosion cracking. To assure the reliability of complex systems such as Si-MEMS, with many potential failure mechanisms operative at the same time, a systematic approach to failure assessment is necessary along with identification of potential failure sites. Fault trees such as Figure 4.15 help model various mechanical, thermal, electrical, and chemical events and relate them to the corresponding failure mechanisms and failure sites. Many failure mechanisms associated with MEMS structures are described in [4].

4.6 SUMMARY

Silicon MEMS devices play a pivotal role in the microtechnology constituting the majority of MEMS devices sold in the market. The superior properties of silicon associated with the well-established industrial fabrication structure of Si-MEMS has made it the material of choice not only for monolithic silicon devices, but also as an important component of nearly all microdevices. Discussion of the properties of silicon was followed by the various microfabrication techniques including surface and bulk micromachining. Applications of Si-MEMS were discussed for various fields of optical, electronic, automotive, and biomedical fields among many others. Recent advances in the application of Si-MEMS were also referenced.

ACKNOWLEDGMENT

The author is grateful to Professors Jim Hughes and Harold Wiebe for their careful review of the manuscript.

REFERENCES

1. Allameh, S. M., Residual stresses in Ni thin films, *Materials and Manufacturing*, in press, 2006.
2. Beeby, S. et al., *MEMS Mechanical Sensors*, Artech House, Inc., Norwood, MA, p. 108, 2004.
3. Gardner, J. W., Varadan, V. K., and Awadelkarim, O. O., *Microsensors MEMS and Smart Materials*, John Wiley and Sons, Ltd, West Sussex, England, 2001.
4. Pecht, M., DasGupta, A., and Evans, J. W., *Quality Conformance and Qualification of Micro-electronic Packages and Interconnects*, Wiley, New York, 1994.
5. Gardner, J. W. and Bartlett, P. N., Book review: Electronic noses, *Nature*, 402(6760), 351, 1999.
6. Greiff, P. et al., Silicon monolithic micromechanical gyroscope, In *1991 International Conference on Solid-State Sensors and Actuators, Jun 24–28, 1991*, San Francisco, CA, IEEE, Piscataway, NJ, pp. 966–968, 1991.
7. Allameh, S. M. et al., Surface topography evolution and fatigue fracture in polysilicon MEMS structures, *Journal of Microelectromechanical Systems*, 12(3), 313–324, 2003.
8. Field, L. A., Micromachined 1*2 optical-fiber switch, *Sensors and Actuators A (Physical) International Solid-State Sensors and Actuators Conference—TRANSDUCERS '95, 25–29 June 1995*, A53(1–3), 311–316, 1996.
9. Abramson, A. R. et al., Fabrication and characterization of a nanowire/polymer-based nanocomposite for a prototype thermoelectric device, *Journal of Microelectromechanical Systems*, 13(3), 505–513, 2004.
10. Maluf, N., *An Introduction to Microelectromechanical Systems Engineering*, ArtechHouse, Boston, MA, 2004.
11. Yang, J. et al., New technique for producing large-area as-deposited zero-stress LPCVD polysilicon films: The MultiPoly process, *Journal of Microelectromechanical Systems*, 9(4), 485–494, 2000.
12. Ni, A. et al., Optimal design of multilayered polysilicon films for prescribed curvature, *Journal of Materials Science*, 38, 4169–4173, 2003.
13. Boyne, D., The Ohio State University, Columbus, OH, 1992.
14. Madou, M. J., *Fundamentals of Microfabrication. The Science of Miniaturization*, 2nd ed., CRC Press, New York, 2002.
15. Feng, G.-H. and Kim, E. S., Piezoelectrically actuated dome-shaped diaphragm micropomp, *Journal of Microelectromechanical Systems*, 14(2), 192–199, 2005.
16. Judy, J. W. and Muller, R. S., Magnetic microactuation of torsional polysilicon structures, *Sensors and Actuators A (Physical) International Solid-State Sensors and Actuators Conference—TRANSDUCERS '95, 25–29 June 1995*, A53(1–3), 392–397, 1996.
17. Gass, V. et al., Micro-torque sensor based on differential force measurement, *Proceedings IEEE Micro Electro Mechanical Systems An Investigation of Micro Structures, Sensors, Actuators, Machines and Robotic Systems, 25–28 Jan. 1994*, IEEE, Oiso, Japan, pp. 241–244, 1994.
18. Despont, M. et al., New design of micromachined capacitive force sensor, *Journal of Micromechanics and Microengineering*, 3(4), 239–242, 1993.
19. Takashima, K. et al., An endoscopic tactile sensor for low invasive surgery, *Sensors and Actuators, A: Physical*, 119(2), 372–383, 2005.
20. Dargahi, J., Parameswaran, M., and Payandeh, S., Micromachined piezoelectric tactile sensor for an endoscopic grasper—theory, fabrication and experiments, *Journal of Microelectromechanical Systems*, 9(3), 329–335, 2000.
21. Usuda, T., Weissenborn, C., and von Martens, H.-J., Theoretical and experimental investigation of transverse sensitivity of accelerometers under multiaxial excitation, *Measurement Science and Technology*, 15(5), 896–904, 2004.
22. Lemkin, M. A. et al., 3-axis force balanced accelerometer using a single proof-mass, *Proceedings of the 1997 International Conference on Solid-State Sensors and Actuators. Part 2 (of 2), 16–19 June 1997*, Chicago, IL, IEEE, Piscataway, NJ, pp. 1185–1188, 1997.

23. Clark, W. A., Howe, R. T., and Horowitz, R., Surface micromachined Z-axis vibratory rate gyroscope, *Technical Digest Solid-State Sensor and Actuator Workshop, 3–6 June 1996*, Tranducer Res. Found., Hilton Head Island, SC, pp. 283–287, 1996.

24. An, S., Dual-axis microgyroscope with closed-loop detection, *Sensors and Actuators, A: Physical, Proceedings of the 11th IEEE International Workshop On Micro Electro Mechanical Systems, MEMS 1998, 25–29 Jan. 1998*, 73(1–2), 1–6, 1999.

25. Fischer, F., *MEMS in Automotive: Driving innovation*, In-Stat Market Research, San Jose, CA, p. 41, 2003.

26. Horsley, D. A. et al., Optical and mechanical performance of a novel magnetically actuated MEMS-based optical switch, *Journal of Microelectromechanical Systems*, 14(2), 274–284, 2005.

27. Ji, C.-H. et al., Electromagnetic 2×2 MEMS optical switch, *IEEE Journal on Selected Topics in Quantum Electronics*, 10(3), 545–550, 2004.

28. Bernstein, J. J. et al., Electromagnetically actuated mirror arrays for use in 3-D optical switching applications, *Journal of Microelectromechanical Systems*, 13(3), 526–535, 2004.

29. Han, J. S., Ko, J. S., and Korvink, J. G., Structural optimization of a large-displacement electromagnetic Lorentz force microactuator for optical switching applications, *Journal of Micromechanics and Microengineering*, 14(11), 1585–1596, 2004.

30. Guerre, R. et al., Fabrication of vertical digital silicon optical micromirrors on suspended electrode for guided-wave optical switching applications, *Sensors and Actuators, A: Physical*, 123–124, 570–583, 2005.

31. Kwon, H. N. and Lee, J.-H., A micromachined 2×2 optical switch aligned with bevel-ended fibers for low return loss, *Journal of Microelectromechanical Systems*, 13(2), 258–263, 2004.

32. Pan, C. T., Silicon-based coupling platform for optical fiber switching in free space, *Journal of Micromechanics and Microengineering*, 14(1), 129–137, 2004.

33. Meng, M. Q.-H. et al., Wireless robotic capsule endoscopy: State-of-the-art and challenges, *Fifth World Congress on Intelligent Control and Automation, 15–19 June 2004*, IEEE, Hangzhou, China, pp. 5561–5565, 2004.

34. Sijie, Z. et al., The research of gastrointestinal wireless endoscope based on MEMS technology, *Chinese Journal of Scientific Instrument*, 26(7), 681–683, 2005.

35. Kim, T. S., et al., Fusion of biomedical microcapsule endoscope and microsystem technology, In *TRANSDUCERS '05. The 13th International Conference on Solid-State Sensors, Actuators and Microsystems. Digest of Technical Papers, 5–9 June 2005*, IEEE, Seoul, South Korea, pp. 9–14, 2005.

36. Moon, S. J., Choi, Y. H., and Lee, S. S., A micromachined biopsy tool for a capsule type endoscope, In *TRANSDUCERS '05. The 13th International Conference on Solid-State Sensors, Actuators and Microsystems. Digest of Technical Papers, 5–9 June 2005*, IEEE, Seoul, South Korea, pp. 1371–1374, 2005.

37. Sendoh, M., Ishiyama, K., and Arai, K.-I., Fabrication of magnetic actuator for use in a capsule endoscope, *IEEE Transactions on Magnetics International Magnetics Conference (INTERMAG 2003), 30 March–3 April 2003*, 39(5, Pt. 2), 3232–3234, 2003.

38. Chiba, A. et al., Magnetic actuator for capsule endoscope navigation system, In *INTERMAG Asia 2005: Digest of the IEEE International Magnetics Conference, 4–8 April 2005*, IEEE, Nagoya, Japan, pp. 1251–1252, 2005.

39. Cheung, E. et al., A new endoscopic microcapsule robot using beetle inspired microfibrillar adhesives, *2005 IEEE/ASME International Conference on Advanced Intelligent Mechatronics, 24–28 July 2005*, IEEE, Monterey, CA, pp. 551–557, 2005.

40. Simons, R. N. and Miranda, F. A., Radiation characteristics of miniature silicon square spiral chip antenna for implantable bio-MEMS sensors, *2005 IEEE Antennas and Propagation Society International Symposium, 3–8 July 2005*, IEEE, Washington, DC, pp. 836–839, 2005.

41. Gomez-Sjoberg, R., Morisette, D. T., and Bashir, R., Impedance microbiology-on-a-chip: Microfluidic bioprocessor for rapid detection of bacterial metabolism, *Journal of Microelectromechanical Systems*, 14(4), 829–838, 2005.

42. Krebs, M. D. et al., Detection of biological and chemical agents using differential mobility spectrometry (DMS) technology, *IEEE Sensors Journal*, 5(4), 696–702, 2005.

43. Wang, X.-N. et al., Fabrication and performance of novel RF spiral inductors on silicon, *Microelectronics Journal*, 36(8), 737–740, 2005.

44. Wu, W. et al., RF inductors with suspended and copper coated thick crystalline silicon spirals for monolithic MEMS LC circuits, *IEEE Microwave and Wireless Components Letters*, 15(12), 853–855, 2005.

45. Wiser, R. F. et al., Polycrystalline silicon-carbide surface-micromachined vertical resonators—Part I: Growth study and device fabrication, *Journal of Microelectromechanical Systems*, 14(3), 567–578, 2005.

46. Nakatani, T. et al., Single crystal silicon cantilever-based RF-MEMS switches using surface processing on SOI, In *18th IEEE International Conference on Micro Electro Mechanical Systems, 30 Jan.–3 Feb. 2005*, IEEE, Miami Beach, FL, pp. 187–190, 2005.

47. Miao, J. and Sun, J., Integrated RF MEMS inductors on thick silicon oxide layers fabricated using SiDeox process, *2004 7th International Conference on Solid-State and Integrated Circuits Technology Proceedings, 18–21 Oct. 2004*, IEEE, Beijing, China, pp. 1683–1686, 2005.

48. Kim, J.-M. et al., A mechanically reliable digital-type single crystalline silicon (SCS) RF MEMS variable capacitor, *Journal of Micromechanics and Microengineering*, 15(10), 1854–1863, 2005.

49. Anagnostou, D. E. et al., Silicon-etched re-configurable self-similar antenna with RF-MEMS switches, In *IEEE Antennas and Propagation Society Symposium, 20–25 June 2004*, IEEE, Monterey, CA, pp. 1804–1807, 2004.

50. Anagnostou, D. E. et al., On the silicon-etched re-configurable antenna with RF-MEMS switches, *2005 IEEE Antennas and Propagation Society International Symposium, 3–8 July 2005*, IEEE, Washington, DC, pp. 417–420, 2005.

51. van Beek, J. T. M. et al., High-Q integrated RF passives and RF-MEMS on silicon, *Materials, Integration and Packaging Issues for High-Frequency Devices Symposium, 1–3 Dec. 2003*, Materials Research Society, Boston, MA, pp. 97–108, 2004.

52. Kranz, M. et al., Performance of a silicon-on-insulator MEMS gyroscope with digital force feedback, *PLANS 2004. Position Location and Navigation Symposium, 26–29 April 2004*, IEEE, Monterey, CA, pp. 7–14, 2004.

53. Kim, J.-M. et al., The SiOG-based single-crystalline silicon (SCS) RF MEMS switch with uniform characteristics, *Journal of Microelectromechanical Systems*, 13(6), 1036–1042, 2004.

54. Kim, J.-M. et al., Design and fabrication of SCS (single crystalline silicon) RF MEMS switch using SiOG process, *17th IEEE International Conference on Micro Electro Mechanical Systems. Maastricht MEMS 2004 Technical Digest, 25–29 Jan. 2004*, IEEE, Maastricht, Netherlands, pp. 495–501, 2004.

55. Liu, Z. et al., The key technologies in silicon based microwave and RF MEMS device fabrication, *2004 4th International Conference on Microwave and Millimeter Wave Technology Proceedings, 18–21 Aug. 2004*, IEEE, Beijing, China, pp. 1–6, 2004.

56. Sharpe, W. N. Jr. et al., Effect of etch holes on the mechanical properties of polysilicon, *Journal of Vacuum Science and Technology B (Microelectronics and Nanometer Structures)*, 15(5), 1599–1603, 1997.

57. Allameh, S. M. et al., On the evolution of surface morphology of polysilicon MEMS structures during fatigue, *Materials Research Society Symposium—Proceedings*, 657, EE231–EE236, 2001.

58. Sharpe, W. N. Jr. and Bagdahn, J., Fatigue Testing of Polysilicon—A Review, *Mechanics and Materials*, 9(1), 3–11, 2004.

59. Sharpe, W. N. Jr. and Turner, K., Fatigue testing of materials used in microelectromechanical systems, In *Fatigue '99: Proceedings of the Seventh International Fatigue Congress*, Wu, X. R., Ed., 8–12th June, Beiging, China, pp. 1837–1844.

60. Shrotriya, P., Allameh, S. M., and Soboyejo, W. O., On the evolution of surface morphology of polysilicon MEMS structures during fatigue, *Mechanics of Materials*, 36(1–2), 35–44, 2004.

61. Muhlstein, C. L., Stach, E. A., and Ritchie, R. O., Mechanism of fatigue in micron-scale films of polycrystalline silicon for microelectromechanical systems, *Applied Physics Letters*, 80(9), 1532–1534, 2002.

62. Muhlstein, C. L. et al., Surface engineering of polycrystalline silicon microelectromechanical systems for fatigue resistance, In *BioMEMS and Bionanotechnology, Symposium*, Materials Research Society, San Francisco, CA, pp. 41–46, 2002.

63. Muhlstein, C. L., Brown, S. B., and Ritchie, R. O., High-cycle fatigue and durability of polycrystalline silicon thin films in ambient air, *Sensors and Actuators, A: Physical*, A94(3), 177–188, 2001.

64. Ando, T., Shikida, M., and Sato, K., Tensile-mode fatigue testing of silicon films as structural materials for MEMS, *Sensors and Actuators, A: Physical*, A93(1), 70–75, 2001.

65. Millet, O. et al., Influence of the step covering on fatigue phenomenon for polycrystalline silicon micro-electro-mechanical-systems (MEMS), *Japanese Journal of Applied Physics, Part 2 (Letters)*, 41(11B), 1339–1341, 2002.

66. Varvani-Farahani, A., Silicon MEMS components: A fatigue life assessment approach, *Microsystem Technologies*, 11(2–3), 129–134, 2005.

67. Chuang, W.-H., Fettig, R. K., and Ghodssi, R., An electrostatic actuator for fatigue testing of low-stress LPCVD silicon nitride thin films, *Sensors and Actuators, A: Physical*, 121(2), 557–565, 2005.

68. Bagdahn, J. and Sharpe, W. N., Fatigue of polycrystalline silicon under long-term cyclic loading, *Sensors and Actuators, A: Physical*, A103(1–2), 9–15, 2003.

69. Choi, G., Horibe, S., and Kawabe, Y., Cyclic fatigue crack growth from indentation flaw in silicon nitride: Influence of effective stress ratio, *Acta Metallurgica et Materialia*, 42(11), 3837–3842, 1994.

70. Horibe, S. and Hirahara, R., Cyclic fatigue of ceramic materials. Influence of crack path and fatigue mechanisms, *Acta Metallurgica et Materialia*, 39(6), 1309–1317, 1991.

71. Kahn, H., Ballarini, R., and Heuer, A. H., Surface oxide effects on static fatigue of polysilicon MEMS, *Nano- and Microelectromechanical Systems (NEMS and MEMS) and Molecular Machines, Symposium*, Materials Research Society, Boston, MA, 2003.

72. Kapels, H., Aigner, R., and Binder, J., Fracture strength and fatigue of polysilicon determined by a novel thermal actuator [MEMS], *IEEE Transactions on Electron Devices*, 47(7), 1522–1528, 2000.

73. Legros, M. et al., Dislocation patterning in fatigued silicon single crystals, *Materials Research Society Symposium—Proceedings*, 779, 201–206, 2003.

74. Allameh, S. M. et al., On the evolution of surface morphology of polysilicon MEMS structures during fatigue, *Materials Science of Microelectromechanical System (MEMS) Devices III*, Boston, MA, pp. EE2.3.1–EE2.3.6, 2000.

75. Allameh, S. M. et al., Surface topography evolution and fatigue fracture in polysilicon MEMS structures, *Journal of Microelectromechanical Systems*, 2002.

76. Bagdahn, J. and Sharpe, W. N., Jr., Static and cyclic fatigue of polysilicon, *Eighth International Fatigue Congress, FATIGUE 2002*, Stockholm, Sweden, EMAS, London, UK, pp. 2197–2212, 2002.

77. Ballarini, R. et al., Fatigue and fracture testing of microelectromechanical systems, *American Society of Mechanical Engineers, Materials Division (Publication) MD*, 84, 151–152, 1998.

78. Frederick, K. M. and Fedder, G. K., Mechanical effects of fatigue and charge on CMOS MEMS, *Proceedings of the SPIE—The International Society for Optical Engineering*, 4180, 108–116, 2000.

79. Muhlstein, C. L., Brown, S., and Ritchie, R. O., High cycle fatigue of polycrystalline silicon thin films in laboratory air, *Materials Science of Microelectromechanical System (MEMS) Devices III*, Boston, MA, pp. EE5.8.1–EE5.8.6, 2000.

80. Brown, S. B., van Arsdell, W., and Muhlstein, C. L., Materials reliability in MEMS devices, *Proceedings of International Solid State Sensors and Actuators Conference (Transducers '97)*, IEEE, New York, NY, 1997.

81. Kahn, H. et al., Electrostatically actuated failure of microfabricated polysilicon fracture mechanics specimens, *Proceedings of the Royal Society of London, Series A (Mathematical, Physical and Engineering Sciences)*, 455(1990), 3807–3823, 1999.

82. Muhlstein, C. L., Brown, S., and Ritchie, R. O., High cycle fatigue of single crystal silicon thin film, *Journal of Microelectromechanical Systems*, 10, 593–600, 2001.

83. Muhlstein, C. L., Howe, R. T., and Ritchie, R. O., *Fatigue of Polycrystalline Silicon for MEMS Applications: Crack Growth and Stability under Resonant Loading Conditions*, Mechanics of Materials, 2002.

84. Kahn, H., Ballarini, R., and Heuer, A. H., Surface oxide effects on static fatigue of polysilicon, *Materials Research Society Symposium—Proceedings*, MRS, Boston, MA, J3.4.1–J3.4.6, 2002.

85. Gardos, M. N., Protective coatings and thin films *Proceedings of NATO Advanced Research Workshop*, Kluwer, Dordrecht, p. 185, 1997.

86. Tanner, D. M. et al., *MEMS Reliability: Infrastructure, Test Structures, Experiments, and Failure Modes*, Sandia National Laboratories, 2000.

87. Tanner, D. M. et al., MEMS reliability in shock environments, *Annual Proceedings—Reliability Physics (Symposium)*, IEEE, Piscataway, NJ, 2000.

88. Tanner, D. M. et al., MEMS reliability in a vibration environment, *Annual Proceedings—Reliability Physics (Symposium)*, IEEE, Piscataway, NJ, pp. 139–145, 2000.

89. Lin, C.-C. et al., Fabrication and characterization of a micro turbine/bearing rig, *Proceedings of 12th International Workshop on Micro Electro Mechanical Systems–MEMS, 17–21 Jan, 1999*, IEEE, Orlando, FL, pp. 529–533, 1999.

90. Epstein, A. H. et al., Power MEMS and microengines, *Proceedings of International Solid State Sensors and Actuators Conference (Transducers '97), 16–19 June 1997*, IEEE, Chicago, IL, pp. 753–756, 1997.

91. Walters, D. S. and Spearing, S. M., On the flexural creep of single-crystal silicon, *Scripta Materialia*, 42(8), 769–774, 2000.

92. Drozhzhin, A. I. and Sidel'nikov, I. V., Distinctive features of creep in silicon whiskers, *Izvestiya Vysshikh Uchebnykh Zavedenii, Fizika*, 24(10), 31–35, 1981.

93. Shmidt, V. A., Negative creep during bending of boron-doped silicon, *Fizika Tverdogo Tela*, 14(12), 3675–3676, 1972.

94. Takashima, K. et al., Fatigue crack growth behavior of micro-sized specimens prepared from an electroless plated Ni–P amorphous alloy thin film, *Materials Transactions*, 42(1), 68–73, 2001.

95. Wagner, U. et al., Mechanical reliability of MEMS-structures under shock load, *Microelectronics Reliability*, 41(9–10), 1657–1662, 2001.

96. Tanner, D. M., Reliability of surface micromachined microelectromechanical actuators, *22nd International Conference on Microelectronics, Proceedings*, IEEE, Piscataway, NJ, pp. 97–104, 1999.

97. Tanner, D. M. et al., Effect of humidity on the reliability of a surface micromachined microengine, In *Annual Proceedings—Reliability Physics (Symposium)*, IEEE, Piscataway, NJ, pp. 189–197, 1999.

98. Tanner, D. M. et al., Frequency dependence of the lifetime of a surface micromachined microengine driving a load, *Microelectronics Reliability*, 39(3), 401–414, 1999.

99. White, C. D. et al., Electrical and environmental reliability characterization of surface-micromachined MEMS polysilicon test structures, *Proceedings of the SPIE—The International Society for Optical Engineering*, 4180, 91–95, 2000.

100. Smith, N. F. et al., Development of characterization tools for reliability testing of Microelectromechanical system actuators, *Proceedings of SPIE—The International Society for Optical Engineering*, 3880, 156–164, 1999.

101. Patton, S. T., Cowan, W. D., and Zabinski, J. S., Performance and reliability of a new MEMS electrostatic lateral output motor, *Annual Proceedings—Reliability Physics (Symposium)*, IEEE, Piscataway, NJ, pp. 179–188, 1999.

102. Marxer, C. et al., Reliability considerations for electrostatic polysilicon actuators using as an example the REMO component, *Sensors and Actuators*, 61, 449–454, 1997.

103. Man, K. F., MEMS reliability for space applications by elimination of potential failure modes through testing and analysis, *Proceedings of SPIE—The International Society for Optical Engineering*, 3880, 120–129, 1999.

104. Ghaffarian, R. et al., Reliability of COTS MEMS accelerometer under shock and thermomechanical cycling, *Proceedings of SMTA International*, Surface Mount Technology Association, Edina, MN, pp. 697–703, 2001.

105. Ramesham, R., Ghaffarian, R., and Kim, N. P., Reliability issues of COTS MEMS for aerospace applications, *Proceedings of SPIE—The International Society for Optical Engineering*, 3880, 83–88, 1999.

106. Lafontan, X. et al., Physical and reliability issues in MEMS microrelays with gold contacts, *Proceedings of the SPIE—The International Society for Optical Engineering*, 4558, 11–21, 2001.

107. Gasparyan, A. et al., Mechanical reliability of surface-micromachined self-assembling two-axis MEMS tilting mirrors, *Proceedings of the SPIE—The International Society for Optical Engineering*, 4180, 86–90, 2000.

108. Arney, S., Designing for MEMS reliability, *MRS Bulletin*, 26(4), 296–299, 2001.

109. Arney, S. et al., Design for reliability of MEMS/MOEMS for lightwave telecommunications, *Proceedings of the SPIE—The International Society for Optical Engineering*, 4558, 6–10, 2001.

110. Kazinczi, R., Mollinger, J. R., and Bossche, A., Reliability of silicon nitride as structural material in MEMS, *Proceedings of the SPIE—The International Society for Optical Engineering Materials and Device Characterization in Micromachining II, 20–21 Sept. 1999*, 3875, 174–183, 1999.

111. Ghaffarian, R. and Ramesham, R., Comparison of IC and MEMS packaging reliability approaches, *Proceedings of SMTA International*, Surface Mount Technology Association, Edina, MN, pp. 761–766, 2000.

112. Wu, J., Pike, R. T., and Wong, C. P., Novel bi-layer conformal coating for reliability without hermeticity MEMS encapsulation, *IEEE Transactions on Electronics Packaging Manufacturing*, 22(3), 195–201, 1999.

113. de Boer, M. P. et al., Role of interfacial properties on MEMS performance and reliability, *Proceedings of the SPIE—The International Society for Optical Engineering*, 3825, 2–15, 1999.

114. Ding, J. N., Meng, Y. G., and Wen, S. Z., Size effect on the mechanical properties and reliability analysis of microfabricated polysilicon thin films, In *Annual Proceedings—Reliability Physics* (Symposium), pp. 106–111, 2001.

115. Tanner, D. M. et al., First reliability test of a surface micromachined microengine using SHiMMeR, *Proceedings of the SPIE—The International Society for Optical Engineering*, 3224, 14–23, 1997.

116. Evans, J. W. and Evans, J. Y., Reliability assessment for development of microtechnologies, *Microsystem Technologies*, 3(4), 145–154, 1997.

5 Porous Metallic Materials

Jikou Zhou

Materials Science and Technology Division, Lawrence Livermore National Laboratory

CONTENTS

5.1 Introduction 103
 5.1.1 Metallic Foams and Applications 103
 5.1.2 Fabrication of Open Cell Metallic Foams 105
5.2 Structure Unit Models 105
 5.2.1 Gibson–Ashby Dimensional Model 106
 5.2.2 Tetrakaidecahedron Unit Cell Model 107
 5.2.3 Four-Strut Unit Cell Model 107
5.3 Stress–Strain Behavior under Monotonic Compression 108
5.4 Multilevel Deformation 110
 5.4.1 Foam Deformation 110
 5.4.2 Strut Deformation 111
 5.4.3 Mechanical Properties of Individual Struts 113
5.5 Fatigue Behavior 116
 5.5.1 Cyclic Strain Accumulation 116
 5.5.2 Multilevel Failure Mechanisms 118
5.6 Creep Behavior 120
References 123

5.1 INTRODUCTION

5.1.1 METALLIC FOAMS AND APPLICATIONS

Metallic foams are porous metals consisting of gas-filled pores with metallic materials between the pores. According to pore connectivity, they are categorized into open cell and closed cell metallic foams. In open cell metallic foams, metallic struts form the structure frame, and open pores are interconnected. In closed cell foams, the pores filled with gas are isolated by cell wall metal. Figure 5.1 shows two representative foam structures of open and closed cell metallic foams. Both types of foam are lightweight structures with a large surface to volume ratio; however, most of the pore surfaces in closed cell foams do not function. They are generally used as structural materials to fill space and support weight in various industries, such as the aerospace, construction, automobile, shipbuilding, and railway industries [1–4].

On the other hand, open cell metallic foams contain an alluring combination of physical, mechanical, biological, chemical, and optical properties. For example, they have well-controlled pore structures, they are lightweight with high stiffness, they have high gas/fluid permeability with a large surface area to volume ratio, and so on. Therefore, open cell metallic foams are frequently

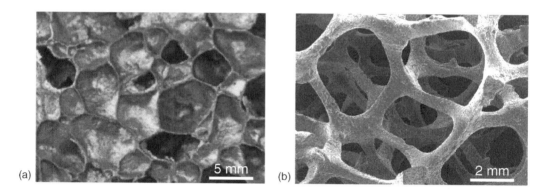

FIGURE 5.1 (a) A closed cell Alporas aluminum foam. (Reproduced from Andrews, E. W., Samders, W., and Gibson, L. J., *Materials Science and Engineering,* A270, 113–124, 1999. With permission of Elsevier Science.) (b) An open cell Duocel aluminum foam.

used to make aerospace structures, automobile components, fluid flow filters/controllers, battery electrodes, heat exchangers, catalyst carriers, bioimplants, medical devices, and optical and semiconductor devices [1–4]. Some interesting applications of open cell aluminum foams are presented in Figure 5.2 [7]. Given that open cell metallic foams have increasing applications and significant impacts on the upcoming high technologies, including biology and biomedicine, energy and environment, electronics and semiconductors, we will focus on the knowledge and mechanical behaviors of open cell metallic foams. In particular, we will present the results of recent studies of open cell Duocel aluminium foams, the most widely used open cell metallic foams in a range of industries. The outcomes of these studies are readily applied to open cell metallic foams made of other metals.

FIGURE 5.2 Applications of Duocel open cell aluminum foams (a) as the isothermalizer and baffle structure in solid cryogenic coolers, (b) as the heat exchange media for the space shuttle, and (c) as the core structure in a solar collector acting as a heat transfer agent. (Reproduced from http://www.ergaerospace.com/index.htm Courtsey of ERG, Duoce®.)

5.1.2 FABRICATION OF OPEN CELL METALLIC FOAMS

Making an open cell metallic foam involves producing open pores and interconnected struts. The first open cell metallic foam was produced in the 1950s [8] by pouring molten aluminum into a preform of compacted rock salt grains. After solidification, the salt grains were dissolved to leave open pores. This idea has been adopted in the fabrication of commercial open cell metallic foams and is generally referred as a casting method. For instance, Duocel aluminum foams (Figure 5.1b) are produced using a casting procedure in which polymeric templates are chosen to prepare preforms of ceramic compact [2]. The pore morphology and strut geometry are largely determined by the selected polymeric templates.

A powder metallurgy (P/M) process can also be used to make open cell metallic foams. In P/M, space-holding materials are mixed with metal powders [4]. After sintering and heat treatment to produce interconnected foam struts, the space-holding materials are removed via chemical dissolution or evaporation at elevated temperatures. The space-holding materials can be polymer spheres and ceramic granules, and their size and shape are crucial in determining cell morphology and pore size. Three-dimensional nanoporous platinum and gold have recently been produced using appropriated ultra-fine polymeric spheres as space-holding materials [9]. The morphology of a nano-scale gold foam is shown in Figure 5.3a.

Open cell metallic foams may also be made by means of deposition processes, including electrodeposition and vapor deposition [2,4]. In both cases, metallic foam materials are deposited onto struts of open cell polymer foam templates. The template struts can be kept or removed after the deposition. A nickel foam produced using the electrodeposition process is shown in Figure 5.3b, in which hollow struts (inset of Figure 5.3b) are obtained after removal of the polymer template.

5.2 STRUCTURE UNIT MODELS

Open cell metallic foams are stochastic structures that bring many challenges to modeling mechanical behavior through a typical mechanics approach using a structure unit. To date, three types of structure unit have been developed. The famous Gibson–Ashby dimensional model [1] is simple and easy to use, and thus, are widely applied to estimate foam strengths and elastic moduli. Representative tetrakaidecahedron unit cells [10] closely resemble the real structure units of

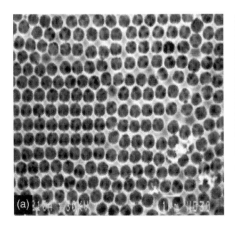
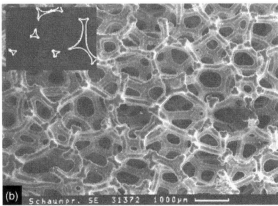

FIGURE 5.3 Open cell metallic foams (a) an open cell gold foam with nanoscale pores. (Reproduced from Velev, O. D., and Kaler, E. W., *Advanced Materials*, 13, 531–534, 1999. With permission of Nature Publishing Group.) (b) A nickel foam with hollow struts. (Reproduced from Ashby, M. F., Evans, A. G., Fleck, N. A., Gibson, L. J., Hutchinson, J. W., and Wadley, H. N. G., *Metal Foams: A Design Guide*, Butterworths, London, 2000. With permission.)

most open cell metallic foams. It is frequently employed in numerical calculation and multiphysics modeling studies. Four-strut structure unit [11] is extracted from open cell foams based on the fact that there are four struts joining at one vertex in open cell metallic foams. It is very useful for studying plastic deformation.

The details of these three structure units are briefly introduced as follows.

5.2.1 GIBSON–ASHBY DIMENSIONAL MODEL

Gibson and Ashby employed a dimensional argument to analyze the deformation of open cell foams. They proposed a simple three-dimensional cubic structure unit, as shown in Figure 5.4a. The cubic unit is constructed with struts that are either parallel with or perpendicular to stress directions. When a compressive stress is applied, the elastic deformation of the foam is achieved via elastic bending of the struts that are orientated in a direction perpendicular to the applied stress. Elastic deformation in the struts parallel with the stress direction is small and negligible (Figure 5.4b). This gives the Young's modulus as a function of the foam density:

$$\frac{E_f}{E_s} = C_1 \left(\frac{\rho_f}{\rho_s}\right)^2 \tag{5.1}$$

where ρ_f is the foam density, ρ_s is the density of the fully dense metal from which the foam is fabricated; E_f and E_s are the Young's moduli of the foam and fully dense metal, respectively; C_1 is a constant related to cell geometry. A numerical analysis using the tetrakaidecahedral unit cell with the cell cross section defined by plateau borders finds $C_1 = 0.98$ [10]. This value is confirmed by data obtained from a wide variety of open cell polymer foams that give $C_1 \sim 1$.

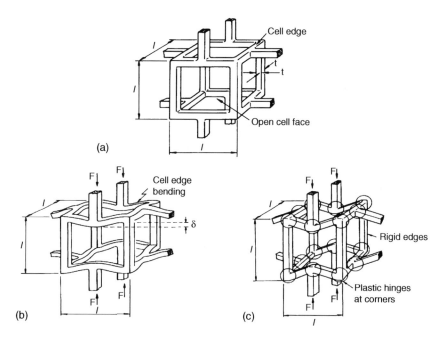

FIGURE 5.4 Gibson–Ashby dimensional cubic model. (a) cubic model without deformation, (b) elastic deformation of the foam is due to elastic bending of the struts perpendicular to the loading axis, and (c) plastic deformation of foams is due to the formation of plastic hinges at the vertex. (Reproduced from Gibson, L. J. and Ashby, M. F., *Cellular Solids: Structure and Properties*, Cambridge University Press, Cambridge, 1997. With permission.)

Plastic deformation of open cell foams starts with the formation of plastic hinges in struts that are orientated in a direction perpendicular to the stress, as indicated by the circles in Figure 5.4c. Corresponding to the stress needed to form plastic hinges, the yield strength of an open cell metallic foam is also a function of foam density:

$$\frac{\sigma_{pl}^{cl}}{\sigma_{YS}} = C_2 \left(\frac{\rho_f}{\rho_s}\right)^{\frac{3}{2}} \tag{5.2}$$

where σ_{pl}^{cl} is plastic collapse strength of a foam under compressive loading, σ_{YS} is the yield strength of the fully dense metal; and the constant C_2 is related to the cell geometry. Data from a wide range of foams suggest that C_2 is ~ 0.3.

5.2.2 TETRAKAIDECAHEDRON UNIT CELL MODEL

The tetrakaidecahedron structure unit is shown in Figure 5.5. It contains fourteen faces including eight regular hexagons and six squares, thirty-six edges, and twenty-four vertices. All edges are the same length. Finite element analysis (FEA) was carried out to calculate the mechanical properties under compression [10]. Comparing FEA results from the tetrakaidecahedron unit cell model with those predicted by Gibson–Ashby dimensional analysis, Kwon [10] found that the latter under-estimated both the Young's modulus and the plastic yield strength of Duocel open cell aluminum foams.

5.2.3 FOUR-STRUT UNIT CELL MODEL

A four-strut unit cell model was extracted from the actual foam structure by Wang et al. [11,12] to study the plastic deformation of three-dimensional linear elastic plastic polymer foams (Figure 5.6a). Using the four-strut unit cell, they constructed three-dimensional foams, as shown in Figure 5.6b. After carefully studying deformation modes in individual struts, the shape of stress–strain curves was obtained for foams deformed under compression using structural mechanics with an energy approach. The predicted stress–strain curve is plotted in Figure 5.6c, in comparison with the measured stress–strain curve. The model prediction is in good agreement with the experimental results particularly in the linear elastic deformation stage and in the plastic deformation plateau

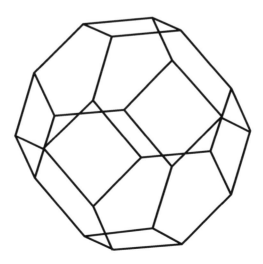

FIGURE 5.5 The ideal tetrakaidecahedron structure unit.

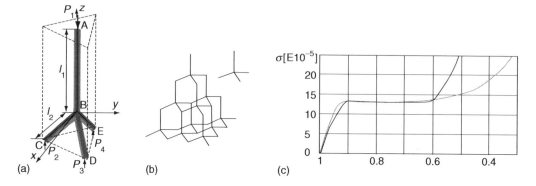

FIGURE 5.6 The four-strut unit cell model (a) a structure unit, (b) an open cell foam constructed by periodic stacking of four-strut unit cells, and (c) experimental (light gray) and predicted stress–strain curves. (Reproduced from Gioia, G., Wang, Y., and Cuitino, A. M., *Proceedings of the Royal Society of London A. Mathematical, Physical, and Engineering Sciences,* A457, 1079–1096, 2001. With permission.)

stage. This suggests that the four-strut unit cell is a good candidate to model the mechanics of open cell metallic foams.

5.3   STRESS–STRAIN BEHAVIOR UNDER MONOTONIC COMPRESSION

Mechanical behavior of metallic foams is fundamental for all types of application. Mechanical data are also needed to form a database for computer-aided modeling and design of foam structure. Therefore, studying the mechanical behavior of open cell metallic foams is important. Various types of mechanical testing have been used to study the mechanical deformation and properties of metallic foams [3]. These include: monotonic and cyclic testing, indentation testing, and dynamic and creep testing at elevated temperatures. Among them, compressive testing is the most frequently used because compression testing is simple and easy, and a lot of information can be obtained from compression testing. Moreover, it is found that the tensile properties of ductile metallic foams are similar to the compressive properties [1]. Hence, the mechanical behavior of open cell metallic foams will be introduced by presenting recent studies carried out on Duocel open cell aluminum foams using compression testing.

An ideal compressive stress–strain curve of ductile metallic foams consists of three stages [3], as shown schematically in Figure 5.7. Compressive deformation starts with linear deformation stage I, in which the loading and unloading curves are not coincident. Deformation in stage II starts from the onset of plastic yielding. It is generally believed that the foam is compressed at a roughly plateau stress level. This is followed by deformation stage III, in which an increase in flow stress is required to achieve further deformation to densify foam.

The real stress–strain behavior of open cell metallic foams strongly depends on the mechanical behavior of the struts, which are determined by the fabrication method, processing, cell, and strut morphologies. Figure 5.8a shows the stress–strain curves of Duocel foams that were subjected to different heat treatments [13]. The details of early-stage deformation are shown in Figure 5.8b at a smaller scale of strain. In the case of the as-fabricated (F) and T6-strengthened (T6) foams, stress peaks and stress valleys are observed in the stress–strain curves, which can be divided into three stages: stage I is a nearly linear deformation stage till to the stress peak; stage II is a stress dropping stage, in which the flow stress continuously drops from the peak stress to the valley stress; and in stage III, the flow stress gradually increases for foam densification.

However, the stress–strain behavior of the annealed (O) foams is significantly different from those of the as-fabricated and T6-strengthened foam. There is no stress peak or stress valley in the

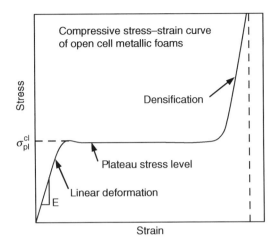

FIGURE 5.7 An ideal stress–strain curve of open cell metallic foams. (Modified from Gibson, L. J. and Ashby, M. F., *Cellular Solids: Structure and Properties*, 2nd ed., Cambridge University Press, Cambridge, 1997.)

stress–strain curve. Upon application of compressive loading, the annealed foam continuously yields to reach a stress plateau, which is sustained in a wide strain range. Stress then rises as foam is densified.

These observations suggest that foam strength cannot be simply defined as a single value. A range with upper and lower bounds is more reflective of the strain behavior of open cell metallic foams. According to the characteristics of the stress–strain curves for foams under three different heat treatment conditions, three strengths can be defined: the upper strength (σ_{upper}) and the lower strength (σ_{lower}) for the as-fabricated and T6-strengthened foams, and the plateau strength ($\sigma_{plateau}$) for the annealed foams. The three strengths are indicated in Figure 5.8b. The difference between the upper and lower strengths could be up to 0.4 MPa, which is more than 10% of the lower strength values.

With the defined foam strengths in accordance with the characteristics of stress–strain behavior, the measured foam strengths are plotted in Figure 5.9. The lower and upper strengths of the T6-strengthened foams rise uniformly, although the measured strength data are scattered over a range

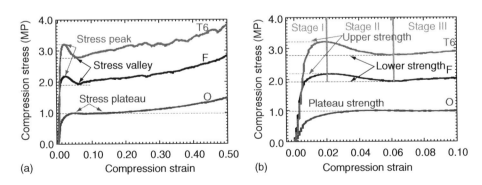

FIGURE 5.8 Stress–strain behavior of Duocel open cell aluminum foams (a) stress–strain curves of foams with similar densities, but subjected to different heat treatment processes and (b) details of the early-stage deformation and definitions of the three types of strengths. (Reproduced from Zhou, J., Gao, Z., Cuitino, A. M., and Soboyejo, W. O., *Materials Science and Engineering*, A386, 118–128, 2004. With permission.)

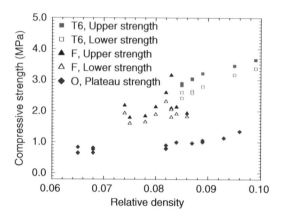

FIGURE 5.9 Effects of heat treatment on foam compressive strength. (Reproduced from Zhou, J. Gao, Z., Cuitino, A. M., and Soboyejo, W. O., *Materials Science and Engineering*, A386, 118–128, 2004. With permission.)

for the as-fabricated foams. The plateau strengths measured from the annealed foam also increase with increasing density.

5.4 MULTILEVEL DEFORMATION

An open cell metallic foam is an architecture constructed by interconnected struts. The mechanical response to an applied load occurs in two levels: macroscale foam-level and microscale strut-level [6,13,14].

5.4.1 FOAM DEFORMATION

Macroscale deformation of open cell aluminum foams was studied using a combination of in situ imaging and digital image correlation (DIC) analysis [13]. During mechanical testing, a foam specimen was deformed by applying a compressive load, and a set of images were taken to record the deformation procedure. DIC analysis was then carried out on these images to study the spatial and temporal strain evolution.

In the case of the as-fabricated foam, plastic deformation starts with the formation of a few discrete regions of strain localization (Figure 5.10a). New regions of strain localization subsequently form with increasing global strain (Figure 5.10b), along with the growth of the deformed regions at the expense of the undeformed regions. When the global strain reaches a level corresponding to the stress peak, the plastic deformation extends completely across the overall specimen. This results in a relatively homogeneous distribution of deformation, with few isolated regions (Figure 5.10c). The local strains in the most fraction of the deformed sample are close to the global strain level. The deformation beyond the peak stress point is localized within a few discrete regions, which then spread out and impinge each other to form a continuous deformation band (Figure 5.10d). Further plastic deformation proceeds by means of the growth of the deformation bands (Figure 5.10e, f) and the formation of new deformation bands (Figure 5.10f, g).

T6-strengthened foams have similar deformation mechanisms as the as-fabricated foams. Even at the so-called linear deformation stage, permanent plastic deformation starts simultaneously from multiple sites as shown in Figure 5.11T_a. The discrete plastic deformation regions grow, along with the formation of new deformation sites, as stress increases. A uniform deformation is formed at a strain corresponding to the stress peak. Beyond this, the deformation is limited in localized regions, leading to decreasing flow stress (Figure 5.11T_c, T_d); and a deformation band is subsequently

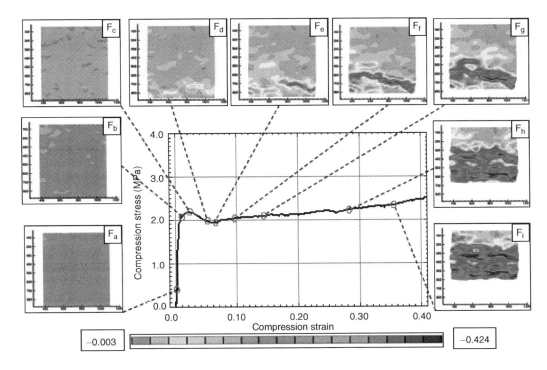

FIGURE 5.10 Macroscale deformation subsequence of a Duocel open cell aluminum foam in the as-fabricated (F) condition.

formed at the valley stress point, as shown in Figure 5.11T_e. Further deformation is carried out by means of growth and formation of new deformation bands.

However, the deformation of the annealed foam is significantly different from those of the as-fabricated and T6-strengthened foams. Upon loading, uniform plastic deformation is achieved through initiation and growth of discrete deformation sites. Further deformation is carried out homogeneously in the overall foam specimen, as shown in Figure 5.11O_c through O_g, which correspond to the deformation in the plateau stage.

The different deformation mechanisms shown by the foams in the T6-strengthened and annealed conditions suggest that the initiation and formation of the first localized deformation band is the cause of the stress drop from peak to valley values. Otherwise, the deformation should proceed under a constant flow stress, as depicted in Figure 5.7 and demonstrated by the annealed foam in Figure 5.8.

5.4.2 STRUT DEFORMATION

Foam deformation is the result of strut/cell deformation. In order to understand how the cells and struts deform in the regions where plastic deformation nucleates and grow, a small open cell metallic foam specimen was progressively loaded and unloaded at various strain levels, and the cell and strut deformation was examined using scanning electron microscopy (SEM) after each unloading. The initiation of plastic deformation in the individual struts was identified by the presence of dislocation slip bands [15].

The incremental deformation of individual struts and single cells in response to the foam deformation is shown in Figure 5.12. Dislocation slip bands are first observed in strut I and strut II in Figure 5.12a, corresponding to a strain of 0.009. At this strain level, there are no observable

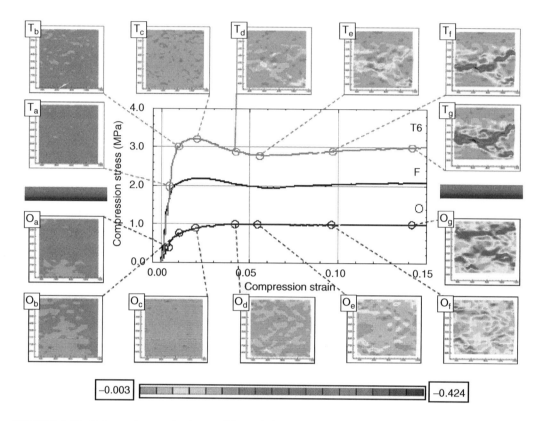

FIGURE 5.11 Deformation mechanisms of Duocel foams in the T6-strengthened (T6) and annealed (O) conditions.

changes in the shape of struts and cells, suggesting that the onset of local micro-plasticity in the ductile metals occurs well before the onset of bulk yielding.

The details of dislocation slip bands in strut I and strut II are shown in images of higher magnification in Figure 5.13. The local regions with slip bands are indicated by rectangles. Instead of a homogeneous distribution of slip bands on the deformed struts, the slip bands are only observed in two discrete regions that are close to the two vertices to which the strut is connected. Subsequent observations reveal that the slip bands are limited to these two regions, even when the strain is increased to about 0.05.

At a strain of about 0.025, dislocation slip bands were observed in strut III and the middle of strut IV. Comparing Figure 5.12b with Figure 5.12a, it is clear that a slight bow has developed in the middle of strut IV.

When strain was increased to 0.039, slip bands are observed on strut V. This strut is deformed to reconcile the cell shape change due to the plastic buckling of strut IV. Before the onset of plastic deformation of the strong strut VI, the cell shape change is achieved by a combination of post-buckling of strut IV and further bending of strut V.

At a strain of 0.053, dislocation slip bands are observed in all the struts within the tested foam block. At a strain of 0.349, most struts are either aligned in a direction perpendicular to the loading direction, or heavily twisted to bring the two vertices together. This is shown in Figure 5.12j. When the strain reaches about 0.433, the vertices and closed cell faces touch each other, and start to participate in the plastic deformation.

In summary, plastic deformation of the foam block was initiated by plastic bending of struts that were not parallel to the loading axis or by plastic buckling of weak struts that are parallel to the

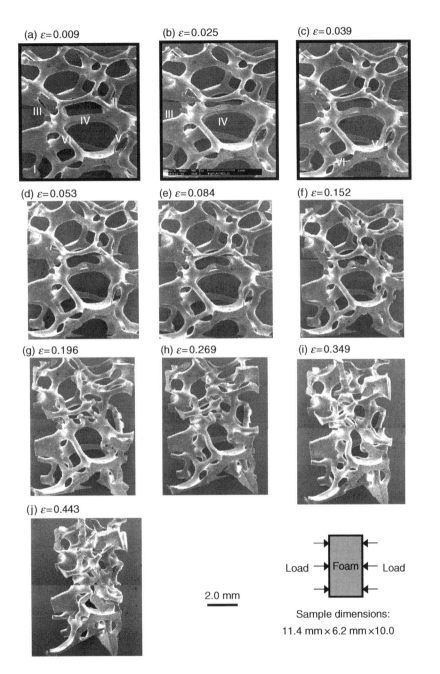

FIGURE 5.12 Deformation of struts and cells at various strain levels. (Reproduced from Zhou, J., Shrotriya, P., and Soboyejo, W. O., *Mechanics of Materials*, 36, 781–795, 2004. With permission.)

loading axis. It is speculated that the occurrence of plastic collapse resulted from the combination of the two failure modes (plastic buckling and plastic bending).

5.4.3 MECHANICAL PROPERTIES OF INDIVIDUAL STRUTS

The three stress–strain curves in Figure 5.8 and the corresponding deformation mechanism were obtained from the foams that had the same pore density, cell morphology, and similar densities of

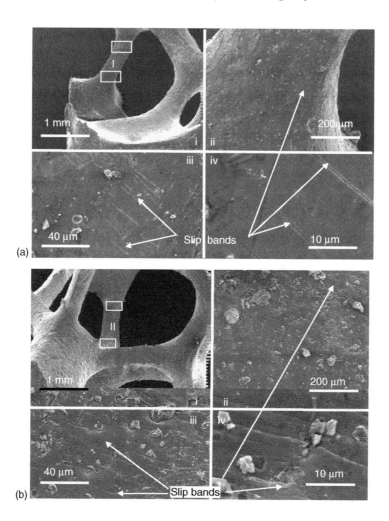

FIGURE 5.13 Plastic deformation of struts I and II (in Figure 12a) at $\varepsilon = 0.009$ (a) dislocation slip bands in strut I and (b) dislocation slip bands in strut II. Each image consists of four sub-images (i–iv) with increasing magnifications. The following sub-image displays the details of the central part of a former sub-image at higher magnification. The magnifications of the four sub-images are 15×, 50×, 200×, and 1000×, respectively. (Reproduced from Zhou, J., Shrotriya, P., and Soboyejo, W. O., *Mechanics of Materials*, 36, 781–795, 2004. With permission.)

~ 0.083. They were produced by means of the same casting processing, and their struts also have the same chemical composition and similar geometry. But the different heat treatment process modified the microstructure and the intrinsic properties of the foam struts. It is the mechanical properties of the struts that are responsible for the significant differences in the stress–strain behavior. Stress–strain behavior of individual struts has been studied using the struts that were extracted from foam blocks subjected to three heat treatment conditions: T6-strengthened (T6), Annealed (O), and As-fabricated (F) [16].

The stress–strain curves are presented in Figure 5.14. The stress and strain of these critical points are summarized in Table 5.1, along with the ultimate tensile strength (UTS) values. Also included in Table 5.1 are values of the ductility, which correspond to the total strain to fracture for each strut.

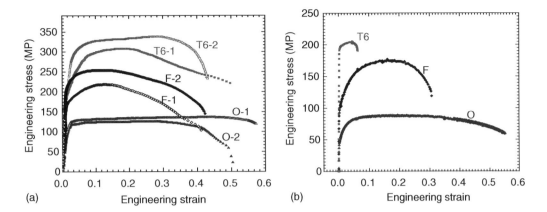

FIGURE 5.14 Stress–strain curves of (a) the struts under the three heat treatment conditions (F, O, and T6), and (b) the fully dense alloys subjected to the same heat treatment conditions. (Reproduced from Zhou, J., Allamah, S., and Soboyejo, W. O., *Journal of Materials Science*, 210, 429–439, 2005. With permission.)

The two as-fabricated struts have similar yielding and plastic flow behaviors. The yield strain levels are ∼0.011 and ∼0.013, and the corresponding yield strengths are 188 and 157 MPa for strut F-1 and F-2, respectively. Similarly, the UTS values obtained for the two struts are 247 and 207 MPa for struts F-1 and F-2, respectively. Struts F-1 and F-2 fail at strain levels of 0.431 and 0.441, respectively.

Compared with the as-fabricated struts, the T6-strengthened struts have significantly higher strengths. The UTS values of struts T6-1 and T6-2 are 307 and 338 MPa, respectively. The yield strength of strut T6-2 is 285 MPa. However, the yield strength of strut T6-1 is only 200.08 MPa. This value is even lower than the strengths of the as-fabricated struts, due to the pre-existing curvatures in this strut. It is worth noting that the T6-strengthening did not impair the ductility of the struts. The fracture strains were 0.400 and 0.431 for struts T6-1 and T6-2, respectively. These are comparable to those of the as-fabricated struts (Table 5.1).

The two annealed struts have similar tensile stress–strain behavior. Corresponding to the critical yielding points, struts O-1 and O-2 have a similar strain of ∼0.023. The yield strengths of the two struts are 111 and 120 MPa, respectively. The UTS values for struts O-1 and O-2 are 125 and 136 MPa, while the fracture strains of the two annealed struts O-1 and O-2 are ∼0.491 and 0.573, respectively. These are significantly greater than those of the as-fabricated and T6-strengthened

TABLE 5.1
Tensile Properties of the Struts under F, O, and T6 Conditions

Specimen ID	Linear Elastic Limit		UTS		Breaking Point	
	Strain	Stress (MPa)	Strain	Stress (MPa)	Strain	Stress (MPa)
F-1	0.011	188.36	0.104	247.37	0.431	140.34
F-2	0.013	156.54	0.109	206.70	0.441	86.48
O-1	0.024	111.23	0.294	124.63	0.491	58.76
O-2	0.023	119.66	0.414	135.71	0.573	118.75
T6-1	0.019	200.08	0.180	306.77	0.400	256.61
T6-2	0.031	284.89	0.270	338.17	0.431	232.87

Source: From Zhou, J., Allamah, S., and Soboyejo, W. O., *Journal of Materials Science*, 40, 429–439, 2005.

TABLE 5.2
Tensile Properties of 6101, the Bulk Alloys under F, O, and T6 Conditions

Specimen ID	0.2% Offset Yield Point		UTS		Breaking Point	
	Strain	Stress (MPa)	Strain	Stress (MPa)	Strain	Stress (MPa)
F-1	0.003	98.00	0.161	174.52	0.305	126.19
F-2	0.003	98.00	0.184	173.84	0.336	143.59
O-1	0.003	45.64	0.194	88.54	0.554	59.26
O-2	0.003	49.00	0.227	87.79	0.304	57.56
T6-1	0.005	192.50	0.046	205.32	0.064	184.01
T6-2	0.005	199.50	0.024	209.87	0.029	186.97

Source: From Zhou, J., Allamah, S., and Soboyejo, W. O., *Journal of Materials Science*, 40, 429–439, 2005.

struts. Furthermore, the annealed struts exhibit nearly elastic-perfectly plastic behavior with very little strain hardening.

It is clear that annealing reduces strut strength, but improves strut ductility. It also results in nearly elastic-perfectly plastic stress–strain behavior. In the case of the T6-strengthened struts, the measured UTS values increase, but strengthening is not associated with degradation in strut ductility. Furthermore, all tested struts have good ductility (Table 5.1).

For a comparison, efforts have been taken to measure the tensile properties of the corresponding fully-dense 6101 aluminum subjected to the same heat treatment conditions (F, O, and T6). It is found that the ~ 0.5 mm thick aluminum are films seriously deformed after being sliced from the bulk alloy using electro-discharged machining (EDM). Thus, macro-scale testing was carried out on dogbone tensile specimens with a gauge cross section of 5.0×10.0 mm and a gauge length of 50.0 mm. The specimens were deformed continuously to fracture at a strain rate of $5 \times 10^{-4} \, \text{s}^{-1}$. The results are summarized in Table 5.2. Typical stress–strain curves are presented in Figure 5.14b. In the case of the as-fabricated bulk alloy, the 0.2% offset yield strength and UTS are ~ 98 and ~ 174 MPa, respectively. The plastic strain to failure is about 30%.

The T6-strengthening process doubles the 0.2% offset yield strength, but leads to only a limited increase in the UTS: from ~ 174 MPa for the as-fabricated alloy to ~ 205 MPa for the T6-strengthened alloy. However, the T6-strengthening process significantly impairs the ductility of the bulk alloy, which fails at a strain less than 6%. Annealing significantly improves the plasticity of the bulk alloy by more than 50%, but both the 0.2% offset yield strength and UTS values of the annealed alloys are reduced to approximately half the corresponding strengths of the as-fabricated alloy.

5.5 FATIGUE BEHAVIOR

5.5.1 CYCLIC STRAIN ACCUMULATION

Under cyclic compressive loading, an open cell metallic foam is progressively shortened. The shortening rate is dependent on the magnitude of maximum load level, $|\sigma_{\text{max}}|$ [17–21]. When the maximum load is smaller than its compressive strength, $\sigma_{\text{pl}}^{\text{cl}}$, the foam is expected to endure a number of cyclic loadings before failure [21]. In this section, the magnitude of the maximum stress will be normalized by the compressive strength to obtain a relative stress level, $|\sigma_{\text{max}}|/\sigma_{\text{pl}}^{\text{cl}}$. The amount of shortening will be named as strain accumulation because it is a result of progressive shortening under cyclic compression.

A typical plot of strain accumulation versus number of cycles is shown in Figure 5.14a. It was obtained from a T6-strengthened foam tested at a relative stress level of 0.8. Upon application of cyclic compressive loading, the foam specimen is instantaneously shortened by a limited amount of

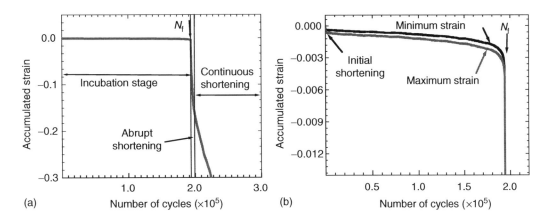

FIGURE 5.15 Plots of accumulated strain versus number of cycles for a T6-strengthened specimen tested at a stress level of 0.80; (a) accumulated strain versus number of cycles, a complete curve; (b) details of the curve in (a) including the abrupt shortening are shown at a smaller strain scale. (Reproduced from Zhou, J. and Soboyejo, W. O., *Materials Science and Engineering*, A369, 23–35, 2004. With permission.)

magnitude, which is followed by progressive shortening. This first stage is an incubation stage, whose details are shown in Figure 5.15b in a smaller strain scale. The maximum and minimum strains in Figure 5.15b correspond respectively to the maximum and minimum compressive stress levels. The dependence of the overall strain accumulation on the relative stress level is presented in Figure 5.16. Generally speaking, increasing load level causes greater strain (Figure 5.16).

The gradual shortening in the incubation stage is followed by dramatic shortening, in which a significant strain accumulation up to ~15% was achieved within a few decades of cycles (within one or a few seconds). This is the abrupt shortening stage (Figure 5.15)a. Beyond that, further deformation depends on the relative stress level. If it is relatively high, i.e., equal to or greater than 0.80, continuous shortening occurs at decreasing rates, as shown in Figure 5.15a. Deformation in this stage is called continuous shortening. However, if the stress level is relatively low, i.e., not

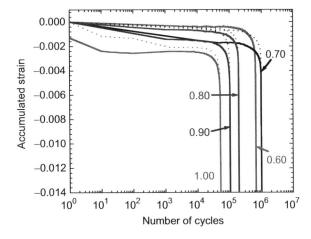

FIGURE 5.16 Plots of accumulated strain versus number of cycles obtained from the T6-strengthened foams that were tested at various stress levels. (Reproduced from Zhou, J. and Soboyejo, W. O., *Materials Science and Engineering*, A369, 23–35, 2004. With permission.)

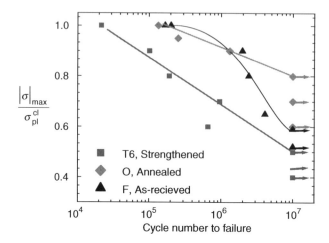

FIGURE 5.17 S–N curves obtained for foams under three conditions as-fabricated, annealed, and T6-strengthened. (Reproduced from Zhou, J. and Soboyejo, W. O., *Materials Science and Engineering*, A369, 23–35, 2004. With permission.)

greater than 0.70, strut powderization occurred after a limited amount of further shortening. This will be further discussed in the next section.

The abrupt shortening is a critical event. It is the onset of foam failure under cyclic loading. The number of cycles corresponding to this point is accordingly defined as the failure point, the number of cycles to failure, N_f, and is indicated in Figure 5.15. The endurance limit is defined as the highest stress level at which the tested specimens survive 10^7 cycles of fatigue loading. The S–N curves obtained for the foams are presented in Figure 5.17.

5.5.2 MULTILEVEL FAILURE MECHANISMS

In the incubation stage, there is no observable change in the shape of struts and cells, nor deformation localization. The abrupt shortening results in the formation of a localized failure band. Subsequent shortening continues through the growth of the local failure band (for the maximum stress level greater than 0.8). The above failure sequence is shown in Figure 5.18 by the three

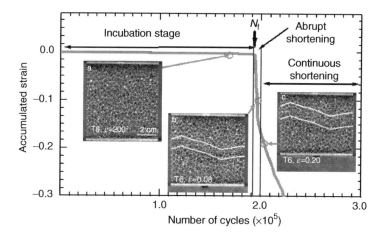

FIGURE 5.18 Macroscale foam failure mechanisms.

representative images corresponding to the three different failure stages of a foam specimen that was tested at a relative stress level of 0.8 [21].

When the maximum stress level is not greater than 0.7, i.e., $|\sigma_{max}|/\sigma_{pl}^{cl} \leq 0.7$, the foam is powderized after a limited amount of continuous shortening. A collapsed foam specimen after powderization is shown in Figure 5.19a. In this case, fractured struts (Figure 5.19b), single four-strut units, and clusters (Figure 5.19c) are observed.

Under cyclic compression, the ultimate failure of a foam specimen is the result of damage accumulation. Before the abrupt shortening, surface cracks are formed on the individual struts, due to the cyclic bending. One surface crack that was formed in a strut is shown in Figure 5.20. The surface cracks grow, leading to the fracture of individual struts. This is evident in Figure 5.21, which was obtained from a foam specimen that underwent the abrupt shortening. Through-thickness cracks are observed in most struts in the images).

Given that a foam block is constructed of individual struts, failure of an individual strut weakens the overall elastic modulus of the foam under testing. In particular, the formation and growth of surface cracks cause foam to loss stiffness. This is verified by gradually decreasing elastic modulus during cyclic compression testing. These are shown in Figure 5.22. A steep drop of elastic modulus was observed right before the onset of the abrupt strain jump. The loss of modulus suggests that a large number of struts failed almost simultaneously, leading to the formation of a localized failure band. This suggests that monitoring of the loss of modulus can be used to evaluate damage accumulation in foams that are in service.

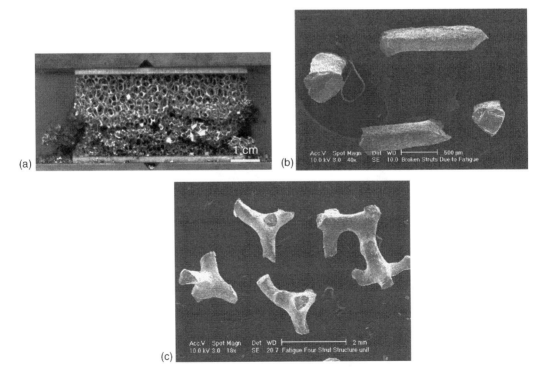

FIGURE 5.19 (a) Powderization of a T6-strengthened foam tested at a stress level of 0.70, (b) fractured individual struts, and (c) four-strut tetrahedral structure units collected from the powderized foam specimen. (Reproduced from Zhou, J. and Soboyejo, W. O., *Materials Science and Engineering*, A369, 23–35, 2004. With permission.)

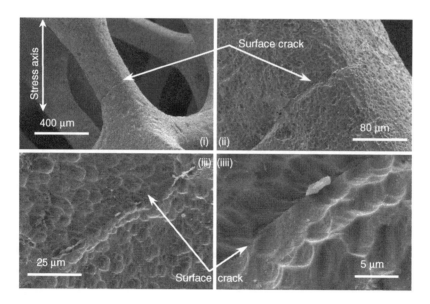

FIGURE 5.20 A crack in a strut of a T6-strengthened foam before the occurrence of the abrupt shortening. It was tested at a stress level of 0.80, and testing was stopped after 1.5×10^5 cycles for scanning electron microscope (SEM) examination. Four sub-images (i–iv) shows a surface crack in the strut. Each of the following sub-images show the details of the central area in the former sub-image. (Reproduced from Zhou, J. and Soboyejo, W. O., *Materials Science and Engineering*, A369, 23–35, 2004. With permission.)

5.6 CREEP BEHAVIOR

Open cell aluminum foam creeps at elevated temperatures. Upon application of a compressive load, a foam specimen starts shortening at a constant strain rate, i.e., the secondary regime. This is followed by the tertiary creep, in which the strain rate increases, leading to significant shortening.

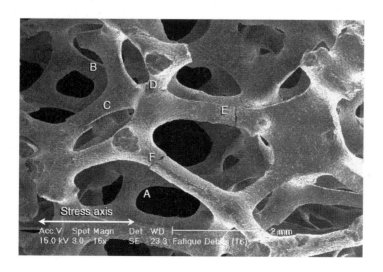

FIGURE 5.21 Cracks in struts of a T6-strengthened foam after abrupt shortening. Through-thickness cracks are labeled. (Reproduced from Zhou, J. and Soboyejo, W. O., *Materials Science and Engineering*, A369, 23–35, 2004. With permission.)

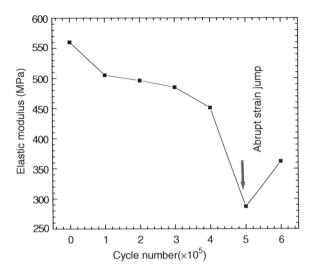

FIGURE 5.22 Loss of elastic modulus as a result of damage accumulation in a foam block subjected to cyclic compressive loading. Data were obtained from a T6-strengthened specimen tested at a stress level of 0.80. The abrupt shortening occurred after 481,000 cycles, as indicated by the arrow. (Reproduced from Zhou, J. and Soboyejo, W. O., *Materials Science and Engineering*, A369, 23–35, 2004. With permission.)

Then the strain rate decreases to a level that is close to the secondary regime. A typical strain–time curve is shown in Figure 5.23 [5]. It was obtained from a test at 300°C. The normalized stress level is about 0.3.

Like deformation under progressive loading, creep deformation of foam is also heterogeneous. It is characterized by the formation of a localized deformation band, which is formed during the tertiary creep stage. Creep deformation continues by forming another localized deformation band in other areas, and this process is repeated during further loading. One localized deformation is shown in Figure 5.24a. Within the band, struts are seriously deformed and twisted, leading to foam densification, as shown in Figure 5.24b.

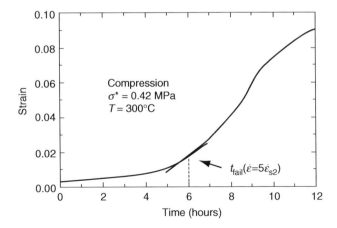

FIGURE 5.23 Strain plotted against time for a creep test of an open cell aluminum foam. (Reproduced from Andrews, E. W., Samders, W., and Gibson, L. J., *Materials Science and Engineering*, A270, 113–124, 1999. With permission.)

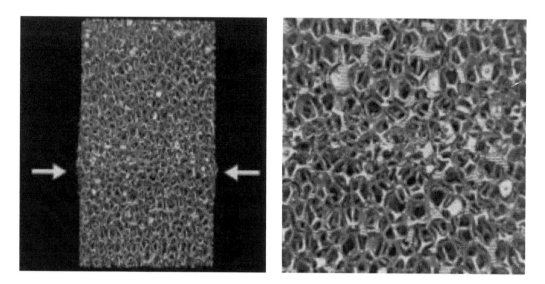

FIGURE 5.24 Multilevel creep mechanisms in open cell aluminum foams. (a) Failure band and (b) foam densification in the band. (Reproduced from Andrews, E. W., Samders, W., and Gibson, L. J., *Materials Science and Engineering*, A270, 113–124, 1999. With permission.)

Assuming that individual struts in foams have the same creep behavior as the make-up metals or alloys, Andrews, Gibson, and Ashby [22] modeled foam creep behavior by means of a dimensional analysis and obtained the steady strain rate $\dot{\varepsilon}_f$ for the secondary regime:

$$\frac{\dot{\varepsilon}_f}{\dot{\varepsilon}_s} = \frac{0.6}{n+2} \left(\frac{1.7(2n+1)}{n} \frac{\sigma}{\sigma_s} \right)^n \left(\frac{\rho_f}{\rho_s} \right)^{-\frac{3n+1}{2}} \qquad (5.3)$$

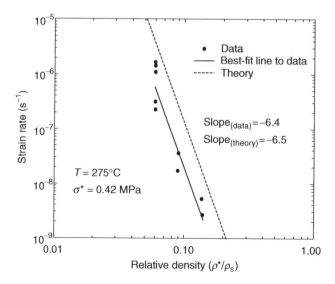

FIGURE 5.25 Secondary creep strain rate plotted against relative density at constant stress and temperature for an open cell aluminum foam (Duocel, ERG). (Reproduced from Andrews, E. W., Gibson, L. J., and Ashby, M. F., *Acta Materialia*, 47, 2583–2863, 1999. With permission.)

where $\dot{\varepsilon}_s \sigma_s$ and n are constants characterizing the creep of the solid metals or alloys of which the struts are made. The parameter $\dot{\varepsilon}_s$ takes the Arrhenius relationship, $\dot{\varepsilon}_s = C \exp(-Q/RT)$, in which C is a constant, Q is the activation energy of creep, R is the gas constant, and T is the absolute temperature.

A comparison of model prediction with experimental data is shown in Figure 5.25. Equation 5.3 predicted a slope value of -6.5 with $n = 4$, measured from the solid alloy from which the foam is made. The slope of the fitting curve is about -6.4. The consistency suggests that the steady creep is a result of strut creep, and the strain rate is determined .by strut creep. It is independent of foam structure parameters, such as sizes of cells and struts, and the shape of cells and struts.

REFERENCES

1. Gibson, L. J. and Ashby, M. F., *Cellular Solids: Structure and Properties*, 2nd ed., Cambridge University Press, Cambridge, 1997.
2. Ashby, M. F., Evans, A. G., Fleck, N. A., Gibson, L. J., Hutchinson, J. W., and Wadley, H. N. G., *Metal Foams: A Design Guide*, Butterworths, London, 2000.
3. Gibson, L. J., Mechanical behavior of metallic foams, *Annual Review of Materials Science*, 30, 191–227, 2001.
4. Banhart, J., Manufacture, characterization and application of cellular metals and metal foams, *Progress in Materials Science*, 46, 559–632, 2001.
5. Andrews, E. W., Samders, W., and Gibson, L. J., Compressive and tensile behavior of aluminum foams, *Materials Science and Engineering*, A270, 113–124, 1999.
6. Zhou, J. and Soboyejo, W. O., An investigation of deformation mechanisms in open cell metallic foams, In *Processing and Properties of Lightweight Cellular Metals and Structures, Third Global Symposium on Materials Processing and Manufacturing*, Ghosh, A. et al., Eds, TMS, Seattle, WA, pp. 209–218, 2002.
7. http://www.ergaerospace.com/index.htm
8. Cellular materials: New concepts provide unique possibilities, feature article, *The Iron Age*, 119–121, February 8, 1962.
9. Velev, O. D. and Kaler, E. W., Structured porus materials via colloidal crystal templating: from inorganic oxide to metals, *Advanced Materials*, 13, 531–534, 1999.
10. Kwon, Y. W., Cooke, R. E., and Park, C., Representative unit-cell model for open cell metal foams with or without elastic filler, *Materials Science and Engineering*, A343, 63–70, 2003.
11. Gioia, G., Wang, Y., and Cuitino, A. M., The Energetics of heterogeneous deformation in open cell solid foams, *Proceedings of the Royal Society of London. Mathematical, Physical, and Engineering Sciences*, A457, 1079–1096, 2001.
12. Gioux, G., McCormack, T. M., and Gibson, L. J., Failure of aluminum foams under multi-axial loads, *International Journal of Mechanical Sciences*, 42, 1097–1117, 2000.
13. Zhou, J., Gao, Z., Cuitino, A. M., and Soboyejo, W. O., Effects of heat treatment on the compressive deformation behavior of open cell aluminum foams, *Materials Science and Engineering*, A386, 118–128, 2004.
14. Zhou, J., Mercer, C., and Soboyejo, W. O., An investigation of the microstructure and strength of open cell 6101 aluminum foams, *Metallurgical and Materials Transactions*, A33, 1413–1427, 2002.
15. Zhou, J., Shrotriya, P., and Soboyejo, W. O., Multi-scale mechanisms of compressive deformation in open cell Al foams, *Mechanics of Materials*, 36, 781–797, 2004.
16. Zhou, J., Allamah, S., and Soboyejo, W. O., Micro-scale strut properties in open cell aluminum foams, *Journal of Materials Science*, 40, 429–439, 2005.
17. Sugimura, Y., Rabiei, A., Evans, A. G., Harte, A. M., and Fleck, N. A., Compression fatigue of a cellular aluminum alloy, *Materials Science and Engineering*, A269, 38–48, 1999.
18. McCullogh, K. Y. G., Fleck, N. A., and Ashby, M. F., The stress-life fatigue behavior of aluminum alloy foams, *Fatigue and Fracture of Engineering Materials and Structures*, 23, 199–208, 2000.
19. Zettl, B., Mayer, H., Stanzl-Tschegg, S. E., and Degischer, H. P., Fatigue properties of aluminum foams at high numbers of cycles, *Materials Science and Engineering*, A292, 1–7, 2000.

20. Banhart, J. and Brinkers, W., Fatigue behavior of aluminum foams, *Journal of Materials Science Letters*, 18, 617–619, 1999.
21. Zhou, J. and Soboyejo, W. O., Compression–compression fatigue of open cell aluminum foams: Macro-/micro- mechanisms and effects of heat treatment, *Materials Science and Engineering* A369, 23–35, 2004.
22. Andrews, E. W., Gibson, L. J., and Ashby, M. F., The creep of cellular solids, *Acta Materialia*, 47, 2583–2863, 1999.

6 A Thermodynamic Overview of Glass Formation Abilities: Application to Al-Based Alloys

Aiwu Zhu and Gary J. Shiflet
Department of Materials Science and Engineering, University of Virginia

CONTENTS

6.1   Introduction 125
6.2   GFA Criterion and Empirical Rules 127
6.3   GFA and Crystallization 127
6.4   General Thermodynamics 128
     6.4.1   Enthalpy Considerations 129
     6.4.2   Entropy Estimation 131
6.5   Thermodynamic Models 131
     6.5.1   CALPHAD Models 131
     6.5.2 T_0-Criterion 132
     6.5.3   Quasi-Kinetics of Primary Crystallization 134
     6.5.4   Atomic Mobility Consideration 136
     6.5.5   Heat of Mixing and SRO in Liquids 136
     6.5.6   Multiple Component Effects 138
6.6   Summary..141
Acknowledgments 141
References 141

Abstract

Glass formability of a solidified alloy depends on the evolution kinetics of thermodynamically viable phases in the undercooled melt. Composition ranges where glass transition may occur under a given processing condition are ultimately controlled by the competing crystallization of the multiple phases including not only the solid solutions but also the intermediate phases, either stable or metastable. Simplified quasi-kinetic calculations can be used to locate the best glass formers for a given system based on the thermodynamic model that includes all the phases concerned. Various empirical rules employed in searching for new glass formers basically address some facets of the comprehensive thermodynamic information.

6.1   INTRODUCTION

As a new structural material, amorphous metals require more effort and improvement in the area of damage tolerance; however, there are many impressive properties with respect to strength, corrosion resistance, elastic limits, etc., (Table 6.1, Figure 6.1). Pursuing improvements in ductility is

TABLE 6.1
Some Physical and Mechanical Properties of Selected Metallic Glasses

Alloys	Density (gm/cm^3)	Fracture/ Yield Strength (MPa)	Vicker's Hardness (GPa)	Young's Modulus (GPa)	Bulk Modulus (GPa)	Poisson Ratio
Titanium-glass	~5.2	1700–1800	~5.48	~100	~105	0.34
Aluminum-glass	3.1–3.7	900–1300	~4	70–75	—	—
Calcium-glass	2.0–2.3	700–750	~2.4	—	—	—
Amorphous steel	7.8–8.0	4000–4400	11–13	180–225	140–215	0.3–0.325
Mg-glass	2.2–3.0	700	2.16	47.6	48.2	0.335

ongoing. When the specific properties are considered, these alloys are further set apart from their crystalline counterparts. Most metallurgical materials usually contain crystalline structures not far from thermodynamic equilibrium. Metallic glasses with amorphous structures similar to the liquid solutions but at temperatures well below the melting points can be synthesized using controlled solidification techniques with sufficient cooling rates to avoid crystallization of at all phases. Over the past few years, extensive worldwide exploration has led to discoveries of new families of metallic glass systems including Zr–, Fe–, Mg–, Cu–, Ca–, Ce(Nd)–, Ti–, Y–, and Ni–, etc. Details of some of these successes are available from the most recent reviews by Inoue [1], Poon et al. [2], and Wang et al. [3]. Various strategies are employed and a number of empirical rules have been inferred. However, in the case of rapid solidification, the competing phase transformation kinetics determines the final product. The controlling factors are principally the atomic mobility and thermodynamic properties of each potential phase, which varies with composition and temperature. The glass forming ability (GFA) of metallic alloys will be analyzed in this chapter with particular attention given to the thermodynamic perspective. Because bulk amorphous metal alloys are now routinely produced, their amorphous, partially crystalline and fully crystalline forms yield new advanced structural alloys. Even in a fully crystalline form, structures and properties can now be obtained with the new thermodynamic and kinetic pathways available from an amorphous precursor that previously could not be attained by any other route. This article will examine some of the thermodynamics and kinetics associated with obtaining these pathways.

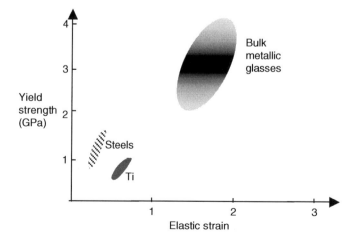

FIGURE 6.1 Plot illustrating strength and elastic properties for BMGs, steels, and titanium alloys.

6.2 GFA CRITERION AND EMPIRICAL RULES

Glass forming ability (GFA) of an alloy can be measured naturally by the minimum cooling rate, R_c, required to vitrify the alloy or by the maximum size (thickness or diameter) of the glass samples that can be synthesized from the α alloy melt. Technologically, characteristic temperatures are often used to define or indicate GFA including glass temperature T_g, solidus temperature T_m, liquidus temperature T_ℓ, onset crystallization temperature T_x, etc.

A practical and useful criterion for GFA is the reduced glass temperature ($T_{rg} = T_g/T_l$). Since T_g is relatively insensitive to changes in alloy composition and T_ℓ generally depends strongly on composition, this criterion indicates that the best glass formers ($T_r > 0.6$) are typically at or near a eutectic reaction with the most depressed T_ℓ. Therefore, phase diagrams, if available, offer a convenient tool for locating the better glass formers. However, among other systems, e.g., Cu–Zr [4] and La–Al–Zr–Cu [5], Al-based systems with rare earth metal (RM) additions is an important exception [6]. The eutectic reaction for Al-rich alloys is rather shallow at the FCC solution end (with a decrement in T_ℓ of about 5% as compared to about 40% for typical easy glass forming eutectic alloys), and the best glass formers in the Al–RM binary systems remarkably have hypereutectic compositions toward the intermediate phase with steep T_ℓ decrements. The eutectic compositions (in mole fraction) range from 0.03 to 0.05 RM, while the glass forming compositions range from 0.08 to 0.16 RM.

GFA has also been related to the supercooled-liquid range $T_{xg}(= T_x - T_g)$ [4–7] and to a parameter $T_x/(T_g + T_\ell)$ [8]. Nevertheless, since T_g or T_x must be known a priori, these T_r-like indicators are not operative for the exploration of new metallic glasses.

When the phase diagrams are unavailable, searching for new bulk metallic glasses (BMGs) has relied upon a trial-and-error approach guided by empirical rules, including Inoue's three rules: (1) consisting of more than two elements, (2) atomic size mismatch greater than 12%, and (3) a negative heat of mixing between the elements [1].

6.3 GFA AND CRYSTALLIZATION

Various crystallization events may occur by nucleation and growth where three factors control the kinetics: chemical driving forces, interfacial energy between the amorphous phase and the crystals, and the atomic mobility for rearrangement or transportation of the partitioning atoms. Addressing the early stage of the transformation, three types of crystallization processes can be characterized from the nucleation to compete against glass transition from an undercooled liquid: polymorphous, primary, or eutectic. Figure 6.2 schematically illustrates the specific reactions that compete with the glass transition during the quenching of an alloy of initial solute concentration C_0: (a) a polymorphous partitionless (PL) reaction from the undercooled liquid to a supersaturated solid solution; (b) primary crystallization (PR) of the solid-solution containing the maximum-driving-force composition or the intermediate phase (Int-P) that is regarded as stoichiometric; and (c) the eutectic reaction (ET) to local equilibrium consisting of the equilibrium primary solid solution and the intermediate phase Int-P. The actual Gibbs energy diagrams for various systems may be different in some details [9].

Conventionally, only the partitionless transformation into the supersaturated solid solution is considered for GFA. It is based on the single effect of the atomic mobility that transformation from the liquid to the corresponding polymorphous crystalline structures requires only the local structural change with the least rearrangement of atoms, and hence, it is the most "kinetically" favored when compared with other possible transformations. Eutectic crystallization involves coordinated growth of coupled zones while the overall local composition is kept invariant with the undercooled liquid. It is reasonable to treat the initial eutectic transformation processes in a manner similar to primary crystallization. Primary crystallization, although requiring both structural and

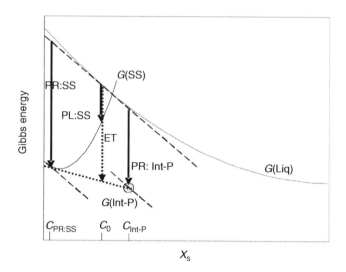

FIGURE 6.2 Schematic Gibbs energy diagram of an alloy system containing undercooled liquid, a primary solid solution (SS), and a line compound (intermediate) phase (Int-P). The arrows PL, PR, and ET denote the partitionless, primary, and eutectic transformations. $C_{\text{Int-P}}$ denotes solute concentration in Int-P, C_0 the (initial) solute concentration in the undercooled liquid, $C_{\text{PR:SS}}$ the maximum-driving-force concentration of the nuclei of the SS in the primary reaction. The dashed tangent lines are parallel and the dotted one is the common tangent for SS and Int-P. The arrow with dotted line indicates the decrease in Gibbs energy for eutectic ET transformation from the liquid phase (From Zhu, A. W., Shiflet, G. J., and Miracle, D. B., *Scripta Mater.*, 50, 987–991, 2004. With permission.)

concentration changes that involve the most extensive atomic rearrangement, has the maximum driving force for nucleation and can also be a dominating factor for GFA, as shown in the recently found Cu–Zr binary bulk glasses [4] as well as for Al–RM–(TM) systems discussed below.

6.4 GENERAL THERMODYNAMICS

The competition between various phases is based on (and partly depends upon) the Gibbs energy G of each concerned phase, e.g., Φ:

$$G^{(\Phi)} = H^{(\Phi)} - TS^{(\Phi)} \tag{6.1}$$

where T is the absolute temperature, $H^{(\Phi)}$ and $S^{(\Phi)}$ the enthalpy and the entropy of the phase. The contributions to the enthalpy that are of principal interest are the atomic heat of mixing ΔH_{mix} for primary solid solutions and liquid solutions or the formation enthalpy ΔH_{mix} for intermediate phases. The entropy consists mostly of configuration entropy S_{conf}, and there are additional contributions from the effects of thermal vibration, magnetic moment distribution, etc.

For a partitionless transformation from a liquid to a solid phase, at a temperature below the melting point, the driving force involving the reaction is related to the heat capacity variation as

$$\Delta G = \Delta H - T\Delta S = \Delta H_{\text{m}}\Delta T_{\text{r}} - \int_{T}^{T_m} (C_{\text{p}}^{\text{Liq}} - C_{\text{p}}^{\text{S}})\mathrm{d}T + T\int_{T}^{T_m} (C_{\text{p}}^{\text{Liq}} - C_{\text{p}}^{\text{S}})/T\mathrm{d}T \tag{6.2}$$

where ΔH_m is the heat of fusion of a solid crystalline phase at its melting point (T_m) $(\Delta H_m = T_m \Delta S_m$ (entropy of fusion)), and $\Delta T_r = (T_\ell - T)/T_\ell$ is the reduced undercooling. The heat capacities of alloy systems under investigation are mostly unknown. When, for the first time, rationalizing GFA from a perspective of nucleation of polymorphous crystal phases, Turnbull [10] obviously used a linear approximation of the Gibbs energy change $\Delta G = \Delta H_m \Delta T_r$, yielding the T_r criterion.

6.4.1 ENTHALPY CONSIDERATIONS

To reduce the driving forces of crystallization, an initial and oftentimes very effective strategy is to raise the enthalpies of the competing crystalline phases to enhance their instability and/or to reduce the enthalpy of the (undercooled) liquid to improve its structural stability. The former is addressed by the empirical rule "atomic size difference" while the "negative heat of mixing" rule is aimed at the latter. The "multicomponent rule" is associated with both ways as will be discussed below.

Heats of mixing of liquid or solid solutions (or formation energies of stoichiometric intermediate phases) are directly related to the interaction among neighboring atoms and more distant atoms. Negative heat of mixing indicates an attractive interaction between unlike atoms and formation of chemical short range ordering (CSRO). It is also significant for efficient atomic packing [11], as well as for high viscosity of the undercooled liquids, as will be discussed.

Atomic interactions should be ultimately described by electronic structure calculations based on quantum mechanics. For a metallic liquid solution, it was argued that the maximum stability occurs when the Fermi level lies at a minimum of the density of states according to a nearly-free-electron perturbation model [12]. It corresponds to a typical valence electronic concentration $e/a = 1.7$ (electron number per atom) of the alloys. This idea has been explored and used to optimize the glass forming compositions of, e.g., Zr–Ni–Al–Cu [13]. For solid solutions with non-spherical Fermi surfaces and structural factors, the electron concentration still plays a similar role for its phase stability as indicated by one of the Hume–Rothery empirical rules. Since it is the relative stability (or enthalpy of mixing) that is relevant to the GFA, the e/a indication should be less significant and at least not physically sound. Another problem for employing this indication is the effective valence electron number of each element that varies in different alloys.

In a semi-empirical way to calculate the mixing enthalpy or phase stabilities of solid solutions and intermediate phases, the Miedema model [14] is employed that considers the electron charge rearrangement upon alloying and treats an unlike-atom as a discontinuous cell of Wigner–Seiz type embedded in a matrix of solvent element cells. The change of chemical enthalpy is represented by the contacts at the boundaries of the corresponding cells and is, therefore, called "interfacial enthalpy". The enthalpy change per mole of dilute solute A atoms in solvent B, depending on the interface area $(\sim V_A^{2/3}$ where V_A is the mole volume of A), has a negative contribution from electronic charge transfer between A and B $(\sim -(\phi_A - \phi_B)^2$ where ϕ_A and ϕ_B represent their electronegativities) and a positive contribution from electronic energy increment due to discontinuity in electron density n_{sw} at the boundary of the cells $(\sim (n_{ws,A}^{1/3} - n_{ws,B}^{1/3})^2)$:

$$\Delta H^\infty(\text{A in B}) = \Delta H^{\text{inter}}(\text{A in B}) = \frac{V_A^{2/3}}{\frac{1}{2}\left(n_{ws,A}^{-1/3} + n_{ws,B}^{-1/3}\right)}\left[-P(\Delta\phi)^2 + Q(\Delta n_{ws}^{1/3})^2\right] \tag{6.3}$$

where the denominator denotes a measure of the electrostatic screen length and P and Q are empirically obtained from a $(\Delta\phi)^2$ versus Δn_{ws} plot for various binary alloys that are distinguished by whether they have either intermetallic compounds or not. The atomic volume change due to electron charge transfer can be estimated; for one mole of dilute A in B:

$$\Delta V_A = \alpha \frac{\phi_A - \phi_B}{n_{ws}^A} \quad \text{and} \quad \Delta V_B = \alpha \frac{\phi_B - \phi_A}{n_{ws}^B} \quad \text{with} \quad \alpha = 1.5 \frac{V_A^{2/3}}{n_{sw,A}^{-1/3} + n_{sw,B}^{-1/3}}.$$

Therefore, for a dilute A-solute in random solid solution of B with c_A and c_B ($= 1 - c_A$), the chemical enthalpy change per mole of atoms: $\Delta H^{\text{chem}} = c_A c_{B,s} \Delta H^{\text{inter}}(\text{A in B})$ where $c_{B,s} = c_B V_B^{2/3} / (c_A V_A^{2/3} + c_B V_B^{2/2})$ is called "surface concentration" of B takes into account the change in contact probability of W–S cells due to atomic size difference between A and B. For intermediate compounds with strict stoichiometry (where all atoms fit in the crystal structures that are usually different from those of terminal elements), the mixing or formation enthalpy consists of only chemical contributions: $\Delta H^{\text{form}} = c_A f_B^A \Delta H^{\text{inter}}(\text{A in B})$ where f_B^A denotes probability of A-atoms in contact with B-atoms and can be estimated as $c_{B,s}(1 + 8(c_{A,s} c_{B,s})^2)$ that is independent of specific crystal structures. The heat of mixing of liquid solutions is regarded to have only a chemical contribution. To distinguish it from the corresponding compounds, the heat of fusion of the compounds estimated using melting points of the elements, referred to as a "topological term," is added. Regarding solid solutions and intermediate phases of metallic bonding characteristics with certain solute solubilities, the major extra contribution is from the strain energy caused by the atomic mismatch between elements when they share any single sub-lattices. Miedema's model considers specific contribution of valence electrons for transition metals, which they call the "structure contribution" ΔH^{struc},

$$\Delta H_{\text{mix}} = \Delta H^{\text{chem}} + \Delta H^{\text{elastic}} + \Delta H^{\text{struc}}$$

with

$$\Delta H_{ij}^{\text{chem}} = c_i c_j [c_j \Delta H^{\text{inter}}(i \text{ in } j) + c_i \Delta H^{\text{inter}}(j \text{ in } i)]$$

$$\Delta H_{ij}^{\text{elastic}} = c_i c_j [c_j \Delta H^{\text{elsctic}}(i \text{ in } j) + c_i \Delta H^{\text{elastic}}(j \text{ in } i)]$$

The atomic size effect in primary solid solutions can be written as the strain enthalpy term $\Delta H^{\text{elastic}}$ in the free energy of the solid solutions using Eshelby's elastic inclusion model, for instance, solid solutions of a binary A–B system [15]:

$$\Delta H^{\text{elastic}} = x_A x_B (x_B \Delta H_{\text{A in B}} + x_A \Delta H_{\text{B in A}})$$

with

$$\Delta H_{\text{A in B}} = \frac{2 K_A G_B (V_B - V_A)^2}{3 K_A V_B + 4 G_B V_A} \tag{6.4}$$

where x_A and x_B are the molar fractions, V_A and V_B the atomic volumes that may take into account the variation (which is usually unknown but can be estimated, e.g., $V_A + \Delta V_A = V_A + \alpha \phi_A - \phi_B / n_{\text{ws,A}}$) due to locally neighboring atom interactions or electron charge transfer between unlike elements, and K_A, K_B and G_A, G_B denote the bulk modulus and shear modulus, respectively, for elements A and B. It explicitly indicates that larger atomic size differences lead to higher elastic strain enthalpy or less stability of solid solutions.

This agrees with the Hume–Rothery's alloy phase stability 15% rule, which has been very successful in the determination of the glass forming composition ranges (GFR) of many binary systems [16]. However, it is basically a geometrical constraint that is actually self-assured for higher order systems. Additionally, it does not consider the intermediate phases with strong and dominating electrochemical interactions between elements while the atomic size difference has, if any, a trivial effect. Last but not least, the atomic "sizes" could be difficult to predetermine since they depend on bonding characteristics with neighbors, or on the local structures, like Fe in Al–Fe–Ce system [17].

In summary, alloying with elements that have strong interactions between unlike elements and compositions having certain electronic concentrations can reduce the negative heat of mixing of the liquids. However, it will, at the same time, reduce the enthalpy of the corresponding solid solutions or intermediate phases at essentially the same level. It is then the elastic strain energy due to atomic size mismatch, that is absent in the liquids, that dominates contributions to the enthalpy of solid solutions and favors relative stability of liquids over the corresponding solid solutions or intermediate phases.

6.4.2 ENTROPY ESTIMATION

The difference of the entropy between liquids and crystal phases is usually neglected explicitly in most GFA analyses, e.g., in Refs. [18,19]. For a simple polymorphous transformation from liquid to a solid phase, the entropy change or the entropy of fusion ΔS is mostly from vibrational contributions that are characterized by the Debye temperature or the characteristic vibration frequencies. Major contributions may also come from differences in the structural or configurational entropy for non-polymorphous crystalline phases. Composition dependence of the entropy can be described by ideal mixing ($\sim -R\sum x_i \ln x_i$) for random solutions. However, for practical systems, the equilibrium configurational entropy depends on the complex interactions between atoms that are indicated in chemical ordering of various degrees. For solid solutions, the cluster variation method, combined with first principles computation can be used to calculate the equilibrium configurational entropy where both the effects of bonding strength and atomic size are taken into account.

The effect of atomic size differences between components can be examined under the hard sphere approximation and hence written in mixing entropy of liquid solutions [20]. A larger difference in atomic size may result in higher entropy of mixing of a liquid [21] but the effect of bonding strength (atomic interactions) could be much more complicated than the hard-sphere description, owing to varied and mostly unknown local structures (e.g., coordination number), and has not yet been treated.

6.5 THERMODYNAMIC MODELS

In order to analyze and predict GFA or GFRs and to search for new glass formers, it is important, as viewed from the discussion above, to develop a quantitative method for alloy systems with multi-components and multiple competing solid phases. All the individual factors, such as those already mentioned should be considered simultaneously. For possible crystallization, not only the partitionless transformations, but also primary reactions could be relevant. Also, competition from the metastable crystalline or quasi-crystalline phases could be significant.

6.5.1 CALPHAD MODELS

The CALPHAD method (CALculation of PHAse Diagrams) principally offers a comprehensive treatment that incorporates all phenomenological factors for complex atomic interactions in each individual phase within an alloy system. It embodies the information about the atomic size, valence electron concentration, and electro-negativity that were considered individually in the previously-mentioned models.

Taking the binary Al–RE (RE = Ce, Gd, Nd, Sm, Y) systems as examples; the most relevant phases are liquid, FCC, and the intermetallic phases Al_3RE or $Al_{11}RE_3$ (only the low temperature phase is considered). The liquid (LIQ) and FCC–Al (FCC) solutions are modeled using the Redlich–Kister–Muggianu formulae and the Al_3RE or $Al_{11}RE_3$ are treated as stoichiometric. The parameters of the Gibbs energies are optimized according to the phase equilibria obtained experimentally. The phases and phase equilibria in the Al-rich corner have experimentally been assessed for Al–Fe–Gd [22] and Al–Ni–Gd [23]. Again the liquid (LIQ) and FCC–Al (FCC) solutions are modeled using the Redlich–Kister–Muggianu formulae. The intermetallic phases $Al_{13}Fe_4$, Al_5Fe_2, Al_2Fe, $Al_{10}Fe_2Gd$ and Al_8Fe_4Gd in Al–Fe–Gd, $Al_{15}Ni_3Gd_2$, Al_4NiGd,

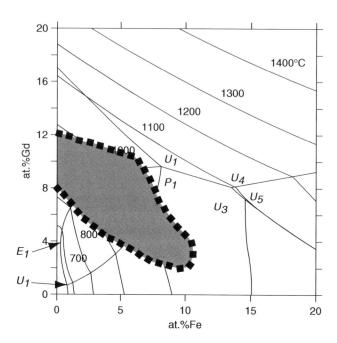

FIGURE 6.3 Liquidus projection of the system at the Al-corner with the univariant equilibrium lines and the isothermals on the surface. The shaded area indicates the composition range where glass ribbons with thickness of about 20 μm can be produced by melt-spinning. (From Zhu, A., Poon, S. J., and Shiflet, G. J., *Scripta Mater.*, 50, 1451–1455, 2004. With permission.)

$Al_7Ni_3Gd_2$, Al_3Ni_2Gd, Al_2NiGd, and AlNiGd in Al–Ni–Gd are treated as stoichiometric. M_3Gd (with DO_{19} structure) and C15-Laves phase M_2Gd in both systems were treated using compound energy formalism with two sub-lattices: Al and Fe (or Ni) mix on the sub-lattice denoted by M. Figure 6.3 shows the liquidus projection of the system at the Al-corner with the univariant equilibrium lines and the isothermals on the surface. The liquid–solid invariant reactions as designated are E_1 at 638°C, U_1 at 647°C, $P_1(U_2)$ at ∼ 1040°C and $U_3(U_4$ or $U_5)$ at ∼ 1100°C. The shaded area in Figure 6.2 indicates the easy glass forming ranges.

6.5.2 T_0-CRITERION

In order to avoid crystallization from undercooled liquids until the glass transition, a straightforward criterion is to keep the liquids thermodynamically favored over the crystal phases, e.g., Φ, that is, $\Delta G = G^{\text{Liq}} - G^{\Phi} \leq 0$. T_0 is the temperature where equilibrium occurs for the intersection composition c_0. A (T_0 versus c_0) line for each phase can thus be constructed by means of Gibbs energy–composition curves.

The T_0-criterion [25,26] was proposed so that the GFR of a system corresponds to the compositions where $T_0 < T_g$ as shown in Figure 6.4. Under this condition, the alloy is constrained not only to avoid polymorphous crystallization but also to prevent segregation during solidification. It is an obviously strict condition for glass formation in a manner that even polymorphous crystallization can also be suppressed kinetically even if it has a non-zero thermodynamic driving force. On the other hand, other possible crystallizations including primary crystallization that usually has a relatively high driving force may prevail and, hence, is more relevant (Figure 6.5).

According to the thermodynamic models for Al–Y, Ce and Nd [27], Al–Gd [23], and Al–Sm [28] the driving forces for the various crystallization reactions illustrated in Figure 6.4 can be calculated. The results show that the polymorphous crystallization to FCC has a zero driving

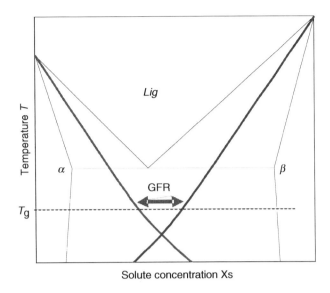

FIGURE 6.4 A metastable phase diagram showing T_0-lines for α and β phases for a eutectic alloy. Glass forming range (GFR) is indicated with the glass transition temperature T_g.

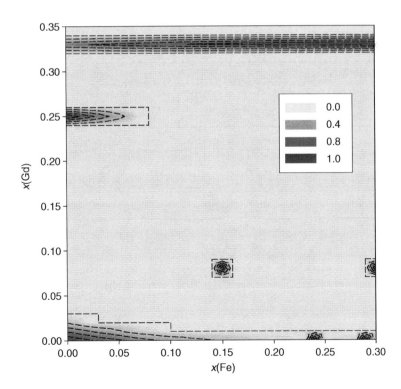

FIGURE 6.5 Driving forces (reduced by the maximum values) contours on mole fractions x(Fe) and x(Gd) for partitionless crystallizations of FCC-A1, $Al_{13}Fe_4$, Al_5Fe_2, Al_2Fe, $Al_{10}Fe_2Gd$, Al_8Fe_4Gd, M_3Gd and M_2Gd from undercooled liquid of Al-Fe-Gd at 600 K."

force beyond $x(\text{RM}) > 0.015-0.025$ (a point on the "T_0 curves") at 500 K. Therefore, even though the kinetics of partitionless FCC polymorphous crystallization is expected to be quite rapid, since no long-range atomic transport is required, the transformation will be confined in these Al-rich corners. A similar analysis is applicable to the stoichiometric compound $Al_{11}RM_3$ or Al_3RM phases. Since the actual GFR($0.08 \leq x(\text{RM}) \leq 0.12$) is well outside of the composition ranges for these partitionless crystallizations, they are not expected to control the GFAs and GFRs of Al–RM glasses.

Figure 6.5 shows the calculated driving forces, in light of the database for the transformations from the undercooled Al–Fe–Gd liquids to various possible crystalline phases of FCC–Al, $Al_{13}Fe_4$, Al_5Fe_2, Al_2Fe, $Al_{10}Fe_2Gd$, Al_8Fe_4Gd, M_3Gd, and M_2Gd. It indicates that the partitionless transformations have driving forces only at the very limited composition ranges corresponding to those of crystalline phases at undercooling as low as 600 K. Similar conclusions can be obtained for crystalline phases including FCC–Al, $Al_{15}Ni_3Gd_2$, Al_4NiGd, $Al_7Ni_3Gd_2$, Al_3Ni_2Gd, Al_2NiGd, $AlNiGd$, M_3Gd, and M_2Gd in Al–Ni–Gd system. It is expected that other metastable phases, possessing higher Gibbs energy than those equilibrium counterparts, could be more limited within certain compositions for their even lower driving forces. Therefore, it can be concluded that the partitionless transformation is not the controlling factor in those system [6].

6.5.3 QUASI-KINETICS OF PRIMARY CRYSTALLIZATION

Consider the primary crystallization of a phase Φ from the liquid solution without phase separation. Assuming homogeneous nucleation without preexisting nuclei and following the simplest treatment [29] based on Johnson–Mehl–Avrami's transformation kinetics, the time t for formation of the phase Φ with a minimal volume fraction f_v (say 10^{-6}) can be estimated using

$$t = \frac{\alpha_\Phi a_0^2 f_v}{D_{R,\text{Liq}}} \cdot \sqrt[4]{\frac{\exp(G_\Phi^*/R_gT)}{(1-\exp(-\Delta G_\Phi/R_gT))^3}} \qquad (6.5)$$

where R_g is the molar gas constant, α_Φ is a structural constant, a_0 the atomic spacing, $D_{R,\text{Liq}}$ the effective diffusivity of the RM solute associated with both nucleation and early growth in the undercooled liquid, and G_Φ^* the nucleation activation energy ($=K\gamma_\Phi^3 v_\Phi^2/\Delta G_\Phi^2$ with the shape factor $K(=16\pi/3)$, the molar volume V_Φ, and interfacial energy γ_Φ of the crystal phase Φ). While the exact values of the interfacial energy γ_Φ are not available, they can be estimated from Turnbull's empirical equation [30] $\gamma_\Phi = (0.35-0.45)\,\Delta H_{f,\Phi}/(N^{1/3}V_\Phi^{2/3})$ where N denotes Avogadro's number and the heat of the fusion $\Delta H_{f,\Phi}$ can be retrieved from the CALPHAD databases. A reduced time-variable $t' = t/t_{\min}$ may be introduced where $t_{\min}$ represents the minimum time needed for the Φ-transformation with respect to the composition. Since the unknown D_{Liq} is usually less dependent on composition, only the variation of 4th root term in Equation 6.4 with composition will be considered. Using the driving forces ΔG_{FCC} and ΔG_{AlxRy} calculated above, the dependence of the reduced time t' is calculated for alloys at various temperatures. The results for Al–Ce, Al–Gd, Al–Sm, and Al–Y alloys at 500 K (Figure 6.6) indicate that the slowest primary crystallization of both FCC and Al_xRM_y phases occur at hypereutectic compositions significantly greater than the eutectic compositions, which generally agrees with the GFRs found experimentally in melt-spun alloys. Thus, these compositional ranges of the slowest primary crystallization kinetics define where metallic glasses may be most easily formed.

For the Al–Gd–Fe ternary system, the reduced times t' for FCC–Al, $Al_{13}Fe_4$, Al_5Fe_2, Al_2Fe, $Al_{10}Fe_2Gd$, Al_8Fe_4Gd, M_3Gd, and M_2Gd from the undercooled liquid of Al–Fe–Gd can be calculated. The overlapped composition contours for all these phases at 600 K are illustrated in Figure 6.7. The relatively large times t', indicating the slowest primary crystallizations of all these phases, clearly occur at compositions that match easy GFR found experimentally.

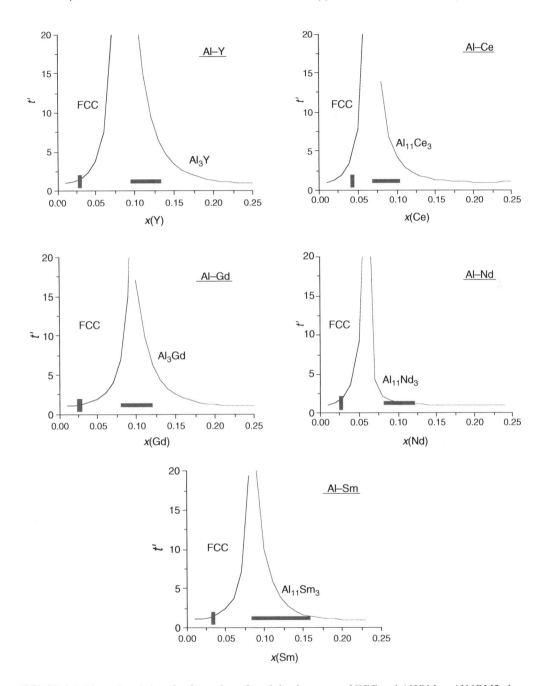

FIGURE 6.6 The reduced time for formation of a minimal amount of FCC and Al3RM or Al11RM3 phases from the undercooled liquid at 500 K. in Al–Sm, –Ce, –Gd, –Nd and –Y systems. The experimentally measured formation ranges of the melt-spun glasses with thickness of 20 mm are indicated by the horizontal bars and the eutectic compositions by the vertical bars.

Similar results have also been obtained for Al–Ni–Gd systems where the reduced times t' for primary crystallization of FCC–Al, $Al_{15}Ni_3Gd_2$, Al_4NiGd, $Al_7Ni_3Gd_2$, Al_3Ni_2Gd, Al_2NiGd, $AlNiGd$, M_3Gd, and M_2Gd from undercooled liquids are considered, c.f. [24] for details.

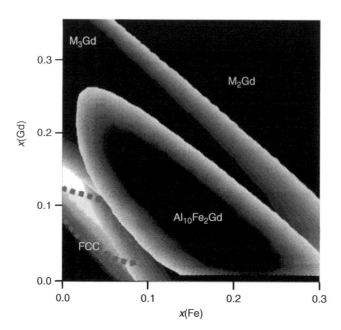

FIGURE 6.7 Overlapped plots of the reduced transformation time t' for the primary crystallizations from the undercooled Al–Fe–Gd liquids at (a) 600 K and (b) 700 K. The area enclosed by dotted line denotes the GFR for melt-sun glass samples with thickness of about 20 μm.

6.5.4 ATOMIC MOBILITY CONSIDERATION

Atomic diffusion process, as well as structural relaxation upon shearing, depends on the atomic interactions (bonding strength) and atomic configurations. Therefore, viscosity and atomic mobility are supposed to be, in some way, connected to the thermodynamic functions of the undercooled liquids. The Moelwyn–Hughes model [31] indicates the correlation between the viscosity and the heat of mixing. As shown in Figure 6.8, the viscosity of an Al–Ce liquid increases with Ce concentration. Similar results were also obtained for Al–Nd, –Sm, –Gd, and –Y systems. Although the results are strictly valid only near or above the melting points, a similar composition dependency of the viscosity might be expected at even deeper undercooling. Therefore, a shift of the t'-curves for Al_3RM or $Al_{11}RM_3$ towards a higher RM composition can be expected. The exact magnitude of this shift may, in general, be different for the different glass systems. The same argument may also be applied to ternary systems; for instance in Al–Ni–Gd, better GFA is obtained for alloys with the more negative heat of mixing among glass formers that are predicted using the quasi-kinetics.

6.5.5 HEAT OF MIXING AND SRO IN LIQUIDS

As already mentioned, the liquid heat of mixing represents the interaction among neighbor atoms and further distant atoms in the liquid solutions. Negative heat of mixing indicates strong interaction between unlike atoms and formation of chemical short range ordering (CSRO). It is also significant for efficient atomic packing, as well as for high viscosity of the undercooled liquids as discussed below.

The heat of mixing ΔH_{mix}, can be estimated from Miedema's "macroscopic atom" model or from the solubilities of solutes in simple phase diagrams when both end solutions have the same structures. Both methods are at best applicable only for binary regular solutions whose interaction parameter Ω is independent of composition. The CALPHAD approach offers a more consistent way

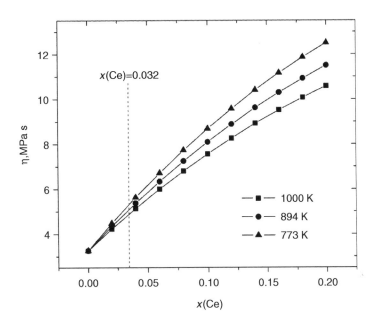

FIGURE 6.8 The calculated composition-dependent viscosity of Al–Ce at $T = 1000$, 894 (the eutectic temperature) and 773 K (extrapolated down from eutectic temperature) according to Moelwyn–Hughes formula. The interaction parameter is retrieved from COST507. The eutectic Ce concentration $x(Ce) = 0.032$ is indicated. (From Zhu, A. W., Shiflet, G. J., and Miracle, D. B., *Scripta Mater.*, 50, 987–991, 2004. With permission.)

to obtain this information without this limitation. The heat of mixing depends on the composition $\{x_i$ (molar fraction of component i)$\}$ and temperature T,

$$\Delta H_{mix} = \Delta G_{mix} - T\frac{\partial \Delta G_{mix}}{\partial T}\Big|_{P, x_i} \tag{6.6}$$

where the interaction parameters for ΔG_{mix} can be retrieved from the appropriate databases. Figure 6.9 shows the heat of the mixing of Al–Fe–Gd liquid at 700°C. The pair interaction (bonding) energy $\Omega (= H_{Al,RM} - 0.5(H_{Al,Al} - H_{RM,RM}))$ between unlike atoms i and j with respect to those of the like-atom interactions can be obtained as $\Omega_{ij} = \Delta H_{mix,(i,j)}/z x_i x_j$ where z is the coordination number of the nearest neighbors for the atomic structure if only the nearest neighbors are considered.

Under the quasi-chemical approximation, the Gibbs energy of a high-order system with n components can be written by extending the formula for binary systems in [32],

$$G = \sum_i N_{ii}E_{ii} + \sum_{i<j} N_{ij}E_{ij} - kT \cdot \ln\left\{\frac{(zN/2)!}{\prod_i N_{ii}! \cdot \left(\prod_{i<j}(N_{ij}/2)!\right)^2} \cdot \frac{\prod_i \left[(z-1)Nx_i\right]!}{[(z-1)N]!}\right\} \tag{6.7}$$

where N denotes the number of atoms, N_{ij} the number of the bonds between component i and j, E_{ij} the bonding energy between i and j. The bonding conservation requires

$$2N_{ii} + \sum_{j>i} N_{ij} = zNx_i, \quad i = 1...n \tag{6.8}$$

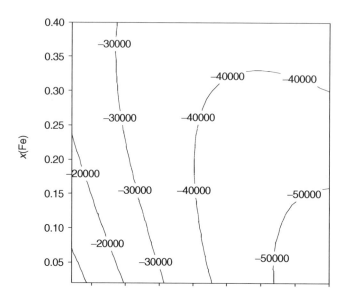

FIGURE 6.9 Composition Contours of Calculated Heat of Mixing (in J/mol) (a) of undercooled LIQ solutions of Al-Fe-Gd at 600 K.

Minimizing G in Equation 6.2 with respect to N_{ij}, equations related to the pair probability $\zeta_{ij} = N_{ij}/zN$ of unlike atoms i and j can be obtained:

$$\frac{\zeta_{ij}}{\left(x_i - \sum_{p=i,q} \zeta_{pq}\right)\left(x_j - \sum_{p,q=j} \zeta_{pq}\right)} = \exp\left(-\frac{2\Omega_{ij}}{kT}\right). \tag{6.9}$$

Solving Equation 6.4, the Warren–Cowley ordering parameters $a(i,j) = 1 - \zeta_{ij}/x_i x_j$ [6] can be obtained. Figure 6.3b and c show the Warren–Cowley parameters of Al–Fe, Al–Gd and Fe–Gd for Al–Fe–Gd at the Al-rich corner at 700 K. For alloys of compositions in the best GFR, Al–Gd and Al–Fe have negative W–C parameters as compared with positive values for Fe–Gd. Particularly for Al–Gd, the W–C parameter is more negative than that for Al–Fe in most of the compositions considered. It indicates a strong tendency of formation of mainly Al–Gd CSRO or aggregates that may act as the backbone in the undercooled amorphous structures (Figure 6.10).

One more example can be illustrated by one of the Al–Cu–Mg eutectics with composition $Al_{75}Cu_{17}Mg_8$. Using conventional melt-spinning (with a cooling rate of about 10^5 C/s), the quenched product is crystallized containing Al FCC solid solution and two intermetallic phases θ (Al_2Cu) and S (Al_2CuMg). It was found [33] that addition of 2–8 at.% Ni makes the alloy a good glass former that can be quenched into a fully amorphous phase under the same processing condition. Using a preliminary thermodynamic model of the liquid phase hierarchically constructed based on CALPHAD databases of Al–Cu–Mg, Cu–Mg–Ni [34] and Al–Ni [35], the effect of Ni addition can be seen from change of the heat of mixing of the alloy. As shown in Figure 6.11, 5 at.% Ni addition reduces the heat of mixing of the system by a factor of two more negative at the eutectic composition ($Al_{75}Cu_{17}Mg_8$).

6.5.6 Multiple Component Effects

New BMGs are usually developed from starting with simple systems such as binary alloys. To make the alloys effectively "complex" for bulk glass forming, selection of specific elements over

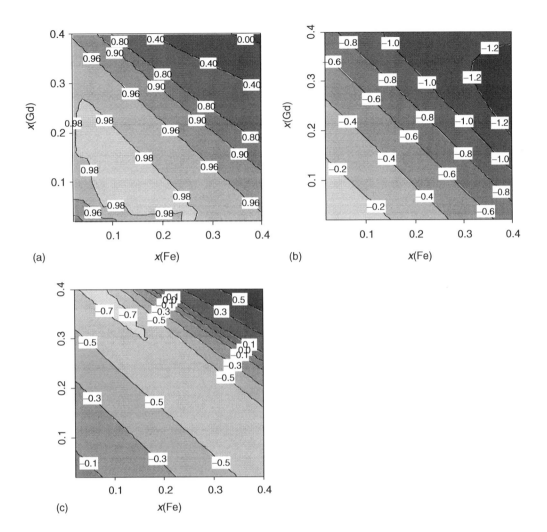

FIGURE 6.10 Composition contours of calculated CSROs of Fe–Gd (a) Al–Gd (b) and Al–Fe (c) in Al–Fe–Gd liquid solutions at 600 K. Glass region is also indicated by the dark shaded areas in the lower left of each plot.

tightly controlled proportions and whether to add or to replace existing components have been the most challenging difficulties in the searching effort.

The multicomponent effects and improved GFA can intuitively be associated with slower transformation kinetics. This could result from more extensive and complex atomic rearrangements required for the solid phase crystallization having a specific composition. The effects may also be viewed from a thermodynamic perspective. This can be illustrated from variation of the chemical driving forces when Ni is added to binary Al–Gd or Gd to Al–Ni. Figure 6.12 shows the calculated driving forces for the FCC–Al, M_3Gd, Al_3Ni, and $Al_{15}Ni_3Gd_2$ from undercooled $Al_{0.8}$–Ni–Gd melts at 600 K where $x(Gd) + x(Ni) = 0.20$. These four phases are the most important for the GFR as shown for the similar Al–Fe–Gd results in Figure 6.3 and Figure 6.5. Replacing Ni with Gd lowers the driving forces for formation of the FCC–Al phase, which has little solubility for Gd, and for the Al_3Ni phase that is considered to be a stoichiometric compound. However, either adding Gd from the Al–Ni end or adding Ni from the Al–Gd end leads to enhanced driving forces for the

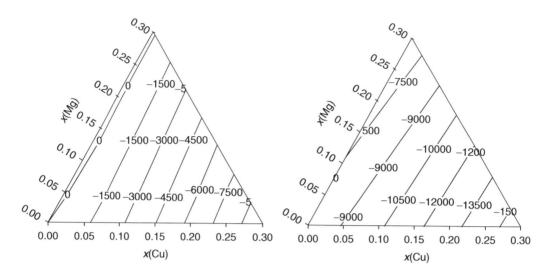

FIGURE 6.11 The calculated heat of mixing of liquid Al–Cu–Mg with 0 and 5 at.% Ni addition (unit in J/mol).

ordered compound M_3Gd, which has solubility for both Gd and Ni, and for the ternary phase $Al_{15}Ni_3Gd_2$.

The changes in the crystallization driving forces, shown in Figure 6.12, represent a general scenario for two types of solid phases when an element is added to complicate the chemistry of a

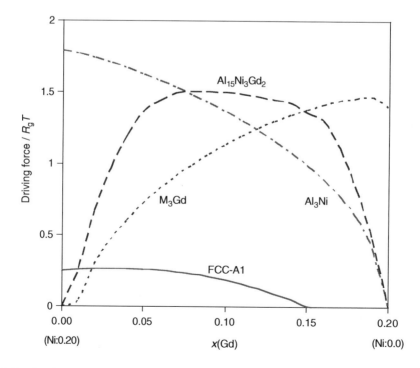

FIGURE 6.12 The driving forces ($/R_gT$) for the nucleation of the solid phases FCC–Al, M_3Gd, Al_3Ni and $Al_{15}Ni_3Gd_2$ from undercooled $Al_{0.80}$–Ni–Gd melts at 600 K where $x(Gd)+x(Ni)=0.20$.

metastable undercooled liquid. The complication of the chemistry is meant to decrease the Gibbs energy of the liquid owing to its increased configuration entropy, as well as it possibly increasing the negative heat of mixing caused by the added element. As an example, the calculation for the Al–Ni–Gd undercooled melts [24] indicates that the heat of mixing of the system decreases when Gd is added to partly replace Ni in Al–Ni. The decrement in Gibbs energy obviously leads to decreased driving forces for the phases with little solubility of the added elements, such as FCC Al and the stoichiometric compound Al_3Ni when adding Gd to replace Ni. On the other hand, added elements may more significantly reduce the Gibbs energy of the phases that incorporate the elements, such as M_3Gd and $Al_{15}Ni_3Gd_2$ for Gd, or the phases with solubility of the elements, such as M_3Gd for Ni as shown in Figure 6.12. For these phases, the driving forces will be enhanced by adding the elements.

6.6 SUMMARY

Future design of bulk structural amorphous metal alloys will require considerable computational materials science, particularly thermodynamics. For the GFA of a solidified alloy, the negative heat of mixing of the system maintains stability of both the undercooled melt and the solid solution, but more importantly, reduces the atomic mobility in the melt. Atomic size factors contribute to the relative instability of the solid solution versus the melt. Additional elements or multiple component effects can be either positive or negative for the GFA depending on a balanced variation of the heat of mixing and configuration entropy induced. These favorite conditions are not necessarily sufficient for glass formation; the kinetics of not only partitionless but also primary/eutectic crystallization of the other competing phases, either stable or metastable, ultimately controls GFR. General guidelines for composition optimization and adding new elements are now given a thermodynamic foundation from known information:

1. Systems with deep eutectics are favored; for usually unsymmetrical reactions, compositions closer to the involving phases with steeper T_l reduction and/or lower chemical variation are preferred; compositions close to the phases with significant composition variation or semi-disordered compounds such as the Laves phase, should be avoided.
2. Additional elements that cause more negative heat of mixing of the system are preferred.
3. Additional elements with little solubility in the phases and yield no new phases in the interested composition ranges are preferred.

ACKNOWLEDGMENTS

The authors of a Multi-University Research Initiative Grant F4960201-1-0352.

REFERENCES

1. Inoue, A., Amorphous, nanoquasicrystalline and nanocrystalline alloys in Al-based systems, *Prog. Mater. Sci.*, 43, 365–520, 1998.
2. Poon, S. J., Shiflet, G. J., Guo, F. Q., and Ponnambalam, V., Glass formability of ferrous- and aluminum-based structural metallic alloys, *J. Non-Cryst. Sol.*, 317, 1–9, 2003.
3. Wang, W. H., Dong, C., and Shek, C. H., Bulk metallic glasses, *Mater. Sci. Eng. R*, 44, 45–89, 2004.

4. Xu, D., Lohwongwatana, B., Duan, G., Johnson, W. L., and Garland, C., Formation and properties of new Ni-based amorphous alloys with critical casting thickness up to 5 mm, *Acta Mater.*, 52, 2621–2624.

5. Tan, H., Zhang, Y., Ma, D., Feng, Y. P., and Li, Y., Optimum glass formation at off-eutectic composition and its relation to skewed eutectic coupled zone in the La based La—Al—(Cu,Ni) pseudo ternary system, *Acta Mater.*, 52, 2621–2624.

6. He, Y., Poon, S. J., and Shiflet, G. J., Synthesis and properties of metallic glasses that contain aluminium, *Science*, 241, 1640–1642, 1988.

7. Waniuk, T. A., Schroers, J., and Johnson, W. L., Critical cooling rate and thermal stability of Zr–Ti–Cu–Ni–Be alloys, *Appl. Phys. Lett.*, 78, 1213–1215, 2001.

8. Lu, Z. P. and Liu, C. T., A new glass-forming ability criterion for bulk metallic glasses, *Acta Mater.*, 50, 3501–3512, 2002.

9. Zhu, A. W., Shiflet, G. J., and Miracle, D. B., Glass forming ranges of Al-rare earth metal alloys: thermodynamic and kinetic analysis, *Scripta Mater.*, 50, 987–991, 2004.

10. Turnbull, D., Formation of crystal nuclei in liquid metals, *J. Appl. Phys.*, 21, 1022–1028, 1950.

11. Miracle, D. B., Sanders, W. S., and Senkov, O. N., The influence of efficient atomic packing on the constitution of metallic glasses, *Phil. Mag.*, 83, 2409–2428, 2003.

12. Nagel, S. R. and Tauc, J., Nearly-free-electron approach to the theory of metallic glass alloys, *Phys. Rev. Lett.*, 35, 380–383, 1975.

13. Chen, W., Wang, Y., Qiang, J., and Dong, C., Bulk metallic glasses in the Zr–Al–Ni–Cu system, *Acta Mater.*, 51, 1899–1907, 2003.

14. Bakker, H., Enthalpies in Alloys, Trans Tech Publications, Switzerland, pp. 1–78, 1998.

15. Van der Kolk, G. J., Miedema, A. R., and Niessen, A. K., *J. Less-Comm. Met.*, 145, 1–77, 1988.

16. Egami, T. and Waseda, Y., Atomic size effect on the formability of metallic galsses, *J. Non-Cryst. Solids*, 64, 113–134, 1984.

17. Mansour, A. N., Wong, C. P., and Brizzolara, R. A., Atomic structure of amorphous Al100-2xCoxCex (x = 8, 9, and 10) and A180Fe10Ce10 alloys: An XAFS study, *Phys. Rev. B50*, 12401–12412, 1994.

18. Lopez, J. M., Alonso, J. A., and Gallego, L. J., Determination of the glass-forming concentration range in binary alloys from a semiempirical theory: Application to Zr-based alloys, *Phys. Rev. B*, 36, 3716–3722, 1987.

19. Shindo, T., Waseda, Y., and Inoue, A., Prediction of glass-forming composition ranges in Zr-Ni-Al alloys, *Mater. Trans.*, 43, 2502–2508, 2002.

20. Mansoori, G. A., Carnahan, N. F., Starling, K. E., and Leland T. Jr., Equilibrium thermodynamic properties of the mixture of hard spheres, 54(4), 1523–1525, 1971.

21. Takeuchi, A. and Inoue, A., Quantitative evaluation of critical cooling rate for metallic glasses, *Mater. Sci. Eng.*, A304–306, 446–451, 2001.

22. Gao, M. C., Hackenberg, R. E., and Shiflet, G. J., Thermodynamic assessment of the Al–Ni–Gd glass forming system, *J. Alloys Compd.*, 353, 114–123, 2003.

23. Hackenberg, R. E., Gao, M., Kaufman, L., and Shiflet, G. J., Thermodynamics and phase equilibria of the Al—Fe—Gd metallic glass-forming system, *Acta Mater.*, 50, 2245–2258, 2002.

24. Zhu, A., Poon, S. J., and Shiflet, G. J., On glass formability of Al—Gd—Ni (Fe), *Scripta Mater.*, 50, 1451–1455, 2004.

25. Boettinger, W. J., Growth Kinetic Limitations During Rapid Solidification, Kear, B. H., Giesson, B. C., and cohen, M., Eds., Vol. 8, MRS Symp. Proc., Elsevier, Amsterdam, p. 15, 1982.

26. Massalski, T. B., Woychik, C. C; Murray, J. L, Alloy Phase Diagrams; Boston, Mass ; U.S.A ; Nov. 1982. pp. 241–247. 1983: Relationships Between Phase Diagrams, the T sub o and T sub n Temperatures, Cooling Rates and Glass-Forming Ability. In *Proc. 4th Int. Conf. Rapidly Quenched Metals*, Masumoto, T. and Suzuki, K., Eds., Japan Inst. of Metals, pp. 203, 1982.

27. COST507, Light Alloys Databases: Round II, European Commission, 1998.

28. Saccone, A., Cacciamiani, G., Maccio, D., Borzone, G., and Ferro, R. J., Saccone A, Cacciamiani G, Maccio D, Borzone G, Ferro RJ, Intermetallics, 1998;6(3):201: Contribution to the study of the alloys and intermetallic compounds of aluminium with the rare-earth metals, *Intermetallics*, 6(3), 201–215, 1997.

29. Uhlmann, D. R., A kinetic treatment of glass formation, *J. Non-Cryst. Sol.*, 7, 337–348, 1972.

30. Turnbull, D., Under what conditions can a glass be formed?, *Contemp. Phys.*, 10, 473–488, 1969.

31. Moelwyn-Hughes, E. A., *Physical Chemistry*, Pergamon, Oxford, 1964.
32. Machlin, E. S., *Thermodynamics and Kinetics*, Giro Press, New York, 1991.
33. Guo, F. Q., Enouf, Poon, S. J., and Shiflet, G. J, Glass formability in Al-based multinary alloys, *Philo. Mag. Lett.*, 81(203), 203–211, 2001.
34. Gorsse, S. and Shiflet, G. J., A thermodynamic assessment of the Cu-Mg-Ni ternary system, *CALPHAD*, 26, 63–83, 2002.
35. SSOL4, ThermoCalc, 2002.

7 NiTi-Based High-Temperature Shape-Memory Alloys: Properties, Prospects, and Potential Applications

Ronald Noebe, Tiffany Biles, and Santo A. Padula II
NASA Glenn Research Center

CONTENTS

7.1 Introduction..146
7.2 Shape-Memory and Superelastic Behavior..147
    7.2.1 Shape-Memory Effect...148
        7.2.1.1 Free Recovery...148
        7.2.1.2 Constrained Recovery 149
        7.2.1.3 Work Production (Actuators)..150
    7.2.2 Superelasticity...150
7.3 Applications 151
    7.3.1 Free Recovery..151
    7.3.2 Constrained Recovery 152
    7.3.3 Work Production (Actuators) 152
    7.3.4 Superelastic Behavior..154
    7.3.5 Damping and Wear Resistant Applications 155
7.4 High-Temperature Ternary NiTi–X Alloys 156
    7.4.1 Background...156
        7.4.1.1 Historical Development of NiTi Alloys.............................156
        7.4.1.2 General Effect of Alloying Additions on NiTi...................157
        7.4.1.3 Important Phase Structures in NiTi Alloys.......................158
    7.4.2 Processing of NiTi-Based Alloys...160
    7.4.3 (Ni,Pd)Ti Alloys 161
    7.4.4 (Ni,Au)Ti Alloys..165
    7.4.5 (Ni,Pt)Ti Alloys 166
    7.4.6 Ni(Ti,Hf) Alloys 170
    7.4.7 Ni(Ti,Zr) Alloys...174
7.5 Conclusions and Future Challenges 177
7.6 Afterword 180
Acknowledgments 180
Abbreviations and Basic Definitions..180
References 181

Abstract

There is a growing demand in the aerospace, automotive, heating and ventilation, systems and controls, and various other industries for shape-memory alloys with transformation temperatures greater than those possible with commercial NiTi systems, which are restricted to use temperatures below 100°C. However, a transformation temperature above 100°C is only the initial requirement for development of a high-temperature commercial alloy. A number of properties including minimum requirements for ductility, thermal stability, dimensional stability, environmental resistance, durability and fatigue life, processability, and functional or shape-memory behavior, especially work output must also be met. Thus, progress to date in the research and development of high-temperature shape-memory alloys based on NiTi with Pd, Pt, Au, Hf, and Zr additions is reviewed, focusing on these properties and with a critical view towards potential commercialization. A brief overview of shape-memory and superelastic behaviors and an update on current and potential applications are also presented, providing the necessary background and context for understanding the challenges involved in developing shape-memory alloys for high-temperature use.

7.1 INTRODUCTION

Shape-memory alloys are an extraordinary group of materials that have been attracting much attention in recent years because of their unique properties. Due to a reversible martensitic transformation, they can display the unusual properties of shape-memory and superelastic behavior. "Shape-memory" is a unique property of certain alloys, where the material can be deformed at a low temperature and recover its original undeformed shape by heating above a given transformation temperature. Superelastic behavior is the ability of the material to exhibit large (on the order of 4–8%) elastic strains at certain temperatures.

These properties are so unique that they actually require an entirely new philosophy in engineering and design. In fact, even the principle properties used to describe shape-memory alloys are different from basic structural materials. In every other chapter in this book, the most important properties are related to strength, modulus, and ductility. While these properties are still important in the development of shape-memory alloys and probably even more so to the development of high-temperature versions, additional properties such as transformation temperatures, transformation or recoverable strain, recovery stress, and work output are critical to design with these materials.

The unique properties of shape-memory alloys are widely exploited, near room temperature, in commercially available NiTi alloys for such products as electrical switches, eyeglass frames, appliance controllers, electrical connectors, couplings for pipes and other tubular products, temperature sensitive valves, actuators, and countless medical and dental devices. However, there are also limitless applications for materials that exhibit shape-memory behavior at higher temperatures within the aerospace, automotive, heating and air-conditioning, chemical processing, energy, and many other industries. In particular, the need for compact, lightweight, but high-force solid-state actuators are critical to the development of morphing and adaptive structures for aerospace and even automotive applications. But no commercial high-temperature shape-memory alloys are currently available to fill this potential need.

Conventional or commercial NiTi alloys can be used at temperatures up to about 100°C but even this has to be done with a great deal of caution because of the reduced life of the alloy and the relatively low loads that can be sustained. The application of high levels of stress can increase the transition temperature of binary NiTi alloys to about 120°C, but this introduces serious problems with creep and reduced cycle life. Therefore, new alloys with higher stress-free transformation temperatures are needed to address the numerous applications that exist at elevated temperatures. Ideally, these high-temperature systems would have the same shape-memory characteristics and properties as the commercial binary NiTi alloys, with only the transformation temperatures shifted to higher temperatures.

The use of ternary alloying additions to NiTi is one way to achieve increased transformation temperatures. Given the experience and knowledge base developed for binary NiTi alloys because of their broad commercialization base, this approach to developing HTSMA has been the most thoroughly studied to date. There are other material systems including CuAlNi and NiAl alloys that display elevated temperature shape-memory characteristics. But there are serious issues with microstructural stability in these materials. Cu-based alloys readily decompose into other phases at intermediate temperatures, making it difficult to stabilize the parent phase necessary for displaying shape memory behavior [1]. NiAl-based systems are brittle and the martensite phase tends to decompose into an Ni_5Al_3 phase, which does not undergo a reverse martensitic transformation [2,3]. Consequently, these systems would appear to have much less promise than NiTi-based alloys. There are also a number of alloys such as platinum-based intermetallics, NbRu, TaRu, and others that have not evolved much beyond laboratory curiosities for various reasons. Therefore, we have decided to limit the scope of this review to NiTi-based systems.

Creating higher transformation temperature alloys with roughly the same functional behavior as binary NiTi is but one of the potential challenges in developing an acceptable high-temperature alloy. In addition to the focus on functional properties (those related to the solid-state phase transformation in the material), there could be potential problems associated with long-term stability of the alloy including its microstructure, phase structure, and resistance to oxidation, and with high-temperature strength, creep resistance, and other mechanical properties. Therefore, high-temperature shape-memory alloys are a functional material but with essentially all the same challenges as other advanced high-temperature structural materials. Meeting these challenges will be difficult, and undoubtedly, tradeoffs will need to be made to develop high-temperature shape-memory alloys with an acceptable balance of properties.

In the remaining sections of this chapter, we will review the NiTi-based alloy systems identified as having potential for use as high-temperature shape-memory materials. Prior to the discussion of specific alloy systems, we will present a brief review of shape-memory and superelastic behavior (Section 7.2). This will be followed by a description of potential applications (Section 7.3), which provides some insight into the wide range of designs where high-temperature shape-memory alloys can be applied. As the potential use of these materials is quite broad, this section is designed to provide enough background to stimulate the reader to explore their own possible uses for these types of materials. The balance of this chapter (Section 7.4) provides a critical review of the various NiTi-based alloys and their realistic potential for commercialization, followed by concluding remarks (Section 7.5) detailing the challenges remaining and the direction needed to develop a commercially viable high-temperature shape-memory alloy.

7.2 SHAPE-MEMORY AND SUPERELASTIC BEHAVIOR

Shape-memory and superelastic behavior are a consequence of the reversible martensite-to-austenite transformation in certain materials. With a few exceptions, the high-temperature austenite phase (A) is a higher-symmetry structure, often cubic and mildly or strongly ordered as in the case of NiTi-based alloys. The austenite phase transforms without long-range diffusion into a lower symmetry martensite (M) structure at some lower temperature. The $A \leftrightarrow M$ transformation is hysteretic and is therefore defined by start and finish temperatures for the reverse austenite and the forward martensite transformations as A_s, A_f, M_s, and M_f, respectively. All shape-memory behavior is subsequently tied to one or more of these temperatures in some way. The width of the hysteresis loop, which is usually defined as $A_f - M_s$ gives an indication of the rapidity of the transformation and can be altered through alloying or thermomechanical processing.

Not all materials that undergo a martensitic transformation display shape-memory or superelastic effects. There are certain properties required of the transformation that are in turn dependent on microstructural features of the parent austenite and resulting martensite phases. The first property is the reversible nature of the *path* for the martensite-to-austenite transformation. A reversible transformation is more likely when the habit plane between the martensite and austenite is essentially coherent and when the parent austenite phase is an ordered compound. Even though there will be multiple lattice variants of martensite that can form from each parent austenite grain, each variant when heated will follow a perfectly reversible path back to the parent austenite phase due to the ordered nature of the alloy. If it did not, it would cause a change in chemical ordering of the superlattice of the original parent phase, which would require a significant input of energy.

The second feature common to materials exhibiting shape-memory behavior, is that easy plastic deformation of the martensite can occur by deformation induced twinning reactions at stresses well below those needed to nucleate and move dislocations. Structurally, the necessary twins are already present in the microstructure, so nucleation is not necessary and deformation simply occurs by the growth of those twins aligned with the largest shear component and the shrinkage and eventual elimination of those twins with the largest negative component. Therefore, as long as dislocation slip does not occur and the only deformation of the martensite is through deformation twinning, there will be no effect on the orientation of the original austenite phase on heating and thus the material will return to its original shape. As will be demonstrated in Section 7.4, strengthening of the martensite phase against slip is actually critical to improving the shape-memory properties of the high-temperature NiTi-based alloys.

7.2.1 Shape-Memory Effect

The shape-memory effect (SME) is a phenomenon where deformation of the martensite phase (a strain introduced below the A_s), is recovered completely by heating the material through the martensite-to-austenite transformation, as long as dislocation slip does not occur in the martensite. This is accomplished when martensite is deformed to some maximum strain only through the movement of twin phase boundaries and the accommodation of martensite variants in the microstructure. The material can then be heated above the A_f temperature and recover the original parent form through a shear-like process. The point where one martensite variant becomes dominant in the structure determines the completely recoverable strain limit for the material and beyond that point, some permanent nonrecoverable strain will be induced through slip.

Once this phenomenon is understood and its limits determined, such as the strain and stresses that can be applied while still maintaining full recovery, the shape-memory effect can be exploited in a number of ways. For example, the transformation can be allowed to proceed unimpeded, i.e., free recovery. The transformation can also proceed with external constraints imposed on the system or as a means of performing work. Applications then can be developed based on each of these processes.

7.2.1.1 Free Recovery

Free recovery of a shape-memory material can be classified into two categories: one-way and two-way. One-way memory is the ability to recover the original austenitic shape upon transformation. Two-way memory involves a reversible shape change on both heating and cooling of the shape-memory material and usually requires prior thermomechanical processing or "training" to stabilize the effect. One-way memory is most commonly used due to the predictability of the transformation and the ability to recover much larger strains. Figure 7.1 schematically illustrates the one-way shape-memory response. The material is cooled from above the A_f, which is the original shape of the element in Figure 7.1a, to below the M_f (Figure 7.1b). This usually occurs without any change in shape except for those due to thermal expansion differences between the two temperatures. The material is then strained from the initial undeformed martensite state in Figure 7.1b with the application of a force (Figure 7.1c). Upon removal of the force, the material will undergo a certain amount of elastic springback but will also exhibit a pseudo-permanent strain (Figure 7.1d). Then upon

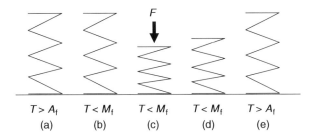

FIGURE 7.1 Free recovery in a one-way shape-memory alloy element.

heating, the material will begin to recover its original high-temperature shape at the A_s temperature and is fully recovered once the A_f transformation temperature is reached (Figure 7.1e). To repeat the process, the material needs to be cooled back down below the M_f temperature and must be redeformed to introduce the strain that is once again to be recovered. For example, for use in a temperature sensitive switch, the shape-memory material is usually coupled with a biasing spring or other device that provides deformation during cooling of the material back below the M_f temperature, so that it is ready to actuate once again on heating.

Two-way memory behavior is shown in Figure 7.2 and does not involve an external force to achieve the strained shape either on heating or cooling. It is, therefore, characterized by a reversible shape change on both heating and cooling of the shape-memory material. Two-way memory is difficult to achieve in most alloys, as an extensive "training regime" must be employed. One training method is to simply repeat the one-way shape-memory effect by deforming the specimen in the martensite condition and then heating it to cause normal recovery. If this process is carefully repeated, deforming the specimen identically each time, stresses will accumulate in the material that will eventually bias the subsequent martensitic transformation during cooling so that only certain variants will form, resulting in a two-way effect. More sophisticated training mechanisms are also possible for creating a two-way shape-memory effect but are beyond the scope of this review. The maximum strain that the material can accommodate through a two-way mechanism is 2–3 times less than what can be accomplished through one-way memory, so practical application of this phenomenon is much more limited. Loss of the shape-memory behavior can also occur more rapidly during this type of process due to any dislocation activity in the material.

7.2.1.2 Constrained Recovery

If a shape-memory material is constrained during recovery, the maximum stress that the element can impart on its surroundings can be up to the yield strength of the austenite phase, due to the

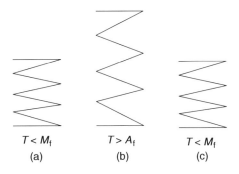

FIGURE 7.2 Free recovery in a two-way shape-memory alloy element. The element transforms between two different shapes on heating and cooling without the need for application of any external load.

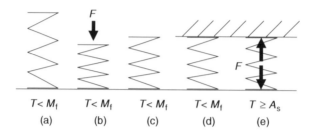

FIGURE 7.3 Constrained recovery in a one-way shape-memory alloy element.

inability of the material to fully recover its shape. Figure 7.3 shows an element (Figure 7.3a) that is strained while in the martensitic state to point (Figure 7.3b), recovers elastically at (Figure 7.3c) and then is inserted into a constrained space at (Figure 7.3d). The element is then heated and once the A_s transformation temperature is exceeded, the material will begin to try to recover its original shape. However, because the element is constrained, its shape cannot change, and forces are generated by the element, which reach a maximum at the A_f temperature (Figure 7.3e). This type of application is most useful in the opposite sense, i.e., when the element shrinks upon heating since it could act as a compression type of fitting.

7.2.1.3   Work Production (Actuators)

Finally, the shape-memory effect can be used to do work against a force, as shown in Figure 7.4. Below the M_f temperature (Figure 7.4a), the material is strained (Figure 7.4b) and allowed to elastically recover (Figure 7.4c). A weight is placed on the shape-memory element while the temperature is still below the M_f (Figure 7.4d), and then the temperature is increased above the A_f (Figure 7.4e). As the temperature is increased above the A_f, the material will then try to recover its original shape, doing work against the weight. Work can be extracted from NiTi shape-memory alloys at energy densities exceeding 10^7 J/m^3 [4]. In fact, most of the recent interest in high-temperature shape-memory alloys resides in this basic property, which essentially allows the shape-memory alloy to act as a solid-state, compact actuator.

7.2.2   Superelasticity

Finally, the isothermal manifestation of the shape-memory effect is a superelastic behavior that occurs at temperatures above the A_f, (but below the M_d). This extreme elastic behavior, illustrated in Figure 7.5, is due to a stress induced martensitic transformation in the material during loading and by the subsequent reverse transformation upon unloading. The maximum amount of strain that a material can recover in the superelastic condition is typically higher than can be achieved through

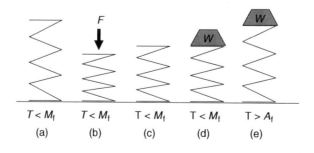

FIGURE 7.4 Work production by a one-way shape-memory alloy element.

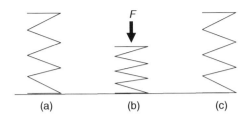

FIGURE 7.5 Superelastic behavior of a shape-memory alloy, which can occur isothermally at a temperature between A_s and M_d.

the shape-memory effect and in certain circumstances can approach 10% in specially prepared NiTi wires, though generally strains on the order of 5–8% can be expected in most other NiTi alloys [5].

7.3 APPLICATIONS

There are a number of diverse applications that involve the use of shape-memory alloys outside the biomedical field [6,7]. Many of these applications and devices, which currently use commercial NiTi alloys, can be retrofitted with a higher temperature alloy to deliver similar results but in a different temperature regime. However, new applications will arise and many are currently being developed for when a successful high-temperature shape-memory alloy emerges on the market. For example, high-temperature shape-memory alloys are considered to be an enabling technology for achieving substantial performance improvements in many systems including aircraft structures and engines, diesel engines, and other components especially in the aerospace, automotive, chemical processing, and heating and cooling industries. While many of the specific designs for use of high-temperature versions of shape-memory alloys are still proprietary, this section provides a brief overview of some of the general applications for these materials and the benefits of their use. For convenience, we have arranged the various applications according to the primary function of the shape-memory material as described in the previous section, since there are common properties, design criteria, and other factors important to each category.

7.3.1 FREE RECOVERY

Free recovery includes any application in which the sole function of the SMA is to cause movement or a shape change while unconstrained. The best example of this is the trade show "giveaway" where a word or picture is formed by a continuous length of SMA wire. The wire can be distorted into any random configuration but when heated will again spell out the desired picture or word. In a more practical sense, SMA-based antenna arrays have been proposed based on the same concept. The antennae for a spacecraft or planetary rover can be folded into a compact unit for storage during travel and then when on station, the antenna can be deployed remotely by sending a small current through the wire, heating the SMA, and causing the antenna to unfold into its deployed state. Even most orbital satellites would require a high-temperature shape memory alloy with capability in excess of 100°C because of heating that occurs when exposed to direct sunlight, and planetary explorers such as those for a mission to Venus, where the environments are quite severe, would need an especially high-temperature material. Another area where free recovery of shape memory alloys would receive considerable attention would be in self-actuating high-temperature switches and safety systems that would actuate passively at a predetermined temperature, making or breaking an electrical or other contact and thus sending some kind of signal.

7.3.2 CONSTRAINED RECOVERY

Under constrained recovery, the SMA element is prevented from changing shape as the temperature moves through the transformation regime, in some cases generating very high stresses. The prototypical example of an application employing this function would be the recovery of an SMA ring onto a rod or tube for the purpose of fastening or joining segments. Applications include elements used as compression fittings to join sections of pipes or tubes, or to apply end caps for tubes [8] and for joining electrical components [9]. In fact, these are some of the most common areas, outside the biomedical field, where conventional SMAs are currently utilized.

7.3.3 WORK PRODUCTION (ACTUATORS)

Most of the prospective applications, leading to the resurgence in interest in high-temperature shape-memory alloys, are based on their use as solid-state actuators. Because they can respond to a temperature change (stimulus) with a mechanical strain (response), it is easy to envision how shape-memory materials can be viewed as solid-state actuators. When this strain or motion moves against a force, work is performed. In fact, SMA elements have a higher energy density than pneumatic actuators or D.C. motors and are equivalent in performance to hydraulic actuators while weighing significantly less and maintaining a much more compact footprint as summarized in Figure 7.6 [10]. SMA actuators also produce a large stroke in relation to their weight compared with other systems such as thermostat metals (bimetallics), magnetic solenoids, and wax actuators [11].

Another advantage of solid-state actuators compared with hydraulic, pneumatic, and motor driven systems, is the reduction in total part count, ease of inspection, and overall reduction in inspection requirements. Shape-memory actuators also are frictionless, quiet, and result in a clean smooth motion. These advantages all equate to lower aftermarket costs for equipment inspection and maintenance. Of course the reduced weight of SMA actuators compared with nonsolid-state systems is also a major benefit in weight critical systems such as jet turbine engines and any other aerospace related applications.

In fact, the development of commercial HTSMAs would represent one of the greatest enabling forces in allowing designers to meet future aircraft efficiency, emissions, and noise goals. The application of high temperature shape-memory alloys for use in clearance control in the compressor and turbine section of an aircraft engine is one of many potential applications being considered

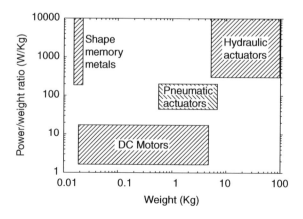

FIGURE 7.6 Comparison of various actuator systems on a power-to-weight ratio versus weight basis indicating the benefits of a solid-state shape-memory alloy system. (From Mavroidis, C., *Res. Nondestr. Eval.* 14, 1, 2002.)

for SMAs [12]. During engine startup, the blades elongate due to centrifugal force and the casing must allow for proper clearance to avoid contact, otherwise the blade tips will wear away. Then as the case warms up, thermal expansion of the materials causes radial expansion of the case and tip clearance increases, decreasing the efficiency of the engine. Clever designs with a high-temperature shape-memory alloy in the compressor and turbine casing would allow for radial contraction upon engine warming and therefore increase engine efficiency.

HTSMAs would also have an enabling impact on the development of variable area or variable geometry inlets for supersonic business jets or military aircraft and designs incorporating HTSMAs are currently being developed [13]. These inlets would be capable of transforming from a geometry optimized for low speed flight to another that optimizes performance during supersonic cruise. Hydraulic systems while effective in performing the needed shape changes are generally too bulky and heavy for such an application. Furthermore, new high-temperature materials are needed for this application since aerodynamic heating would increase the temperature of the inlet beyond the capability of conventional SMAs.

Shape-memory alloys are also being considered in designs for a variable area fan nozzle, which requires area changes of up to 20% during flight [14]. The optimum nozzle area is dependant on many different variables such as altitude and speed, so the area must be modified many times during flight, requiring innovative actuator designs and complicated controller systems. Similarly, HTSMAs are being considered for application in vectored exhaust nozzles and variable area bypass nozzles. Other applications include shape changing or articulating blades, various flow control devices, and fuel nozzles.

The advanced actuator designs proposed for the aerospace systems mentioned in the previous paragraphs mostly involve active control by self-resistance heating (Joule heating) of the SMA and feedback through stress or strain measurements for precision movement. However, this requires a material with a rather narrow transformation hysteresis. Therefore, not all the alloy systems summarized in this chapter would be appropriate for such applications.

Other applications for shape-memory alloys simply take advantage of the ability to both "sense" temperature and actuate in a passive manner. Safety switches in many electrical appliances, thermal control valves in automobiles, and heating and air-conditioning equipment, and even scald free plumbing devices all make use of conventional shape-memory alloys in a passive manner. However, many if not all of these applications could also make use of higher-temperature SMAs.

Another prominent application for shape-memory alloys is in the area of fire safety. In particular, there is a need for devices that would automatically, upon sensing an abnormally high temperature, trigger or turn off some mechanical or electrical system. Depending on the specific application and normal operating range of the system, however, the "abnormally" high temperature can in actuality be several hundred degrees centigrade and thus another reason for the development of higher temperature shape-memory alloys. In this type of application, SMAs generally compete with thermostatic bimetals. However, SMAs are capable of performing much more work over a narrower range of temperatures, since the actuation is dependent on the temperature change required to complete a first order phase transformation and not the thermal expansion differences between two materials.

There are also aerospace applications that would benefit from a passively controlled shape-memory alloy actuator. Noise reduction during aircraft takeoff has become a major focus of new engine technologies in order to meet strict noise ordinances within communities. It has been found that a 3.5 dB reduction in noise can be achieved by disrupting airflow past the core stream with a chevron nozzle [15]. The chevrons mix the cold bypass air with the hot exit exhaust gas, thereby reducing the acoustic energy at the back of the engine. While this is beneficial for noise reduction during take off, the effect of the chevron nozzle in the core stream is detrimental to fuel efficiency during cruise. Designs such as those in Figure 7.7, using high-temperature shape-memory alloys to actuate the core exhaust chevrons and conventional shape-memory alloys to actuate the fan bypass chevrons, are currently being developed to passively remove the chevrons from the engine flow

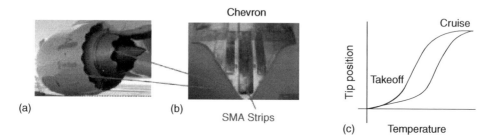

FIGURE 7.7 (a) Jet engine with chevrons on both the fan bypass and core exhaust nozzles for noise reduction. (b) Shape-memory modified chevron for passive actuation and (b) temperature dependent behavior of 2D or 3D shape-memory alloy flaps as a function of the flight characteristics that would allow passive actuation of the chevron so that it would move into the exhaust stream during take-off for noise reduction and would move out of the stream during cruise so as not to effect engine efficiency. (From Smith, R. C., *Smart Material Systems: Model Development and Control Design*, SIAM, To appear. With permission.)

path during cruise in order to maintain engine efficiency [15]. This allows for maximum benefit from the chevron nozzle during each flight cycle without added duties to the pilot.

7.3.4 SUPERELASTIC BEHAVIOR

In an operating window between the A_f and M_d temperatures, shape-memory alloys can be used in applications requiring significant levels of elastic or recoverable strain under isothermal conditions, which can often exceed 6–8%. There are a number of current applications for superelastic materials that operate near room temperature. Many of these involve biomedical applications for the insertion or deployment of devices into the body such as arterial stents. Some of the more prominent nonmedical applications include virtually indestructible eyeglass frames or cell phone antennas capable of significant deflections without permanent deformation. At high temperatures, superelastic properties could be used to advantage in high-temperature seals, which can be deformed to large strains without permanent deflection under high operating temperatures.

A unique application that currently takes advantage of the superelastic properties of high-temperature shape-memory alloys is in a pressure transducer (Figure 7.8), recently developed by Orbital Research, Inc. [16]. This pressure transducer, was developed for diesel and jet turbine engine applications and uses a NiTiPd or NiTiPt thin film strain gauge deposited on a silicon substrate in order to measure pressure pulses at high temperatures. These HTSMA-based pressure transducers have a 6–10× sensitivity improvement over standard metal thin film gages, coupled with high frequency response and high use temperatures.

In fact, this device has a frequency response on the order of 60–80 kHz. The reason for the high frequency response is that the operation of the sensor is not based on the thermal transformation of martensite to austenite (as in actuators) but is based instead on the resistance change in the material when the austenite is transformed to martensite through a stress-induced transformation. In other words, the sensor is relying on the superelastic properties of the material in order to achieve both good resolution and fast response time as a sensor. Consequently, the sensor employs a heating unit to keep the alloy within the A_f to M_d temperature range, enabling the pressure variations to induce martensite. Since the martensite phase has significantly different electrical properties than austenite, variances between the two can be calibrated back to the external pressure. These sensors can currently be used in diesel engines replacing existing in-cylinder pressure transducers and next generation transducers are being developed for aero-turbine engines to control combustion in order to increase fuel efficiency, and reduce emissions, and screech.

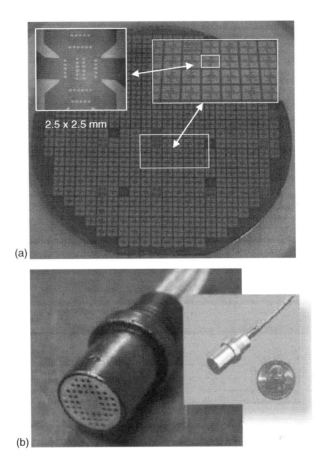

(a)

2.5 x 2.5 mm

(b)

FIGURE 7.8 (a) (Ni,Pd)Ti HTSMA thin film sputtered on a Si substrate followed by etching to form a high-temperature strain gage for use in (b) MEMS-based high-temperature pressure transducer for diesel and turbine engine applications. (Figure courtesy of Joe Snyder, Orbital Research Inc., Cleveland, OH.)

7.3.5 DAMPING AND WEAR RESISTANT APPLICATIONS

Shape-memory alloys naturally have good damping capabilities due to the energy absorption properties of the twinned martensitic structure, which is attributed to the high mobility of the martensite interfaces including twin variant interfaces and twin planes [11]. This leads to several other potential aerospace applications including the use of SMAs in (a) acoustic liners to dampen acoustic energy from the aircraft engine, (b) self-damping components or connectors in mechanical systems, such as fuel line clamps, to help reduce fatigue damage, and (c) flutter control or damping of fan blades.

The ability of SMAs to absorb and dissipate energy also could be useful in protecting systems during impact events [17]. Consequently, conventional SMAs are currently being investigated for use in fan containment systems to capture debris from a catastrophic engine failure. Depending on the type of engine and operating conditions, higher-temperature shape-memory alloys may be needed for such applications.

Finally, while applications involving wear resistance are not usually considered when discussing shape-memory alloys, they do exhibit excellent wear behavior due to the unique properties of the martensitic phase transformation. Strain energy from sliding wear is dissipated into the material, causing a localized temperature increase and phase transition [18]. Therefore, shape-memory alloys would also appear to have potential as a multifunctional coating, possibly for fan or compressor blades, that would provide not only wear resistance but also be used for their damping or shape change abilities.

7.4   HIGH-TEMPERATURE TERNARY NiTi–X ALLOYS

7.4.1   BACKGROUND

7.4.1.1   Historical Development of NiTi Alloys

Like most ordered intermetallic systems, NiTi was originally investigated as a potential light-weight, high-temperature structural material for advanced missile and aircraft applications [19]. However, Buehler et al. also noted some curious behaviors in binary NiTi, including high vibration damping capability [19] and an unusual ability to recover its shape [20]. In fact, Buehler et al. [20] were the first to demonstrate the free recovery of a room-temperature coiled NiTi spring back to a straight wire upon heating and the opposite case, the coiling of a shape-memory wire spring that was straightened at room temperature after heating. However, it took a few more years before anyone began to realize that the root cause of these unusual behaviors was related to a martensitic transformation [21] and it was several more years before Otsuka et al. [22] were able to finally identify the correct structure of the low temperature martensite phase in binary NiTi alloys. Nevertheless, given these startling shape changes, subsequent emphasis on this material quickly focused on the shape-memory and other unusual properties displayed by NiTi alloys around room temperature [23,24].

NiTi has never been utilized as a high-temperature structural material given its relatively poor strength at elevated temperatures [25,26] (Figure 7.9) and marginal oxidation behavior even at intermediate temperatures [27,28]. In contrast, use as a functional material has progressed steadily, due to excellent shape-memory properties including large recoverable strains and stresses, high corrosion resistance, and biocompatibility. However, issues such as oxidation resistance and

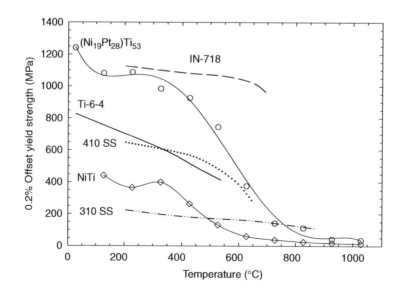

FIGURE 7.9 Strength of NiTi and a (Ni,Ti)Pt HTSMA alloy as a function of temperature compared with conventional structural materials. (From NASA Glenn Research Center, High-Temperature Shape-Memory Alloy Group, unpublished research, 2003.)

elevated temperature strength will return, once alloys are developed for functional use at elevated temperatures. Yet with a few exceptions, the majority of the work performed on high temperature shape-memory alloy systems, as reviewed in the following sections, has focused on processing, transformation temperatures, and microstructural stability. Also, the current limitations of conventional NiTi alloys, such as a thermal hysteresis and fatigue degradation of the recovery properties, which have generally been addressed through ternary alloying additions and advanced thermomechanical processing cycles, will also have to be addressed anew in these high-temperature systems.

Transformation temperatures for binary NiTi alloys are essentially managed through control of stoichiometry [29] and thermomechanical processing so that M_s temperatures can be adjusted from cryogenic to about 100°C [30]. As this essentially straddles room temperature and given their superior shape-memory effect, superelasticity, and compatibility with the human body, NiTi has become the most common commercial shape-memory alloy system. Therefore, it is not surprising that the major direction research in high-temperature systems has taken, starting primarily in the early 1990s, has been in identifying and characterizing ternary NiTi-based alloys with increased transformation temperatures.

7.4.1.2　General Effect of Alloying Additions on NiTi

The transformation temperatures for NiTi are very strongly dependent on composition, especially on the Ni-rich (Ti-lean) side of stoichiometry. Ni contents even slightly greater than stoichiometry lead to a rapid decrease in transformation temperature [29,31], while the transformation temperatures for Ti-rich alloys are much less sensitive to composition primarily as a result of the precipitation of Ti_2Ni particles, which leave the composition of the matrix relatively unchanged [30].

For the most part, ternary alloying additions, at levels of less than about 10 at.% either decrease or have little effect on the M_s temperature of NiTi. For example, Fe, or Co, substituted for Ni; or Al, Mn, V, or Cr substituted for Ti will severely depress the transformation temperatures such that only 1–6 at.% ternary addition is necessary to drive the M_s temperature below $-100°C$ [32,33]. Slight Fe additions have also been found to lower the transformation temperature of NiTiNb alloys without significantly changing the transformation characteristics of the alloy [34]. Other alloying additions such as Cu in place of Ni or Mn when substituted evenly for Ti and Ni [35] have relatively little effect on the transformation temperature of NiTi even at large concentrations. Then there is a final group of alloying additions, namely Hf, Zr, Au, Pd, and Pt, which have been found to increase the transformation temperature of NiTi proportional to their concentration in the alloy as summarized in Figure 7.10, but only in amounts greater than about 10 at.%. Details of these high-temperature shape-memory alloy systems are presented in the balance of this chapter.

Alloying additions also have a dramatic effect on the thermal hysteresis between the forward and reverse transformation in NiTi alloys, by controlling the type of martensite phase that occurs. For example, binary NiTi alloys have a thermal hysteresis on the order of 20–40°C with the low temperature martensite phase having a monoclinic structure [34]. When Cu additions, on the order of 10%, are added to NiTi (at the expense of Ni) the hysteresis is lowered to 10–15°C and the martensite phase has an orthorhombic structure. Through combinations of work hardening and low-temperature heat treatment it is possible to get a rhombohedral martensite phase (commonly referred to in the literature as R-phase) in NiTi, NTiCu, and NiTiFe alloys, which has a hysteresis of only few degrees. On the other hand, Nb additions will increase the hysteresis to nearly 100°C [36].

Hysteresis in turn, has a large impact on thermomechanical fatigue life with fatigue life increasing dramatically as the hysteresis is lowered. Conversely, the recoverable strain level in superelastic NiTi alloys, and the temperature range over which superelastic behavior is observed

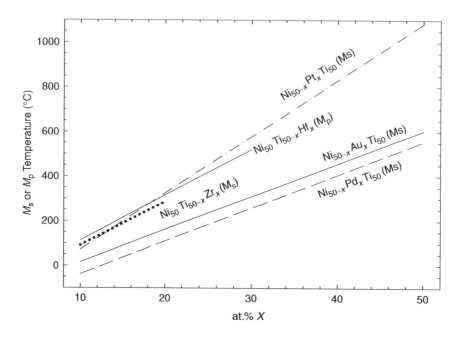

FIGURE 7.10 Effect of ternary alloying additions on the M_s (or M_p) temperature for NiTi-based high-temperature shape-memory alloy systems.

increases with increasing hysteresis. Therefore, in the case of NiTiNb alloys, the temperature window over which superelasticity can occur is over 160°C [36]. Hysteresis is also an important factor when trying to actively control shape-memory-based actuator systems, with a smaller hysteresis being greatly preferred. Therefore, the composition of NiTi alloys can be optimized depending on whether the application requires shape-memory or superelastic behavior by controlling hysteresis.

7.4.1.3 Important Phase Structures in NiTi Alloys

Early research on the phase structure of binary NiTi alloys around the stoichiometric composition, and for the phase diagram in general, has been fraught with numerous inconsistencies and mistakes. One can refer back to earlier articles for historical development in this area [21,29,37]. Unfortunately, some of the uncertainties concerning the phase relationships in Ni-Ti persist. To this end, Otsuka and Ren [38] have provided an excellent review of the NiTi system that includes an updated phase diagram and description of the resulting phases in near-stoichiometric NiTi alloys, including the Ni–Ti based martensites. Given the important foundation that the binary NiTi phase diagram serves in our understanding of the microstructural behavior of ternary NiTi-based alloys, we have included the phase diagram in Figure 7.11 (with permission of the publisher).

From this work it is clear that there is little solubility for Ti (or probably other Ti group elements) in excess of 50 at.%, resulting in the formation of Ti_2Ni, and that it is not possible to heat treat Ti-rich alloys to control microstructure, with the exception of amorphous NiTi thin films [39]. There is more room for microstructural control on the Ni-rich side of stoichiometry. On this side of stoichiometry, precipitation proceeds through a series of phases: $Ti_3Ni_4 \rightarrow Ti_2Ni_3$ until finally the stable $TiNi_3$ phase is reached with increasing aging time. Unfortunately, while there is room for microstructural control in Ni-rich alloys, excess Ni sharply lowers the transformation temperatures of binary alloys.

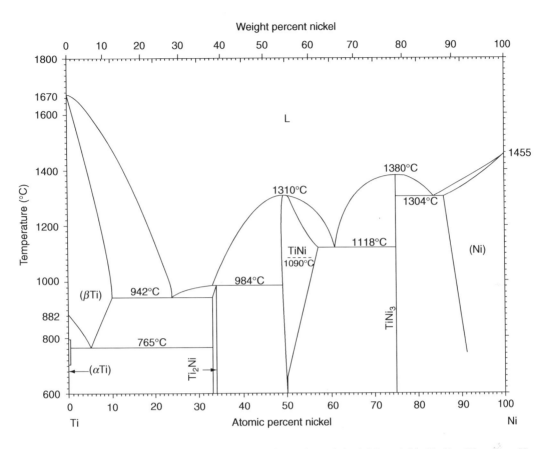

FIGURE 7.11 Ni–Ti phase diagram. (Modified from the original Massalski, T. B., Okamoto, H., Subramanian, P. R., and Kacprzak, L. Eds., In *Binary Alloy Phase Diagrams*, 2nd ed., Vol. 3, ASM International, Metals Park, OH, p. 2875, 1990 by Otsuka, K. and Ren, X., *Intermetallics*, 7, 511, 1999.) Note that the original eutectoid decomposition at 630°C was deleted and a dotted line is added at 1090°C representing an order–disorder transition. (Reprinted from Otsuka, K. and Ren, X., *Intermetallics*, 7, 511, 1999. With permission.)

The parent B2-NiTi, or austenite phase as it is often referred to, can undergo a number of different martensitic transformations depending on minor alloying additions and thermomechanical treatments [40,41]. These include

- B2 ↔ B19
- B2 ↔ B19′
- B2 ↔ B19 ↔ B19′
- B2 → R → B19′

All of these transformations are reversible, except the B2 → R transformation, with the B19′ phase reverting directly to B2 on heating. The B19 phase is an orthorhombic structure formed by $\{110\}\langle 1\bar{1}0\rangle$ shear of the B2 parent structure (Figure 7.12). The B19′ is a monoclinic structure with a space group of $P2_1/m$ [37], formed by an additional $\{001\}\langle 110\rangle_{B2}$ shear of the B19 structure (Figure 7.12c). The R-phase is rhombohedral in structure and occurs by elongating the cubic B2 cell along $\langle 111\rangle$. The space group is $P3$, which has no center of symmetry, and the structure of the R-phase is described in detail in [42]. In the case of the high-temperature shape-memory alloy systems discussed below, the most commonly observed martensitic phases are the

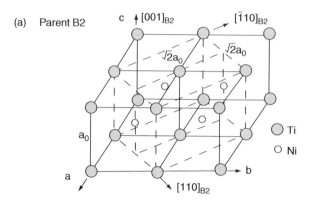

(a)   Parent B2

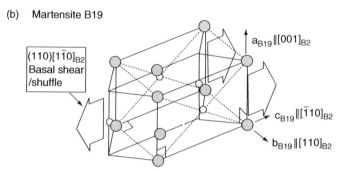

(b)   Martensite B19

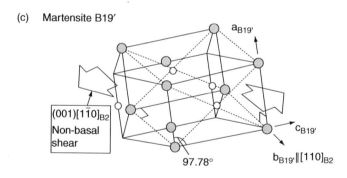

(c)   Martensite B19′

FIGURE 7.12 Structural relationship between the cubic B2-parent phase and B19 and B19′ martensites. (a) four parent B2 cells with a face-centered tetragonal cell outlined by dashed lines. (b) orthorhombic B19 martensite formed by shear/shuffle of the basal plane $(110)_{B2}$ along $[1\bar{1}0]_{B2}$ direction, close packed (basal) planes are indicated by thin dotted lines. (c) monoclinic B19′ martensite formed by a non basal shear (001) $[1\bar{1}0]_{B2}$ to the B19 structure to produce a monoclinic β angle, close packed (basal) planes are indicated by thin dotted lines. (Reprinted from Otsuka, K. and Ren, X., *Intermetallics*, 7, 511, 1999. With permission.)

B19 and B19′. The structural relationship between these martensites and the parent B2-NiTi structure is shown in Figure 7.12 [38].

7.4.2   PROCESSING OF NiTi-BASED ALLOYS

Because of the narrow compositional window for the processing of NiTi-based shape-memory alloys and the high price of certain ternary additions (Pd, Pt, Au), thin film processing techniques such as magnetron sputtering and other physical vapor deposition techniques have been strongly championed and recently reviewed by Grummon [4]. Therefore, thin film SMAs will not be covered in any detail in this chapter.

For the development of other than thin film components, these techniques are impractical and in most cases stoichiometry control even for binary alloys is still a considerable challenge [43], let alone the problem of maintaining compositional control for three or more constituents. The other major challenge is that most thin films will be used in some sort of micromachine, and therefore, the substrate will probably be Si. The problem that has to be addressed is how to avoid the constraint or large stresses that build up due to the large difference in thermal expansion coefficient between the NiTi thin film and the Si substrate [44]. Yet, significant progress has been made in all these areas by Orbital Research, Inc. [16].

Conventional metallurgical processing, i.e., casting followed by thermomechanical processing (hot rolling, forging, or extrusion), and appropriate heat treatment is capable of producing NiTi-based shape-memory alloys with good shape-memory properties and minimal amount of undesirable second phase. The major concerns with conventional processing of NiTi-based alloys are those generally encountered with any Ti-based alloy, namely reaction with crucible materials during melting and interstitial pickup during melting and heat treatment. Fortunately, solutions to these problems are quite tractable and routine [45].

The majority of the work on ternary alloys discussed below has been performed on material that has been initially prepared by nonconsumable arc melting of high purity constituents using copper crucibles. In this case there is minimal reaction with the mold materials. The arc-melted castings are then heat treated and/or thermomechanically processed by typical metallurgical processes. In a few studies induction melting is used, resulting in the additional complication that reaction with mold materials is possible. However, induction melting is also quite common in the production of commercial NiTi products using graphite crucibles [46] and for the most part does not seem to introduce a serious problem.

7.4.3 (Ni,Pd)Ti Alloys

Boriskina and Kenina [47] originally determined that PdTi and NiTi form a continuous solid solution with a high-temperature B2 phase and that transformation of the B2 phase to a martensitic structure occurred at temperatures that constantly decreased from the binary PdTi to a minimum temperature for ternary compositions containing 5–10 at.% Pd (40–45 at.% Ni). Therefore, Pd additions have since been used to increase the transformation temperatures of NiTi alloys through substitution for Ni. Unfortunately, as the data by Boriskina and Kenina [47] and others have since borne out, it takes a significant amount of ternary addition to generate an increase in transformation temperature. Initial additions of Pd to NiTi actually shift the M_s to lower temperatures with a minimum of $-26°C$ found at 10 at.% Pd [48], and it takes nearly 20 at.% Pd just to get an M_s near 100°C. Between 10 and 50 at.% Pd the M_s temperature is essentially proportional to Pd content finally reaching a value somewhere between 510 and 563°C for 50 at.% Pd (or TiPd) [47–51]. This data is summarized in Figure 7.13.

It is interesting to note that alloys with 10% or less Pd transformed to a monoclinic martensite on cooling [47]. Lindquist and Wayman [48] observed a similar result but also observed an intermediate structure during the transformation so that the following two stage process: B2→ R→B19' was observed on cooling, while a one stage B19'→B2 transformation was observed on heating for alloys with 10 at.% or less Pd. In contrast, in the region where the transformation temperature is found to increase with further additions of Pd to NiTi, the structure of the martensite phase was found to be B19 (orthorhombic) with a one stage B2↔B19 transformation observed on heating and cooling [47,48].

Shimizu et al. [52] also examined the effect of stoichiometry (Ni+Pd:Ti≠50:50) on the transformation temperatures of a series of $Ti_{50-x}Pd_{30}Ni_{20+x}$ alloys (where $x = -0.6$ to 1.5%). These results are presented in Figure 7.14, and show an extremely steep drop in transformation

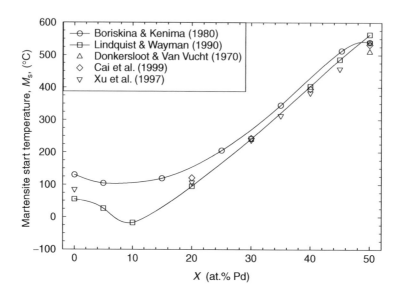

FIGURE 7.13 Martensite start temperature, M_s, as a function of Pd concentration for $Ni_{50-x}Pd_xTi_{50}$ alloys.

temperature for Ti-deficient alloys but only a moderate change in transformation temperature on the Ti-rich side where Ti_2Ni type precipitates were observed. This trend is consistent with binary NiTi alloys where there is a very steep drop in transformation temperatures for Ti-deficient compositions and where the transformation temperatures are much less sensitive to composition for Ti-rich alloys [30,31].

The unconstrained shape recovery in (Ni,Pd)Ti alloys can be quite good [48], approximately 6% for an alloy containing 30 at.% Pd, and at least 3.5% (the tensile test samples fractured at that point limiting measurements) for samples prestrained at room temperature on either side of 30 at.% Pd.

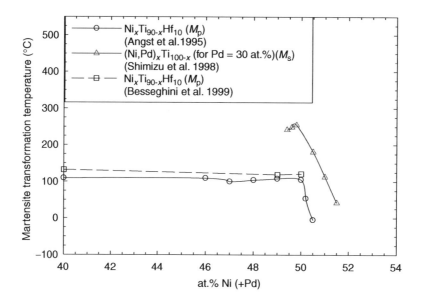

FIGURE 7.14 Martensite transformation temperature as a function of Ni concentration for high-temperature shape-memory alloys containing Pd or Hf.

Khachin et al. [53] also reported the complete recovery of 4% strain introduced by a 200 MPa stress in torsion for a $Ni_{13}Pd_{37}Ti_{50}$ alloy. However, for alloys containing more than 40 at.% Pd, only about 0.5% strain could be recovered in alloys prestrained in tension [48]. Poor shape-memory properties were also reported for stoichiometric TiPd alloys, which was attributed to a low critical stress for slip in the alloy, so that significant slip occurred in addition to twinning in those materials [50].

Consequently, a high transformation temperature is just the minimum requirement for developing HTSMAs since it does not guarantee good shape-memory properties. In the case of TiPd [50], deformation at room temperature is accommodated by both slip and martensite reorientation so that the recovered strain in the material is rather low even for samples deformed at room temperature. However, as the deformation temperature is increased (although still kept below the A_s) the amount of recoverable strain in TiPd quickly approaches zero. This problem, which stems from a low critical stress for slip, is only exacerbated at high temperatures in some shape-memory alloy systems. With increasing temperature, the critical stress for slip generally decreases allowing dislocation processes to become more dominant. This in turn reduces the amount of strain accommodated only by martensite reorientation, severely degrading the shape-memory behavior of the alloy.

In the case of NiTiPd (and probably for most NiTi-based HTSMA systems), it is necessary to increase the critical stress for slip in the martensite phase in order to improve shape-memory characteristics [50]. This can be accomplished through a number of conventional metallurgical processes including: (1) solid solution alloying through the addition of quaternary elements, (2) precipitate strengthening, and (3) thermomechanical processing. Strengthening of these materials against slip in an effort to improve high-temperature shape-memory behavior has thus been a major area of emphasis for TiPd and NiTiPd alloys.

To date, attempts at improving the shape-memory characteristics of HTSMAs through solid solution strengthening have only been moderately successful. Arguments can be made that Ni additions improve the shape-memory characteristics of binary TiPd alloys [50]. The effect of boron additions on the shape-memory characteristics of NiTiPd alloys has been investigated by a number of researchers [54–56]. Yang and Mikkola [54] examined the effect of boron on shape-memory characteristics of a $Ni_{22.3}Pd_{27}Ti_{50.7}$ alloy by determining the amount of recovered deformation in samples originally deformed in compression at room temperature. However, no real effect of the boron was observed. They did, however, attribute an increase in the room temperature tensile ductility of their material to the grain refining effect of the boron addition.

Suzuki et al. [55] also found that 0.2 at.% B additions to a $Ni_{20}Pd_{30}Ti_{50}$ alloy had relatively no effect on the transformation temperature of the alloy or its shape-memory characteristics. This was attributed to the fact that the boron precipitated out as a distribution of course (greater than 1 μm in size) TiB_2 precipitates with apparently little effect on the slip behavior of the alloy. However, they did find that the boron addition nearly doubled the ultimate tensile strength and ductility of the alloy at 170°C. Consistent with the findings of Yang and Mikkola [54], the grain size of the boron-doped material was much finer than that of the undoped material.

Shimizu et al. [52] used a different approach for improving the shape-memory properties of NiTiPd alloys. They used slight deviations from stoichiometry in order to generate a homogeneous distribution of fine Ti_2Ni precipitates as a method for improving the shape-memory characteristics of a $Ti_{50.6}Pd_{30}Ni_{19.4}$ alloy. This was demonstrated by measuring the recovery rate of samples deformed at various temperatures below the M_f. The recovery rate is the percent of the total strain imparted to the alloy below the M_f that could be recovered when cycled through the transformation temperature to a temperature of about $A_f + 100°C$ and back to the original deformation temperature. For the $Ti_{50.6}Pd_{30}Ni_{19.4}$ alloy the recovery rate was about 90% for samples originally deformed to a total strain of 6% at 100°C. This was about 10% higher than the recovery rate for a comparable $Ti_{50}Pd_{30}Ni_{20}$ alloy as seen in Figure 7.15. The improvement in shape-memory properties was attributed to the hardening effect due to a homogeneous distribution of fine Ti_2Ni-type precipitates in the Ti-rich alloy.

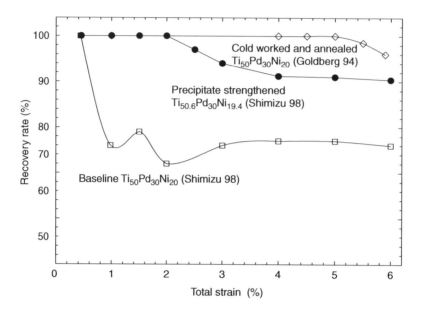

FIGURE 7.15 Effect of various strengthening strategies on the recovery rate of a (Ni,Pd)Ti alloy.

The effect of thermomechanical processing on the shape-memory characteristics of Ni–Ti–Pd alloys has been explored by Goldberg et al. [57]. They used cold rolling followed by a recovery anneal at 400°C (which is below the recrystallization temperature for the alloy [51]), to enhance the shape-memory behavior of a $Ni_{20}Pd_{30}Ti_{50}$ alloy and were able to achieve total recoverable strains in tensile samples in excess of 5% (Figure 7.15) with 3–3.5% of that strain attributed to shape-memory effect. Samples annealed at 900°C (fully recrystallized) did not display complete recovery of strains greater than 2%. The difference in behavior was attributed to the high dislocation density observed in the B19-martensite phase in 400°C annealed samples resulting in strengthening of the martensite phase against slip deformation. The cold-worked material also displayed superelastic behavior when tested at 262°C, the first time superelastic behavior has been reported in a NiTiPd alloy.

The benefits of thermomechanical processing, have also been investigated by Cai et al. [58] by thermal cycling samples under load. In a $Ti_{50.6}Pd_{30}Ni_{19.4}$ alloy, thermal cycling under stress was found to increase the M_s temperature slightly and had a fairly significant effect on the transformation strain with increasing number of cycles for a given stress. This was attributed to the introduction of dislocations during the cycling. The increase in M_s or transformation strain also increased as the level of applied stress increased. However, the increase in M_s or transformation strain occurred essentially over the first 20–40 thermal cycles and then saturated after that. In this way thermal cycling could be used to improve and stabilize the transformation characteristics of an alloy before they are put into use.

The effects of thermal cycling on M_s temperature are quite unusual and deserve additional comment at this point. Under no-load (stress-free) conditions, thermal cycling generally results in a *decrease* in the M_s temperature of NiTi [59], and NiTiHf and NiTiZr alloys, as will be discussed in following sections. An explanation for this behavior is that under stress-free conditions, the stress field formed by the dislocations generated during cycling has the effect of suppressing the martensitic transformation resulting in a decrease in M_s [59]. But under a constant stress condition during thermal cycling through the transformation temperature, the internal stress field caused by the generated dislocations assists the nucleation of preferred variants of the stress-induced martensite during cooling and thus increases the M_s temperature [60], as in the case of the NiTiPd alloy [58].

Another phenomenon that can influence transformation temperatures in shape-memory alloys is the martensite aging effect, which is a concern especially for high-temperature shape-memory alloys. The martensite aging effect refers to changes that occur in the material during aging of the martensite phase. These include an increase in the reverse transformation temperatures (A_s and A_f), also referred to as martensite stabilization, and the occurrence of a rubber-like pseudoelasticity of the martensite phase. The most peculiar thing about the rubber-like behavior of the martensite is that martensite deformation involves only twinning and there is no phase transformation involved, unlike that of superelasticity of the austenite phase due to a stress-induced martensitic transformation. Therefore, there must be some unique mechanism causing a rubber-like, pseudoelastic restoring force, independent of the typical transformation process. According to Otsuka and Ren [38], this restoring force is due to a symmetry-conforming short-range order process (SC-SRO), where the short-range order of point defects tries to conform to the symmetry of the crystal. Details of the theory are presented in [38].

From a mechanistic standpoint, the necessary conditions for the occurrence of these phenomena (or martensite aging effects) are (a) the existence of point defects and (b) the possibility of diffusion in the martensite phase [61]. The existence of point defects is a common feature of ordered alloys, with the possibility for numerous substitutional and vacancy-type defects. Diffusion in the martensite phase depends on the reduced martensitic transformation temperature, M_s/T_m, where T_m is the melting temperature of the alloy on an absolute scale. The higher the reduced temperature the faster diffusion in martensite becomes. However, observation of martensite aging effects generally occurs only in a critical range of reduced transformation temperatures of about 0.23–0.5 [62]. At higher values, aging is so fast that it completes almost immediately after the martensitic transformation and at values much less than this range, aging is too slow to be observed.

The transformation temperatures for (Ni,Pd)Ti alloys can range from ambient to about 540°C resulting in M_s/T_m ratios of 0.21–0.49 for alloys containing 20–50 at.% Pd. Consequently, Cai et al. [63] examined martensite aging effects in (Ni,Pd)Ti alloys as a function of composition. They found that significant martensite aging effects were observed for alloys aged just below the A_s temperature with 40 and 50 at.% Pd but in alloys containing 30 at.% Pd or less no effects were observed.

Another advantage of the Pd-doped alloys is the narrow hysteresis in these materials, on the order of 15–30°C [48], which make them very amenable to active control, through various temperature or strain feedback mechanisms. Furthermore, thermal cycling under stress can increase the M_s temperature and possibly decrease the thermal hysteresis further [58,60]. In most applications these types of changes would be advantageous if they were consistent and well characterized. Another benefit of the NiTiPd alloys is their very stable microstructure. The transformation temperatures for $Ni_{49-x}Pd_xTi_{51}$ alloys with Pd contents from 15 to 40 at.% were essentially unaffected by thermal cycling to $A_f+50°C$ (Figure 7.16) or long term aging at 500°C [64].

Finally, while Pd is obviously an expensive metal, it is considerably less expensive than gold or platinum, and the material costs in many applications would be offset by the system level payoffs that would include such advantages as weight savings and benefits in performance, such as increased maneuverability, larger optimum flight envelope, increased fuel efficiency, or noise and emissions reductions. Furthermore, components made from NiTiPd HTSMAs can be recycled once the part is no longer required or needs to be replaced for some reason.

7.4.4 (Ni,Au)Ti Alloys

Compared with the NiTiPd system, almost no work has been performed on (Ni,Au)Ti alloys, even though the transformation temperatures for the NiTiAu alloys are slightly higher than the NiTiPd alloys for a given concentration of precious metal (Figure 7.10). The effect of up to 2 at.% Au additions, substituted for Ni, on transformation temperatures was first studied by Eckelmeyer [32]. At the other extreme Donkersloot and Van Vucht [49] determined the transformation temperatures of binary AuTi alloys and found that similar to the other precious metal-titanium systems (PtTi and

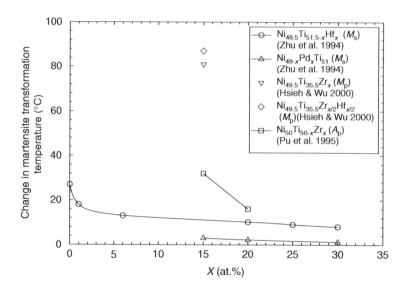

FIGURE 7.16 Change in martensite transformation temperature for various compositions of NiTiX alloys after 100 thermal cycles between room temperature and $A_f + \sim 50°C$.

PdTi), binary AuTi alloys consist of a high-temperature B2 phase and lower temperature B19-type orthorhombic structure.

Wu and Wayman [65,66] studied compositions within these extremes, primarily investigating premartensitic effects and the structural qualities of the transformation products in $Ti_{50}Ni_{50-x}Au_x$ alloys with $x = 5, 10, 40,$ and 50 at.% Au. At levels of 5 and 10 at.% Au the martensite was of the B19′ monoclinic version. In agreement with Donkersloot and Van Vucht [49], alloys containing 40 and 50% Au transformed to a B19 orthorhombic martensite. The latter two alloys were shown to qualitatively exhibit a one-way shape-memory effect, by bending and recovering a bar of material, but they were not capable of two-way shape-memory behavior [65]. The inability to exhibit a two-way shape-memory effect was probably due to the accumulated bias stresses being relaxed (annealed out) from the material when the temperature was raised above the M_s (440°C) during the training process. Wu and Wayman determined phase transformation temperatures using an electrical resistivity measurement technique [65,66]. The M_s temperatures along with data from the other investigators appears in Figure 7.17.

Even though the high Au containing alloys did not display two-way shape-memory behavior, the strong one-way effect demonstrated has significant implications for this material as a HTSMA. Unfortunately, not much else is known about the shape-memory characteristics of these materials. All the studies performed to date were essentially limited to the investigation of transformation temperatures or the structure of the martensitic phases. No research has been performed on the basic mechanical behavior or any detailed characterization of the functional properties of these ternary alloys. Thus, other than a reasonable characterization of the transformation temperature and transformation products, relatively little is known about ternary NiTiAu alloys.

7.4.5 (Ni,Pt)Ti Alloys

(Ni,Pt)Ti alloys, like the (Ni,Au)Ti alloys discussed in the previous section, have received very little attention in the past, though they are starting to generate interest, since for a given at.% ternary addition the Pt containing alloys exhibit significantly higher transformation temperatures compared with any of the other NiTi-based systems (Figure 7.10). Consequently, they have the greatest potential temperature capability.

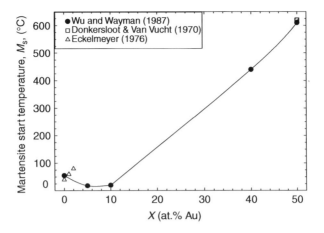

FIGURE 7.17 Martensite start temperature, M_s, as a function of Au concentration for $Ni_{50-x}Au_xTi_{50}$ alloys.

As in the other NiTi-based alloys, Pt added at the expense of Ni increases the martensite transformation temperature only after a threshold value of approximately 10 at.% Pt is reached [48]. Between 0 and 10 at.% Pt, the M_s decreases slightly, though the magnitude varies depending on the study. Using resistivity measurements on rolled sheet, Lindquist and Wayman [48] found a minimum M_s at 10 at.% Pt of $-10°C$, while others using a DSC technique on cast and homogenized material have observed a minimum M_s at 5 at.% Pt of 45°C [26]. However, at higher levels of Pt the results are quite similar with M_s increasing linearly to approximately 1040°C for 50 at.% Pt additions (TiPt). These and other results [26,48,49,67] are compared in Figure 7.18.

At levels of 10% Pt or less, the martensite phase is the monoclinic B19$'$ phase and the transformation path is either a one step $B2 \rightarrow B19'$ or a two step $B2 \rightarrow R \rightarrow B19'$ on cooling depending on how the material is processed [26]. Cast and homogenized $Ni_{45}Pt_5Ti_{50}$ was found to follow the two step process while rolled sheet was observed to follow a one stage transformation sequence.

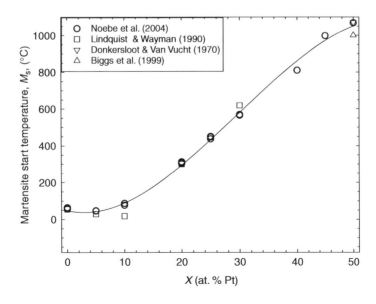

FIGURE 7.18 Martensite start temperature, M_s, as a function of Pt concentration for $Ni_{50-x}Pt_xTi_{50}$ alloys.

In either case a one step $B19' \rightarrow B2$ transformation occurs on heating. While the low Pt alloys do not have suitable transformation temperatures for use at high temperatures, there is growing interest in alloys with 5–12% Pt in the biomedical device field for stents and related hardware since the alloys are radiopaque, and will therefore, show up more readily in x-rays. For alloys with higher levels of Pt, at least 16% or greater, the martensite formed is the B19 (orthorhombic) type and the transformation path is essentially a single stage $B2 \leftrightarrow B19$ on heating or cooling [26,68].

Lindquist and Wayman [48] recorded a rather narrow hysteresis for all the Pt alloys studied, similar to the Pd alloys, except for the 30 at.% alloy, which had a hysteresis of over 80°C. Data by NASA [26] revealed a similar trend with a rather large hysteresis (on the order of 40–80°C) for 30 at.% Pt but with considerable scatter. A hysteresis of generally less than 20°C was observed for alloys with 25 at.% or less Pt. In addition, Meisner and Sivokha [68] observed very little hysteresis (less than 5°C) for a $Ni_{34}Pt_{16}Ti_{50}$ alloy under applied stresses up to 600 MPa. The nature of the anomalously large hysteresis at 30 at.% Pt is completely unknown at this time but could operationally represent an upper limit in Pt level for use in actively controlled devices. Preliminary data also indicates that these alloys, like the Pd alloys, are very stable against thermal cycling [26]. A $Ni_{25}Pt_{25}Ti_{50}$ alloy exhibited essentially no change in M_s temperature when cycled 150 times from above the A_f to room temperature.

Lindquist and Wayman attempted to measure the recoverable strains in NiTiPt alloys but were limited by the extremely low room temperature ductility of these alloys. However, as seen in Figure 7.19, $Ni_{30}Pt_{20}Ti_{50}$ alloys are capable of significant shape recovery in bending. Other than these minor observations, the shape memory or work characteristics of these alloys are essentially unresolved.

Yet compared with some other ternary-NiTi-based systems much more basic mechanical testing has been performed. Hosoda et al. [69] have measured the room temperature tensile properties of (Ni,Pt)Ti alloys and found that ductility generally decreased as the amount of Pt in the alloys increased, but that yield stress followed a much more complicated dependence on composition. Ni actually appears to harden PtTi alloys considerably, while Pt additions to NiTi soften the monoclinic martensite phase. This behavior is summarized in Figure 7.20. The tensile ductility values of Hosoda et al. [69] are very consistent with those of Lindquist and Wayman [48], who observed about 3.5% and 0.5% tensile ductility for 20 and 30 at.% Pt alloys, respectively.

The yield strength of a $Ni_{19}Pt_{28}Ti_{53}$ alloy as a function of temperature is included in Figure 7.9, and as can be seen in the figure the strength of this ternary alloy is far superior to binary NiTi and most other common structural materials at intermediate temperatures. Consequently, the Pt addition not only serves as a potent aid in increasing the transformation temperatures of NiTi but is a potent

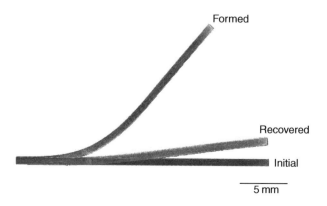

FIGURE 7.19 Qualitative demonstration of the one-way shape-memory effect in a $Ni_{30}Pt_{20}Ti_{50}$ alloy, prestrained to a 38° angle at room temperature and recovered to an 8° angle by heating to 350°C.

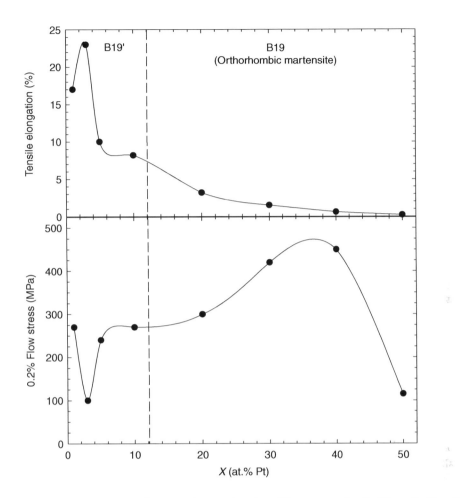

FIGURE 7.20 Room temperature tensile properties as a function of Pt concentration for $Ni_{50-x}Pt_xTi_{50}$ alloys. (From Hosoda, M., Tsuji, M., Takahashi, Y., Inamura, T. Wakashima, K., Yamabe-Mitaria, Y., Miyazaki, S., and Inoue, K., *Mater. Sci. Forum*, 426–432, 2333, 2003.)

strengthening agent. However, since the alloy was technically Ti-rich and contained $Ti_2(Pt,Ni)$ type precipitates, a portion of the strength may be attributed to precipitate strengthening. The compressive creep behavior of a Ti-rich NiTiPt alloy at 827°C is also shown in Figure 7.21. Interestingly, the creep strength for alloys containing 18.5 and 28 at.% Pt was quite similar and additional alloying with 0.1 at.% B had no further effect on the strength of either alloy. Furthermore, both alloys were considerably more creep resistant than binary NiTi, which at equivalent stress levels had a creep rate that was over four orders of magnitude greater than the Pt-containing alloys.

As with the other precious metal additions to NiTi, the high initial material costs will have to be offset by the increased benefits in using HTSMAs at the system level, seriously limiting bulk material applications to such high-performance and demanding applications as exist in the aerospace industry. While the data in most cases is still cursory, the NiTiPt alloys with 25 at.% Pt or less show promise for such demanding applications, since they have very high transformation temperatures, low hysteresis, apparently good thermal stability, and reasonable high-temperature strength. The major questions that still need to be resolved are whether these alloys have acceptable shape-memory properties and work behavior, and if they do, whether tensile ductility will be a limiting factor in the utilization of these alloys.

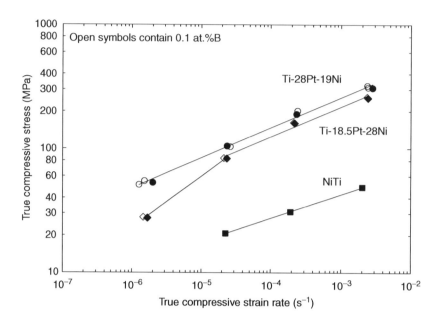

FIGURE 7.21 Compressive creep strength of (Ni,Pt)Ti and NiTi alloys at 827°C.

However, materials like the Pd-, Au-, and Pt-containing alloys will probably gain wider acceptance over a much broader range of applications as part of MEMS type devices as thin film actuators or sensors. For this reason, other low cost alternative high-temperature systems have also received considerable attention such as the Ni(Ti,Hf) and Ni(Ti,Zr) alloys discussed below.

7.4.6 Ni(Ti,Hf) ALLOYS

NiTiHf alloys are a possible alternative to the much more costly precious metal containing systems discussed previously [70]. They have higher transformation temperatures than the Pd and Au containing alloys for an equivalent amount of ternary addition (Figure 7.10), though the maximum level of Hf that can be added is limited, since there is no continuous solid solution between NiTi and NiHf. While these alloys are much more difficult to process than binary NiTi alloys [71], they have at least sufficient ductility to undergo most thermomechanical treatments [72,73]. Therefore, they have been studied rather significantly.

Transformation temperatures have been determined [73,74] as a function of Hf content (Figure 7.22) and as a function of stoichiometry (or Ni content) for 10 at.% Hf containing alloys (Figure 7.14). As with the precious metal containing systems, transformation temperatures do not begin to significantly exceed that of binary NiTi until Hf concentrations greater than about 10 at.% and from there the transformation temperature increases linearly with Hf concentration until an M_p temperature of approximately 525°C is reached for alloys containing 30 at.% Hf. Sputtered Ni(Ti,Hf) films generally have slightly lower transformation temperatures than reported for similar bulk compositions, especially for Hf compositions above 20 at.% [75].

Surprisingly, as the Ni content is varied from 40 to 50 at.% for alloys containing 10 at.% Hf (balance Ti), the transformation temperatures remain essentially independent of Ni content as shown in Figure 7.14. This was the case even though alloys with a composition of $Ni_{40}Ti_{50}Hf_{10}$ contained approximately 50 volume percent second phase, while the $Ni_{50}Ti_{40}Hf_{10}$ alloy was essentially single phase [74]. Also, consistent with binary NiTi and the other ternary alloys is a steep drop in transformation temperature for these same 10 at.% Hf alloys when the Ni content even slightly exceeds 50 at.%, falling to below zero centigrade for a $Ni_{50.2}Ti_{49.8}Hf_{10}$ composition.

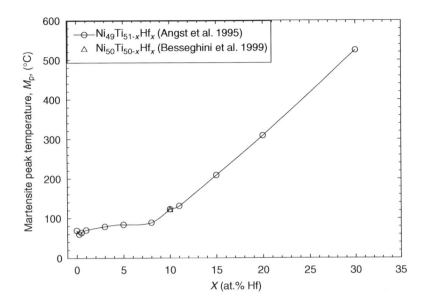

FIGURE 7.22 Martensite peak temperature, M_p, as a function of Hf concentration for Ni(Ti,Hf) alloys.

Alloys containing at least 15 at.% Hf still have a monoclinic, B19′ martensite structure as observed in binary NiTi alloys [76,77]. At higher Hf levels, the B2 phase transforms by a single stage transformation into a B19, orthorhombic martensite [78], which is the same as the precious metal containing alloys with higher Pd, Pt, or Au content.

There is also the possibility that a number of additional phases could appear in these alloys, especially as one deviates from stoichiometry [79], which could have implications for the stability of the alloy and its transformation characteristics. However, not a lot of quantitative analysis has been performed on the microstructure, microstructural evolution, or resulting effects on the properties of NiTiHf alloys. Zhu et al. [64] attributed a decrease in the transformation temperatures of $Ni_{48.5}Ti_{36.5}Hf_{15}$ and $Ni_{48.5}Ti_{21.5}Hf_{30}$ alloys during thermal cycling at $A_f + 50°C$ and aging at 500°C to precipitation of a second phase, though no attempt was made to identify this phase. However, the effects did saturate fairly quickly and the material was considered to be much more stable than binary NiTi alloys treated under similar conditions, although not as stable as the (Ni,Pd)Ti alloys. These results are summarized in Figure 7.16.

Meng et al. [80] examined the effect of aging at 700°C on the microstructure, transformation temperatures, and room temperature tensile properties of a $Ni_{49}Ti_{36}Hf_{15}$ alloy. They found that this treatment resulted in precipitation of a $(Ti,Hf)_2Ni$ phase that increased in size and volume fraction with increased aging time. The M_s temperature decreased significantly, almost 60°C, after the first 20 hours of aging but stabilized after that. Tensile properties showed a slight peak in yield strength after aging for 20 hours while tensile elongation was found to be insensitive to aging time.

Olier et al. [71] identified the presence of a cubic $(Ti,Hf)_4Ni_2O_x$ phase in a cast $Ni_{50}Ti_{38}Hf_{12}$ alloy, but the impact of this phase on properties was not determined. In general, a $(Ti,Hf)_4Ni_2O_x$ phase would be indistinguishable from $(Ti,Hf)_2Ni$ and the actual role of oxygen in stabilizing the microstructure of these intermetallic phases in either binary NiTi alloys (Ti_2Ni versus $Ti_4Ni_2O_x$) or ternary alloys such as NiTiHf is unclear.

In contrast, Han et al. [76] have identified a face-centered orthorhombic structured precipitate in a $Ni_{48.5}Ti_{36.5}Hf_{15}$ alloy after aging at 600°C for 150 hours. These precipitates were formed during the aging treatment and were not observed in as-homogenized samples. The precipitates were relatively fine, lenticular-shaped, and coherent with the B2-matrix. The lattice parameters and space group for this precipitate phase have been reported in [76] and are not consistent with a Ti_2Ni type structure, which would normally be expected in slightly Ni-lean alloys [80].

Another possible concern with the Hf-containing ternary alloys is the potentially large hysteresis between the forward and reverse transformations. This hysteresis is on the order of 60–80°C [73,80], which makes these systems less attractive for active control purposes. However, other studies [71,64] report smaller values of hysteresis, on the order of 40–50°C, for similar alloy compositions. The reason for the discrepancy in these values is unknown, but could be due to the amount of residual work in the material, as a result of varied thermomechanical treatments. Still, these latter values for hysteresis are still too large for certain applications requiring precision control.

Regardless, NiTiHf alloys would still be viable for applications where hysteresis may not be as much of a concern. This would include applications in heating and cooling, where the nature of the application involves a low frequency response or in the case of fire suppression, which would usually require a single operational cycle. In addition, these applications would rely on passive control with the material simply responding to the temperature change in the environment. Thus, NiTiHf alloys could still be attractive for these types of applications, if the shape-memory characteristics are acceptable.

However, poor shape-memory properties, approximately 80% shape recovery rate for samples deformed at room temperature to a strain of 2.5%, were originally reported for a $Ni_{50}Ti_{38}Hf_{12}$ alloy by Olier et al. [71]. The poor SME was attributed to the large stress to begin reorientation of the martensite phase (535 MPa), which was probably close to the plastic deformation limit of the material due to the onset of dislocation processes. However, full recovery of up to 4% strain was possible, when the same alloy was strained in the austenite phase, producing stress-induced martensite (SIM) in the material. The SIM apparently did not revert to austenite after unloading, providing permanent strain in the material, which could be recovered by heating above the A_f.

Firstov et al. [81] observed recoverable strains on the order of 3%, measured in compression, for a $Ni_{49.42}Ti_{35.95}Hf_{14.63}$ alloy and suggested that about another 0.8% recoverable strain would be possible if necessary strengthening to avoid plastic deformation by slip could be achieved. Similarly, Meng et al. [82] determined the shape-memory characteristics of a $Ni_{49}Ti_{36}Hf_{15}$ alloy in bending and found about 3% completely reversible strain when the samples were deformed at room temperature. The shape recovery rate was also constant at about 92% for samples deformed to 4.5% strain at deformation temperatures below 184°C (the M_s and A_s temperatures for the alloy were 179 and 216°C, respectively). A two-way shape-memory effect was also observed but it was unstable, decreasing in strain with increasing number of thermal cycles. In general, the relatively low shape recovery rate of the NiTiHf alloys has been attributed to a high critical stress for reorientation or detwinning of the martensite combined with a low critical stress for dislocation slip [71,82].

Mechanical data on a thin film $Ni_{48.9}Ti_{36.6}Hf_{14.5}$ alloy [75] is consistent with the previous data for bulk alloys, exhibiting full recovery of 3% strain at an applied stress of 250 MPa in constant load tensile experiments. However, at 300 MPa and above, strain recovery was incomplete indicating that this level of stress was sufficient to cause dislocation activity in the material. The strain–temperature behavior of this alloy as a function of applied stress is shown in Figure 7.23. This same material was capable of as much as 4% superelastic strain at 190°C ($\sim$6°C above the A_f temperature), a behavior not typically observed in the bulk alloys. It is interesting to note that the stresses required to deform the thin film material by twinning are less than half of that required to initiate martensite reorientation in bulk alloys. It is therefore possible that stress state differences between the bulk and thin film samples could have affected the behavior of the alloy as well.

The general tensile behavior of NiTiHf alloys has been studied by Meng et al. [80,77]. They examined the room temperature tensile properties of a $Ni_{49}Ti_{36}Hf_{15}$ alloy and found that unlike

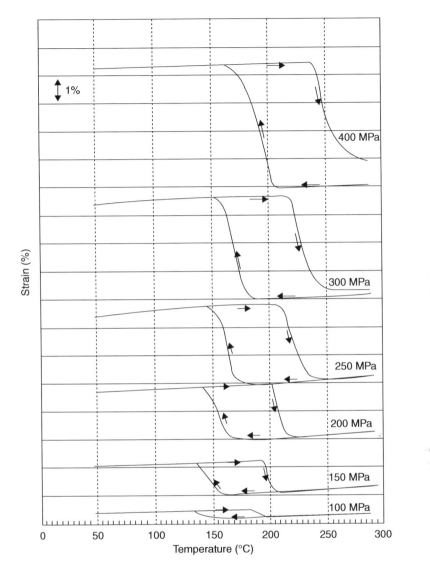

FIGURE 7.23 Strain–temperature curves for a Ni$_{48.9}$Ti$_{36.6}$Hf$_{14.5}$ thin film during thermal cycling under various constant stresses (From Zhang, J., Ph.D. Dissertation, Michigan State University, 2002.)

similarly treated binary NiTi alloys [83], no stress plateau was observed that would have been indicative of easy martensite reorientation, even though samples exhibited tensile elongations on the order of 7%. Instead, the stress–strain curves were characterized by continuous yielding and high work hardening rate, with ultimate tensile strengths between 1000 and 1200 MPa. Meng et al. [77,84] also examined the tensile behavior of the Ni$_{49}$Ti$_{36}$Hf$_{15}$ alloy at 250°C ($A_f = 234$°C) and higher temperatures [84]. The substructure of the stress-induced martensite was found to be composed of (001) compound twins, the same as for the thermally produced martensite, and did not change with deformation temperature or strain. Again, no stress plateau was observed during tensile testing of the material in the austenite state, no superelastic behavior was observed, and the stress-induced martensite formed during testing apparently did not revert during unloading. The reason for this behavior has been attributed to the simultaneous occurrence of stress induced martensite and dislocation slip in the alloy during deformation at 250°C [84]. This results in strong work hardening and decreases the mobility of the martensite variant interfaces, consequently, leading to deterioration of the SME at high temperature and preventing the occurrence of superelasticity in the alloy [77].

There also have been a handful of studies looking at the effect of quaternary alloying additions on the shape-memory properties of NiTiHf alloys. Hsieh et al. [85,86] looked at the combined effects of Hf + Zr additions to NiTi alloys and for the most part found that the transformation temperatures fell between those of the comparable ternary NiTiHf and NiTiZr alloys, while the stability of the transformation temperatures during thermal cycling were more comparable to those of NiTiZr alloys.

Meng et al. [87] added 3–5 at.% Cu to a $Ni_{49-x}Cu_xTi_{36}Hf_{15}$ alloy with an intent to decrease the transformation hysteresis and improve the thermal stability of the alloy, as was originally reported by Liang et al. [88]. However, Meng et al. [87] found that there was no effect of Cu on the stability of the alloy, with the transformation temperatures stabilized after 15–20 thermal cycles, similar to the ternary $Ni_{49}Ti_{36}Hf_{15}$ alloy. Furthermore, instead of shrinking, the hysteresis was slightly expanded due to the Cu addition. Moreover, 5 at.% Cu additions had almost no effect on the transformation temperatures while 3 at.% Cu additions decreased the transformation temperatures of the base NiTiHf alloy by about 20°C. The structure and lattice parameters of the martensite phase were also unchanged. In the end, no real benefit was attributed to the Cu-additions; however, no mechanical or shape-memory properties were determined.

Besides the limited work on quaternary alloys, no other attempts have been made to optimize or improve the shape-memory characteristics of Ni(Ti,Hf) alloys. Given the success of thermomechanical treatments and the judicious use of second phases in improving the strain recovery rate of (Ni,Pd)Ti alloys, a similar approach would seem to be worth investigating for the Ni(Ti,Hf) system.

7.4.7 Ni(Ti,Zr) Alloys

Ecklemeyer [32] was apparently the first person to look at the effect of Zr additions substituted for Ti, on the transformation temperatures of NiTi base alloys. However, he only investigated total alloying additions of up to 2 at.%. His results are fascinating in that while he correctly determined that Zr additions for Ti (and Au additions for Ni) should increase the transformation temperature of NiTi, he worked in that range of compositions where all other investigators find a decrease in transformation temperature for similar levels of Zr [89,90] and other additions. Ecklemeyer's results notwithstanding, NiTiZr alloys are similar to the other alloys discussed in this chapter in that significant increases in transformation temperature only occur above 10 at.% of the ternary addition (Figure 7.24). Above these levels Mulder et al. [89] found that Zr increases the M_s temperature for Ni(Ti,Zr) alloys at a rate of about 18°C/at.%. Also, consistent with the behavior of NiTi alloys in general, the Zr-containing materials only exhibit elevated temperature transformation characteristics as long as the alloys are slightly Ni-deficient [89]. Unlike the precious metal containing (Ni,X)Ti alloys, where there exists a continuous solid solution between the NiTi and XTi (X = Pd, Pt, Au) with a high-temperature B2 structure, solubility for Zr in Ni(Ti,Zr) alloys is approximately 30 at.% at 700°C [91]. Instead, a two phase equilibrium forms between the NiTiZr phase and the NiZr solid solution at these higher Zr concentrations. In contrast, Meisner et al. [92] reported that the structure of the 30 at.% Zr alloy was NiZr plus $Ni_{10}(Ti,Zr)_7$ at room temperature.

The structure of the martensite phase in NiTiZr alloys is apparently in dispute. Meisner et al. [92,93] and Hsieh et al. [86] claim that the martensite phase is B19' in all materials containing up to at least 20 at.% Zr. However, Pu et al. [90] claim that the martensite phase is the B19, orthorhombic structure for alloys containing 15% or more Zr, which would be similar to the other ternary systems discussed in this review that undergo a monoclinic to orthorhombic change in martensite structure

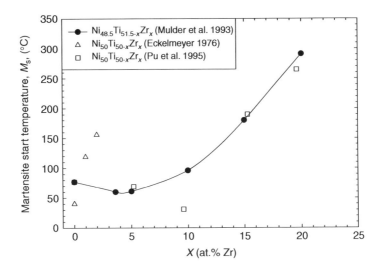

FIGURE 7.24 Martensite start temperature, M_s, as a function of Zr concentration for Ni(Ti,Zr) alloys.

with increasing third element addition. Everyone is in agreement that at lower Zr contents, the B2 phase has been observed to transform into the monoclinic B19′ martensite phase through a single stage (B2→B19′), or two stage (B2→R→B19′) process.

Thermal hysteresis in these alloys is generally broader than that of binary NiTi alloys, but similar to the NiTiHf alloys. A number of investigators [71,89,90] have shown that the thermal hysteresis in NiTiZr alloys with Zr contents up to 20 at.% Zr is on the order of 50–60°C, but that small B additions (0.1 at.%) can reduce the hysteresis by about 10°C [71].

Microstructural control and stability is especially a concern in the Ni(Ti,Zr) system. As expected, any deviation toward Ti(Zr)-rich compositions will result in formation of a $(Ti,Zr)_2Ni$ phase. Additionally, Mulder et al. [89] have reported the presence of a NiTiZr Laves phase in alloys containing greater than about 7 at.% Zr. Hsieh and Wu [94] also have reported an unknown TiNiZr intermetallic phase besides the $(Ti,Zr)_2Ni$ phase in alloys with Zr contents between 10 and 20 at.%. However, it is likely that Mulder et al. and Hsieh et al. are actually referring to the same phase. On the other side of stoichiometry, Ni-rich alloys, consist of some combination of the $Ni_7(Ti,Zr)_2$, $Ni_{10}(Ti,Zr)_7$, NiZr, and B2-Ni(Ti,Zr) phases depending on Zr content [89,93].

Precipitation and growth of any of these phases over time can result in a change in the composition of the austenite phase, which ultimately affects the transformation temperature and shape-memory characteristics of the alloy. For example, Mulder et al. [89] have reported a decrease in the transformation temperatures of a $Ti_{31.5}Ni_{48.5}Zr_{20}$ alloy during thermal cycling that was attributed to precipitation in the alloy, but cycling was stopped after only a few cycles and long before the transformation temperatures stabilized. The transformation temperatures and hardness of Ti(Zr)-rich Ni(Ti,Zr) alloys were also studied by Hsieh et al. [85]. The transformation temperatures were found to decrease with increasing number of thermal cycles with a 50% reduction in M_p after 100 thermal cycles between 0 and 300°C (Figure 7.25). Also, hardness was found to increase by 200–300% after thermal cycling, which would seem to be consistent with precipitation in the alloy. While still significant, Pu et al. [90] found only about a 20–30°C, change in transformation temperature after 100 thermal cycles, compared with the over 80°C change observed by Hsieh et al. [85]. Both of these sets of data for the Ni(Ti,Zr) alloys are compared with the thermal cycling response of Hf and Pd containing alloys in Figure 7.16. From Figure 7.16, it is obvious that stability of the transformation temperatures is much more of a concern in the NiTiZr alloys than any of the other

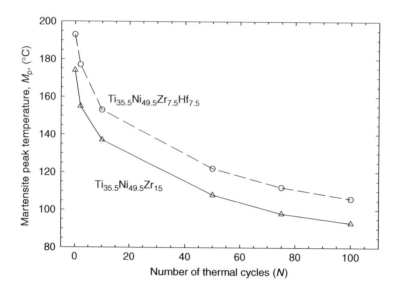

FIGURE 7.25 Martensite peak temperature, M_p, as a function of the number of thermal cycles between 0 and 300°C for Ni(Ti,Zr,Hf) alloys. (From Hsieh, S. F. and Wu, S. K., *Mater. Charact.*, 45, 143, 2000.)

systems discussed. The reason for the large discrepancy in results for the Zr-containing alloys is not clear; however, it is probably related to differences in the way the alloys were processed and heat treated before testing.

The mechanism affecting the stability of the transformation temperatures in the Zr-containing alloys is also open to debate. Hsieh et al. [85] have proposed that the dramatic changes in transformation temperature for NiTiZr and NiTiZrHf alloys, shown in Figure 7.25, and the corresponding

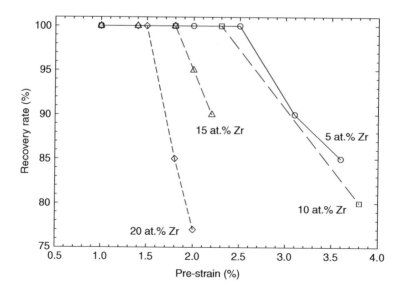

FIGURE 7.26 Recovery rate versus pre-strain in bending at room temperature for Ni(Ti,Zr) alloys of various Zr content. (From Pu, Z., Tseng, H., and Wu, K., In *Smart Structures and Materials 1995: Smart Materials*, Vol. 2441, SPIE Proceedings, pp. 171–178, 1995. With permission.)

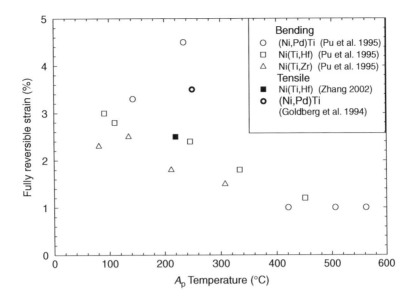

FIGURE 7.27 Fully reversible plastic strain limit as a function of transformation temperature (A_p) for high-temperature NiTi-based systems.

increases in hardness are solely due to dislocation generation during thermal cycling. However, precipitation of undesirable phases has been shown to be a major concern in these materials and is probably responsible for the observed behaviors, especially given the large changes in hardness. But regardless of the root cause of such effects, the major changes observed in Figure 7.25 make design with these materials quite challenging, especially if the transformation temperature needs to be correlated to a specific use temperature.

A one-way shape-memory effect was qualitatively confirmed in a NiTiZr alloy containing 15 at.% Zr; however, alloys with 20% Zr were too brittle to deform in tension at room temperature in order to determine the shape-memory characteristics of higher Zr-containing alloys [89]. Similarly, Meisner et al. [92,93] used torsion testing to determine that a one-way shape-memory effect exists in alloys containing up to 20 at.% Zr. Unfortunately, alloys with 20 at.% or more Zr were very brittle and contained a significant fraction of intermetallic phase other than the B2/martensite, while alloys containing greater than 30 at.% Zr displayed no shape-memory effect. Bend testing of alloys with intermediate levels of Zr, by Pu et al. [90], indicates that the amount of fully recoverable strain decreases with increasing Zr content from about 2.5% to 1.5% for alloys containing 5–20 at.% Zr, respectively, as shown in Figure 7.26.

Consequently, the shape-memory characteristics of the Zr-containing alloys, at least in terms of fully reversible strain, are inferior to both the (Ni,Pd)Ti and Ni(Ti,Hf) alloys (Figure 7.27). But given the limited work to date on Ni(Ti,Zr) alloys, it is obvious that there has been no attempt to try to improve or optimize the shape-memory characteristics of this system. However, as with the other systems discussed in this review, tensile ductility may ultimately be the deciding factor in the viability of these high-temperature shape-memory alloys.

7.5 CONCLUSIONS AND FUTURE CHALLENGES

For applications beyond 100°C, alloys other than binary NiTi need to be developed. Even at temperatures near this limit, binary NiTi alloys are not strong enough to be used as high-force actuators. Fortunately, there are ternary alloy compositions based on NiTi with transformation temperatures from below room temperature to over 1000°C. However, exhibiting a high martensite transformation temperature is only the very first characteristic in a long list of properties that a HTSMA needs to possess in order to be considered commercially viable.

While the origin and mechanism of shape-memory is rather well understood in the high-temperature NiTi-based alloys and the transformation temperatures well defined, most engineering aspects of these materials are not. Almost no research as been done in identifying the operational limits of HTSMAs including the effects of temperature and stress on strain recovery, fatigue life, and work output. Furthermore, these new alloys will need to be more than just functional materials (exhibiting typical shape-memory transformational characteristics). They will have to be robust to handle the challenges associated with an elevated operating temperature. These include long-term stability of the alloy including its microstructure, phase structure, and resistance to oxidation. The shape-memory process depends on the diffusionless transformation of the martensite phase to the higher temperature austenite phase. For materials such as NiTi, where this diffusionless transformation occurs near room temperature, alloy stability is not an issue. However, when this transformation occurs at elevated temperatures, where decomposition, recovery, recrystallization, and other thermal processes are prevalent, alloy and microstructural stability can become major concerns. These diffusional processes can retard the shape recovery process and affect transformation temperatures. Then there are environmental concerns. In general, titanium-based alloys including NiTi have poor oxidation resistance above about 700°C. But serious concerns exist even at lower temperatures especially for thin film components. Therefore, mapping the metal recession rates of alloys as a function of composition and temperature is critical in defining durability limits for these alloys under both isothermal and cyclic conditions.

Since actuators are usually highly loaded, relaxation and creep is also a concern. But the concern lies deeper than just holding dimensional tolerances. The recoverable strain and the amount of work that a shape-memory actuator can perform depend directly on the materials' resistance to dislocation-mediated deformation. The higher the resistance the material has to dislocation motion, the greater the amount of work that can be performed during the martensite-to-austenite transformation. Additionally, this strengthening needs to be accomplished without affecting the properties giving rise to the shape-memory effect, in other words, without significantly increasing the stress for deformation twinning. This requirement also has to be tempered by the need for the material to possess adequate ductility and toughness to survive handling and typical damage during use. Furthermore, a growing concern is that poor tensile ductility may ultimately limit the amount of strain recovery in these materials more than the transformation properties of the alloy.

Then there are the effects of thermal cycling on the transformation characteristics of the material. Cycling under no-load conditions tends to result in a decrease in the M_s temperature, while cycling under load can be used to increase the M_s temperature slightly and can significantly increase the transformation strains. These contradictory behaviors are both attributed to dislocation generation during cycling. Nevertheless, these materials will need to be stabilized against such drifts prior to operation, particularly in applications requiring precision control. Fortunately, this can be accomplished by "training" (cycling) the material under device realistic conditions prior to service.

Finally, as with all materials, the issue of processing must be addressed. Given the often steep dependence of transformation temperature with ternary alloying additions and overall stoichiometry, compositional control, which influences the transformation temperatures of the alloy as well as the shape-memory properties, will be a challenge. Of course each heat of material could be tested for actual transformation (use) temperature, as is the practice with commercial NiTi alloys, but this would be a prohibitively costly option with the precious metal containing alloys. Limited ductility in some of these high-temperature systems also makes thermomechanical processing an ordeal, which is a concern since one of the most common forms that these HTSMAs will be used is wire.

Given all these concerns, it is no wonder that the design and development of shape-memory alloys with an acceptable balance of properties for elevated temperature application is a daunting metallurgical challenge. Of the systems featured in this review, the (Ni,Pd)Ti system has been the most thoroughly studied and would seem to hold some promise. Alloys with 20–40 at.% Pd have a range of transformation temperatures from about 100 to 400°C so that compositions can be tailored for a given application. Alloys in this composition range also exhibit good shape-memory behavior including significant strain recovery under constrained and unconstrained conditions. The alloys can be stabilized against M_s temperature drift through training and are generally not susceptible to martensite aging effects. In addition to alloy stability, these materials have a narrow thermal hysteresis and therefore would make good candidate materials for actuators requiring active control.

In contrast, very little is known about the (Ni,Au)Ti alloys except that they have similar but slightly higher transformation temperatures compared with the (Ni,Pd)Ti alloys. Issues concerning long-term stability, durability, and even basic shape-memory characteristics of these alloys are completely unknown. Much more research also needs to be performed on the (Ni,Pt)Ti alloys. However, this latter group does seem promising in that they have the highest transformation temperatures of any of the NiTi-based systems, appear to be stable against thermal cycling effects, have a narrow hysteresis for compositions with less than 30 at.% Pt, and seem to have relatively good high-temperature strength. The most important question that remains to be answered is whether the NiTiPt alloys have adequate shape-memory behavior, but even if they did, tensile ductility may be an Achilles' heel for this material.

Ni(Ti,Hf) alloys have also been studied in some detail but primarily for alloys containing around 15 at.% Hf. At Hf levels greater than about 15 at.% the already marginal shape-memory characteristics of the ternary alloys drop off considerably. The bottom line is that these alloys have recoverable strains on the order of 3% and can be used primarily at temperatures up to about 200°C, but they also have such a low M_s that there is relatively little advantage over binary NiTi alloys for applications requiring thermal cycling. These materials are not as competitive as the precious metal containing alloys in terms of potential use temperatures, shape-memory characteristics, or alloy stability, and they have a wide hysteresis. Though it is believed that the shape-memory properties of the NiTiHf alloys can be improved by strengthening against slip, no consistent studies have been performed along these lines, as has been done for the Pd-containing alloys. The main advantage of Ni(Ti,Hf) alloys is the low cost of the constituent materials, since they do not contain any precious metals, making them competitive for passive applications requiring low frequency response, primarily in the 100–200°C range.

The transformation temperatures for Ni(Ti,Zr) alloys are similar to the Ni(Ti,Hf) alloys, but they possess slightly worse shape-memory properties and ductility compared with the latter. The Ni(Ti,Zr) alloys are also the most microstructurally unstable of any of the NiTi-based high-temperature systems, which results in extremely large changes in transformation temperature, shape-memory characteristics, and strength with aging or thermal cycling. Consequently, these alloys probably have the least potential of any of the NiTi-based high-temperature systems.

Based on the literature reviewed to date, success in creating shape-memory alloys for use at 100–200°C seems possible. However, development of shape-memory alloys for very high temperature applications, 700°C or greater, at least from the systems currently identified, seems unlikely given the limited transformation temperatures for the majority of these alloys and the breadth of the challenges including microstructural stability, environmental resistance, creep, toughness, thermo-mechanical fatigue, and durability. Probably the biggest challenge over the next decade will be in developing systems for use at intermediate temperatures, in the range of 200–600°C. This entire temperature range would seem to be ideally suited for the NiTiPt alloys, while NiTiPd alloys can be used to address applications in the bottom half of this temperature range.

As large as the material challenges are, this is just part of the problem in creating successful SMA devices. One would expect that many early uses for HTSMAs will be in applications requiring passive control and low frequency response. But the major payoffs will most likely occur only after synergistic designs that integrate these new alloys with active closed-loop control systems incorporating the latest developments in control electronics, sensors, and software are developed.

7.6 AFTERWORD

When this chapter was originally written in early 2003, there were two primary objectives beyond a simple review of the literature. The first was to determine whether a commercially viable HTSMA could be developed from ternary NiTi alloys, specifically for the development of high-force solid-state actuators needed for the fabrication of adaptive components for aeropropulsion systems, and to begin to identify the most promising compositions. The second objective was to identify explicit areas of research that needed immediate attention to hasten the commercialization of such materials.

As a result, the primary focus of high-temperature shape-memory alloy research at NASA Glenn over the last few years (since this review was written) has been on determination of the basic work characteristics of NiTiPt [95,96] and NiTiPd [97] alloys and the effect of alloying and thermomechanical processing, including fabrication of fine wire, on work output and dimensional stability in these systems [98,99]. The overall results of these recent studies have been extremely promising and indicate that it is commercially feasible to develop shape memory alloys for use between 100 and 300°C [100]. But beyond that temperature range the challenges increase exponentially, for the reasons described in this chapter.

ACKNOWLEDGMENTS

RDN would like to thank Anita Tenteris and Nichaela Noebe for their patience and support during this project. Thoughtful discussions with Michael Nathal, Anita Garg, Steve Arnold, Bernie Carpenter, and personnel from Continuum Dynamics, Orbital Research, and Johnson Matthey are also appreciated. This work was supported by the NASA Aeronautics Program under the UEET-IPSFT and QAT projects and by the NASA Glenn IR&D fund.

ABBREVIATIONS AND BASIC DEFINITIONS

A:	austenite, the high-temperature and higher symmetry phase in shape-memory alloys usually consisting of an ordered B2 structure
A_f:	austenite finish temperature, the temperature at which the reverse transformation is complete upon heating
A_p:	austenite peak temperature, the peak temperature observed in DSC or DTA curves during heating associated with the martensite-to-austenite transformation
A_s:	austenite start temperature, the temperature at which the reverse transformation starts upon heating
DSC:	differential scanning calorimetry, a thermal analysis technique used to determine the transformation temperatures in metals and alloys
DTA:	differential thermal analysis, a thermal analysis technique used to determine the transformation temperatures in metals and alloys
HTSMA:	high-temperature shape-memory alloy
Hs:	thermal hysteresis between the forward and reverse $M \leftrightarrow A$ transformations, usually defined as $A_f - M_s$. Alternatively, many studies define Hs as $A_p - M_p$
M:	martensite, the low temperature, and lower symmetry phase in shape-memory alloys usually consisting of a monoclinic or orthorhombic crystal structure

M_d: maximum temperature that martensite can be stress induced

M_f: martensite finish temperature, the temperature at which the forward transformation is complete upon heating

M_p: martensite peak temperature, the peak temperature observed in DSC or DTA curves during cooling associated with the austenite-to-martensite transformation

M_s: martensite start temperature, the temperature at which the forward transformation starts upon heating

SIM: stress induced martensite

SMA: shape-memory alloy

SME: shape-memory effect, where an SMA deformed below the M_f will recover its original shape when heated through the transformation temperature

TWSME: two-way shape-memory effect, where a change in shape occurs on both heating and cooling through the transformation temperature

REFERENCES

1. Perez-Landazabal, J. I., Recarte, V., No, M. L., and San Juan, J., Determination of the order in γ1 intermetallic phase in Cu-Al-Ni shape memory alloys, *Intermetallics*, 11, 927, 2003.
2. George, E. P., Liu, C. T., Horton, J. A., Sparks, C. J., Kao, M., Kunsmann, H., and King, T., Characterization, processing, and alloy design of NiAl-based shape memory alloys, *Mater. Charact.*, 32, 139, 1994.
3. Yang, J. H. and Wayman, C. M., On the formation mechanism of Ni_5Al_3 in NiAl-base alloys: Part I. microstructures, *Intermetallics*, 2, 111, 1994.
4. Grummon, D. S., Thin-film shape-memory materials for high-temperature applications, *JOM*, 55(12), 24, 2003.
5. Duerig, T. W. and Zadno, R., An engineer's perspective of pseudoelasticity, In *Engineering Aspects of Shape-Memory Alloys*, Duerig, T. W., Melton, K. N., Stockel, D., and Wayman, C. M., Eds., Butterworth-Heinemann, London, pp. 369–393, 1990.
6. Van Humbeeck, J., From a seed to a need: The growth of shape memory applications in Europe, In *Shape-Memory Materials and Phenomenon—Fundamental Aspects and Applications*, Liu, C. T., Kunsmann, H., Ostuka, K., and Wutting, M., Eds., MRS Proceedings, Vol. 246, pp. 377–387, 1992.
7. Van Humbeeck, J., Non-medical applications of shape memory alloys, *Mater. Sci. Eng.*, A273–275, 134, 1999.
8. Kapgan, M. and Melton, K. N., Shape memory alloy tube and pipe couplings, In *Engineering Aspects of Shape-Memory Alloys*, Duerig, T. W., Melton, K. N., Stockel, D., and Wayman, C. M., Eds., Butterworth-Heinemann, London, pp. 137–148, 1990.
9. Cydzik, E., The design of electrical interconnection systems with shape memory alloys, In *Engineering Aspects of Shape-Memory Alloys*, Duerig, T. W., Melton, K. N., Stockel, D., and Wayman, C. M., Eds., pp. 149–157, 1990.
10. Mavroidis, C., Development of advanced actuators using shape memory alloys and electrorheological fluids, *Res. Nondestr. Eval.*, 14, 1, 2002.
11. Van Humbeeck, J., Damping capacity of thermoelastic martensite in shape memory alloys, *J. Alloys Compd.*, 355, 58, 2003.
12. Schetky, L. McD., The industrial applications of shape memory alloys in North America, *Mater. Sci. Forum*, 327–328, 9, 2000.
13. Quackenbush, T. R., Smart materials for high speed adaptive inlet nozzle design, Phase I SBIR Final Report, NASA Contract NNC04CA45C, 2004.
14. Barooah, P., and Rey, N., Closed loop control of a shape memory alloy actuation system for variable area fan nozzle, In *Smart Structures and Materials 2002: Modeling, Signal Processing, and Control*, SPIE Proceedings, Vol. 4693, pp. 384–395, 2002.
15. Smith, R. C., *Smart Material Systems: Model Development and Control Design*, SIAM, Philadelphia, PA, Frontiers in Applied Mathematics Series, 32, 2005.

16. Huff, M. A., Bernard, W. L., Lisy, F. J., and Prince, T. S., Method and sensor for detecting strain using shape memory alloys, U.S. Patent No. 6622558, 2003.

17. Chen, Y.-C. and Lagoudas, D., Impact induced phase transformations in shape memory alloys, *J. Mech. Phys. Solids*, 48, 275, 2000.

18. Lin, H. C., Liao, H. M., He, J. L., Chen, K. C., and Lin, K. M., Wear characteristics of TiNi shape memory alloys, *Metall. Mater. Trans.*, 28A, 1871, 1997.

19. Buehler, W. J. and Wiley, R. C., TiNi-ductile intermetallic compound, *Trans. ASM*, 55, 269, 1962.

20. Buehler, W. J., Gilfrich, J. V., and Wiley, R. C., Effect of low-temperature phase changes on the mechanical properties of alloys near composition TiNi, *J. Appl. Phys.*, 34, 1475, 1963.

21. Wang, F. E., Buehler, W. J., and Pickard, S. T., Crystal structure and a unique Martensitic transition of TiNi, *J. Appl. Phys.*, 36, 3232, 1965.

22. Otsuka, K., Sawamura, T., and Shimizu, K., Crystal structure and internal defects of equiatomic TiNi martensite, *Phys. Status Solidi A*, 5, 457, 1971.

23. Wasilewski, R. J., Elastic-modulus anomaly in TiNi, *Trans. Metall. Soc. AIME*, 233, 1691, 1965.

24. Wasilewski, R. J., The effects of applied stress on the martensitic transformation in TiNi, *Metall. Trans.*, 2, 2973, 1971.

25. Suzuki, T. and Shigeaki, U., Mechanical properties of FeTi, CoTi, and NiTi at elevated temperatures, In *Titanium 80, Science and Technology, Proceedings of the 4th International Conference on Titanium*, Kimura, H. and Izumi, O., Eds., The Metallurgical Society of AIME, Warrendale, PA, pp. 1255–1263, 1980.

26. NASA Glenn Research Center, High-Temperature Shape-Memory Alloy Group, unpublished research, 2003.

27. Chu, C. L., Wu, S. K., and Yen, Y. C., Oxidation behavior of equiatomic TiNi alloy in high temperature air environment, *Mater. Sci. Eng.*, A216, 193, 1996.

28. Firstov, G. S., Vitchev, R. G., Kumar, H., Blanpain, B., and Van Humbeeck, J., Surface oxidation of NiTi shape memory alloy, *Biomaterials*, 23, 4863, 2002.

29. Wasilewski, R. J., Butler, S. R., Hanlon, J. E., and Worden, D., Homogeneity range and the martensitic transformation in TiNi, *Metall. Trans.*, 2, 229, 1971.

30. Otsuka, K. and Ren, X., Factors affecting the M_s temperature and its control in shape-memory alloys, *Mater. Sci. Forum*, 394–395, 177, 2002.

31. Melton, K. N., Ni-Ti based shape memory alloys, In *Engineering Aspects of Shape-Memory Alloys*, Duerig, T. W., Melton, K. N., Stockel, D., and Waymam, C. M., Eds., Butterworth-Heinemann, London, pp. 21–35, 1990.

32. Eckelmeyer, K. H., The effect of alloying on the shape memory phenomenon in Nitinol, *Scripta Metall.*, 10, 667, 1976.

33. Hwang, C. M. and Wayman, C. M., Phase transformations in TiNiFe, TiNiAl and TiNi alloys, *Scripta Metall.*, 17, 1345, 1983.

34. Suzuki, Y. and Horikawa, H., Thermal hysteresis in Ni-Ti and Ni-Ti-X alloys and their applications, In *Shape-Memory Materials and Phenomenon—Fundamental Aspects and Applications*, Liu, C. T., Kunsmann, H., Ostuka, K., and Wuttig, M., Eds., MRS Proceedings, Vol. 246, pp. 389–398, 1992.

35. Honma, T., Matsumoto, M., Shugo, Y., and Yamazaki, I., Effects of 3d transition elements on the phase transformation in TiNi compound, In *ICOMAT-79: Proceedings of the International Conference on Martensitic Transformations*, Cambridge, MA, pp. 259–264, 1979.

36. Yang, J. H. and Simpson, J. W., Stress-induced transformation and superelasticity in Ni-Ti-Nb alloys *J. Phys. IV*, 5, C8–771, 1995.

37. Kudoh, Y., Tokonami, M., Miyazaki, S., and Otsuka, K., Crystal structure of the martensite in Ti-49.2 at.% Ni analyzed by the single crystal X-ray method, *Acta Metall.*, 33, 2049, 1985.

38. Otsuka, K. and Ren, X., Recent developments in the research of shape memory alloys, *Intermetallics*, 7, 511, 1999.

39. Zhang, J. X., Sato, M., and Ishida, A., On the Ti_2Ni precipitates and Guinier-Preston zones in Ti-rich Ti-Ni thin films, *Acta Mater.*, 51, 3121, 2003.

40. Hwang, C. M. and Wayman, C. M., Compositional dependence of transformation temperatures in ternary TiNiAl and TiNiFe alloys, *Scripta Metall.*, 17, 381, 1983.

41. Miyazaki, S. and Otsuka, K., *Metall. Trans.*, 17A, 53, 1986.

42. Hara, T., Ohba, T., and Otsuka, K., Structural study of R-phase in Ti-50.23 at. %Ni and Ti-47.75 at.%Ni-1.50 at.%Fe alloys, *Mater. Trans. JIM*, 38, 11, 1997.

43. Grummon, D. S. and Pence, T. L., Thermotractive Titanium-Nickel thin films for microelectromechanical systems and active composites, In *Materials for Smart Systems II*, Vol. 459, George, E. P., Gotthardt, R., Otsuka, K., Trolier-Mckinstry, S., and Wun-Fogle, M., Eds., MRS Proceedings, Vol. 459, pp. 331–343, 1997.

44. Grummon, D. S. and Zhang, J., Stress in sputtered films of near-equiatomic TiNiX on (100) Si: intrinsic and extrinsic stresses and their modification by thermally activated mechanisms, *Phys. Status Solidi A*, 186, 17, 2001.

45. Rusell, S. M., NITINOL melting and fabrication, In *SMST-2000: International Conference on Shape-memory and Superelastic Technologies*, Russell, S. M. and Pelton, A. R., Eds., Int, Org. on SMST, Pacific Grove, CA, pp. 1–9, 2001.

46. Pelton, A. R., Russel, S. M., and DiCello, J., The physical metallurgy of Nitinol for medical applications, *JOM*, 55(5), 33, 2003.

47. Boriskina, N. G. and Kenina, E. M., Phase equilibria in the Ti-TiPd-TiNi system alloys, In *Titanium 80, Science and Technology, Proceedings of the 4th International Conference on Titanium*, Kimura, H, and Izumi, O., Eds., The Metallurgical Society of AIME, Warrendale, PA, pp. 2917–2927, 1980.

48. Lindquist, P. G. and Wayman, C. M., Shape memory and transformation behavior of martensitic Ti-Pd-Ni and Ti-Pt-Ni alloys, In *Engineering Aspects of Shape-Memory Alloys*, Duerig, T. W., Melton, K. N., Stockel, D., and Waymam, C. M., Eds., Butterworth-Heinemann, London, pp. 58–68, 1990.

49. Donkersloot, H. C. and Van Vucht, J. H. N., Martensitic transformations in gold-titanium, palladium-titanium and platinum-titanium alloys near the equiatomic compositions, *J. Less Common Met.*, 20, 83, 1970.

50. Otsuka, K., Oda, K., Ueno, Y., Piao, M., Ueki, T., and Horikawa, H., The shape memory effect in a $Ti_{50}Pd_{50}$ alloy, *Scripta Metall. Mater.*, 29, 1355, 1993.

51. Xu, Y., Shimizu, K., Suzuki, Y., Otsuka, K., Ueki, T., and Mitose, K., Recovery and recrystallization processes in Ti-Pd-Ni high-temperature shape memory alloys, *Acta Mater.*, 45, 1503, 1997.

52. Shimizu, S., Xu, Y., Okunishi, E., Tanaka, S., Otsuka, K., and Mitose, K., Improvement of shape memory characteristics by precipitation-hardening of Ti-Pd-Ni alloys, *Mater. Lett.*, 34, 23, 1998.

53. Kachin, V. N., Matveeva, N. M., Sivokha, V. P., Chernov, D. B., and Yu, K., Koveristyi, High-temperature shape memory effects in alloys of the TiNi-TiPd system, *Doklady Akad. Nauk SSSR*, 257, 167, 1981.

54. Yang, W. S. and Mikkola, D. E., Ductilization of Ti-Ni-Pd shape memory alloys with boron additions, *Scripta Metall. Mater.*, 28, 161, 1993.

55. Suzuki, Y., Xu, Y., Morito, S., Otsuka, K., and Mitose, K., Effects of boron addition on microstructure and mechanical properties of Ti-Pd-Ni high-temperature shape memory alloys, *Mater. Lett.*, 36, 85, 1998.

56. Qiang, D. S., Ying, Q. G., Bo, Y. H., and Ming, T. S., Phase transformation and shape memory effect of the high temperature shape memory alloy $Ti_{49}Ni_{25}Pd_{26}B_{0.12}$, In *Proceedings of the International Symposium on Shape Memory Materials*, Youyi, C, and Hailing, T. U., Eds., International Academic Publishers, Beijing, China, pp. 248–252, 1994.

57. Goldberg, D., Xu, Y., Murakami, Y., Morito, S., Otsuka, K., Ueki, T., and Horikawa, H., Improvement of a $Ti_{50}Pd_{30}Ni_{20}$ high temperature shape memory alloy by thermomechanical treatments, *Scripta Metall. Mater.*, 30, 1349, 1994.

58. Cai, W., Tanaka, S., and Otsuka, K., Thermal cyclic characteristics under load in a $Ti_{50.0}Pd_{30}Ni_{19.4}$ alloy, *Mater. Sci. Forum*, 327–328, 279, 2000.

59. Miyazaki, S., Igo, Y., and Otsuka, K., Effect of thermal cycling on the transformation temperature of Ti-Ni alloys, *Acta Metall.*, 34, 2045, 1986.

60. Stachowiak, G. B. and McCormick, P. G., Shape memory behavior associated with the R and martensitic transformations in a NiTi alloy, *Acta Metall.*, 36, 291, 1988.

61. Ren, X. and Otsuka, K., Origin of rubber-like behaviour in metal alloys, *Nature*, 389, 579, 1997.

62. Otsuka, K. and Ren, X., Origin of aging effect and rubber-like behavior in martensite, In *Pacific Rim International Conference on Advanced Materials and Processing (PRICM 3)*, Imam, M. A., DeNale, R., Hanada, S., Zhong, Z, and Lee, D. N., Eds., The Minerals, Metals, and Materials Society, Warrendale, PA, pp. 1173–1180, 1998.

63. Cai, W., Otsuka, K., and Asai, M., Martensite aging effect in Ti-Pd and Ti-Pd-Ni high temperature shape memory alloys, *Mater. Trans. JIM*, 40, 895, 1999.

64. Zhu, Y. R., Pu, Z. J., Li, C., and Wu, K. H., The stability of NiTi-Pd and NiTi-Hf high temperature shape memory alloys, In *Proceedings of the International Symposium on Shape Memory Materials*, Youyi, C, and Hailing, T. U., Eds., International Academic Publishers, Beijing, China, pp. 253–257, 1994.

65. Wu, S. K. and Wayman, C. M., Martensitic transformations and the shape memory effect in $Ti_{50}Ni_{10}Au_{40}$ and $Ti_{50}Au_{50}$ alloys, *Metallography*, 20, 359, 1987.

66. Wu, S. K. and Wayman, C. M., TEM studies of the martensitic transformation in a $Ti_{50}Ni_{40}Au_{10}$ alloy, *Scripta Metall.*, 21, 83, 1987.

67. Biggs, T., Cortie, M. B., Witcomb, M. J., and Cornish, L. A., Martensitic transformations, microstructure, and mechanical workability of TiPt, *Metall. Mater. Trans.* 32A, 1881, 2001.

68. Meisner, L. L. and Sivokha, V. P., The effect of applied stress on the shape memory behavior of TiNi-based alloys with different consequences of martensitic transformations, *Physica B*, 344, 93, 2004.

69. Hosoda, M., Tsuji, M., Takahashi, Y., Inamura, T., Wakashima, K., Yamabe-Mitaria, Y., Miyazaki, S., and Inoue, K., Phase stability and mechanical properties of Ti-Ni shape memory alloys containing platinum group metals, *Mater. Sci. Forum*, 426–432, 2333, 2003.

70. Mulder, J. H., Beyer, J., Donner, P., and Peterseim, J., On the high temperature shape memory capabilities of Ni-(TiZr) and Ni-(TiHf) alloys, In *SMST-94: The First International Conference on Shape-Memory and Superelastic Technologies*, Pelton, A., Hodgson, D., and Duerig, T., Eds., Int, Org. on SMST, Pacific Grove, CA, pp. 55–61, 1995.

71. Olier, P., Brachet, J. C., Bechade, J. L., Foucher, C., and Guenin, G., Investigation of transformation temperatures, microstructure and shape memory properties of NiTi, NiTiZr and NiTiHf alloys, *J. Phys. IV*, 5, C8–741, 1995.

72. AbuJudom, D. N., Thoma, P. E., and Kao, M., Angst, D. R., High transformation temperature shape memory alloy, U.S. Patent No. 5,114,504, 1992.

73. Angst, D. R., Thoma, P. E., and Kao, M. Y., The effect of hafnium content on the transformation temperatures of $Ni_{49}Ti_{51-x}Hf_x$ shape memory alloys, *J. Phys. IV*, 5, C8–747, 1995.

74. Besseghini, S., Villa, E., and Tuissi, A., Ni-Ti-Hf shape memory alloy: effect of aging and thermal cycling, *Mater. Sci. Eng.*, A273–275, 390, 1999.

75. Zhang, J., Processing and characterization of high-temperature nickel-titanium-hafnium shape memory thin films, Ph.D. Dissertation, Michigan State University, 2002.

76. Han, X. D., Zou, W. H., Wang, R., Zhang, Z., Yang, D. Z., and Wu, K. H., The martensite structure and aging precipitates of a TiNiHf high temperature shape memory alloy, *J. Phys. IV*, 5, C8–753, 1995.

77. Meng, X. L., Cai, W., Wang, L. M., Zheng, Y. F., Zhao, L. C., and Zhou, L. M., Microstructure of stress-induced martensite in a Ti-Ni-Hf high temperature shape memory alloy, *Scripta Mater.*, 45, 1177, 2001.

78. Wu, K. H., Pu, Z. J., Tseng, H. K., and Biancaniello, F. S., The shape memory effect of the Ni-Ti-Hf high temperature shape memory alloy, In *SMST-94: The First International Conference on Shape-memory and Superelastic Technologies*, Pelton, A., Hodgson, D., and Duerig, T., Eds., Int, Org. on SMST, Pacific Grove, CA, pp. 61–66, 1995.

79. Nishida, M., Wayman, C. M., and Honma, T., Precipitation processes in near-equiatomic TiNi shape memory alloys, *Metall. Trans.* 17, 1505, 1986.

80. Meng, X. L., Zheng, Y. F., Wang, Z., and Zhao, L. C., Effect of aging on the phase transformation and mechanical behavior of $Ti_{36}Ni_{49}Hf_{15}$ high temperature shape memory alloy, *Scripta Mater.*, 42, 341, 2000.

81. Firstov, G. S., Van Humbeeck, J., and Kval, Y. N., Comparison of high temperature shape memory behavior for ZrCu-based, Ti-Ni-Zr and Ti-Ni-Hf alloys, *Scripta Mater.*, 50, 243, 2004.

82. Meng, X. L., Zheng, Y. F., Wang, Z., and Zhao, L. C., Shape memory properties of the $Ti_{36}Ni_{49}Hf_{15}$ high temperature shape memory alloy, *Mater. Lett.*, 45, 128, 2000.

83. Eucken, S. and Duerig, T. W., The effects of pseudoelastic prestraining on the tensile behavior and two-way shape memory effect in aged NiTi, *Acta Metall.*, 37, 2245, 1989.

84. Meng, X. L., Cai, W., Zheng, Y. F., Tong, Y. X., Zhao, L. C., and Zhou, L. M., Stress-induced martensitic transformation behavior of a Ti-Ni-Hf high temperature shape memory alloy, *Mater. Lett.*, 55, 111, 2002.

85. Hsieh, S. F. and Wu, S. K., Martensitic transformation of quaternary $Ti_{50.5-x}Ni_{49.5}Zr_{x/2}Hf_{x/2}$ (X = 0-20 at.%),, *Mater. Charact.*, 45, 143, 2000.

86. Hsieh, S. F. and Wu, S. K., Lattice parameters of martensite in $Ti_{50.5-x}Ni_{49.5}Zr_{x/2}Hf_{x/2}$ quaternary shape memory alloys, *J. Alloy Compd.*, 312, 288, 2000.

87. Meng, X. L., Tong, Y. X., Lau, K. T., Cai, W., Zhou, L. M., and Zhao, L. C., Stress-induced martensitic transformation behavior of a Ti-Ni-Hf high temperature shape memory alloy, *Mater. Lett.*, 57, 452, 2002.

88. Liang, X. L., Chen, Y., Shen, H. M., Zhang, Z. F., Li, W., and Wang, Y. N., Thermal cycling stability and two-way shape memory effect of Ni-Cu-Ti-Hf alloys, *Solid State Commun.*, 119, 381, 2001.

89. Mulder, J. H., Maas, J. H., and Beyer, J., Martensitic transformations and shape memory effects in Ni-Ti-Zr alloys, In *ICOMAT-92: Proceedings of the International Conference on Martensitic Transformations*, Wayman, C. M., and Perkins, J., Eds., Monterey Institute for Advanced Studies, Carmel, CA., pp. 869–874, 1993.

90. Pu, Z., Tseng, H., and Wu, K., Martensite transformation and shape memory effect of NiTi-Zr high temperature shape memory alloys, In *Smart Structures and Materials 1995: Smart Materials*, SPIE Proceedings, Vol. 2441, pp. 171–178, 1995.

91. Eremenko, V. N., Semenova, E. L., Tret'yachenko, L. A., and Domtyrko, Z. G., The structure of Ni-Zr-Ti alloys in the region of 0-50% (at) of Ni at 700°C, *Dopovidi Akademii Nauk Ukrains'koi RSR, Seriya A: Fiziko-Matematichni Ta Tekhnichni Nauki*, 50(2), 76 (1988).

92. Meisner, L. L., Grishkov, V. N., and Sivokha, V. P., The Martensitic transformations and the shape memory effect in the $Ni_{50}Ti_{50-x}Zrx$ alloys, In *Proceedings of the International Symposium on Shape Memory Materials*, Youyi, C and Hailing, T. U., Eds., International Academic Publishers, Beijing, China, pp. 263–266, 1994.

93. Meisner, L. and Sivokha, V., Deformation of crystal lattice in the process of martensitic transformations in alloys of $Ni_{50}Ti_{50-x}Zr_x$,, *J. Phys. IV*, 5, C8–765, 1995.

94. Hsieh, S. F. and Wu, S. K., Room-temperature phases observed in $Ti_{53-x}Ni_{47}Zr_x$ high-temperature shape memory alloys, *J. Alloy Compd.*, 266, 276, 1998.

95. Noebe, R., Gaydosh, D., Padula, S., Garg, A., Biles, T., and Nathal, M., Properties and potential of two (Ni,Pt)Ti alloys for use as high-temperature actuator materials, In *Smart Structures and Materials 2005: Active Materials: Behavior and Mechanics*, Vol. 5761, SPIE Conference Proceedings, pp. 364–375, 2005.

96. DeCastro, J. A., Melcher, K. J., and Noebe, R. D., System-level design of a shape memory alloy actuator for active clearance control in the high-pressure turbine, In *Proceedings of the 41st Joint Propulsion Conference and Exhibit, American Institute of Aeronautics and Astronautics*, Paper #AIAA-2005-3988, 2005.

97. Noebe, R., Padula II, S., Bigelow, G., Rios, O., Garg, A., and Lerch, B., Properties of a $Ni_{19.5}Pd_{30}Ti_{50.5}$ high-temperature shape memory alloy in tension and compression, In *Smart Structures and Materials 2006: Active Materials: Behavior and Mechanics*, Vol. 6170, SPIE Conference Proceedings, 2006.

98. Bigelow, G., Padula II, S., Noebe, R., Garg, A., and Olson, D., Development and characterization of improved NiTiPd high-temperature shape-memory alloys by solid-solution strengthening and thermomechanical processing, In *SMST-2006: International Conference on Shape-Memory and Superelastic Technologies*, Mitchell, M. R., and Berg, B., Eds., ASM International, Metals Park, OH, 2006.

99. Noebe, R., Draper, S., Gaydosh, D., Garg, A., Lerch, B., Penney, N., Bigelow, G., Padula, S., and Brown, J., Effect of thermomechanical processing on the microstructure, properties, and work

behavior of a $Ti_{50.5}Ni_{29.5}Pt_{20}$ high-temperature ahape memory alloy, In *SMST-2006: International Conference on Shape-Memory and Superelastic Technologies*, Mitchell, M. R., and Berg, B., Eds., ASM International, Metals Park, OH, 2006.

100. Padula, S., Bigelow, G., Noebe, R., and Gaydosh, D., Challenges and progress in the development of high-temperature shape memory alloys based on NiTiX compositions for high-force applications, In *SMST-2006: International Conference on Shape-Memory and Superelastic Technologies*, Mitchell, M. R., and Berg, B., Eds., ASM International, Metals Park, OH, 2006.

8 Cobalt Alloys and Composites

M. Freels and Peter K. Liaw
Department of Materials Science and Engineering, The University of Tennessee

L. Jiang
Corporate Research and Development Center, General Electric Company

D. L. Klarstrom
Haynes International, Inc.

CONTENTS

8.1 Introduction 188
8.2 Structure and Phase Equilibria of Cobalt and its Alloys 188
8.3 Occurrence, Extraction, and End Uses of Cobalt...191
    8.3.1 Occurrence..191
    8.3.2 Extraction and Processing 191
    8.3.3 End Uses of Cobalt 191
8.4 Mechanical and Physical Properties 192
    8.4.1 Pure Cobalt 192
    8.4.2 Cobalt-Based Wear-Resistant Alloys..194
    8.4.3 Cobalt-Based High-Temperature Alloys 198
    8.4.4 Cobalt-Based Corrosion-Resistant Alloys 204
    8.4.5 Fatigue Behavior of Selected Cobalt-Based Alloys 208
        8.4.5.1 High-Cycle Fatigue Behavior 208
        8.4.5.2 Low-Cycle Fatigue..209
        8.4.5.3 Temperature Evolution During Fatigue...212
8.5 Applications of Cobalt and its Alloys 219
    8.5.1 High-Temperature Applications...219
    8.5.2 Magnetic Devices 221
    8.5.3 Biomedical Applications 221
8.6 Cemented Carbides 222
References 223

Abstract

This chapter discusses the use of cobalt as the basic element in alloys and composites, including the basics of the cobalt-alloy design and physical metallurgy. The topics discussed include the structure and phase equilibria of pure cobalt and its alloys, the occurrence, extraction, and processing of cobalt, the mechanical and physical properties of pure cobalt, cobalt-based alloys, and cemented carbides, as well as some applications. Also discussed is the fatigue behavior of selected cobalt-based alloys.

8.1   INTRODUCTION

In 1907, Elwood Haynes developed a series of cobalt-chromium alloys suitable for tableware, surgical instruments, laboratory equipments, analytical weights, and cutting tools to be used on relatively soft metals [1]. In 1913, these cobalt-chromium alloys were placed on the market [1]. Determination of phase diagrams for various alloys systems was very active at the start of the twentieth century, which allowed for the development of new alloys with many useful properties [2]. After the initial application of the cobalt–chromium alloys and advances in phase diagrams, various wrought cobalt-based alloys were developed that found extensive usage for such components as forged turbine blades, combustor liners, and afterburner tailpipes [3]. The "cobalt crisis" in the late 1970s increased the price of cobalt tremendously, hindering their continued development and use. However, cobalt-based alloys offer unique combinations of properties, such as high-temperature creep and fatigue strength, as well as resistance to aggressive corrosion and various forms of wear. Properties such as these should secure the use of cobalt-based alloys now and in the future.

Today, cobalt-based alloys are usually categorized as: (a) wear-resistant alloys with high-carbon contents, (b) high-temperature alloys with medium-carbon contents, and (c) corrosion-resistant alloys or simultaneous corrosion- and wear-resistant alloys with low carbon contents.

8.2   STRUCTURE AND PHASE EQUILIBRIA OF COBALT AND ITS ALLOYS

There are two allotropic modifications of cobalt; a hexagonal close-packed (HCP) form, ε, stable at temperatures below 422°C [3], and a face-centered-cubic (FCC) form, α, which is stable between 422°C and its melting point of 1,495°C [4]. It should be noted that these temperatures depend critically on the purity and the rate of the temperature change. There is some controversy as to the stability of the latter FCC form [5]. Some investigators believe that there is a reversion from the cubic to hexagonal form near the Curie temperature (1,121°C). However, these reports are not always confirmed and are probably due to the presence of interstitial impurities [5]. Table 8.1 summarizes the variation of the lattice parameters with the temperature of cobalt for both allotropic forms [5].

The mechanism of the phase transformation occurs by shear with the following crystallographic relationships between the two phases [6]:

$$\{111\}_\alpha \parallel \{0001\}_\varepsilon : \langle 110 \rangle_\alpha \parallel \langle 11\text{--}20 \rangle_\varepsilon$$

Therefore, the transformation is classified as martensitic, arising from the mobility of partial dislocations along the close-packed planes. The stacking-fault energy of both allotropes is low and

TABLE 8.1
Lattice Parameters of Cobalt (nm)

Temperature (°C)	ε Co			α Co
	a	c	a/c	a
24	0.25071 ± 0.00005	0.40695 ± 0.00005	1.6233	0.35446 ± 0.00005
520	—	—	—	0.35688
598	—	—	—	0.35720
1,003	—	—	—	0.35900
1,398	—	—	—	0.36214

Source: From Betteridge, W., *Cobalt and its Alloys*, Ellis Horwood Limited, Chichester, 1982.

TABLE 8.2
Variation of the Stacking-Fault Energy of the ε Cobalt and α Cobalt with Temperature

Form	Temperature (°C)	Stacking-Fault Energy (J/m²)
ε Co	20	31×10^{-3}
	150	24.5×10^{-3}
	370	20.5×10^{-3}
α Co	500	13.5×10^{-3}
	710	18.5×10^{-3}

Source: From Betteridge, W., *Cobalt and its Alloys*, Ellis Horwood Limited, Chichester, 1982.

some reported values are given in Table 8.2 [5]. Grain size plays a major role in the stability of the two allotropes [5]. Finer grain sizes tend to favor the cubic form. Cobalt powders or thin films have been known to retain an essentially cubic structure at normal temperatures. Even in the normal solid hot or cold-worked cobalt that has been annealed and slowly cooled to room temperature, a portion of the metastable cubic phase will always be present.

There have been very few published phase diagrams that indicate the presence of a two-phase $\alpha + \varepsilon$ field, as is required for compliance with the Phase Rule [5]. This is probably due to the sluggishness of the transformation. Although pure cobalt transforms from one structure to the other at a fixed temperature, alloys should transform gradually over a temperature range with both phases present in the intervening field.

Alloying elements can be classified according to their influence on the transformation temperature [5]. The alloying elements that raise the transformation temperature are described as leading to a restricted α-field (or the HCP-phase stabilizing elements) such as chromium or other refractory elements. Those lowering the temperature are described as giving an enlarged α-field (or FCC phase stabilizing elements) such as nickel, iron, or manganese. Most alloying additions make the transformation more sluggish, some to the extent of raising the $\varepsilon \rightarrow \alpha$ temperature but lowering the $\alpha \rightarrow \varepsilon$ temperature. Such elements are described as a "combined" type. The effects of alloying additions according to the above description are given in Table 8.3 [5].

The effects of individual alloying elements have been systematically studied [7], and Figure 8.1 is a simplified representation relating the solubility with the effect of an element on the $\varepsilon \rightarrow \alpha$ transformation temperature (designated here as A_s). The A_s temperature of cobalt-based alloys may be calculated, based on the information provided in Figure 8.1 by

$$A_s = 422 + \sum_{i=1}^{n} \Delta T_i \times C_i \tag{8.1}$$

where 422 is the allotropic phase transformation temperature in °C of the pure cobalt, ΔT_i is the temperature change per atomic percent of a solute element, i, and C_i is the atomic percentage of the solute element, i. The effect of alloying elements on the $\alpha \rightarrow \varepsilon$ transformation temperature is not well understood.

Cobalt-based alloys that have a metastable FCC phase tend to transform to a HCP phase under the action of a mechanical stress (or strain) at room temperature, aging in the intermediate temperature range (below the $\varepsilon \rightarrow \alpha$ transformation temperature), or mechanical stress and aging at intermediate temperatures [6]. This strain-induced FCC to HCP phase transformation process involves the coalescence of stacking faults, which is achieved by the passage of a Shockley partial dislocation, resulting from the dissociation of a full dislocation due to the low stacking-fault energy on every second {111} plane in an FCC crystal

ABCABC $\Rightarrow$ ABABAB

TABLE 8.3
Classification of Solute Elements in Cobalt-Rich Binary Alloys

Enlarged α-Field (or FCC Phase-Stabilizing Elements)	Restricted α-Field (or HCP Phase-Stabilizing Elements)	Combined
Aluminum	Silicon	Beryllium
Boron	Germanium	Lead
Copper	Arsenic	Vanadium
Titanium	Antimony	Palladium
Zirconium	Chromium	Gallium
Carbon	Molybdenum	Gold
Tin	Tungsten	—
Niobium	Tantalum	—
Manganese	Rhenium	—
Iron	Ruthenium	—
Nickel	Osmium	—
—	Rhodium	—
—	Iridium	—
—	Platinum	—

Source: From Betteridge, W., *Cobalt and its Alloys*, Ellis Horwood Limited, Chichester, 1982.

where A, B, and C are the stacking sequences of an FCC crystal. However, the passage of a Shockley partial dislocation on every adjacent {111} close-packed plane results in the formation of an FCC twin

$$ABCABC \Rightarrow ACBACB$$

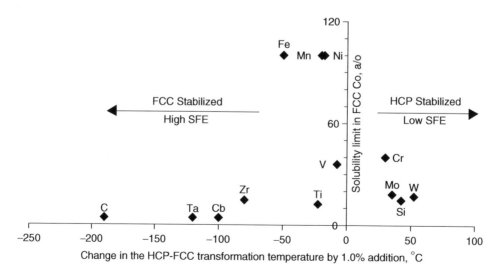

FIGURE 8.1 Effects of alloying additions on the HCP to FCC transformation temperature in cobalt as a function of solubility in the FCC cobalt. Note that SFE represents stackingfault energy. (From Beltran, M., Cobalt-base alloys, In *Superalloys II*, Sims, C. T., Stoloff, N. S., and Hagel, W. C., Eds., pp. 135–164, 1987. With permission.)

The formation of the HCP phase or FCC twins of the macroscopic size may be considered as a nucleation and growth process. Once the HCP phase or FCC twin embryo is formed by dislocation interactions, it may then thicken by the overlapping of other nuclei formed on the {111} planes parallel to its habit planes. The dislocation density also plays an important role for the formation of the HCP phase and FCC twins.

8.3 OCCURRENCE, EXTRACTION, AND END USES OF COBALT

8.3.1 OCCURRENCE

Cobalt occurs naturally in widespread form, and is found in small quantities in many rocks, coals, soils, plant lives, and manganese-rich marine nodules [5]. However, the main source of cobalt arises as a byproduct of other more abundant metals, mainly copper and nickel, but also zinc, lead, and platinum-group metals [5]. Table 8.4 lists the principal cobalt-containing ores, along with the composition and location [8]. The largest supplier of cobalt is found in Africa in the Democratic Republic of Congo (formerly Zaire) and Zambia. Canada, Russia, Australia, Finland, and Norway are other countries where the mining and refining of cobalt are important.

8.3.2 EXTRACTION AND PROCESSING

Since cobalt production is usually subsidiary to that of copper and nickel and other more abundant metals, extraction procedures vary and are usually developed individually to suit the particular characteristics of the ore being dealt with [5]. Processes such as hydrometallurgical, pyrometallurgical, vapormetallurgical, electrolytic, or combinations of these, are employed to extract and refine cobalt.

8.3.3 END USES OF COBALT

In the U.S., superalloys (Co-based, Ni-based, and Fe-based) make up 47% of all cobalt consumed, versus 24% worldwide [8]. Metallurgical uses account for more than 70% of the cobalt consumed in the U.S. In some countries, most notably Japan, nonmetallurgical uses account for more than 50% of the cobalt consumed. One of the fastest growing markets for cobalt use is in rechargeable batteries used for portable electronic devices, such as mobile phones, camcorders, and computers.

TABLE 8.4
The Principal Cobalt-Containing Ores

Ore	Composition	Location
Skutterudite	$(Co, Ni)As_3$	Bou-Azzer, Morocco; Cobalt, Ontario, Canada
Carrolite	$CuCo_2S_4$ or $CuSCo_2S_3$	Congo; Zambia
Linnaeite	$Co_3S_4(+Ni, Cu, Fe)$	Congo; U.S. (Mississippi Valley); Zambia
Cattierite	CoS_2	Congo
Nickel cattierite	$(Co, Ni)S_2$	Congo
Cobaltite	$CoAsS$	U.S. (Idaho); Cobalt, Ontario, Canada
Heterogenite	$2Co_2O_3CuO \cdot 6H_2O$	Congo
Asbolite	Mixed manganese–iron oxides + Co	Congo; Zambia; New Caledonia
Pyrrhotite	Ni, Cu, (Co) sulfides	Sudbury, Ontario, Canada
Latterites	Weathered igneous rock containing silicates, magnesium oxide, limonite, and 2–3% Ni + Co	Moa Bay, Cuba; Various locations in the U.S., Australia, Brazil, and Russia

Source: From Davis, J. R., *Nickel, Cobalt, and their Alloys—ASM Specialty Handbook*, ASM International, Metals Park, OH, pp. 343–406, 2000.

8.4 MECHANICAL AND PHYSICAL PROPERTIES

8.4.1 PURE COBALT

Wide variations exist among the reported values of the physical and mechanical properties of cobalt. Many of the variations are due to the sluggish nature of the $\varepsilon \rightarrow \alpha$ transformation. The fundamental reason for the sluggishness, and its sensitivity to experimental conditions, is due to the very low free-energy change (about 500 J mol^{-1} for $\varepsilon \rightarrow \alpha$ and about 360 J mol^{-1} for $\alpha \rightarrow \varepsilon$) associated with the FCC to HCP transformation [5]. In addition, physical and mechanical properties are greatly influenced by the orientation of the crystals in the HCP structure, as well as the purity of the material.

Cobalt has an atomic number of 27, which puts it between iron and nickel in the periodic table. The density of cobalt is 8.85 g/cm^3, which is similar to that of nickel (8.902 g/cm^3). Cobalt is ferromagnetic, exhibiting a room-temperature initial permeability of 68, a max permeability of 245, and a Curie temperature of 1,121°C [8].

The elastic moduli of cobalt are 211 GPa for Young's modulus, 82 GPa for shear modulus, and 183 GPa for compressive modulus [8]. Figure 8.2 shows the variations of the moduli with temperature, exhibiting an inflection in the region near the transformation temperature [5]. Cobalt has a 0.2% yield stress of 305–345 MPa, a tensile strength of 800–875 MPa, and an elongation of 15–30% [5]. Figure 8.3 shows the tensile properties of the pure cobalt as a function of temperature [5]. At around 500°C, a maximum in ductility is seen, which, consequently, is the ideal temperature for working cobalt. The Poisson's ratio ranges from 0.29 to 0.32 [5].

The elastic moduli of HCP single crystals of cobalt are dependent on the orientation of the crystal and the crystallographic direction in which they are measured. Table 8.5 gives values of Young's moduli and shear moduli at room temperature for different directions, along with the expected moduli for randomly oriented polycrystalline materials based on the single crystal values [5].

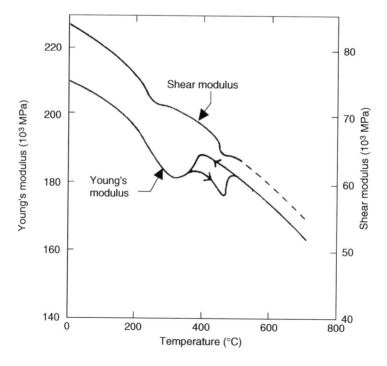

FIGURE 8.2 Variation of moduli with temperature. (From Betteridge, W., *Cobalt and its Alloys*, Ellis Horwood Limited, Chichester, 1982. With permission.)

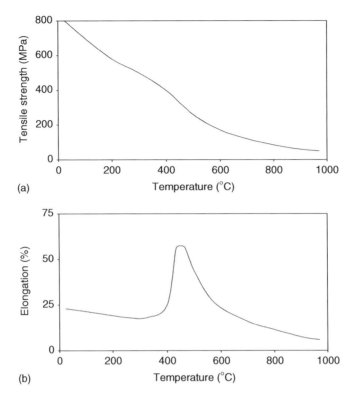

FIGURE 8.3 Tensile properties of pure cobalt as a function of temperature, (a) tensile strength as a function of temperature, (b) elongation as a function of temperature. (From Betteridge, W., *Cobalt and its Alloys*, Ellis Horwood Limited, Chichester, 1982. With permission.)

Creep properties of the pure cobalt are of little interest. However, it has been established that the activation energy for creep is higher for the hexagonal cobalt than for the cubic cobalt [5]. The corrosion rate of cobalt is found to be of the same order as that of nickel in many dilute media [5].

For the well-annealed and high-purity samples, hardness values fall between 140 and 160 HV [5]. Figure 8.4 shows how hardness varies with temperature [5]. No marked inflection is seen at the transformation temperature. Microhardness measurements on single crystals present results that are dependent on the orientation, with values varying from 81 to 250 HV [5]. The low values are related to the deformation by the slip on the basal planes of the hexagonal form, and the high values result from twinning. Other important physical properties of cobalt are given in Table 8.6 [8].

TABLE 8.5
Elastic Moduli at Room Temperature for Different Crystallographic Directions

Crystallographic Direction	Young's Modulus (103 MPa)	Shear Modulus (103 MPa)
0001	213	62.4
10–12	169	74.1
10–10	175	62.2
11–20	174	62.2
Polycrystal	174	64.8

Source: From Betteridge, W., *Cobalt and its Alloys*, Ellis Horwood Limited, Chichester, 1982.

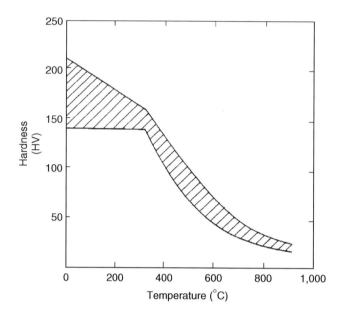

FIGURE 8.4 Hardness of cobalt as a function of temperature. (From Betteridge, W., *Cobalt and its Alloys*, Ellis Horwood Limited, Chichester, 1982. With permission.)

8.4.2 COBALT-BASED WEAR-RESISTANT ALLOYS

Cobalt-based alloys have seen use in wear-related engineering applications for more than 70 years due to their inherent high-strength, corrosion resistance, and ability to retain hardnesses at elevated temperatures. The pioneering work on cobalt-based wear-resistant alloys was carried out by Elwood Haynes in the early 1900s, and the cobalt-based alloys of today have changed only slightly from the early alloys developed by Haynes.

Table 8.7 lists compositions of the popular cobalt-based STELLITE wear-resistant alloys [8]. The typical microstructure of these alloys consists of chromium-rich M_7C_3 carbides, and in higher

TABLE 8.6
Important Physical Properties of Pure Cobalt

Property	Value
Boiling point	3,100°C
Coefficient of thermal expansion	13.8 μm/m K (at room temperature)
Specific heat	0.0414 kJ/kg K (at room temperature)
Thermal conductivity	69.04 W/m K (at room temperature)
Electrical conductivity	27.6% IACS (at 20°C)
Electrical resistivity	52.5 nΩ m (at 20°C)
Coercive force	8.9 Oe (708 A/m)
Saturation magnetization	1.87 T (18.7 kG)
Residual magnetization	0.49 T (4,900 G)

Source: From Davis, J. R., *Nickel, Cobalt, and their Alloys—ASM Specialty Handbook*, ASM International, Metals Park, OH, pp. 343–406, 2000.

TABLE 8.7
Compositions of Various Cobalt-Based Stellite Wear-Resistant Alloys

Alloy Tradename	Co	Cr	W	Mo	C	Fe	Ni	Si	Mn	Others
						Nominal Composition (wt.%)				
STELLITE1	Bal.	30	13	0.5	2.5	3	1.5	1.3	0.5	—
STELLITE 3 (P/M)	Bal.	30.5	12.5	—	2.4	5 (max.)	3.5 (max.)	2 (max.)	2 (max.)	1 B (max.)
STELLITE 4	Bal.	30	14	1 (max.)	0.57	3 (max.)	3 (max.)	2 (max.)	1 (max.)	—
STELLITE 6	Bal.	29	4.5	1.5 (max.)	1.2	3 (max.)	3 (max.)	1.5 (max.)	1 (max.)	—
STELLITE 6 (P/M)	Bal.	28.5	4.5	1.5 (max.)	1	5 (max.)	3 (max.)	2 (max.)	2 (max.)	1 B (max.)
STELLITE 12	Bal.	30	8.3	—	1.4	3 (min.)	1.5	0.7	2.5	—
STELLITE 21	Bal.	27	—	5.5	0.25	3 (max.)	2.75	1 (max.)	1 (max.)	0.007 B (max.)
STELLITE 98M2 (P/M)	Bal.	30	18.5	0.8 (max.)	2	5 (max.)	3.5	1 (max.)	1 (max.)	4.2 V, 1 B (max.)
STELLITE 703	Bal.	32	—	12	2.4	3 (max.)	3 (max.)	1.5 (max.)	1.5 (max.)	—
STELLITE 706	Bal.	29	—	5	1.2	3 (max.)	3 (max.)	1.5 (max.)	1.5 (max.)	—
STELLITE 712	Bal.	29	—	8.5	2	3 (max.)	3 (max.)	1.5 (max.)	1.5 (max.)	—
STELLITE 720	Bal.	33	—	18	2.5	3 (max.)	3 (max.)	1.5 (max.)	1.5 (max.)	0.3 B
STELLITE F	Bal.	25	12.3	1 (max.)	1.75	3 (max.)	22	2 (max.)	1 (max.)	—
STELLITE STAR J (P/M)	Bal.	32.5	17.5	—	2.5	3 (max.)	2.5 (max.)	2 (max.)	2 (max.)	1 B (max.)
STELLITE STAR J	Bal.	32.5	17.5	—	2.5	3 (max.)	2.5 (max.)	2 (max.)	2 (max.)	—
TRIBALOY T-400	Bal.	9	—	29	—	—	—	2.5	—	—
TRIBALOY T-800	Bal.	18	—	29	—	—	—	3.5	—	—
STELLITE 6B	Bal.	30	4	1.5 (max.)	1	3 (max.)	2.5	0.7	1.4	—
STELLITE 6K	Bal.	30	4.5	1.5 (max.)	1.6	3 (max.)	3 (max.)	2 (max.)	2 (max.)	—

Source: From Davis, J. R., *Nickel, Cobalt, and their Alloys–ASM Specialty Handbook*, ASM International, Metals Park, OH, pp. 343–406, 2000.

tungsten alloys, tungsten-rich M_6C type carbides are dispersed in a cobalt-rich solid-solution matrix. Chromium is the main carbide former in these alloys, and plays a crucial role in providing added strength to the matrix, as well as increasing resistance to corrosion and oxidation. Due to their large atomic sizes, additions of tungsten and molybdenum provide additional strengths to the matrix, as well as providing additional corrosion resistances. Figure 8.5 shows the typical microstructure of STELLITE alloys 1, 6, 12, and 21 [8].

Hardness and resistances to the abrasive wear in these alloys are strongly influenced by the carbide-volume fraction and morphology, which is determined by the cooling rate and subtle chemistry changes. The matrix composition is, however, equally important for resistances to the adhesive and erosive wear, and is determined by the chromium, tungsten, molybdenum, and carbon content levels.

STELLITE alloys 6B and 6K (Table 8.7) are wrought versions of the previously described alloys [9]. Chemical homogeneity, ductility, and abrasion resistance (due to the formation of

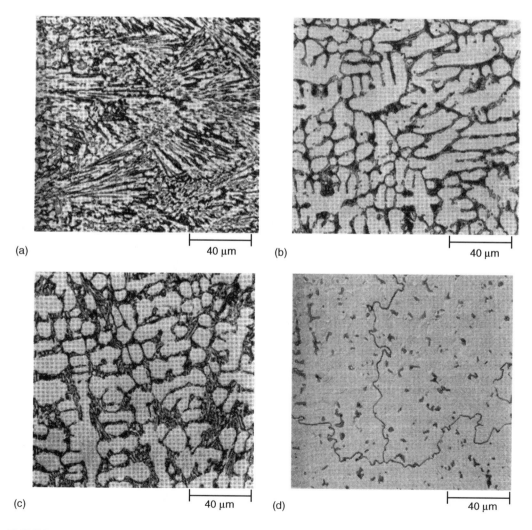

(a)   40 μm   (b)   40 μm

(c)   40 μm   (d)   40 μm

FIGURE 8.5 Typical microstructures of (a) Stellite 1, (b) Stellite 6, (c) Stellite 12, and (d) Stellite 21. (From Davis, J. R., Introduction to Cobalt and Cobalt Alloys, *Nickel, Cobalt, and their Alloys-ASM Specialty Handbook*, ASM International, Metals Park, OH, pp. 343–406, 2000. With permission.)

coarse, blocky carbides) are all improved by the wrought process. Along with excellent wear resistances, these alloys have good resistances to a variety of corrosive media. Again, chromium is the main carbide former and provides strength and corrosion resistances to the solid-solution, while tungsten provides the additional solid-solution strengths. STELLITE alloys 703, 706, 712, and 720 (Table 8.7) employ molybdenum to provide strength and to assist in dividing the carbides to enhance wear resistances further [8]. These alloys have excellent corrosion-resistances in reducing environments [8]. The powder-metallurgy (P/M) versions of several STELLITE alloys (Table 8.7) that have been developed have excellent wear resistances and mechanical properties. These alloys also provide a cost-effective means of producing small, simple shapes in large quantities [8]. It has been shown that the hot-isostatically-pressed (HIPed) STELLITE alloy 6 possesses superior erosion and erosion–corrosion resistances when compared with that of the cast STELLITE alloy 6 [10].

The TRIBALOY alloys, such as T-400 and T-800 (Table 8.9), are from an alloy family developed by DuPont in the early 1970s [11]. These alloys employ excessive amounts of molybdenum and silicon to promote the formation during the solidification of a hard and corrosion-resistant intermetallic compound known as a Laves phase [11]. This intermetallic compound can be of the MgZn type, or could possibly have a more complex composition such as CoSiMo or Co_2W, depending on the composition of the alloy. The microstructure of the Tribaloy T-800 is shown in Figure 8.6 [9]. These alloys are usually fabricated by means of casting, hardfacing, powder metallurgy, or plasma spraying. Unlike the STELLITE alloys, the Laves phase gives these alloys their hardnesses and excellent abrasion resistances, rather than carbides. To prevent carbides from forming, the carbon content is held as low as possible. Because the Laves phase makes up 35–70 weight percent (wt.%) of the microstructure, it determines all of the properties. As a result, these alloys have limited ductilities and impact strengths.

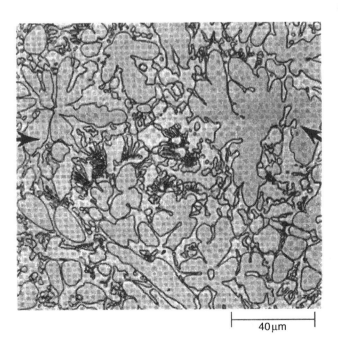

40 μm

FIGURE 8.6 Typical microstructure of Tribaloy T-800, showing the Laves precipitates (the largest continuous precipitates some of which are indicated with arrows). (From Crook, P., Cobalt and Cobalt Alloys, *Metals Handbook*, Vol. 2, pp. 446–454, 1990. With permission.)

TABLE 8.8
Properties of Selected Cobalt-Based Wear-Resistant Alloys

Property	STELLITE 21	STELLITE 6	STELLITE 12	STELLITE 1	Tribaloy T-800
Density (g/cm^3)	8.3	8.3	8.6	8.6	8.6
Ultimate compressive strength (MPa)	1295	1515	1765	1930	1780
Ultimate tensile strength (MPa)	710	834	827	620	690
Elongation (%)	8	1.2	1	1	<1
Coefficient of thermal expansion (°C^{-1})	14.8×10^{-6}	15.7×10^{-6}	14×10^{-6}	13.1×10^{-6}	12.3×10^{-6}
Hot hardness, HV, at:					
445°C	150	300	345	510	659
540°C	145	275	325	465	622
650°C	135	260	285	390	490
760°C	115	185	245	230	308
Unlubricated sliding wear,[a] (mm^3) at:					
670 N	5.2	2.6	2.4	0.6	1.7
1,330 N	14.5	18.8	18.4	0.8	2.1
Abrasive wear,[b] (mm^3)					
OAW	—	29	12	8	—
GTAW	86	64	57	52	24
Unnotched Charpy impact strength (J)	32	23	5	5	1.4
Corrosion resistance[c]:					
65% nitric acid at 65°C	U	U	U	U	S
5% sulfuric acid at 65°C	E	E	E	E	—
50% phosphoric acid at 400°C	E	E	E	E	E

[a] Wear measured from tests conducted on Dow-Corning LFW-1 against 4620 steel ring at 80 rev/min for 200 rev varying the applied load.
[b] Wear measured from dry sand rubber wheel abrasion tests. Tested for 2,000 rev at a load of 135 N using a 230 mm diam. rubber wheel and American Foundrymen's Society test sand. OAW, oxyacetylene welding; GTAW, gas-tungsten arc welding.
[c] E, less than 0.05 mm/yr; S, 0.5 to less than 1.25 mm/yr; U, more than 1.25 mm/yr.
Source: From Crook, P., *Metals Handbook*, Vol. 2, pp. 446–454, 1990.

Properties of selected cobalt-based wear-resistant alloys are presented in Table 8.8 [9].

8.4.3 COBALT-BASED HIGH-TEMPERATURE ALLOYS

Today, cobalt-based high-temperature alloys are not as widely used as nickel and nickel–iron superalloys. However, due to unique properties such as high-melting temperatures and correspondently flatter stress-rupture curves, superior hot-corrosion resistances to contaminated gas-turbine atmospheres, and superior fatigue resistances and weldabilities, cobalt-based alloys still find significant uses in the gas-turbine industry [7].

STELLITE and TRIBALOY are trademarks of Deloro Stellite, Inc.

Cobalt-based alloys high-temperature strength stems from a careful balance of alloying elements (such as molybdenum, tungsten, tantalum, and niobium) that strengthen the solid-solution, and along with carbon, promote the precipitation of carbides. Carbides that precipitate are in the form of MC, M_6C, M_7C_3, $M_{23}C_6$, and occasionally, M_2C_3. Forms present depend on the composition, carbon content, and thermal history of the alloy. To achieve optimal strengthening, the precipitation of carbides at both grain boundaries and within grains is necessary. The strengthening effect is accomplished by (a) increasing temperatures of the effective recovery and recrystallization processes, (b) reducing the stacking-fault energy, (c) increasing the lattice friction and causing the chemical pinning of dislocations, (d) retarding dislocation glides and climbs, (e) increasing the stability of the solid solution, and (f) increasing the valence [6]. The roles of various alloying elements in cobalt-based high-temperature alloys are summarized in Table 8.9 [12].

High-temperature cobalt-based alloys are typically divided into two categories: wrought and cast. Typical compositions of wrought cobalt-based alloys are given in Table 8.10 [8]. Alloys such as these are considerably more ductile, oxidation resistant, and microstructurally stable than the wear-resistant wrought cobalt-base alloys. Creep and stress-rupture properties can be improved by a solution treatment, followed by rapid cooling. An aging or precipitation treatment is not usually done, since this trend would lead to rapid overaging during service. HAYNES 25 alloy possesses excellent high-temperature strengths (particularly for long-term applications at temperatures of 650–980°C), good resistances to oxidizing environments up to 980°C for prolonged exposures, excellent resistances to sulfidation, as well as excellent resistances to metal galling [13]. Figure 8.7a shows the microstructure of alloy 25 [8]. HAYNES 188 alloy has excellent high-temperature strengths (particularly effective at temperatures of 650°C or more), is resistant to oxidizing environments up to 1,095°C for prolonged exposures, as well as excellent resistances to sulfate-deposit hot corrosion, molten chloride salts, and gaseous sulfidation [13]. Figure 8.7b presents the microstructure of the 188 alloy [8].

TABLE 8.9
The Roles of Various Alloying Elements in Cobalt-Based High-Temperature Alloys

Element	Effect
Chromium	Improves oxidation and hot corrosion resistance; produces strengthening by the formation of M_7C_3 and $M_{23}C_6$ carbides
Nickel	Stabilizes the fcc form of matrix; produces strengthening by the formation of the intermetallic compound, Ni_3Ti; improves forgeabilities
Molybdenum, tungsten	Solid-solution strengtheners; produces strengthening by the formation of intermetallic compounds Co_3M; formation of M_6C carbides
Tantalum, niobium	Solid-solution strengtheners; produces strengthening by formation of intermetallic compounds, Co_3M and MC carbide; formation of M_6C carbides
Carbon	Produces strengthening by the formation of carbides MC, M_7C_3, $M_{23}C_6$, and possibly M_6C
Aluminum	Improves oxidation resistances; formation of carbides, MC, M_7C_3, $M_{23}C_6$, and possibly M_6C
Titanium	Produces strengthening by the formation of the MC carbides and intermetallic compound Co_3Ti with sufficient nickel produces strengthening by the formation of the intermetallic compound, Ni_3Ti
Boron, zirconium	Produces strengthening by the effect on grain boundaries and by precipitate formation; zirconium produces strengthening by the formation of MC carbides
Yttrium, lanthanum	Increase oxidation resistance

Source: From Sullivan, C. P., Donchie, M. J., and Morral, F. R., *Cobalt-Base Superalloys-1979*, Cobalt Information Center, Brussels, Belgium, 1980.

TABLE 8.10
Typical Wrought Cobalt-Based Heat Resistant Alloy Compositions

Alloy Tradename	Nominal Composition (wt.%)									
	Co	Cr	W	Mo	C	Fe	Ni	Si	Mn	Other
HAYNES 25 (L605)	Bal.	20	15	—	0.1	3 (max.)	10	0.4 (max.)	1.5	—
HAYNES 188	Bal.	22	14	—	0.1	3 (max.)	22	0.35	1.25	0.03 La
INCONEL 783	Bal.	3	—	—	0.03 (max.)	25.5	28	0.5 (max.)	0.5 (max.)	5.5 Al, 3 Nb, 3.4 Ti (max.)
UMCo-50	Bal.	28	—	—	0.02 (max.)	21	—	0.75	0.75	—
S-816	40 (min.)	20	4	4	0.37	5 (max.)	20	1 (max.)	1.5	4 Nb

Source: From Davis, J. R., *Nickel, Cobalt, and their Alloys—ASM Specialty Handbook*, ASM International, Metals Park, OH, pp. 343–406, 2000.

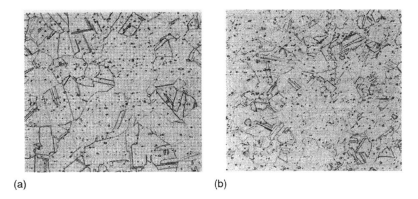

(a)   (b)

FIGURE 8.7 Microstructure of two wrought cobalt-base high-temperature alloys etched with hydrochloric acid/oxalic acid solution. (a) Alloy 25. (b) Alloy 188. Both at 200x. (From Davis, J.R., Introduction to Cobalt and Cobalt Alloys, *Nickel, Cobalt, and their Alloys-ASM Specialty Handbook*, ASM International, Metals Park, OH, pp. 343-406, 2000. With permission.)

Wrought cobalt-based high-temperature alloys exhibit rapid hardening with respect to cold working when their composition is not fully stabilized relative to the FCC to HCP formation, as illustrated for 25 (10 wt.% Ni) and 188 (22 wt.% Ni) alloys in Figure 8.8 [3]. These alloys tend to have relatively high yield strength properties over a broad temperature range, and good high-temperature stress-rupture strengths when compared with other solid-solution alloys (such as X alloy), as shown in Figure 8.9 [3]. Though the properties seem to be better for the γ'-strengthened alloy, such as the R-41 alloy (Figure 8.9), many times solid-solution-strengthened alloys are preferred because of their ease of fabrication. The stress to produce 1% creep in 1,000 hours for 188 alloy is given in Figure 8.10 [13]. Data for some nickel-based alloys are shown for comparative purposes.

Deleterious intermetallic compounds, such as the σ, μ, and Laves phases, can precipitate in these alloys [3]. The precipitation of these phases can cause embrittlement, mainly at low temperatures. Some success in retarding the formation of the Laves phase was achieved by the development of 188 alloy [3]. However, if the material is held in the 760–870°C temperature range for sufficient

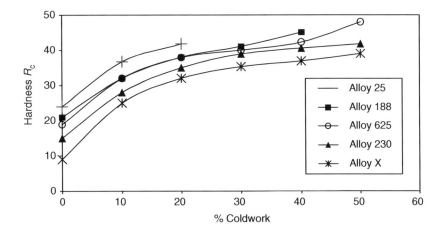

FIGURE 8.8 Hardness versus cold work for cold reduced sheet of various alloys. (From Klarstrom, D.L., Wrought Cobalt-Base Superalloys, *J. Mater. Eng. Perform.*, 2, pp. 523-530, 1993. With permission.)

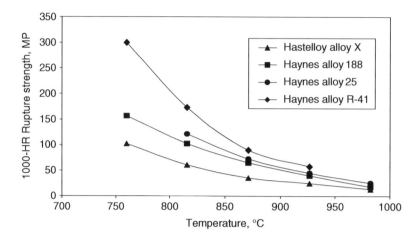

FIGURE 8.9 Comparative 1,000-h stress-rupture strengths for high-temperature sheet alloys. (From Klarstrom, D. L., Wrought Cobalt-Base Superalloys, *J. Mater. Eng. Perform.*, 2, pp. 523–530, 1993. With permission.)

time, it has been found to cause enough precipitations of the Laves phase to cause severe embrittlement [3].

Compositions of typical cast cobalt-based high-temperature alloys are given in Table 8.11 [8]. These alloys have chromium and tungsten-rich carbides dispersed in a cobalt–chromium matrix, with the chromium and tungsten providing additional solid-solution strengthening. The distribution of the carbides is determined by the solidification conditions. Most cast alloys do not undergo any further heat treatment, other than stress relieving at temperatures too low to affect the distribution of the carbides, so that the initial microstructure remains that of the as-cast condition. However, some additional carbide precipitation can occur during service. The resistances to oxidation, hot corrosion, and sulfidation are almost entirely dependent on the chromium content in these alloys.

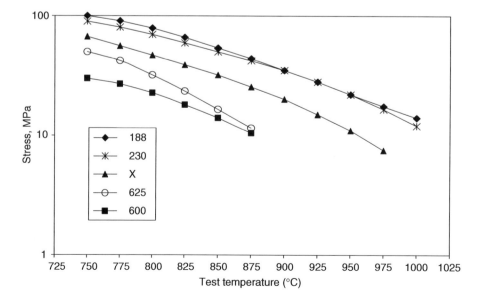

FIGURE 8.10 Stress to produce 1% creep in 1,000 h. (From Haynes International, Kokomo, IN. http://www. haynesintl.com With permission.)

TABLE 8.11
Typical Cast Cobalt-Based Heat-Resistant Alloy Compositions

Alloy Tradename	Nominal Composition (wt.%)												
	C	Ni	Cr	Co	Mo	Fe	Al	B	Ti	Ta	W	Zr	Other
AiResist 13	0.45	—	21	62	—	—	3.4	—	—	2	11	—	0.1Y
AiResist 213	0.20	0.5	20	64	—	0.5	3.5	—	—	6.5	4.5	0.1	0.1Y
AiResist 215	0.35	0.5	19	63	—	0.5	4.3	—	—	7.5	4.5	0.1	0.1Y
FSX-414	0.25	10	29	52.5	—	1	—	0.010	—	—	7.5	—	—
HAYNES 25 (L-605)	0.1	10	20	54	—	1	—	—	—	—	15	—	—
J-1650	0.20	27	19	36	—	—	—	0.02	3.8	2	12	—	—
MAR-M 302	0.85	—	21.5	58	—	0.5	—	0.005	—	9	10	0.2	—
MAR-M 322	1.0	—	21.5	60.5	—	0.5	—	—	0.75	4.5	9	2	—
MAR-M 509	0.6	10	23.5	54.5	—	—	—	—	0.2	3.5	7	0.5	—
NASA Co-W-Re	0.40	—	3	67.5	—	—	—	—	1	—	25	1	2 Re
S-816	0.4	20	20	42	—	4	—	—	—	—	4	—	4 Mo, 4 Nb, 1.2 Mn, 0.4 Si
V-36	0.27	20	25	42	—	3	—	—	—	—	2	—	4 Mo, 2 Nb, 1 Mn, 0.4 Si
WI-52	0.45	—	21	63.5	—	2	—	—	—	—	11	—	2 Nb + Ta
STELLITE 23	0.40	2	24	65.5	—	1	—	—	—	—	5	—	0.3 Mn, 0.6 Si
STELLITE 27	0.40	32	25	35	5.5	1	—	—	—	—	—	—	0.3 Mn, 0.6 Si
STELLITE 30	0.45	15	26	50.5	6	1	—	—	—	—	—	—	0.6 Mn, 0.6 Si
STELLITE 31 (X-40)	0.50	10	22	57.5	—	1.5	—	—	—	—	7.5	—	0.5 Mn, 0.5 Si

Source: From Davis, J. R., *Nickel, Cobalt, and their Alloys—ASM Specialty Handbook*, ASM International, Metals Park, OH, pp. 343–406, 2000.

Stress-rupture properties of commonly used cast cobalt-based alloys are given in Figure 8.11 [8]. Mechanical and physical property data of selected alloys are given in Table 8.12 [9].

8.4.4 COBALT-BASED CORROSION-RESISTANT ALLOYS

The electrochemical behavior of cobalt is similar to that of nickel, and, in general, the corrosion resistance of cobalt is slightly inferior to that of nickel [14]. Moreover, since cobalt is more expensive than nickel, its use is only justified for special applications (requiring erosion, cavitation, and galling resistances) where nickel-based alloys would not perform well. Table 8.13 lists some of the typical cobalt-based corrosion-resistant alloys and their respective compositions [8]. These Co–Ni–Cr–Mo alloys still exhibit some characteristics typical of other cobalt-based alloys, which are resistant to wear and have strengths at elevated temperatures. These alloys have outstanding resistances to aqueous corrosion, yet still have some of the advantages (wear resistances and high-temperature strength) that stem from using cobalt as the base alloy.

The influence of alloying elements in cobalt-based corrosion-resistant alloys are similar to that for nickel-based alloys [14]. Increasing molybdenum and tungsten contents tends to increase resistances to reducing conditions (such as, hydrochloric and sulfuric acids), while the addition

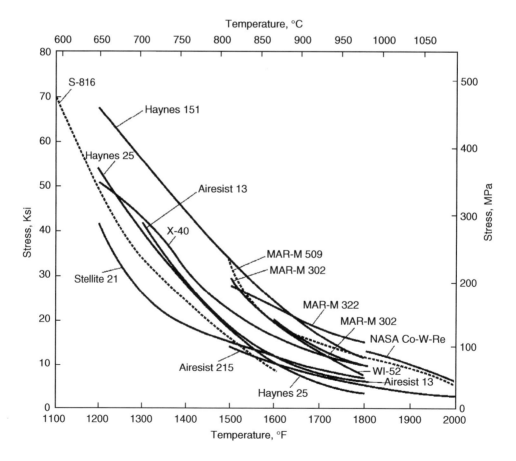

FIGURE 8.11 Stress-rupture properties of commonly used cast cobalt-based alloys. (From Davis, J. R., Introduction to Cobalt and Cobalt Alloys, *Nickel, Cobalt, and their Alloys-ASM Specialty Handbook*, ASM International, Metals Park, OH, pp. 343–406, 2000. With permission.)

TABLE 8.12

Mechanical and Physical Properties of Selected Cobalt-Based High-Temperature Alloys

	Alloy		
Property	25	188	MAR-M 509
	Yield strength (MPa)		
At 21°C	445[a]	464[b]	585[c]
At 540°C	—	305[d]	400[c]
	Tensile strength (MPa)		
At 21°C	970[a]	945[b]	780[c]
At 540°C	800[c]	740[d]	570[c]
	1,000 h rupture strength (MPa)		
At 870°C	75	70	140
At 980°C	30	30	90
Elongation (%)	62[a]	53[b]	3.5[c]
	Thermal expansion coefficient (μm/m K)		
21–93°C	12.3[e]	11.9	—
21–540°C	14.4	14.8	—
21–1,090°C	17.7	18.5	—
	Thermal conductivity (W/m K)		
At 20°C	9.8[f]	10.8	—
At 500°C	18.5[g]	19.9	—
At 900°C	26.5[h]	25.1	—
Specific gravity	9.13	8.98	8.86
Electrical resistivity (μΩ m)	0.89	1.01	—
	Melting range (°C)		
Solidus	1,329	1,302	1,290
Liquidus	1,410	1,330	1,400

[a] Sheet 3.2 mm thick.

[b] Sheet 0.75–1.33 mm thick.

[c] As-cast.

[d] Sheet heat treated at 1,175°C for 1 h with rapid air cool.

[e] Sheet, heat treated at 1,230°C for 1 h with rapid air cool.

[f] At 38°C.

[g] At 540°C.

[h] At 815°C.

Source: From Crook, P., *Metals Handbook*, Vol. 2, pp. 446–454, 1990.

of chromium gives these alloys resistances to oxidizing conditions (such as, nitric acid) [14]. Carbon contents are held within the soluble range to improve resistances to the heat-affected-zone sensitization during welding [8].

A careful balance of the chromium, molybdenum, tungsten, and nickel contents determines the resistances to the localized corrosion, although their mechanism of protection is not well understood [14]. These alloys are more resistant to stress–corrosion cracking than nickel-based alloys. However, due to their metastable FCC structure with the low stacking-fault energy that promotes the planar slip, these alloys are susceptible to transgranular stress–corrosion cracking [14].

Low-carbon ULTIMET corrosion-resistant alloy exhibits outstanding wear resistances that are comparable to those of the STELLITE alloys, which makes this alloy suitable for applications in environments that contain the combined effects of corrosion and wear [14]. This alloy does not

TABLE 8.13
Compositions of Typical Cobalt-Based Corrosion-Resistant Alloys

Alloy Tradename	Nominal Composition (wt.%)									
	Co	Cr	W	Mo	C	Fe	Ni	Si	Mn	Others
ULTIMET	Bal.	26	2	5	0.06	3	9	0.3	0.8	0.08 N
MP159	Bal.	19	—	7	—	9	25.5	—	—	3 Ti, 0.6 Nb, 0.2 Al
MP35N	35	20	—	10	—	—	35	—	—	—
Duratherm 600	41.5	12	3.9	4	0.05(max.)	8.7	Bal.	0.4	0.75	2 Ti, 0.7 Al, 0.05 Be
Elgiloy	40	20	—	7	0.15	Bal.	15.5	—	2	1 Be (max.)
Havar	42.5	20	2.8	2.4	0.2	Bal.	13	—	1.6	0.06 Be (max.)

Source: From Davis, J. R., *Nickel, Cobalt, and their Alloys—ASM Specialty Handbook*, ASM International, Metals Park, OH, pp. 343–406, 2000.

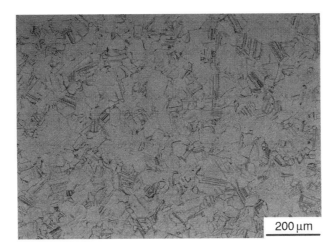

FIGURE 8.12 Characteristic microstructure of ULTIMET alloy. (From Jiang, L., *Fatigue Behavior of ULTIMET Alloy: Experiment & Theoretical Modeling*, Ph.D. Thesis, The University of Tennessee, Knoxville, 2000. With permission.)

exhibit the intragranular or grain-boundary carbide precipitation, but does show a sparse dispersion of nitrides [8]. The typical microstructure of the ULTIMET alloy is shown in Figure 8.12 [6]. It is resistant to localized corrosion and many aggressive environments, and possesses excellent resistances to slurry erosion, cavitation erosion, and galling [8]. This alloy also possesses excellent ductilities and resistances to weld cracking, which makes it suitable for welding applications [14].

Table 8.14 is a comparison of the corrosion rates for selected cobalt-, iron-, and nickel-based alloys in various solutions [9]. Table 8.15 lists mechanical and physical properties of selected cobalt-based corrosion-resistant alloys [9].

HAYNES and ULTIMET are trademarks of Haynes International, Inc.

TABLE 8.14
Comparison of Corrosion Rates for Selected Cobalt-, Iron-, and Nickel-Based Alloys in Various Solutions

	Corrosion Rate (mm/year)							
Alloy	Boiling 99% Acetic Acid	Boiling 65% Nitric Acid	Boiling 1% Hydro-chloric Acid	Boiling 2% Hydro-chloric Acid	54% P205 at 116°C	Boiling 10% Sulfuric Acid	Boiling ASTM-G28A Solution	Boiling ASTM-G28B Solution
ULTIMET	<0.01	0.15	0.01	13.49	0.19	2.52	0.20	0.02
C-276	<0.01	21.51	0.52	1.90	0.58	0.51	8.05	0.86
625	0.01	0.51	0.03	14.15	0.30	0.64	0.43	71.08
20Cb-3	0.11	0.19	1.80	5.77	0.92	0.40	0.25	69.08
316L	0.19	0.24	13.31	25.15	5.11	47.46	0.94	80.51

Source: From Crook, P., *Metals Handbook*, Vol. 2, pp. 446–454, 1990.

TABLE 8.15
Mechanical and Physical Properties of Selected Cobalt-Based Corrosion-Resistant Alloys

Property	ULTIMET[a]	MP35N
Hardness	28 HRC	90 HRB[b]
Yield strength (MPa)	558	380
Ultimate tensile strength (MPa)	1,020	890[b]
Elongation (%)	33	65[b]
Thermal expansion coefficient (μm/mK)		
21–93°C	—	12.8[c]
21–315°C	—	14.8[c]
21–540°C	—	15.7[c]
Thermal conductivity (W/mK)	—	11.2[c]
Electrical resistivity (μΩ m)	—	1.03[c]
Melting range (°C)		
Solidus	1,333	1,315
Liquidus	1,355	1,440

[a] 13 mm Plate, solution annealed.
[b] Cold-drawn bar, solution annealed.
[c] Work-strengthened and aged.
Source: From Crook, P., *Metals Handbook*, Vol. 2, pp. 446–454, 1990.

8.4.5   FATIGUE BEHAVIOR OF SELECTED COBALT-BASED ALLOYS

8.4.5.1   High-Cycle Fatigue Behavior

Liang et al. have performed extensive studies on the high-cycle fatigue behavior of the ULTIMET alloy [15,16]. In one study [15], the effects of the test frequency, the temperature increase during fatigue, and the change of crack-initiation sites from the surface to subsurface on fatigue lives were investigated. It was concluded that both 1,000 and 20 Hz high-cycle fatigue tests had comparable fatigue lives, as shown in Figure 8.13. During fatigue testing, plateaus were found in the stress versus number of cycles to failure (*S–N*) curves at an applied maximum stress level near the yield strength of the alloy, 586 MPa, regardless of the testing frequency. The plateau region was determined to be due to an FCC to HCP phase transformation induced by the plastic deformation, and changed the crack-initiation site from the surface to subsurface (Figure 8.13).

An advanced, high-speed, and high-sensitivity infrared (IR) imaging system was also employed in this study to record the temperature changes during high-cycle fatigue [17–27]. Because the damping energy per cycle and the accumulation rate of the dissipated energy was greater at 1,000 Hz than at 20 Hz, the temperature increase at 1,000 Hz was considerably higher than at 20 Hz.

The crack-initiation site changed from the surface to the subsurface, depending on the applied maximum stress level and environment, as shown in Figure 8.14 and Figure 8.15. The fracture surface exhibited typical two-stage fatigue-crack-growth process including (a) stage I fatigue-crack initiation and (b) stage II fatigue-crack propagation. In a second study [16], using a uniform design method, the effects of the *R* ratio, environment and maximum stress level on the fatigue life were systematically studied. It was found that there was no interaction among these three factors. A linear statistical model was presented to predict the effects of these factors. This model is

$$\log(N) = 8.2743 + 3.0968 \times R - 0.0050 \times \sigma + 0.6325 \times \text{Env}$$

where N is the fatigue life defined as the number of cycles to failure, R is the R ratio, σ is the applied maximum stress (MPa), and Env is 1 for the testing in a vacuum environment, and 0 for an air

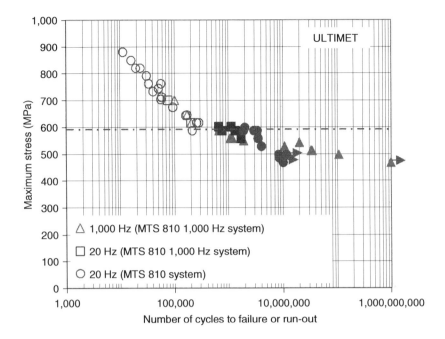

FIGURE 8.13 High-cycle fatigue testing results, SN data, of ULTIMET alloy with R ratio of0.05 in air at room temperature. Note that the opened symbol represents the fatigue crack initiated from the surface, the solid symbol stands for the fatigue crack initiated from the subsurface, the arrow indicates a run-out without failure, and the dashed line represents the applied maximum stress of the plateau region. (From Jiang, L., Brooks, C. R., Liaw, P. K., Wang, H., Rawn, C. J., and Klarstrom, D. L., High- Frequency Metal Fatigue: The High-Cycle Fatigue Behavior of ULTIMET Alloy, *Mat. Sci. Eng. A-Struct.*, 314(1–2), pp. 162–175, 2001. With permission.)

environment. It was found that the vacuum environment increased the fatigue life about four times over that in air. The *R* ratio also significantly affected the fatigue life.

It was found that the cracks initiated at the specimen surface when the maximum stress level was greater than 600 MPa. However, the cracks initiated at the subsurface of the specimen when the maximum stress level was lower than 600 MPa. All crack-initiation sites were located at the subsurface of the specimen when tested in a vacuum.

8.4.5.2 Low-Cycle Fatigue

Chen et al. [28] studied the effects of the tensile hold time on the cyclic deformation behavior and the low-cycle fatigue lifetime of HAYNES 188 alloy. It was concluded that the introduction of a hold time at the maximum tensile strain was found to lead to a significant reduction in the fatigue lifetime of 188 alloy. Table 8.16 shows the summary of the cyclic data obtained, and Figure 8.16 presents the relationship between the fatigue life and the hold time. The well-known Coffin–Mason equation of the type

$$\Delta\varepsilon_{in} \propto N_f^{-\beta}$$

was used to correlate the fatigue-life data. It was determined that the increased inelastic strain due to the stress relaxation during strain holding was responsible for the reduction in the fatigue life.

HAYNES 188 alloy also exhibited cyclic hardening, followed by softening or stability during low-cycle fatigue deformation, which mainly depended on the test temperature and the duration of the hold time. Stress relaxation occurred rapidly within the initial several decades of seconds of a

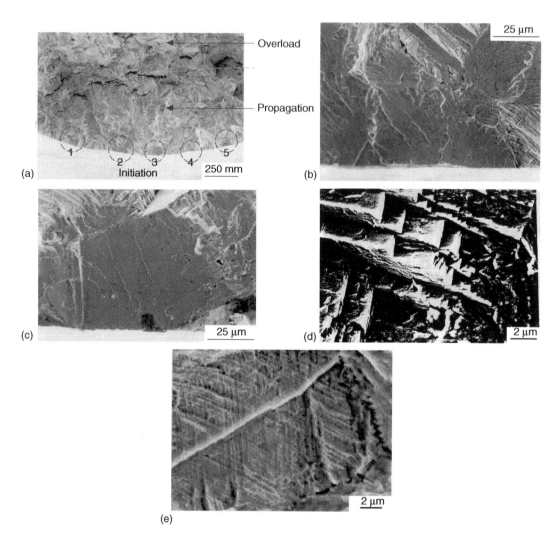

FIGURE 8.14 Crack initiation from the specimen surface, (a) Overview of the fractured surface and the crack initiation sites noted by the circles with numbers, (b) SEM fractograph of initiation site 2 from (a), (c) SEM fractograph of initiation Site 3 from (a), (d) 'Stair Step' feature of the crack-propagation region, and (e) Dimpled fracture surface of the final overload region. (σ_{max}=821 MPa, R=0.05, Room Temperature, in Air, Fatigue Life of 23,568 Cycles, and 20 Hz). (From Jiang, L., Brooks, C. R., Liaw, P. K., Wang, H., Rawn, C. J., and Klarstrom, D. L., High- Frequency Metal Fatigue: The High-Cycle Fatigue Behavior of ULTIMET Alloy, *Mat. Sci. Eng. A-Struct.*, 314(1–2), pp. 162–175, 2001. With permission.)

strain hold period, and then it continued at a diminishing rate in the remainder of the period. In the tests with a 60-minute hold time, the creep deformation was unconstrained at the test temperature of 816°C, while it became strain-limited at 927°C. For the hold-time tests, crack propagation occurred in a mixed transgranular and intergranular mode.

Jiang et al. [29] also studied the low-cycle fatigue behavior of the ULTIMET alloy at temperatures of 21°C, 600°C, and 900°C under isothermal conditions. It was concluded that, in general, the material possessed the longest fatigue life at 600°C due to the greater ductility at this temperature. The interaction of fatigue and oxidation at 900°C resulted in a significant reduction of the fatigue life.

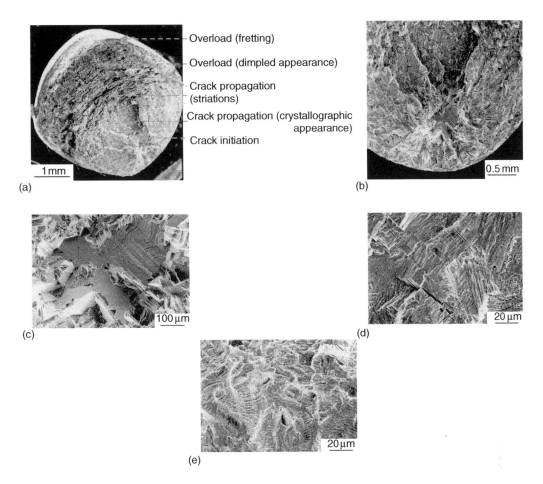

FIGURE 8.15 Crack initiation from the subsurface at an applied maximum stress level of 528 MPa at 1,000 Hz with an R Ratio of 0.05 at room temperature in air, (a) Overview, (b) Crack-initiation and propagation regions, (c) Cleavage-like crack-initiation site, (d) Crack-propagation region with crystallographic appearance, and (e) Crack-propagation region with striations. (From Jiang, L., Brooks, C. R., Liaw, P. K., Wang, H., Rawn, C. J., and Klarstrom, D. L., High- Frequency Metal Fatigue: The High-Cycle Fatigue Behavior of ULTIMET Alloy, *Mat. Sci. Eng. A-Struct.*, 314(1–2), pp. 162–175, 2001. With permission.)

The evolution of the peak stresses during low-cycle fatigue was strongly related to the test temperature, as shown in Figure 8.17. At 21°C, the material initially hardened, then softened moderately. At 600°C, the material initially hardened, then stabilized for a short period, followed by extensive hardening until the onset of macrocrack initiation and propagation. This difference was due to the greater amount of the HCP phase that formed at 600°C as compared with that at 21°C. At 900°C, the material experienced significant hardening until macrocracks formed, which was due to the precipitation of carbides.

At both 21°C and 600°C, a transgranular mode of fracture was observed, and was categorized as typical stage I crack initiation and stage II crack propagation. At 900°C, the alloy experienced moderate oxidation. Below strain ranges of 1.5%, fracture modes were similar to those at lower temperatures. However, at strain ranges of 1.5% or greater, the cracks initiated intergranularly, and fracture was of a mixed mode nature. At 21°C and 600°C, the plastic strain induced the FCC to HCP phase transformation. At 900°C, no phase transformation was observed.

TABLE 8.16
Summary of Cyclic Test Data on Haynes 188 Alloy

Test Temperature (°C)	Hold Time (min)	Number of Cycles to Failure	Time to Failure (min)	Strain Ranges (%) $\Delta\varepsilon_{in}$	$\Delta\varepsilon_{cp}$
816	0	359	5.98	0.497	—
	2	163	328.72	0.668	0.168
	10	251	2514.2	0.642	0.211
	60	97	5821.6	0.789	0.179
927	0	465	7.6	0.632	—
	2	276	556.6	0.789	0.184
	10	158	1582.6	0.781	0.208
	60	112	6721.9	0.829	0.220

Note that $\Delta\varepsilon_{in}$ and $\Delta\varepsilon_{cp}$ represent the inelastic strain range and the creep-strain component, respectively [17].

8.4.5.3 Temperature Evolution During Fatigue

Jiang et al. [30] studied the temperature variations of the ULTIMET alloy subjected to low-cycle fatigue using high-speed, high-resolution infrared thermography. Figure 8.18 presents the mean temperature and peak stress evolution of the specimen during low-cycle fatigue. It was found that the temperature evolution could be divided into three stages. Initially, the temperature increased due to the hysteresis energy dissipated during each fatigue cycle. The temperature then reached a

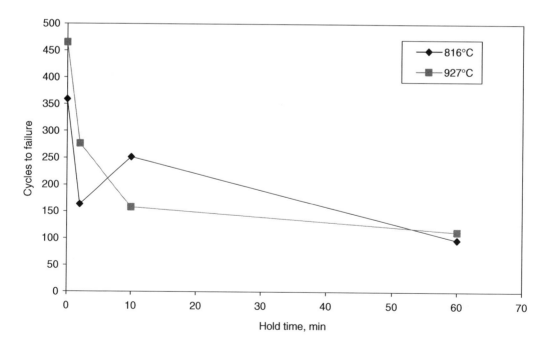

FIGURE 8.16 Relationship between the fatigue life and hold time for HAYNES 188. (From Chen, L. J., Liaw, P. K., He, Y. H., Benson, M. L., Blust, J. W., Browing, P. F., Seeley, R. R., and Klarstrom, D. L., Tensile Hold Low-Cycle Fatigue Behavior of Cobalt-Based Haynes 188 Superalloy, *Scripta Mater.*, 44, pp. 859–865, 2001. With permission.)

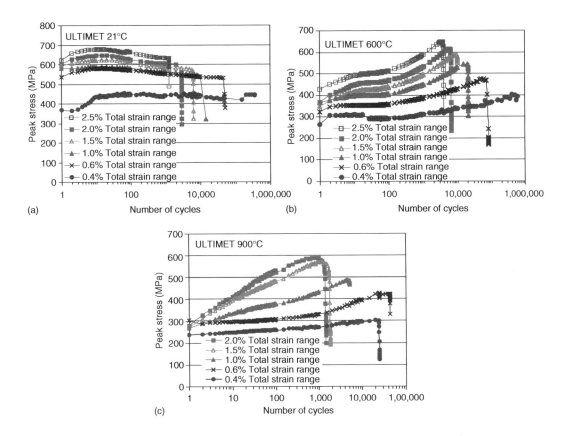

FIGURE 8.17 The Evolution Curves of the Peak Stresses of ULTIMET Alloy during Low-Cycle Fatigue at (a) 21° C, (b) 600° C, and (c) 900° C. Note that the valley stresses of the ULTIMET alloy were approximately symmetrical to the peak stresses during tensioncompression low-cycle fatigue. (From Jiang, L., Brooks, C. R., Liaw, P. K., Dunlap, J., Rawn, C. J., Peascoe, R., and Klarstrom, D. L., Low-Cycle Fatigue Behavior of ULTIMET alloy, *Metall. Mater. Trans. A*, 35(3), pp. 785–796, 2004. With permission.)

steady-state condition that resulted from the balance of the heat dissipation and the hysteresis energy associated with the stabilized state of the stress–strain response. Finally, the temperature dropped until the specimen failed due to the formation of microcracks, and since most of the plastic strain was concentrated in front of the microcracks, the generated hysteresis energy was decreased.

Figure 8.19 shows the typical material responses of the ULTIMET alloy during a fatigue test. Initially, the strain was in the elastic range for which Hooke's law applied, and the stress or strain–temperature relation can be described by the thermoelastic effect. At the initial stress-free stage, the temperature was 26.09°C, but the temperature dropped to 25.69°C when the stress increased from 0 MPa to the yield stress level of 586 MPa, owing to the thermoelastic effect.

Once the stress level was above the yield stress, the temperature started to increase because of the heat dissipation due to the irreversible plastic deformation. When the stress elastically decreased from the maximum level, 660–0 MPa, the temperature continued to increase because of the thermoelastic effect. When the stress turned into a compressive stress, the temperature increase extended because of the plastic deformation until the stress reached the minimum stress level, 650 MPa. When the stress elastically increased from 650 to 200 MPa, the temperature decreased due to the thermoelastic effect. After the stress went over 200 MPa, the temperature increased again because of the plastic deformation.

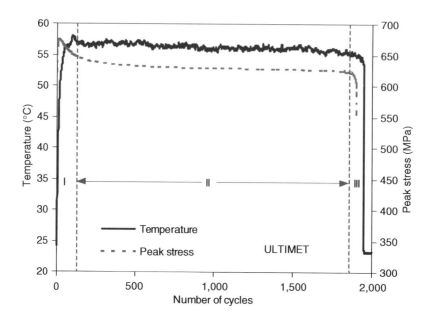

FIGURE 8.18 The mean temperature and peak stress evolution of the specimen during lowcycle fatigue ($\Delta\varepsilon = 2.5\%$, $R = -1$, 21° C, in air, fatigue life of 1,903 cycles, strain rate of $7.5\times10^{-3}\,\mathrm{s}^{-1}$, and 0.15 Hz). Note that the temperature profile was taken at an IR camera speed of 0.15 Hz. (From Jiang, L., Wang, H., Liaw, P. K., Brooks, C. R., and Klarstrom, D. L., Temperature Evolution During Low-Cycle Fatigue of ULTIMET alloy: Experiment and Theoretical Modeling, *Mech. Mater.*, 36 (1–2), pp. 73–84, 2004. With permission.)

For subsequent cycles, the strain and temperature responses repeated, and the mean temperature continuously increased. As indicated by the dashed line in Figure 8.19, the lowest temperature of each cycle corresponded to the tensile yield stress level of each cycle, i.e., 586 MPa for the first cycle and 200 MPa for the rest of the fatigue cycles. Starting from the lowest temperature of each cycle, there were three different mechanisms causing the temperature increase: (a) the temperature increase between the dashed line and dotted line was due to the plastic deformation in tension, (b) the temperature increase between the dotted line and dash-dotted line resulted from the thermo-elastic effect, and (c) the temperature increase between the dash-dotted line and solid line was owing to the plastic deformation in compression. Another interesting observation is that the highest temperature of each cycle was coincident with the lowest stress level of each cycle, as indicated by the solid line in Figure 8.19. Thus, in the low-cycle fatigue process, the thermoelastic effect caused the temperature to oscillate, and the irreversible inelastic deformation caused the mean temperature to increase. The thermoelastic effect caused the temperature to oscillate during the low-cycle fatigue process, and the irreversible inelastic deformation increased the mean temperature. Figure 8.20 shows the experimental and predicted temperature oscillations of the ULTIMET alloy at the initial stage of low-cycle fatigue. It can be seen that the predicted results, with the consideration of different heat-transfer mechanisms, such as conduction, convection and radiation, were in good agreement with the experimental results.

A constitutive model was formulated for predicting the mechanical and thermal responses of the ULTIMET alloy subjected to cyclic deformation. The model was restricted to consideration of an isotropic cylindrical bar subjected to a homogeneously applied uniaxial infinitesimal deformation field. Only thermal and mechanical properties of materials were considered (magnetic, electric, and other factors were neglected). The inelastic deformation behavior was described by a set of internal-state variables. All the mechanical, thermodynamic, and internal-state variables were referred to a

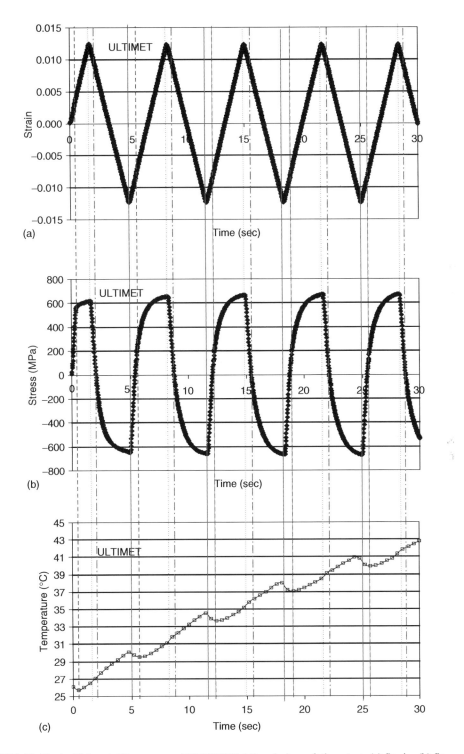

FIGURE 8.19 Typical Material Responses of ULTIMET Alloy during a fatigue test: (a) Strain, (b) Stress, and (c) Temperature Evolution versus Time Curves. ($\Delta\varepsilon = 2.5\%$, R $= -1.21°$ C, in Air, Fatigue Life of 1,903 Cycles, Strain Rate of 7.5×10^{-3} s^{-1}., and 0.15 Hz) Note that the temperature profile was taken at an IR camera speed of 2.1 Hz. (From Jiang, L., Wang, H. ,Liaw, P. K., Brooks, C. R., and Klarstrom, D. L., Temperature Evolution During Low-Cycle Fatigue of ULTIMET alloy: Experiment and Theoretical Modeling, *Mech. Mater.*, 36 (1–2), pp. 73–84, 2004. With permission.)

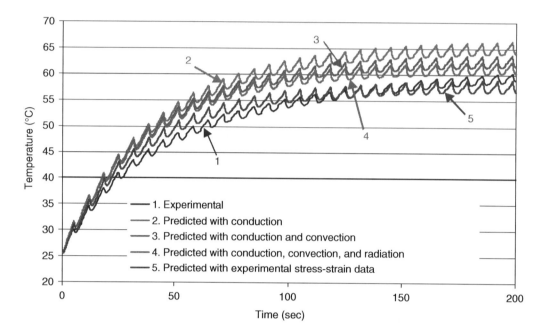

FIGURE 8.20 Experimental and Predicted Temperature Oscillations of ULTIMET Alloy at Initial Cycles during Low-Cycle Fatigue. ($\Delta\varepsilon = 2.5\%$, R = - -, Room Temperature, in Air, Fatigue Life of 1,903 Cycles, Strain Rate 7.5^{-3} s^{-1}., and 0.15 Hz) Note that the temperature profile was taken at an IR camera speed of 2.1 Hz. (From Jiang, L., Wang, H., Liaw, P. K., Brooks, C. R., and Klarstrom, D. L., Temperature Evolution During Low-Cycle Fatigue of ULTIMET alloy: Experiment and Theoretical Modeling, *Mech. Mater.*, 36 (1–2), pp. 73–84, 2004. With permission.)

one-dimensional system with the axis, x. The x axis was designated along the longitudinal direction of the cylindrical bar, and the midpoint of the specimen gage section represents the origin of the x axis. As usual, the time was designated as t. Thus, all the state variables described herein are functions of x and t.

The state of the bar was characterized by independent observable state variables, which include the total strain, ε, absolute temperature, T, temperature gradient, ΔT, and internal-state variables, α_i, where i is the number of internal variables ranging from 1 to n. The total strain was assumed to be composed of the elastic strain, ε_E, and the inelastic strain, ε_I.

$$\varepsilon = \varepsilon_E + \varepsilon_I \tag{8.3}$$

The state of the bar must satisfy thermodynamic laws, i.e., the conservation of energy, and the second law of thermodynamics. The coupled heat-transfer equation can be written as

$$\rho\frac{\partial H}{\partial \alpha_i}\dot{\alpha}_i - \rho T\frac{\partial^2 H}{\partial \alpha_i \partial T}\dot{\alpha}_i - \rho T\frac{\partial^2 H}{\partial \varepsilon \partial T}\dot{\varepsilon} - \rho T\frac{\partial^2 H}{\partial T^2}\dot{T} - k\frac{\partial^2 T}{\partial x^2} + h\frac{p}{A}(T - T_r) + \frac{p}{A}GFs(T^4 - T_r^4) = \rho r \tag{8.4}$$

where $(\cdot) = d/dt(\ )$, ρ is the mass density, H is the Helmholtz free energy per unit mass, h is the heat transfer coefficient, p is the circumference of the cylindrical bar, A is the cross-sectional area of the cylindrical bar, T_r is the temperature of the environment, G is the geometric factor between the specimen and surroundings, F (≤ 1) is the factor termed the emissivity, and s is the Stefan–Boltzman constant. On the left-hand side of Equation 8.4, the first three terms are related to the change of the mechanical state, the fourth term is the accumulation of the temperature, the fifth term is the heat conduction, the sixth term is the heat convection, and the seventh term is the heat radiation. Equation 8.4 is the governing equation describing the thermal and mechanical responses of a material subjected to cyclic fatigue. In addition, there was no internal heat source other than the material plastic energy dissipation.

Unified theories contain time and temperature dependent constitutive equations that describe the elastic and inelastic behavior of materials subjected to thermal and mechanical loads. In these theories, all mechanisms of the inelastic deformation are represented with kinetic equations that define the inelastic strain rate, $\dot{\alpha}_i$. The Bodner–Partom unified model is one of the unified theories. The three basic elements that comprise the Bodner–Partom unified model are: (a) a flow law relating the inelastic strain rate to deviatoric stresses, (b) a kinetic equation that relates the inelastic strain rate to stress invariant through a function of temperature and internal state variables, and (c) evolution equations defining the growth of internal state variables [31].

The internal state variable, α_1, can be represented by the kinetic equation [31–36]

$$\dot{\alpha}_1 = \frac{2}{\sqrt{3}} D_0 \exp\left[-\left(\frac{1}{2}\right)\left(\frac{Z}{\sigma}\right)^{2n}\right] \tag{8.5}$$

where D_0 and n are experimentally obtained material constants, and Z is a measure of the strain hardening, which is a function of the plastic work. Z is composed of isotropic and kinematic hardening,

$$Z = Z^I + Z^D \tag{8.6}$$

where Z^I is the isotropic variable and Z^D is the directional hardening variable. The evolution equations for the isotropic and the directional hardening variables without thermal recovery are described by Equation 8.6 and Equation 8.7, respectively,

$$\dot{Z}^I = m_1(Z_1 - Z^I)\dot{W}_p; \quad Z^I(0) = Z_0 \tag{8.7}$$

$$\dot{Z}^D = \frac{m_2}{2}[1 + \exp(-m_3 Z^D)](Z_2 - Z^D)\dot{W}_p; \quad Z^D(0) = 0 \tag{8.8}$$

where, m_1, m_2, m_3, Z_1, Z_2, and Z_0 are material constants. m_1, m_2, m_3, Z_1, and Z_2 are strain rate and temperature independent for material modeling purposes. Z_0 is strain-rate independent, but, is, however, temperature dependent. In addition, the plastic work, W_p, is determined by

$$\dot{W}_p = \sigma\dot{\alpha}_1 \tag{8.9}$$

With the consideration of thermal expansion and plastic strain, α_1, the elastic strain can be described as

$$\varepsilon_E = \varepsilon - \varepsilon_1 = \varepsilon - \alpha_1 - \bar{\alpha}\Delta T \tag{8.10}$$

where $\bar{\alpha}$ is the coefficient of the thermal expansion in the x direction; ΔT is the temperature difference between the current temperature, T, and the initial temperature, T_0, at the strain-free state, $\Delta T = T - T_0$. Herein, the inelastic strain is composed of the plastic strain, and thermal expansion, i.e., $\varepsilon_1 = \alpha_1 + \bar{\alpha}\Delta T$.

With the above development of constitutive equations, Equation 8.5 through Equation 8.10, and a proper description of the Helmholtz free energy function, the evolution of Equation 8.4 can be resolved. Following Allen's postulation [37], the Helmholtz free energy is expanded in the form of Taylor's series in terms of the elastic strain and temperature as

$$H = H_0 + \frac{E}{2\rho}\varepsilon_E^2 - \frac{C_v}{2T}\Delta T^2 \tag{8.11}$$

where H_0 is the free energy in the initial state, a constant, and E is the elastic modulus. The specific heat at a constant volume, C_v, is

$$C_v = -T\frac{\partial^2 H}{\partial T^2} \tag{8.12}$$

Note that in Equation 8.11, although the first-order terms have been neglected, the coupling among the total strain, inelastic strain, and temperature is retained [37]. Note also that the energy dissipation due to microstructural changes has been neglected in the free-energy Equation 8.11, because this mechanism was shown to contribute only a small portion of energy to the dissipation process [37,38]. Further, the fracture energy loss resulting from the microvoid growth and transgranular macrofracture is neglected.

Substitution of Equation 8.10 and Equation 8.11 into the energy balance Equation 8.4 results in the coupled heat-transfer formula:

$$[(E_e\varepsilon - E_e\alpha_1 + E_e\bar{\alpha}T_0)\dot{\alpha}_1] - [E_e\bar{\alpha}T\dot{\varepsilon} + \rho C_v\dot{T}] + k\frac{\partial^2 T}{\partial x^2} - h\frac{p}{A}(T - T_r) - \frac{p}{A}GFs(T^4 - T_r^4)$$

$$= 0 \tag{8.13}$$

where the terms in the first bracket arise due to the inelastic response, and the terms in the second bracket are the classical thermoelastic coupling terms for an adiabatic condition [39]. The stress–strain relation was in the form of

$$\sigma = E(\varepsilon - \alpha_1 - \bar{\alpha}\Delta T) \tag{8.14}$$

The temperature and stress–strain response of the uniaxial bar subjected to uniaxial homogeneous low-cycle fatigue loading can be characterized using Equation 8.13 and Equation 8.14 together with the constitutive equations Equation 8.5 through Equation 8.10.

Jiang et al. [40] studied the temperature evolution of the ULTIMET alloy subjected to high-cycle fatigue as well. The cumulative fatigue-damage process was revealed by the temperature evolution during high-cycle fatigue, as shown in Figure 8.21. The high-cycle fatigue process could be divided into four stages. Initially, the extension of the specimen due to the uniformly distributed plastic deformation corresponded to a sharp rise in the mean temperature. Next, a stabilized state in the stress–strain response was observed, which resulted from the localized plastic deformation in persistent slip bands, and was associated with a steady-state mean temperature. The shakedown of the stabilized stress–strain state due to crack propagation produced an abrupt rise in the mean temperature. Finally, the failure of the specimen resulted in a final drop of the mean temperature.

Material response typical of the ULTIMET alloy during a fatigue test is shown in Figure 8.22. When the load was initially applied, the strain was elastic. At the initial stress-free stage, the temperature was 22.45°C, but the temperature dropped to 22.12°C when the stress increased from 0 MPa to a mean stress level of 386 MPa, owing to the thermoelastic effect, as shown in Figure 8.22. During the initial short hold time with the mean stress level of 368 MPa, as presented in Figure 8.22a, the strain remained constant at 0.16 pct, as exhibited in Figure 8.22b, and the temperature remained at 22.12°C without any obvious heat loss, as indicated in Figure 8.22c.

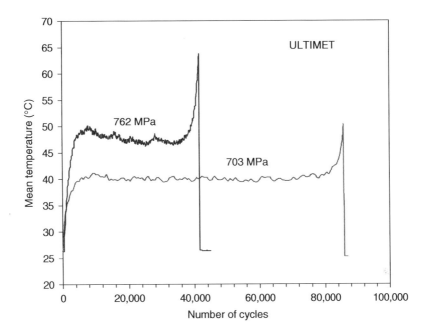

FIGURE 8.21 Mean temperature evolutions of ULTIMET alloy at σ_{max} levels of 703 and 762 MPa, *R* ratio of 0.05, room temperature, in air, and 20 Hz. Note that the IR camera was operated at a speed of 1/8 Hz. (From Jiang, L., Wang, H., Liaw, P. K., Brooks, C. R., and Klarstrom, D. L., Characterization of the Temperature Evolution During High-Cycle Fatigue of the ULTIMET superalloy: Experiment and Theoretical Modeling, *Metall. Mater. Trans. A*, 32 (9), pp. 2279–2296 2001. With permission.)

As stress and strain were continuously increased, the temperature decreased from 22.12°C to 21.87°C. After the stress reached the yield-strength level of 586 MPa, the temperature started to increase because of the heat dissipation due to the irreversible plastic deformation beyond yielding. During the unloading cycle, the strain decreased, and the temperature continued to increase. For subsequent cycles, the strain and temperature responses were repeated, but the residual level of plastic strain grew, and the mean temperature continuously increased. It is clear that the temperature reached its highest point in each cycle when the stress and strain were at their minima. However, when the temperature reached its cyclic minimum, the stress and strain were not at their maxima. This trend results from the fact that the plastic yielding occurred prior to reaching the maximum stress, which initiated the temperature rise. Thus, in the fatigue process, the thermoelastic effect caused the temperature to oscillate, and the irreversible inelastic deformation caused the mean temperature to increase.

A constitutive model, similar to that which is described above for low-cycle fatigue, was developed for predicting the mechanical and thermal responses of the ULTIMET alloy subjected to high-cycle fatigue. The experimental and predicted results were found to be in good agreement.

8.5 APPLICATIONS OF COBALT AND ITS ALLOYS

8.5.1 HIGH-TEMPERATURE APPLICATIONS

When selecting materials for high-temperature applications, one must consider mechanical properties, such as creep and stress-rupture strength, resistance to high-temperature degradation phenomena, fabricability characteristics, and cost. The proper alloy selection is important for safety and economic reasons. Although cobalt-based alloys meet many of these requirements,

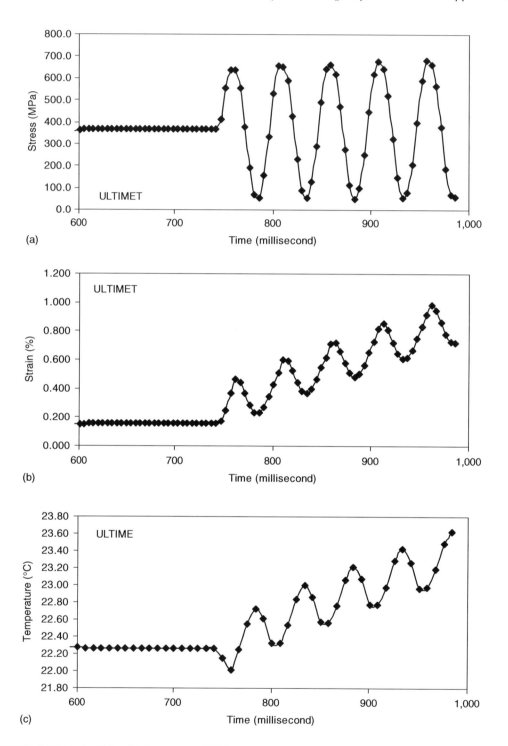

FIGURE 8.22 Typical Material Responses of ULTIMET Alloy during a fatigue test: (a) Stress, (b) Strain, and (c) Temperature Evolution versus Time Curves. ($\sigma_{max} = 703$ MPa, $R = 0.05$, Room Temperature, in Air, Fatigue Life of 97,542 Cycles, and 20 Hz) Note that the temperature profile was taken at an IR camera speed of 120 Hz. (From Jiang, L., Wang, H., Liaw, P. K., Brooks, C. R., and Klarstrom, D. L., Characterization of the Temperature Evolution During High-Cycle Fatigue of the ULTIMET superalloy: Experiment and Theoretical Modeling, *Metall. Mater. Trans. A*, 32(9), pp. 2279–2296 2001. With permission.)

the high cost of cobalt and the need for higher service temperatures have hindered their use and development.

HAYNES 25 and 188 alloys (Table 8.12) are the most frequently employed cobalt-based alloys for high-temperature applications [8]. HAYNES 25 alloy has been used for the hot sections of gas turbines, nuclear-reactor components, and for fasteners and wear pads in the cold-worked condition [8]. HAYNES 188 alloy has found use in combustors and transition ducts in gas turbines [8]. INCONEL 783 alloy (Table 8.12) has been used in gas-turbine casings, rings, and seals, while UMC0-50 (Table 8.12) has been extensively employed in Europe for furnace parts and fixtures [8]. Cast cobalt-base alloys, such as MAR-509 and FSX-414 (Table 8.13), are commonly used for complex shapes, such as first- and second-stage vanes/nozzles in gas-turbine engines [8]. MP35N (Table 8.15) was employed as fasteners for highly-loaded joints in the space-shuttle Columbia because of its resistances to salt-water corrosion and the massive stresses generated by the rocket blast [41].

8.5.2 MAGNETIC DEVICES

Cobalt is a naturally ferromagnetic material, and provides resistances to the demagnetization in several groups of magnetic materials. Cobalt was first introduced into magnet steels in 1917 at a level of 35 wt.%, along with additions of chromium and tungsten [5]. This combination led to an increase in coercive force to over 16 kA m^{-1}, however, these materials are now considered obsolete.

Additions of cobalt to the high-coercive-force Ni–Fe–Al alloys (from 5 to 35 wt.%) in the 1940s and 1950s produced major improvements in all magnetic properties [5]. Today, these alloys are known as Alnico alloys, and are still widely used in both cast and sintered forms [8]. Generally, Alnico alloys are superior to other permanent magnet materials in resisting temperature effects on the magnetic performance.

INCONEL is a trademark of Special Metals Corporation.

MP35N is a trademark of SPS Technologies, Inc.

Platinum–cobalt (23 wt.% Co) permanent magnets are isotropic, ductile, easily machined, resistant to corrosion and high temperatures, and have magnetic properties superior to the Alnico alloys [8]. Except in very specialized applications, the high cost of these alloys has resulted in their replacement by rare earth-cobalt alloys, and attempts to cheapen these alloys by the partial substitution of cobalt by iron and of platinum by other metals have been unsuccessful [5].

Rare-earth-permanent magnets were developed around 1970, and include the samarium–cobalt compounds, $SmCo_5$ and Sm_2Co_{17} [8]. Initially, these materials found applications in small, high-performance devices operating between 175°C and 350°C [8]. In 1983, less expensive, more powerful Nd–Fe–B alloys were developed that largely replaced the rare-earth permanent magnets [8]. In some cases where corrosion resistances are necessary, small amounts of cobalt are added to these Nd–Fe–B alloys.

Many soft magnetic materials used today consist of iron or iron-silicon alloys. However, cobalt can be alloyed with iron to produce alloys that exhibit a high positive-magnetostrictive coefficient [8]. The properties of these alloys depend on their thermo-mechanical production procedure. Typical applications include laminations for motors and generators, transformer laminations and tape torroids, magnetic pole caps, and in extremely accurate positioning devices [8]. Amorphous cobalt-based alloys containing various amounts of metalloids are now utilized as soft magnetic materials [8]. These alloys are known for their low hysteresis loss and low coercive force, and are typically used as core materials for power-distribution transformers.

8.5.3 BIOMEDICAL APPLICATIONS

The American Society for Testing and Materials (ASTM) lists four types of cobalt-based alloys, which are recommended for surgical-implant applications: (1) cast Co–Cr–Mo alloy (F75), (2) hot-forged Co–Cr–Mo (F799), (3) wrought Co–Cr–W–Ni alloy (F90), and (4) wrought Co–Ni–Cr–Mo alloy (F562). However, only the castable Co–Cr–Mo alloy and the wrought Co–Ni–Cr–Mo alloy are used extensively as implant materials.

The castable Co–Cr–Mo alloy (STELLITE alloy 21) has been used for many decades in dentistry, and relatively recently, in making artificial joints. Its employment is due primarily to its corrosion resistance in chloride environments, which is related to its bulk composition and the surface oxide (nominally Cr_2O_3) [42]. Variations in the casting process can produce three micro-structural features that can strongly influence properties [42]. First, nonequilibrium cooling can produce a cored microstructure, which causes Cr-depleted regions to become anodic with respect to the rest of the microstructure. Second, large grain sizes can be produced, which can reduce the yield strength. Small additions of molybdenum can help reduce the grain size, and therefore, increase the strength. Third, casting defects, such as inclusions or porosities, may develop. ASTM F799 is a modified version of the F75 alloy that has been mechanically processed by hot forging after casting [42]. The microstructure has a more worked grain structure than the as-cast F75, and has an HCP phase that forms via a shear-induced transformation of the FCC matrix to HCP platelets. The fatigue, yield, and ultimate tensile strengths of this alloy are about twice that of the as-cast F75 [42].

Co–Cr–W–Ni alloy (STELLITE alloy 25) employs tungsten and nickel to improve machin-ability and fabrication properties. In the annealed state, its mechanical properties are close to those of the F75 alloy, but cold working can more than double the properties.

The Co–Ni–Cr–Mo alloy (MP35N) is highly corrosion resistant to seawater (containing chloride ions) under stresses [43]. Thermal treatments and cold working can produce a controlled microstructure and increase the strength of the alloy considerably [42]. The processing of this alloy includes 50% cold work, which induces extensive twin formation. The dislocation motion is impeded, resulting in a significantly strengthened structure. However, it is difficult to cold work this alloy when making large devices, such as hip-joint stems [43]. Only hot forging can be used to fabricate a large implant. An aging treatment (430–650°C) strengthens this alloy even further by forming Co_3Mo precipitates. Superior fatigue and ultimate tensile strengths of this alloy make it suitable for applications that require long service lives, such as the stems in hip-joint prostheses [43].

8.6 CEMENTED CARBIDES

One of the most important uses of cobalt is its use as bonding material in cemented carbides. Cemented carbides constitute a group of extremely hard and wear-resistant materials, which also show a high compressive strength. They are essentially made of hard carbide particles of a refrac-tory metal (usually tungsten) bound together, or cemented, by a soft and ductile metal binder. The binder is usually cobalt, but nickel is sometimes used to improve resistances to oxidation and corrosion. They are extensively used as cutting tools for steels and other high-strength metals and alloys, wire-drawing or extrusion dies, metal shaping and forming tools, rock drills in mining, and many other applications where high hardnesses, wear resistances, and rigidities are required.

WC–Co alloys, commonly referred to as straight grades, have a microstructure of two phases: a phase of the angular WC grains and a cobalt-binder phase [44]. Hard-carbide particles provide the cutting surface. However, they are extremely brittle and incapable of withstanding the cutting stresses independently. Surrounding the carbides with a ductile-metal matrix enhances toughness by preventing particle-to-particle crack propagation. Both the matrix and particulate are capable of withstanding the high temperatures generated during cutting of extremely hard materials. Too much or too little carbon can be detrimental to performance; therefore, strict control of carbon content is necessary [44].

Properties depend on the type and amount of carbides present, carbide grain size, and the amount of the binder material. By adding small amounts of TaC, NbC, VC, or CrC to WC–Co straight grade alloys, submicron grain sizes can be developed [44], which gives the alloy higher toughness and edge strengths. WC–TiC–Co alloys have an increased resistance to chemical attack and diffusion wear, which are problems with WC–Co alloys when cutting steels [44]. Other alloys, known as complex grades (WC, TiC, TaC, and NbC) have been developed that possess a wide variety of properties [44].

REFERENCES

1. American Society of Metals (ASM), Past times, *Adv. Mater. Processes*, 8, 90, 1999.
2. Cottrell, A., A centennial report, *Mater. Res. Bull.*, 25(2), 2000. http://www.mrs.org/publications/bulletin/2000/feb/cottrell.html
3. Klarstrom, D. L., Wrought cobalt-base superalloys, *J. Mater. Eng. Perform.*, 2, 523–530, 1993.
4. Cobalt Monograph, Centre d'Information du Cobalt, Brussels, 1960.
5. Betteridge, W., *Cobalt and its Alloys*, Ellis Horwood Limited, Chichester, 1982.
6. Jiang, L., *Fatigue Behavior of ULTIMET Alloy: Experiment & Theoretical Modeling*, Ph.D Thesis, The University of Tennessee, Knoxville, 2000.
7. Beltran, M., Cobalt-base alloys, In *Superalloys II*, Sims, C. T., Stoloff, N. S., Hagel, W. C., Eds., pp. 135–164, 1987.
8. Davis, J. R., Introduction, to cobalt and cobalt alloys, *Nickel, Cobalt, and their Alloys—ASM Specialty Handbook*, ASM International, Metals Park, OH, pp. 343–406, 2000.
9. Crook, P., Cobalt and cobalt alloys, *Metals Handbook*, Vol. 2, pp. 446–454, 1990.
10. Malayoglu, U. and Neville, A., *Wear*, 255, 181–194, 2003.
11. Cameron, C. B. and Ferriss, D. P., Tribaloy intermetallic materials—new wear-resistant and corrosion-resistant alloys, *Anti-Corros. Methods Mater.*, 22(4), 5–8, 1975.
12. Sullivan, C. P., Donchie, M. J., Jr., and Morral, F. R., *Cobalt-Base Superalloys-1979*, Cobalt Information Center, Brussels, Belgium, 1980.
13. Haynes International, Kokomo, IN. http://www.haynesintl.com
14. Cahn, R. W., Haasen, P., and Kramer, E. J., *Materials Science and Technology, A Comprehensive Treatment Corrosion and Environmental Degradation*, Vol. 2, Wiley, New York, 2000.
15. Jiang, L., Brooks, C. R., Liaw, P. K., Wang, H., Rawn, C. J., and Klarstrom, D. L., High-frequency metal fatigue: The high-cycle fatigue behavior of ULTIMET alloy, *Mat. Sci. Eng. A-Struct.*, 314(1–2), 162–175, 2001.
16. Jiang, L., Wang, H., Liaw, P. K., Klarstrom, D. L., Rawn, C. J., and Muenchen, B., Phenomenological aspects of the high-cycle fatigue of ULTIMET alloy, *Mat. Sci. Eng. A-Struct.*, 316(1–2), 66–79, 2001.
17. Yang, B., Liaw, P. K., Wang, H., Jiang, L., Huang, J. Y., Kuo, R. C., and Huang, J. G., Thermographic investigation of the fatigue behavior of reactor pressure vessel steels, *Mat. Sci. Eng. A-Struct.*, 314(1–2), 131–139, 2001.
18. Jiang, L., Wang, H., Liaw, P. K., Brooks, C. R., and Klarstrom, D. L., Characterization of the temperature evolution during high-cycle fatigue of the ULTIMET superalloy: Experiment and theoretical modeling, *Metall. Mater. Trans. A*, 32(9), 2279–2296, 2001.
19. Wang, H., Jiang, L., Liaw, P. K., Brooks, C. R., and Klarstrom, D. L., Infrared temperature mapping ULTIMET alloy during high-cycle fatigue tests, *Metall. Mater. Trans. A*, 31, 1307–1310, 2000.
20. Jiang, L., Wang, H., Liaw, P. K., Brooks, C. R., and Klarstrom, D. L., *Trans. Nonferrous Metals Soc. China*, 12, 734, 2002.
21. Wang, H., Jiang, L., He, Y. H., Chen, L. J., Liaw, P. K., Seeley, R. R., and Klarstrom, D. L., Infrared imaging during low-cycle fatigue of HR-120 alloy, *Metall. Mater. Trans. A*, 33, 1287–1292, 2002.
22. Liaw, P. K., Yang, B., Tian, H., Jiang, L., Wang, H., Huang, J. Y., Kuo, R. C., Huang, J. G., Fielden, D., Strizak, J. P., and Mansur, L. K., Fatigue and fracture mechanics, In *ASTM STP 1417*, Reuter, W. G. and Piascik, R. S., Eds., Vol. 33, ASTM International, West Conshohocken, PA, p. 524, 2002.
23. Liaw, P. K., Wang, H., Jiang, L., Yang, B., Huang, J. Y., Kuo, R. C., and Huang, J. G., Thermographic detection of fatigue damage of pressure vessel steels at 1,000 and 20 Hz, *Scripta Mater.*, 42, 389–395, 2000.

24. Yang, B., Liaw, P. K., Wang, G., Peter, W. H., Buchanan, R. A., Yokoyama, Y., Huang, J. Y., Kuo, R. C., Huang, J. G., Fielden, D. E, and Klarstrom, D. L., Thermal-imaging technologies for detecting damage during high-cycle fatigue, *Metall. Mater. Trans. A*, 35, 15–23, 2004.

25. Yang, B., *J. Mater. Eng. Perform.*, 12(3), 345, 2003.

26. Chen, L. J., Liaw, P. K., Wang, H., He, Y. H., McDaniels, R. L., Jiang, L., Yang, B., and Klarstrom, D. L., Cyclic deformation behavior of HAYNES HR-120 superalloy under low-cycle fatigue loading, *Mech. Mater.*, 36, 85–98, 2004.

27. Jiang, L., Wang, H., Liaw, P. K., Brooks, C. R., Chen, L., and Klarstrom, D. L., Temperature evolution and life prediction in fatigue of superalloys, *Metall. Mater. Trans. A*, 35, 839–848, 2004.

28. Chen, L. J., Liaw, P. K., He, Y. H., Benson, M. L., Blust, J. W., Browing, P. F., Seeley, R. R., and Klarstrom, D. L., Tensile hold low-cycle fatigue behavior of cobalt-based Haynes 188 superalloy, *Scripta Mater.*, 44, 859–865, 2001.

29. Jiang, L., Brooks, C. R., Liaw, P. K., Dunlap, J., Rawn, C. J., Peascoe, R., and Klarstrom, D. L., Low-cycle fatigue behavior of ULTIMET alloy, *Metall. Mater. Trans. A*, 35(3), 785–796, 2004.

30. Jiang, L., Wang, H., Liaw, P. K., Brooks, C. R., and Klarstrom, D. L., Temperature evolution during low-cycle fatigue of ULTIMET alloy: Experiment and theoretical modeling, *Mech. Mater.*, 36(1–2), 73–84, 2004.

31. Rowley, M. A. and Thornton, E. A., Constitutive modeling of the visco-plastic response of hastelloy-X and aluminum alloy 8009, *J. Eng. Mater Technol.*, 118, 19–27, 1996.

32. Bodner, S. R. and Partom, Y., Constitutive equations for elastic–viscoplastic strain-hardening materials, *J. Appl. Mech.*, 42(6), 385–389, 1975.

33. Chan, K. S., Bonder, S. R., and Lindholm, U. S., Phenomenological modeling of hardening and thermal recovery in metals, *J. Eng. Mater. Technol.*, 110, 1–8, 1988.

34. Chan, K. S., Lindholm, U. S., Bonder, S. R., and Walker, K. P., High temperature inelastic deformation under uniaxial loading: Theory and experiment, *J. Eng. Mater. Technol.*, 111, 345–353, 1989.

35. Li, K. and Sharpe, W. N., Jr., Viscoplastic behavior of a notch root at 650 C: ISDG Measurement and finite element modeling, *J. Eng. Mater. Technol.*, 118, 88–93, 1996.

36. Skipor, A. F., Harren, S. V., and Botsis, J., On the constitutive response of 63/37 Sn/Pb eutectic solder, *J. Eng. Mater. Technol.*, 118, 1–11, 1996.

37. Allen, H., A prediction of heat generation in a thermoviscoplastic uniaxial bar, *Int. J. Solids Struct.*, 21(4), 325–342, 1985.

38. Fine, M. E. and Davidson, D. L., Fatigue mechanisms, advances in quantitative measurement of physical damage, In *ASTM STP 811*, Lankford, J., Davidson, D. L., Morris, W. L., and Wei, R. P., Eds., American Society for Testing and Materials, pp. 350–370, 1983.

39. Stanley, P. and Chan, W. K., Quantitative stress analysis by means of the thermoelastic effect, *J. Strain Anal.*, 20(3), 129–137, 1985.

40. Jiang, L., Wang, H., Liaw, P. K., Brooks, C. R., and Klarstrom, D. L., Characterization of the temperature evolution during high-cycle fatigue of the ULTIMET superalloy: Experiment and theoretical modeling, *Metall. Mater. Trans. A*, 32(9), 2279–2296, 2001.

41. Boyer, H. E., *Exotic Alloy Helps Hold Space Shuttle Together Selection of Materials for Service Environments*, ASM International, Metals Park, OH, pp. 222–223, 1987.

42. Brunski, J. B., Classes of materials used in medicine: Metals, In *Biomaterials Science*, Ratner, B. D., Hoffman, A. S., Schoen, F. J., and Lemons, J. E., Eds., Academic Press, London, 1996.

43. Park, J. B. and Kim, Y. K., Metallic biomaterials, In *Biomaterials: Principles and Applications*, Park, J. B. and Bronzino, J. D., Eds., CRC Press, Boca Roton, 2003.

44. Davis, J. R., Tool materials, *ASM Specialty Handbook*, pp. 36–58, 1995.

9 The Science, Technology, and Applications of Aluminum and Aluminum Alloys

T. S. Srivatsan and Satish Vasudevan
Department of Mechanical Engineering, University of Akron

CONTENTS

9.1   Introduction 226
9.2   Availability of Aluminum 227
9.3   Production of Aluminum 227
9.4   Early Alloys of Aluminum 228
    9.4.1   Durability and Damage Tolerant Alloys 229
    9.4.2   Fracture Resistant Alloys 230
9.5   Alloys of Aluminum 231
9.6   Physical Metallurgy of Aluminum Alloys 231
    9.6.1   Alloying Additions to Aluminum and the Role of Impurities 232
9.7   Mechanical Deformation and its Influence on Strength of Aluminum Alloys ... 234
    9.7.1   Solid Solution Strengthening 235
    9.7.2   Precipitation Hardening 236
    9.7.3   Presence and Role of Trace Elements in Governing Precipitation Kinetics 242
9.8   Properties of Aluminum 242
9.9   Wrought Aluminum Alloys 243
9.10 Wrought Non-Heat-Treatable Aluminum Alloys 244
9.11 The Powder Metallurgy Alloys 245
    9.11.1 Advantages and Barriers to Aluminum Powder Metallurgy Alloys ... 247
9.12 Temper Designation System 249
    9.12.1 System for Heat-Treatable Aluminum Alloys 250
9.13 Mechanical Behavior of Aluminum Alloys: Role of Microstructure 251
    9.13.1 Tensile Behavior (Properties) 252
    9.13.2 Fracture Toughness 254
    9.13.3 Fatigue Response 256
    9.13.4 Fatigue Crack Initiation 257
    9.13.5 Fatigue Crack Growth 258
    9.13.6 Corrosion Fatigue 261
    9.13.7 Creep Characteristics 261
    9.13.8 Properties at Elevated-Temperatures 261
    9.13.9 Low-Temperature Properties 263
9.14 Corrosion Behavior of Wrought Aluminum Alloys 263
9.15 Recent Developments on Aluminum Alloys: Aluminum–Lithium Alloys 263

9.16 Aluminum Macro-Laminates 266
9.17 The Engineered Products from Aluminum Alloys 266
    9.17.1 Castings 266
    9.17.2 Extrusions 268
    9.17.3 Forgings 268
    9.17.4 Powder Metallurgy 268
9.18 Classification of Aluminum Alloy Products 268
    9.18.1 Building and Construction Application 268
    9.18.2 Electrical Applications: Conductor Alloys 269
    9.18.3 Household Appliances 269
9.19 Concluding Remarks 270
References 271

Abstract

The technology of aluminum seems to grow stronger with competition, stronger in the physical properties of the alloys themselves, and stronger in the dominant role they play in a variety of performance-critical and non-performance critical applications. Aluminum alloys have been the primary material of choice for the structural components of aircraft since the 1930s and continue to be an outstanding choice for the performance-critical structures of commercial airliners, military cargo, and transport aircraft. Well-established performance characteristics, known fabrication costs, sound and established experience and practices in design, and established manufacturing methods and facilities are the reasons for continued confidence in aluminum alloys that will ensure their sustained use in the years ahead. In the preliminary stages of alloy development, they were developed by trial and error. However, time has resulted in significant advances in our understanding of the relationships between composition, processing, microstructural development, properties, and performance. This knowledge base led to continuous efforts to achieve improvements in properties that are essential for a spectrum of applications. This chapter covers the basic science, availability, production, types, temper designation, physical metallurgy, mechanical behavior with specific reference to microstructural influences on properties, and recent developments in emerging alloys. Finally, engineered products from aluminum alloys and classification of the product forms are discussed.

9.1 INTRODUCTION

Aluminum is the second most plentiful metallic element on earth. Through the years, it has grown in use to become an economic competitor to conventionally used materials in a spectrum of engineering applications as recently as the end of the nineteenth century. It has progressively grown in use and acquired the status of a well-placed metal of its time. Important industrial developments in the field of materials science and engineering that have emerged while keeping pace with technological evolution, and the demanding material characteristics consistent with the unique qualities of aluminum and its alloys, have led to growth in both the production and use of the metal [1]. It is predicted that most of the aluminum and its advanced variations produced in the years ahead will find use for applications that did not exist two or three decades ago [2]. The present markets for aluminum are greatly affected by shortages of other metal counterparts coupled with a combination of economical, environmental, and safety factors [3]. In recent years, the technology related to aluminum has grown stronger with competition, with an emphasis on the physical properties of the alloys themselves and the dominant role they play in structural applications. The characteristics that make aluminum and its emerging variations, i.e., rapidly solidified alloys, new ingot alloys, new tempers, aluminum–lithium alloys, and metal matrix composites, a potent and dominant material for structural, aerospace, and automotive applications, are its high strength, low density, recyclability, availability, workability, work experience, and cost effectiveness.

The first commercial application of aluminum was for novelty items such as mirror frames, house numbers, and serving trays [1]. Cooking utensils were also an early market for the metal. Aluminum has found use in aerospace structures since the 1930s. As more alloys are developed and emerge in the commercial market, they make it difficult for composite materials to move into primary structures [4]. Until now, the metal has been increasingly used in a spectrum of applications, spanning both performance-critical and nonperformance-critical, to the extent that virtually every aspect of modern life is either directly or indirectly affected by its usage [1].

The objective of this technical chapter is to present and discuss the availability and the production of early and subsequent alloys, physical metallurgy, mechanical behavior, and corrosion response. In the next part of this paper an overview of the newer variations of this alloy family is provided. Finally, a brief review of the applications of these alloys is presented.

9.2 AVAILABILITY OF ALUMINUM

Industrialized nations leading in the world production of aluminum are (a) United States of America, (b) Soviet Union, and (c) Japan, with Soviet Union having the largest smelter in the world (rated at 600,000 tons per annum) [1]. A study undertaken during the past decade predicted the production of aluminum to expand at an average annual rate of 5% during the subsequent two to three decades when compared to a growth rate of 5% for plastics, 4% for steel, and 4% for copper [5]. Of the major structural materials chosen for use in commercial products, only the consumption of aluminum and plastics is expected to increase at a rate greater than the predicted rate of growth of the Gross National Product (GNP) of the developed countries [5]. At the same time, large increases in energy costs have been responsible for the closure of several aluminum smelters in both the United States and Japan. The production of aluminum is being transferred to countries were cheaper energy is available. A representative example being Australia, where the output of aluminum is expected to triple current production levels [5].

In the developed world and including a large number of developing countries, aluminum and its alloys have found use in four major areas: (a) building and construction, (b) containers and packaging, (c) transportation, and (d) electrical conductors, which are commonly ranked in this order. However, the major growth market for present and future projects is expected to be ground and air transportation systems as more alloys of aluminum find use in emerging automotive vehicles and the newer generation of military and civilian aircraft [1,6].

9.3 PRODUCTION OF ALUMINUM

Although aluminum was isolated in small quantities early in the nineteenth century, it remained an expensive curiosity until 1886 when independent discoveries by Hall in America and Heroult in France led to the development of an economical method for its electronic extraction. Since that time, the emergence of aluminum as a safe, practical, and commercially viable metal has been primarily dependent on the availability of bulk quantities of electricity at affordable prices [5].

Aluminum is obtained from bauxite, which is the name given to ores generally containing about 40–60% hydrated alumina together with impurities such as iron oxides, silica, and titania [5]. The name originates from Les Baux the district in Provence, France where the ore was first mined. Bauxite is formed by the gradual surface weathering of the aluminum-bearing rocks such as granite and basalt under tropical weather conditions. The largest known reserves are found in Northern Australia, Guyana, and Brazil [5]. Up until now, and in the near future, the major suppliers of the ore will be the countries of Australia (30%), Jamaica (15%), and Guyana (15%) [2]. Considering the current demand and consumption levels, it is expected that high-grade bauxite with low silica content will soon become depleted in the years ahead [5].

Production of aluminum from bauxite involves two distinct processes, which are often operated at different locations. First, pure alumina (Al_2O_3) is extracted from bauxite almost exclusively by the Bayer process. This process essentially involves digesting crushed bauxite with a caustic soda solution at temperatures up to 240°C. Most of the alumina is carefully extracted leaving behind an insoluble residue known as "red mud." The "red mud" consists of a combination of iron oxide and silica, which is removed by filtration. Upon cooling, the liquor is seeded with crystals of alumina trihydrate to reverse the chemical reaction. The trihydrate is precipitated and the caustic soda is recycled. The entire process is represented by the chemical reaction

$$Al(OH)_3 + NaOH = NaAlO_2 + H_2O \qquad (9.1)$$

The alumina trihydrate is subsequently calcined in a rotary kiln at 1200°C to remove the water of crystallization resulting in the production of alumina as a fine powder. Alumina has a high melting point (2040°C) and is a poor conductor of electricity [5]. A key to the successful production of aluminum lies in dissolving the alumina in molten cryolite (Na_2AlF_6). A typical electrolyte contains 80–90% of this compound coupled with 2–8% alumina together with additives such as aluminum and calcium fluoride. An electrical current is passed through a bath containing the mixture to dissolve the alumina with the occurrence of the following: (a) oxygen forming at and reacting with the carbon anode, and (b) aluminum collecting as a metal pad at the cathode. The separated metal is periodically removed by either siphon or vacuum methods into crucibles [5].

The major impurities present in smelted aluminum are iron and silicon [1]. However, zinc, gallium, titanium, and vanadium are also present as minor contaminants. Internationally, minimum aluminum purity is the primary criterion for defining both composition and value [5]. In the United States, there exists a convention for considering the relative concentrations of iron and silicon. The reference to grades of the unalloyed aluminum metal is often based on metal purity [7,8].

9.4　EARLY ALLOYS OF ALUMINUM

The need for new and improved materials that offer attractive savings in weight while concurrently providing high durability and damage tolerance for both performance-critical and nonperformance critical airframe structures has existed since the advent of powered flight. A cast aluminum block provided the Wright brothers with an engine having sufficient ratio of thrust-to-weight [7]. However, the strength of wrought alloys available at that time was low. Thus, the alloys of aluminum were not used as structural materials on the early aircraft. With time and gradual advances in research and progressive evolution of technology the discovery of precipitation hardening led to the development and emergence of Duralumin (Alloy 2017-T4), which had much higher ratio of strength-to-weight (σ/ρ) relative to conventional wood and cloth construction [7].

Alloy was initially used as structural material in airframe in 1916. At that time, the traditionally used material was desirable primarily because it saved weight. However, its corrosion resistance was less than optimal [7]. The development of ALCLAD sheet, a product in which thin layers of unalloyed aluminum are metallurgically bonded to an aluminum core, provided the level of corrosion resistance desired by the aircraft industry [7]. Continued progress in research and development efforts on lightweight alloys led to the emergence of super Duralumin (2024-T3), which was significantly stronger than Duralumin. By the 1930s, Alclad 2024-T3 became the standard material for airframe construction and was an ideal replacement to the widely used aluminum alloys and other companion materials. With time and advances in research, the discoveries of the intrinsic influence of cold working and artificial aging led to the development of the stronger T36 and T8 tempers for aluminum alloy 2024. However, these found selective use only because of their reduced ductility when compared to 2024-T3. The intrinsic difficulty to forge aluminum alloy 2024 led to the development and emergence of aluminum alloy 2014-T6. This alloy was initially used both in the form of extrusions and flat rolled products in Europe [8].

Over time and aided by incremental advances in processing science and the gradual evolution of technology, newer materials were developed to improve the corrosion resistance of the widely chosen and used aluminum–copper alloys. The newer material consisted of a high-strength core, coated on one or both surfaces with a metallurgically bonded, thin layer of pure aluminum or aluminum alloy, which is anodic relative to the core. When a corrosive solution is exposed to the cladding, it corrodes moderately and uniformly while concurrently aiding in protecting the high-strength core from selective corrosive attack [9,10].

During World War II, the need for improved performance led to the development and emergence of alloy 7075-T6 as sheet and thin extrusions. The first 7XXX alloy product was resistant to stress corrosion cracking in the long-transverse (LT) direction. It was initially used for parts of the structure, which are sized by strength. Alloy 7075 was followed by the development and emergence of the following: (a) alloy 7079-T6 for forgings and thick plate and (b) alloy 7178-T6 for thin sheet and extrusions loaded in compression. This was because the newer alloys offered higher strengths in such products. With gradual growth in the size of civilian aircraft and a concurrent change in design philosophy from sheet stringer construction to the extensive machining of thick plate and forgings, experience revealed the 7XXX alloys in the T6 temper to be susceptible to stress corrosion cracking when stressed in the short transverse (ST) direction [7].

Over time and influenced by the gradual evolution of technology, design concepts leading to safer, cheaper and more durable aircraft were emphasized and put into practice. In the 1930s, the concept of "safe-life" was introduced. This is similar to the "engineering safety factor" that most engineers are familiar with. The goal of designing for no failure in four lifetimes is heavily dependent on previous service experience and rigorous mechanical testing [11,12]. The late 1960s and early 1970s saw the emergence of the concept of damage tolerant design. This concept relied upon an irreversible slow growth of undetected flaws in the structure, which eventually culminated in failure of the part. Once the growth rate of cracks and/or flaws can be predicted in service parts, then regularly scheduled maintenance and replacement can be carried out to prevent catastrophic failure [11]. This design philosophy led to the onset of the "fail-safe" concept. The goal of the "fail-safe" concept was to ensure a redundant load path, such that in the case of a part or component failure, despite detection and maintenance efforts, another part was readily available to prevent critical loss of air-worthiness while facilitating a safe return to the ground [7]. All this led to current design efforts to control, predict, and prevent failure due to widespread and/or multi-site damage [12]. A material that is currently available to meet the challenges of emerging aircraft structures and airframe design is aluminum and its alloy counterparts.

9.4.1 Durability and Damage Tolerant Alloys

As airplanes continued to increase in size, designers needed thicker materials. Experience with aluminum alloy 7075-T6 revealed that it developed significantly lower strengths in thick sections because of the slower quench rate. Furthermore, the thick products were easily susceptible to stress-corrosion cracking (SCC) in the short transverse (ST) direction. In an attempt to obviate this problem, alloy 7079-T6 was imported from Germany because it provided higher strength than 7075-T6 in thick sections. While it did perform well in accelerated SCC tests, the 7079-T6 alloy developed severe stress-corrosion problems while in service and was found to be unacceptable [6,13].

The stress corrosion cracking (SCC) problems in 7XXX alloy products were solved by the development of the T73 temper for aluminum alloy 7075. Development of this temper resulted in a penalty in strength. Even with reduced strength, it was safely applied on the Douglas DC10, which was the first aircraft to feature forgings of alloy 7075-T73. Continued and ongoing research efforts led to the development of the T76 temper, which showed good resistance to exfoliation corrosion while retaining some of the strength loss, and concurrently enabling its application as plate on the Lockheed L-1011 aircraft [6].

The key companies involved in the production and processing of aluminum alloys pursued conventional ingot metallurgy product development for specific applications such as the upper wing, the lower wing, the fuselage, and other components. The first major technical success was the T77 temper for aluminum alloy 7150 [6]. This alloy, i.e., 7150-T77, provided corrosion resistance with minimal sacrifice in strength. Further, unlike the unconventionally produced alloys, such as the powder metallurgy (PM) processed aluminum alloys or aluminum–lithium (Al–Li) alloys, aluminum alloy 7150-T77 was cost effective because it was produced using conventional ingot metallurgy (IM) techniques. This was followed by the development and emergence of (i) the high strength 7055-T77 plate and extrusions and (ii) high toughness 2524-T3 sheet and plate stock [6]. The family of alloys developed and offered to customers has been growing since the 1920s up until the year 2000 (Figure 9.1).

Over a 75-year period, since the early 1920s, aluminum has shown a record of continuous improvement and cost effectiveness. The use of derivative alloys and tempers offers low-risk replacement of parts on existing airframes. Newer alloys provided added design flexibility coupled with the potential for a higher payoff. A real-life example of the continuing improvement of aluminum alloys paralleling airframe design can be seen in the BOEING 777. Newer aluminum alloys and tempers (7055-T7751, 2224-T3511, 7150-T77511, and C188) have been used in this aircraft to attain higher combinations of strength, durability, and damage tolerance than the previously designed aircraft [6] (Figure 9.2).

9.4.2 FRACTURE RESISTANT ALLOYS

Loss of military aircraft as a direct consequence of the unstable propagation of relatively small flaws caused aircraft designers to become more concerned about the fracture toughness of materials used in airframe construction. In the late 1960s and up until the 1980s, specially processed versions of the standard aircraft aluminum alloys containing low levels of the impurity elements, i.e., iron and silicon, were developed to meet the need for higher toughness in military aircraft. These included alloy 2124-T8 and alloy 7475 in various tempers. Strength levels were the same as those of the standard alloys in equivalent tempers. Further, the maximum impurity content and ancillary alloying element additions to alloy 7050 were selected with the idea of achieving high toughness [7].

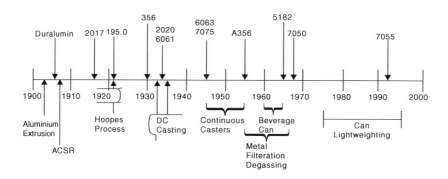

FIGURE 9.1 A timeline showing aluminum product development.

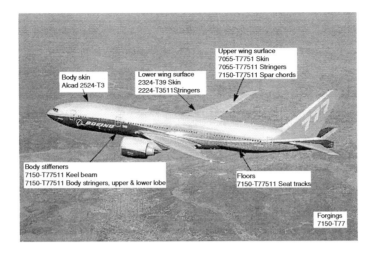

FIGURE 9.2 Aluminum alloys used in a Boeing 777 aircraft.

9.5 ALLOYS OF ALUMINUM

It is convenient to divide the alloys of aluminum into two major categories: (i) cast compositions and (ii) wrought compositions [13–15]. Further differentiation for each category is based on the primary mechanism of property development. Many of the alloys respond to thermal treatments based on the solubilities of the phases present. The thermal treatments include (a) solution heat treatment, (b) quenching, and (c) artificial aging to facilitate or promote precipitation. The cast and wrought aluminum alloys responsive to these treatments are termed as heat treatable [15]. A number of wrought alloy compositions rely on work hardening through mechanical deformation, usually in combination with various annealing procedures for property development. These alloys are referred to as the work hardened alloys. Some casting alloys are not heat treatable and are used in the as-cast condition or a thermally modified condition unrelated to solution heat treatment or precipitation effects [5,15,16].

The Aluminum Association system is the most widely recognized in the United States. Their alloy identification system employs different nomenclatures for wrought alloys and cast alloys, but divides the alloys into families for the purpose of simplification. For the wrought alloys, a four-digit designation system is used to produce a list of wrought composition families. The designation system for the wrought aluminum alloys is shown in Table 9.1.

Casting compositions are described by a three-digit system followed by a decimal value. The decimal 0.0 in all cases pertains to casting alloy limits. Decimals 0.1 and 0.2 concern ingot compositions, which after melting and processing result in alloy chemistries conforming to the requirement of casting specifications. The designation system for the cast alloys is provided in Table 9.2.

9.6 PHYSICAL METALLURGY OF ALUMINUM ALLOYS

While most metals tend to easily alloy with aluminum, comparatively few have sufficient solid solubility to serve as a major alloying addition. Of the most commonly used elements, only zinc, magnesium (both greater than 10 atomic %), copper (2.4 atomic %), and silicon (1.6 atomic %) have significant solid solubility in molten aluminum (Table 9.3). Several other elements having a solubility below 1 atomic % confer important improvements to the properties of aluminum alloy. Notable examples being the transition metals, i.e., chromium, manganese, and zirconium, which are

TABLE 9.1
Designation System for Wrought Aluminum Alloy

1XXX	Controlled unalloyed compositions	Not age-hardenable
2XXX	Alloys in which copper is the principal alloying element, though other elements notably magnesium, may be specified	Age-hardenable
3XXX	Alloys in which manganese is the principal alloying element	Not age-hardenable
4XXX	Alloys in which silicon is the principal alloying element	Age-hardenable if magnesium is present
5XXX	Alloys in which magnesium is the principle-alloying element	Not age-hardenable
6XXX	Alloys in which magnesium and silicon are the principal alloying elements	Age-hardenable
7XXX	Alloys in which zinc is the principle-alloying element, but other elements such as copper, magnesium, chromium, and zirconium may be present	Age-hardenable
8XXX	Alloys including the tin and lithium composition	Age-hardenable
9XXX	Alloys reserved for emerging ones	

used primarily to form compounds that control the grain structure [2,15,16]. With the exception of hydrogen, elemental gases have no detectable solubility in either liquid or solid aluminum. Apart from tin, which is sparingly soluble, the maximum solid solubility in binary alloys occurs at the eutectic and peritectic temperature [2]. A typical eutectic and peritectic binary phase diagram for the Al–Cu and Al–Cr alloy systems is as shown in Figure 9.3 and Figure 9.4. Further, aluminum is a soft metal and the fact that high strength-to-weight (σ/ρ) ratio can be achieved only in certain alloys arises because these alloys show a marked response to precipitation hardening or age hardening.

9.6.1 ALLOYING ADDITIONS TO ALUMINUM AND THE ROLE OF IMPURITIES

Alloying elements are normally present as

  (a) Solid solution with aluminum, and
  (b) Micro-constituents comprising (i) the element itself, (e.g., silicon) or (ii) as a compound between one or more elements and aluminum (e.g., Al_2CuMg).

TABLE 9.2
Designation System for Cast Aluminum Alloys

1XX.X	Controlled unalloyed (pure) composition	Not age-hardenable
2XX.X	Alloys in which copper is the principal alloying element, but other alloying elements may be specified	Age-hardenable
3XX.X	Alloys in which silicon is the principal alloying element, but other alloying elements such as copper and magnesium are specified	Some are age-hardenable
4XX.X	Alloys in which silicon is the principal alloying element	Not age-hardenable
5XX.X	Alloys in which magnesium is the principle alloying element	Not age-hardenable
6XX.X	Unused	
7XX.X	Alloys in which zinc is the principal alloying element, but other alloying elements such as copper and magnesium may be specified	Age-hardenable
8XX.X	Alloys in which tin is the principal alloying element	Age-hardenable
9XX.X	Unused	

TABLE 9.3
Solid Solubility of Elements in Aluminum

Element	Temperature (°C)	Maximum Solid Solubility	
		(Wt. %)	(At. %)
Cadmium	649	0.4	0.09
Cobalt	657	<0.02	<0.01
Copper	548	5.65	2.4
Chromium	661	0.77	0.4
Germanium	424	7.2	2.7
Iron	655	0.05	0.025
Lithium	600	4.2	16.3
Magnesium	450	17.4	18.5
Manganese	658	1.82	0.9
Nickel	640	0.04	0.02
Silicon	577	1.65	1.59
Silver	566	55.6	13.8
Tin	228	~0.06	~0.01
Titanium	665	~1.3	~0.74
Vanadium	661	~0.4	~0.21
Zinc	443	70	28.8
Zirconium	660.5	0.28	0.08

Note: Maximum solid solubility occurs at eutectic temperatures for all elements except chromium, titanium, vanadium, zinc, and zirconium for which it occurs at peritectic temperatures.

Source: From Van Horn, K. R., Ed., *Aluminum*, Vol. 1, American Society of Metals, Ohio, U.S., 1967; Mondolfo, L. F., *Aluminum Alloys, Structure, and Properties*, Butterworth, London, 1976.

Solid solubility at 20°C is estimated to be approximately 2 wt.% for magnesium and zinc, 0.1–0.2 wt.% for germanium, lithium and silver and below 0.1 wt.% for all other elements.

In a typical commercial aluminum alloy, any one or all of the two conditions may tend to exist [15,16]. The solid solution is the most corrosion resistant form in which an aluminum alloy can exist. Magnesium when dissolved in aluminum tends to make it more anodic although dilute. Alloys of Al–Mg retain a relatively high resistance to corrosion, particularly to the seawater

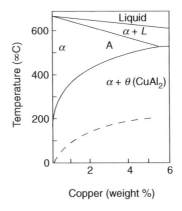

FIGURE 9.3 Section of the aluminum–copper eutectic phase diagram. (From Polmear, I. J., *Light Alloys: Metallurgy of the Light Metals*, Edward Arnold Publishers, London, 1981.)

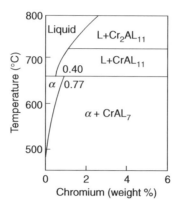

FIGURE 9.4 Section of the aluminum–chromium peritectic phase diagram. (From Polmear, I. J., *Light Alloys: Metallurgy of the Light Metals*, Edward Arnold Publishers, London, 1981.)

environment and alkaline solutions. Chromium, silicon, and zinc in solid solution with aluminum, have only a minor influence on corrosion resistance, although zinc does cause a significant increase in electrode potential. Thus, the aluminum–zinc alloys find use as (a) clad coatings for aluminum alloys and (b) galvanic anodes for the cathodic protection of steel structures in seawater environment. Copper reduces the corrosion resistance of aluminum more than any other alloying element and this arises because of its presence in micro-constituents.

9.7 MECHANICAL DEFORMATION AND ITS INFLUENCE ON STRENGTH OF ALUMINUM ALLOYS

Commercial usefulness of pure aluminum is limited by its strength. An important step towards improving the strength of the pure metal is by alloying to form a solid solution. This is followed by supersaturating the solid solution and by suitable heat treatment. This favors the excess solute to precipitate as second phases. The wrought aluminum-base alloys can be classified as being heat treatable or non-heat treatable depending on their response to precipitation hardening [15–17]. Heat treatable alloys derive their strength from a homogeneous distribution of fine precipitates developed by solutionizing followed by quenching and aging treatments. This process is known as precipitation hardening or age hardening [2]. Important alloying elements in this group have extensive solid solubility at high temperatures and include copper, magnesium, zinc, and lithium. Non-heat treatable alloys derive their strength from a combination of [17]

(i) A fine dispersion of intermetallic phases that form during solidification and
(ii) The dislocation substructure produced during mechanical deformation, i.e., strain hardening.

The phenomenon of recovery and recrystallization during subsequent processing plays an important role in (a) the strength levels achieved, (b) accompanying ductility, (c) formability, and (d) texture developed in the final wrought product [18]. Alloys in this group contain silicon, iron, manganese, and magnesium either singularly or in various combinations. During mechanical deformation, the wrought alloys are strengthened when the movement of dislocations is hindered such that a higher stress is needed to continue the deformation. Obstacles to dislocation movement that important are (a) solute atoms, (b) other dislocations, (c) particles of a second phase, and (d) grain and sub-grain boundaries [2,15].

9.7.1 SOLID SOLUTION STRENGTHENING

Magnesium provides an effective solid solution strengthening effect, partly because of its high solubility in the aluminum matrix. The solid solubility of magnesium in aluminum varies from 1.9 wt.% at 1000°C to 15.35 wt.% at 450°C [19]. The hardening contribution of the alloying element in solid solution is dependent on the (a) lattice structure of the solvent, (b) concentration of the solute addition, (c) type of solution, and (d) the degree of order achieved in the solution [20]. Strengthening due to the solid solution becomes significant at high solute concentrations where either short-range order (SRO) or long-range order (LRO) may develop and the passage of the first dislocation reduces order [21]. Short-range order (SRO) tends to confine the dislocations to a relatively small number of slip planes primarily because the leading dislocations create considerable disorder [22]. An offset of the sheared portions of an ordered region prevents complete reordering from occurring due to the passage of the second dislocation. Consequently, successive dislocations following on the same slip plane encounter less resistance to movement either because of the energy released upon partial re-creation of order or because of the reduced order on the slip plane [23].

When the atoms in the solid solution rearrange themselves into solute-rich clusters, strains are generated because of the difference in equilibrium lattice spacing, or atomic diameter of the solute rich regions and the FCC parent aluminum matrix. These clusters are known as Guinier–Preston zones (GP zones). In silver-containing aluminum alloys, the structure of the Guinier–Preston (GP) zones was determined by transmission electron microscopy and x-ray diffraction studies [13]. A typical Guinier–Preston (GP) zone in an Al-5.0 atomic percent silver (Ag) alloy is shown in Figure 9.5. Gregg and Cohen [25] observed that upon aging below 443 K, the GP zones in an Al-5 atomic percent Ag alloy were octahedral in shape with no internal order. At temperatures

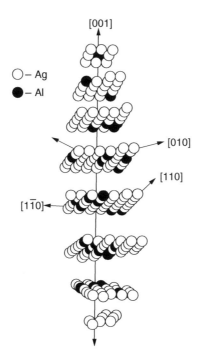

FIGURE 9.5 Exploded view of a typical Guinier–Preston (GP) zone in an aluminum- 5 atomic percent silver alloy formed on aging at 110°C. (From Cottrell, A. H., *Dislocations and Plastic Flow in Crystals*, Oxford University Press, Oxford, 1953.)

above 443 K, the GP zones were of spherical morphology and contained about one-half the silver of the zones formed at the low temperature [26]. During mechanical deformation, the Guinier–Preston zones should be sheared by the moving dislocations rather than be bypassed. For example, in the aluminum–silver (Al–Ag) system, Gregg and Cohen [25] observed that the destruction of order in the Al–Ag zones because of cutting/shearing of the zones by the moving of dislocations is not a source of strengthening. As the zones are not ordered, a difference in stacking fault energy between the zones and the matrix strengthening precipitates plays an important role. Alternatively, strengthening arises from an increase in surface area when the moving dislocations [27] cut the fine precipitates.

In solid solutions with short-range order, the dislocations tend to form co-planar groups with the lead dislocation being pushed by the trailing dislocations. In solutions with long-range order (LRO), the dislocations tend to agglomerate. However, there is no net amount of disorder created by the passage of two equal dislocations through a slip plane where the sum of the Burger's vectors equals the repeat distance of the structure [28]. Consequently, there is little tendency for groupings of greater than two dislocations. The distribution of such dislocation pairs (known as super lattice dislocations) is homogeneous on the primary slip planes having the greatest critical resolved shear stress (CRSS) in a solution having long-range order (LRO). A concentration of super lattice dislocations occurs on relatively few of the potentially active slip planes when short-range order (SRO) exists [28].

Fleischer [30] for the case of nonuniform solute concentration considered the elastic interaction between dislocations and solute atoms, known as the Cottrell interaction [29]. The magnitude and intrinsic severity of this interaction was found to depend on the change in concentration across a region containing the stress field of the dislocation. Steep concentration gradients resulted in interaction effects of appreciable magnitude [30].

9.7.2 PRECIPITATION HARDENING

An aluminum alloy is said to be precipitation hardenable when its yield strength or hardness increases with time at a constant temperature (the aging temperature) following rapid cooling from a higher temperature, i.e., the solution heat treatment temperature [2,24]. The primary requirement for a heat-treatable aluminum alloy to be amenable to age hardening is a gradual decrease in solid solubility of one or more of the alloying elements with decreasing temperature, and in concentrations that exceed their equilibrium solubility at ambient and moderately higher temperatures [31]. The heat treatment sequence involves the following steps:

(a) Solution heat treatment to a relatively high temperature within a single-phase region in order to dissolve the alloying elements.
(b) Rapid cooling or quenching to room temperature to obtain a supersaturated solid solution of the elements in the aluminum matrix, thereby, minimizing the precipitation of solute atoms as secondary phases.
(c) Controlled decomposition of the supersaturated solid solution (SSSS) to form a finely dispersed phase.

The final step in this three-step sequence is accomplished below the equilibrium solvus temperature, and below a metastable miscibility gap called the GP zone solvus [31]. The aluminum alloys having slow precipitation reactions at the ambient temperature must be heat-treated to attain the high strengths that are generated through precipitation [31]. Depending on the kinetics of a particular precipitation reaction, some alloys can be aged at room temperature (T4 temper). A few alloys require a thermal driving force, provided by elevated temperatures, in order to achieve their peak strength (maximum strength). In certain aluminum alloys, a considerable increase in strength is obtained by imposing a controlled amount of cold working immediately following quenching.

A fraction of the increase in strength due to this procedure can be attributed to strain hardening. However, when cold working is followed by precipitation heat treatment, the precipitation kinetics is greatly accentuated [31]. This is a result of the effectiveness of the dislocation substructure generated during the cold working step, which lowers the free energy required for heterogeneous nucleation.

The phenomenon of precipitation hardening in aluminum-base alloys was first discovered by Alfred Wilm [32] who observed that the hardness of an Al-4% Cu-0.5% Mg-0.5% Mn alloy to increase with time at ambient temperature after having been quenched from a higher temperature. Subsequently, Merica and co-workers [33] postulated that age hardening occurred in alloys whose solid solubility increased with an increase in temperature, thereby, enabling the second-phase particles to form at a lower temperature by precipitation from an initially supersaturated solid solution (SSSS). The primary focus of the research work during the 1920s and 1930s was on the mechanism of precipitation or aging rather than the mechanisms of strengthening. The earliest attempted explanation of age hardening using the concept of dislocations was that of Mott and Nabarro [34]. These researchers suggested the observed strengthening to be the result of an interaction between dislocations and the internal stresses produced by the misfitting coherent precipitates. Subsequently, Orowan [35] developed and put forth a relationship relating the strength of an alloy containing hard particles to the ratio of shear modulus of dislocation and the average planar spacing of the particles. The Orowan relationship serves as the basis for the theory of dispersion strengthening of alloys by non-deformable particles. The Orowan relationship is expressed as [35]

$$\tau = \alpha [Gb/\lambda] \tag{9.2}$$

where τ is the shear stress, G is the matrix shear modulus, b is the Burgers vector, and λ is the inter-particle spacing.

During mechanical deformation, interaction of the moving dislocations with the non-deformable particles results in the generation of new dislocations necessary to accommodate the strain. In this way, the non-deformable particles increase, at least initially, the work hardening rate of the aluminum alloy. The extent and/or severity of work hardening depend upon the mutually interactive influences of the following: (a) dislocation-particle interaction, (b) the dispersion parameters, and (c) the possible recovery processes occurring.

Increase in strength of precipitation-hardenable aluminum alloys over that of pure metals or the solid solution alloys is due to an interaction of the moving dislocations with the pre-precipitates and precipitates, which are formed during aging below the Guinier–Preston zone solvus temperature (T_C) [24,31]. Decomposition of the supersaturated solid solution (SSSS), below T_C, normally occurs in the following sequence:

$$\text{SSSS} \quad \rightarrow \quad \text{solute clusters} \quad \rightarrow \quad \text{intermediate precipitate} \quad \rightarrow$$
$$\text{equilibrium precipitates (Transition structure)} \tag{9.3}$$

The relationship depicting the decomposition of the super-saturated solid solution (SSSS) involves a complex sequence of both time-dependent and temperature-dependent changes. At relatively low temperatures and during the initial stages of artificial aging at elevated temperatures, the principal change occurring is a redistribution of solute atoms within the solid solution lattice to form clusters, which are referred to as Guinier–Preston (GP) zones. The GP zones are solute rich clusters of atoms. They retain the structure of the matrix and are coherent with it. Besides, they do produce appreciable elastic strain as depicted in Figure 9.6. The strain is dependent on their volume fraction. As the number and density of the GP zones increases, the degree of disturbance of the regularity and periodicity of the lattice increases.

Matrix planes

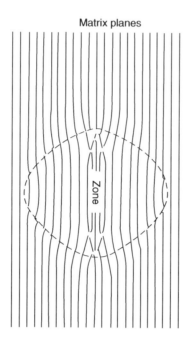

FIGURE 9.6 Representation of the distortion of matrix lattice planes near the coherent Guinier–Preston zone.

Obstacles to the motion of dislocations in age-hardened aluminum alloys can be regarded as being either localized or diffuse. Localized obstacles are those, which interact with the dislocations only when they are in direct physical contact. On the other hand, diffuse obstacles interact with the dislocations at a distance characterized by a range (d) and interaction energy (U). When the interaction energy is greater than zero, the diffuse obstacles are repulsive, and when the interaction energy is less than zero, they are attractive [36]. The precipitate particles, which lie along the slip planes on which the dislocations move, cause the dislocations to behave in one of two ways: (i) cut or shear through the precipitate particles, or (ii) take a path around the particles [24].

With respect to coherent GP zones, maximum hardening occurs when spacing between the particles (GP zones) is equal to the limiting radius of curvature of the moving dislocation lines, which is approximately 10 nm [36]. During mechanical deformation, the moving dislocations shear the coherent GP zones. The observed increase in yield strength caused by the presence of these zones arises because of their high volume fraction. The strengthening effect of the GP zones can be interpreted as resulting from an interference offered to the motion of dislocations when they cut the zones. This is due to chemical strengthening, which results from

(A) The production of new particle-matrix interfaces by the dislocation when it shears a coherent particle, and

(B) An increase in the stress required to move a dislocation through a region distorted by coherency stresses.

Strengthening also arises from an increase in surface area when the moving dislocations cut the precipitates. The progressive increase in strength of an age-hardened aluminum alloy with natural aging can be attributed to the conjoint and interactive influences of (i) an increase in the size of the GP zone and (ii) an increase in their number density.

With aging at higher temperatures (known as artificial aging), the GP zones are replaced by second-phase particles having a crystal structure distinct from that of the solid solution and also

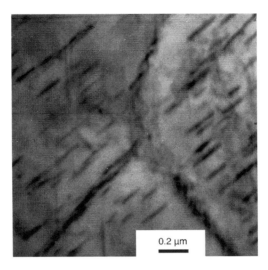

0.2 µm

FIGURE 9.7 Bright field transmission electron micrograph showing the intermediate θ' (Al$_2$Cu) precipitates in an Al–Cu–Li–Mn alloy. 2020-T651 alloy. (From Srivatsan, T. S., Coyne, E. J. Jr., and Starke, E. A. Jr., *Journal of Materials Science*, 21, 1553–1560, 1986.)

different from the structure of the equilibrium phase. These particles are referred to as the transition precipitates [24,31]. The transition or intermediate precipitates (Figure 9.7) have specific crystallographic orientation relationships with the solid solution such that the phases remain either coherent or partially coherent with the parent FCC aluminum matrix [39,40]. For example in an Al–Cu system, the metastable phases, which form prior to the equilibrium precipitates, are denoted as θ'' and θ' (Al$_2$Cu). The θ'' and θ' (Al$_2$Cu) phases are coherent with the alpha-aluminum matrix along the (001) planes [24,36]. The strengthening contribution of the transition precipitate structures is related to the impedance offered to the motion of dislocations by the presence of lattice strains. As long as the mobile dislocations continue to cut the strengthening precipitates, the strength increases with aging time in which the average size of the precipitate particle increases while the volume fraction of precipitates remains essentially constant. The strength increases up to a certain critical size of the particle. Beyond the critical diameter, the dislocations are forced to loop or bow around the second-phase particles resulting in a concomitant loss in strength. There are other factors such as (a) morphology of the second-phase particles and (b) their distribution through the alloy microstructure, which must also be taken into consideration. Blazer and co-workers [37] calculated the critical resolved shear stress (CRSS) for the glide of dislocations through a random array of δ' (Al$_3$Li) precipitates in a lithium-containing aluminum alloy and concluded that the more sharply peaked the distribution, the more effective is the strengthening effect. Variation of strength of a precipitation hardenable aluminum alloy with particle diameter is shown in Figure 9.8.

In the maximum strength (peak-aged) condition, the Guinier–Preston (GP) zones and the intermediate (metastable) precipitates are the principal strengthening agents in commercial wrought aluminum-base alloys [24,34,36]. GP zones, which are solute-rich clusters of atoms, are only one or two atom planes in thickness. For example, in the aluminum–copper alloys, Guinier [41] and Preston [42] reported the discovery of solute-rich clusters (GP zones), which were fully coherent with the aluminum alloy metal matrix. The GP zones, which form at ambient temperature, are known to be thin {100} platelets of copper enriched regions one or two atomic diameters in thickness and about 15 atomic diameters in radius [41,42]. However, their structure and composition was not known. The copper atoms are 13% larger than the aluminum atoms, giving rise to

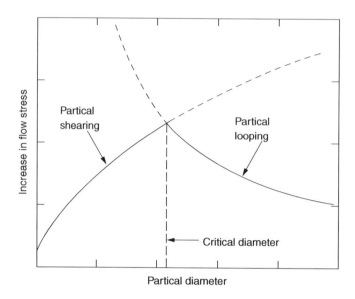

FIGURE 9.8 Schematic showing the variation of flow stress with particle diameter for the shearing and dislocation looping mechanism. (From Ardell, A. J., *Metallurgical Transactions*, 16A, 2131, 1985.)

strain energy. In order to minimize the strain energy, the GP zones form as thin platelets on the {100} planes of aluminum [43,44].

Intermediate precipitates are normally much larger than the GP zone. In most precipitation hardened aluminum alloys, the intermediate precipitates tend to nucleate at the sites of the stable GP zones. In a few other alloys, the precipitates nucleate heterogeneously at lattice defects such as dislocations and grain boundaries [23,31,40]. Gradual shearing of the coherent GP zones by the moving dislocations, during mechanical deformation, increases the number of solute-solvent bonds across the slip planes as shown in Figure 9.9. As a result, the process of clustering, or agglomeration, tends to be removed. Additional work must be done by the applied stress in order for this to occur. The magnitude of work is controlled by factors such as (a) the relative sizes of the atoms concerned and (b) the difference in stacking fault energy between the alloy matrix and the precipitate [43].

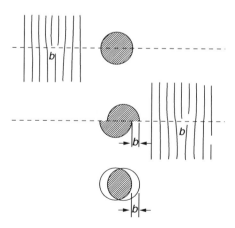

FIGURE 9.9 Schematic showing cutting of a Guinier–Preston zone by a moving dislocation. (From Starke, Jr., E. A., *Materials Science and Engineering*, 29, 40, 1977.)

Maximum age hardening in an aluminum alloy occurs in the presence of a critical dispersion of (a) the GP zones, (b) the intermediate precipitates, or (c) both. In deforming the metal into its plastic zone the coherent GP zones as well as the fully coherent and partially coherent intermediate precipitates are penetrated or sheared by the moving dislocations. For the shearing/cutting mechanism, the critical resolved shear stress resulting from an interaction of gliding dislocations with the dispersion is expressed by the relationship

$$\tau = C f^m r^p \tag{9.4}$$

where f is the particle volume fraction, r is the particle radius, and m and p are positive empirical constants. Parameter C is an alloy constant. Strength increases with both volume fraction and particle size. Particle size is an important factor when considering dislocation-second-phase particle interactions [42].

When the precipitate particles cease to be coherent, its crystal structure is very different from that of the FCC matrix and the particles are large and widely spaced. At this point, the dislocations no longer cut the precipitate particles but readily bow out between them, or bypass them and rejoin by a mechanism originally proposed by Orowan [35]. A schematic of this mechanism is shown in Figure 9.10. Wavy slip lines observed on the surface of aluminum-base alloys suggest that the bypassing process essentially involves cross-slip [31]. Loops of dislocations are left around the precipitate particles that have been bypassed [44,45]. The yield strength of the alloy is low but the rate of work hardening is high. As a result, plastic deformation tends to be spread more uniformly throughout the grains, as is typically the case with the overaged (OA) aluminum alloys. A typical age hardening curve, in which the strength initially increases and subsequently decreases with aging time has been correlated with a transition from dislocation shearing of the precipitate particles to dislocation bypass of the precipitates as is shown in Figure 9.11. In the mechanism of Orowan hardening, [35] the strengthening effect is independent of the properties of the particles. Consequently, the yield stress σ_Y can be described by the modified Orowan relationship:

$$\sigma_X = \sigma_o + 0.8\,(Gb/L) \tag{9.5}$$

where σ_o is the flow stress of the unstrengthened matrix, which may include a component of solid solution strengthening, G is the shear modulus, b is the Burger's vector and L is the inter-particle spacing.

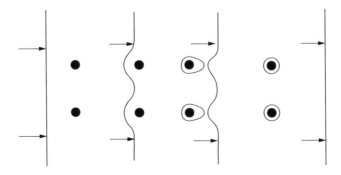

FIGURE 9.10 Schematic showing a dislocation bypassing the incoherent, widely spaced particles. (From Orowan, E., *Symposium on Internal Stress in Metals and Alloys*, The Institute of Metals, London, United Kingdom, p. 451, 1948.)

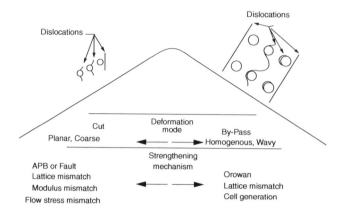

FIGURE 9.11 Schematic showing the effect of aging time and particle size on strengthening mechanisms and the deformation mode of a precipitation hardened aluminum alloy. (From Orowan, E., *Symposium on Internal Stress in Metals and Alloys*, The Institute of Metals, London, United Kingdom, p. 451, 1948.)

9.7.3 PRESENCE AND ROLE OF TRACE ELEMENTS IN GOVERNING PRECIPITATION KINETICS

Quite common with other nucleation and growth processes, precipitation reactions are often influenced by the presence of minor amounts or traces of certain elements [15,31,38]. These changes arise due to the concurrent and competing influence of the following:

1. Preferential interaction with vacancies, which reduces the rate of nucleation of the GP zones;
2. Raising the GP zone solvus, which alters the temperature range over which the various phases are stable;
3. Stimulating the nucleation of an existing precipitate by reducing the interfacial energy between the matrix and the precipitate; and
4. Promoting the formation and presence of a different precipitate.

9.8 PROPERTIES OF ALUMINUM

Among the most striking characteristics of aluminum is its versatility. The range of physical and mechanical properties that can be developed, from refined high purity aluminum to the most complex alloys, is truly remarkable [15]. More than three hundred alloy compositions are currently recognized, and many additional variations are continuously being developed internationally and in supplier-consumer relationships. The key properties of aluminum that make the metal and its alloy counterparts an affordable, attractive and viable choice for a wide variety of uses are appearance, light-weight, fabricability, physical properties, mechanical properties, and corrosion resistance [2,14,15].

Aluminum has a density of 2.7 g/cm^3, approximately one-third that of steel (7.8 g/cm^3), one-third of copper (8.93 g/cm^3) or one-third of brass (8.53 g/cm^3) [14,15]. One cubic inch of aluminum weighs one-tenth of a pound [14]. Pure aluminum melts at 1220°F, which is considerably lower than the melting point of other structural materials. Further, it displays excellent corrosion resistance in most environments, including ordinary atmosphere (relative humidity of 55%), water (including salt water), petro-chemicals, and many other chemical systems. Aluminum surfaces are most often highly reflective. Radiant energy, visible light, radiant heat, and electromagnetic waves are efficiently reflected, while anodized and dark anodized surfaces can be either reflective or absorbent [14]. The reflectance of polished aluminum, over a broad range of wavelengths, leads to its selection for a variety of decorative and functional uses [2,14,15].

Pure aluminum and its alloys display excellent electrical and thermal conductivity. However, sustained research efforts over a period have resulted in the development and emergence of alloys having a high degree of electrical resistivity. On a volume basis, the electrical conductivity of pure aluminum is roughly 65% of the International Annealed Copper Standard [14]. However, pound-for-pound aluminum is a better conductor than copper, surpassed only by sodium. The alloys of aluminum also find use in high torque electric motors. Aluminum is often selected for its electrical conductivity, which is nearly twice that of copper on an equivalent weight basis [14]. The requirements of high conductivity and mechanical strength can be effectively met by use of long-line, high voltage aluminum steel cord reinforced transmission cable. The thermal conductivity of aluminum alloys, at about 50–60% that of copper, is advantageous in varying applications ranging from (a) heat exchangers, (b) evaporators, (c) electrically heated appliances, (d) utensils, (e) automotive cylinder heads, and (f) radiators [14]. As far as mechanical properties are concerned, pure aluminum is relatively weak but highly ductile. Its modulus of elasticity is 10 million psi compared to 30 million psi for steel. Further, aluminum has a tendency to deform elastically about three times more than steel under a comparable loading condition. This property is not substantially improved by the addition of alloying elements to aluminum. In addition, aluminum is tough and unlike steel and titanium, aluminum and its alloy counterparts do not lose their ductility and become brittle at cryogenic temperatures [14]. It is a unique metal having varying applications since cold working, or strain hardening, to a significant extent, can strengthen it. However, the greatest strengthening is obtained by alloying aluminum with a variety of elements. Pure aluminum (99.9% content) finds use as electrical conductors, chemical equipment, and sheet metal working. The alloys of aluminum find use where strength is an important design consideration [15,17,23].

9.9 WROUGHT ALUMINUM ALLOYS

The earliest approach to alloy development was strictly empirical. Alloying elements were added and the products were evaluated by the potential end user. If an alloying element provided an improvement in a particular property, more was added until such time other properties of interest became unsuitable for the concerned application. With time, a gradual evolution of the following occurred:

(a) Customer needs became more sophisticated in that a combination of properties was often required.
(b) The science of metals grew.
(c) The instruments and procedures to characterize microstructure and properties were improved.
(d) Producers developed many tests in an attempt to predict their performance during service.

As science related to the metal progressed, alloy developers began to understand why certain alloying elements produced the desired changes in properties. However, the scientific community still did not understand the reasons for the effects they observed for many years after the initial observation. The concept of "if a little is good, add more until something bad happens" was in use. The realization that properties are controlled by microstructure, which is in turn controlled by (a) composition, (b) method of primary processing, i.e., casting, (c) secondary processing or fabrication, and (d) heat treatment practices, came considerably before the ability to identify the microstructural features that controlled properties [46]. This led to the alloy developer creating a list of heuristic rules relating alloy composition and processes to properties. The relationship between microstructure and mechanical properties is better known today than it has ever been. However, it is still insufficient to serve as a guideline to alloy developers and manufacturers. The progressive development and emergence of aluminum alloys was proceeded by a process of "enlightened empiricism [46]." The alloy developer determines which existing alloy-processing combination develops the properties closest to the target. Then, a judgment is made as to which microstructural features are controlling the property of interest. Changes in alloy composition and processing (primary + secondary) designed to provide the desired microstructure were then tried using experience and science as the guiding tools.

In a few to many instances, aluminum and its alloys have replaced another material in an existing application. The driving force for the replacement is largely dependent on application. The first alloys of aluminum were developed when the characteristics of unalloyed aluminum were inadequate to meet the needs of the end user. For purposes of structural use, the strongest alloy that meets the minimum requirements for properties such as corrosion resistance, ductility, and toughness is usually selected provided it is cost-effective [46,47]. The continuing pursuit for a cost-effective and viable solution to materials was the driving force for alloy development. The properties of an aluminum alloy depend on microstructure, which, in turn, is controlled by both composition and processing to include both primary and secondary [46–49]. Consequently, the development of tempers is equally important, as is the establishment of a chemical composition. The different temper designations are covered in a following section. The chemical composition of selected wrought heat treatable aluminum alloys that find selection and use in aerospace is provided in Table 9.4.

9.10 WROUGHT NON-HEAT-TREATABLE ALUMINUM ALLOYS

Manufacture of strain hardened or non-heat treatable (NHT) aluminum alloy sheet represented a major portion of the aluminum industry since the late 1980s. The Aluminum Association has designated the 1XXX-, 3XXX-, and 5XXX- series alloy designations to be the major non-heat treatable aluminum alloys [50]. Unlike the heat treatable counterparts, which derive their strength from the mechanism of precipitation hardening, [23,49] the NHT alloys are strengthened by the conjoint and mutually interactive influences of elements in solid solution and the dislocation structures introduced by cold rolling, i.e., the phenomenon of strain hardening [50]. The 1XXX alloys are defined as those that are 99.0 wt.% or greater aluminum with trace amounts of iron and silicon as the major alloying elements. The 3XXX alloys have manganese as the primary alloying element, while the 5XXX alloys are alloyed principally with magnesium. A few 8XXX-series NHT alloys, the Al–Fe–X compositions, have been developed during the last three decades but are not used in large volumes [51].

Microstructure development during work hardening (strain hardening), recovery, and recrystallization were found to be strongly influenced by the interactive influences of the following: (a) the elements in solid solution and (b) the phases formed during casting (the constituents) and preheating (the dispersoids). For the 1XXX (Al–Fe–Si) alloys the presence and role of iron-bearing constituent particles and silicon in solid solution on strain hardening, and nucleation of recrystallized grains must be considered. For the 3XXX (Al–Mn–Fe–Si) alloys the distribution of the $Al_{12}(Fe, Mn)_3Si$ dispersoids was found to be critical in controlling grain size and mechanical properties [50]. In the 5XXX alloys a mutual interaction between the magnesium atoms and the moving dislocations aids in inhibiting dynamic recovery while concurrently facilitating high strength in the alloy without heat treatment. The precipitation and presence of the metastable phase β' (Al_3Mg_2) does not provide useful strengthening to a commercial 5XXX alloy but tends

TABLE 9.4
Nominal Compositions of Aluminum Aerospace Alloys

Alloy	Zn	Mg	Cu	Mn	Cr	Zr	Fe	Si	Li	O	Other
1420		5.2				0.1			2		
2004			6			0.4					
2014		0.5	4.4	0.8			0.7[a]	0.8			
2017		0.6	4	0.7			0.7[a]	0.5			
2020			4.5	0.55			0.4[a]	0.4[a]	1.3		0.25Cd
2024		1.5	4.4	0.6			0.5[a]	0.5[a]			
X2080	1.85	3.7			0.2	0.20[a]	0.10[a]			0.2	
2090			2.7			0.1	0.12[a]	0.10[a]	2.2		
2091		1.5	2.1			0.1	0.30[a]	0.20[a]	2		
X2095	0.25[a]	0.4	4.2			0.1	0.15[a]	0.12[a]	1.3		0.4Ag
2219			6.3	0.3		0.2	0.3[a]	0.2[a]			0.1V
2224		1.5	4.1	0.6			0.15[a]	0.12[a]			
2324		1.5	4.1	0.6			0.12[a]	0.10[a]			
2519		0.2	5.8	0		0.2	0.3[a]	0.2[a]			0.1V
6013		1	0.8	0.35			0.30[a]	0.8			
6113		1	0.8	0.35			0.30[a]	0.8		0.2	
7010	6.2	2.35	1.7			0.1	0.15[a]	0.12[a]			
7049	7.7	2.45	1.6		0.15		0.35[a]	0.25[a]			
7050	6.2	2.25	2.3			0.1	0.15[a]	0.12[a]			
7055	8	2.05	2.3			0.1	0.15[a]	0.1[a]			
7075	5.6	2.5	1.6		0.23		0.4[a]	0.4[a]			
7079	4.3	3.2	0.6	0.2	0.15		0.4[a]	0.3[a]			
X7093	9	2.5	1.5			0.1	0.15[a]	0.12[a]		0.2	
7150	6.4	2.35	2.2			0.1	0.15[a]	0.12[a]			
7178	6.8	2.8	2		0.23		0.5[a]	0.4[a]			
7475	5.7	2.25	1.6		0.21		0.12[a]	0.10[a]			
8009							8.65	108		0.30[a]	1.3V
X8019							8.3	0.2[a]		0.2	4.0Ce
8090		0.9	1.3			0.1	0.30[a]	0.20[a]	2.4		

Most ingot metallurgy aluminum alloys contain about 0.05–1% Ti to refine the ingot grain size.

[a] Maximum

to exacerbate problems due to corrosion because of its tendency to precipitate both at and along the grain boundaries.

For the non-heat treatable (NHT) aluminum alloys the temper designations used are based on the amount of strengthening achieved due to work hardening that remains in the sheet subsequent to processing. The cold working operations, recovery and recrystallization treatments are manipulated during fabrication to provide the desired combination of final properties in the sheet product. A summary of the Aluminum Association NHT alloy temper designations are provided in Table 9.5.

9.11 THE POWDER METALLURGY ALLOYS

Use of aluminum as powders began when flake-type aluminum powder was manufactured in the early 1900s by either milling or attriting thin sheets into tiny flakes used as an additive pigment in paint formation. Introduction of the atomization process in the 1920s provided for an efficient production process and marked a new era in aluminum powder production [51]. This process

TABLE 9.5
Temper Designations for the Non-Heat Treatable Aluminum Alloys

F-	*As-fabricated*: no special control over thermal or working operations; no mechanical property limits
O-	*Annealed*: to obtain the lowest strength temper
H-	*Strain-hardened*: applies to products that have had their strength increased by strain-hardening, with or without subsequent thermal operations; always followed by two digits
H1X-	Strain-hardened only, no subsequent thermal operation
H2X-	Strain-hardened and partially annealed
H3X-	Strain-hardened and stabilized

X indicates the degree of strain hardening remaining after rolling and/or partially annealing; except for H1X tempers. It does not directly correspond to the cold-reduction experienced by the metal.

Examples:

H18-	Corresponds to UTS achieved by cold-reduction of 75% after full annealing
H14-	Applies to material strengthened to one-half UTS level of H18
H34-	Applies to material with strength of one-half UTS level of H18 but Achieved by stabilizing after cold rolling
H19-	Applies to material with UTS of 2.0 ksi (14 MPa) greater than that of H18
H26-	Applies to material strengthened to three-quarters UTS level of H18 by cold rolling and partial annealing

aided by the initiation of World War II and the concurrent development of aluminized explosives created a major market for the atomized aluminum powder. From a technological viewpoint, atomized aluminum powder has found application in both the commercial and defense-related areas. The use varies from pigments for paints, roofing, commercial blasting materials, reducing agents for the manufacture of metals and alloys, pharmaceuticals, weld conductors, cellular concrete, pyrotechnics, chemicals, and rocket propellants. The commercial viability of low-cost pressed and sintered powder metallurgy parts emerged in the 1960s resulting in an expansion of the powder metal market into near net shape components [51].

Aluminum powders were initially investigated for use in high temperature applications. Sintered aluminum products (SAP) utilized aluminum oxide (Al_2O_3) dispersions to improve the high temperature strength and creep response of parts consolidated from milled aluminum powder [52–55]. During subsequent years experimental work [56,57] on atomized aluminum powders containing high solute content was initiated and provided the basis for research that gradually evolved into the development of the current PM aluminum alloys. These alloys utilized the benefits derived from rapid solidification, which included (i) alloying flexibility, (ii) increased solid solubility limits, (iii) compositional homogeneity, and (iv) fine scale microstructure. The benefits of rapid solidification are highlighted in Table 9.6.

TABLE 9.6
Benefits of Rapid Solidification

1.	Unique composition
2.	Fine-grain, constituent and dispersoid sizes
3.	Homogeneous distribution of alloy elements
4.	Superior combinations of strength, toughness, and resistance to corrosion and fatigue
5.	Potential for improved modulus, thermal stability, and low density

TABLE 9.7
Extension of Solid Solubility for Binary Aluminum Alloys

Elements	At. % (Equilibrium)	At. % (Reported Maximum)
Ce	0.01	1.9
Co	0.01	0.5–5
Cr	0.44	5.0–6.0
Cu	2.5	17–8
Fe	0.025	4.0–6.0
Mg	18.9	37–40
Mn	0.7	6.0–9.0
Mo	0.07	>1.0
Ni	0.023	1.2–7.7
Si	1.59	10–16
Ti	0.15	0.22–2
V	0.2	1.4–2
Zn	66.5	38
Zr	0.083	1.15–1.5

Techniques used for the processing of powder metallurgy alloys have gradually evolved [51,55]. Rapidly solidified particulates, in the form of powders, ribbon, and attrited flakes, must be consolidated to full density for maximizing mechanical properties [49]. Processing procedures were carefully designed to maintain the refined structure while concurrently obtaining the benefits inherent in the powder alloys. Advances in powder processing such as vacuum degassing, consolidation, hot isostatic pressing, direct spray deposition, extrusion, forging, and rolling have aided in the development of aluminum powder metallurgy alloys for structural applications [57]. Rapidly solidified particulates, produced by atomization, mechanical alloying, or attriting melt spun ribbon, have been consolidated to full density with excellent mechanical properties. Since the mid 1950s, rapid solidification of aluminum alloys has been based on gas atomization techniques [11,55]. The other techniques that were tried for the production of aluminum alloy powders were centrifugal atomization, rotating cup, rotating electrode process, plasma rotating electrode process, single and double-roll quenching, ultrasonic gas atomization, the Osprey process, and the soluble gas atomization process [11,55]. Following the production of alloy powders, secondary processes, such as mechanical alloying combining the carbide and oxide reinforcements, provide opportunities for achieving and retaining strength at high temperatures and improving the elastic modulus. The refinement of microstructure through rapid solidification coupled with an extension of solid solubility are important physical features of the process. Table 9.7 lists the maximum extended solid solubility reported by various investigators. The resulting benefits demonstrate the utility in the use of aluminum powder metallurgy alloys for aerospace applications.

9.11.1 ADVANTAGES AND BARRIERS TO ALUMINUM POWDER METALLURGY ALLOYS

The ability to achieve a refined microstructure coupled with economical powder processing techniques have created the desired potential for powder metallurgy alloys to be applied where (a) high strength and stiffness, (b) elevated temperature performance, (c) corrosion, and (d) wear resistance are required [50,51]. Powder metallurgy alloys offer an attractive combination of mechanical and physical properties that are far superior to the cast alloy counterparts. The benefits of rapid solidification are highlighted in Table 9.6. With time, several advances in the use of powder metallurgy aluminum alloys have been achieved in the areas spanning (a) alloy design and (b) atomization and

TABLE 9.8
Barriers Limiting the Application and Acceptance of Powder Metallurgy Alloys in the Marketplace

1.	High cost
2.	Product availability and size limitation
3.	Quality-control issues

consolidation processes [51]. With the gradual evolution of technology, the challenges in the future would involve the development of improved levels of properties coupled with assured reliability and reproducibility in powder properties and processing.

Although the powder metallurgy alloys offer significant potential in terms of mechanical properties, there are several barriers that limit their application and acceptance in the market place (Table 9.8). As with most materials, the cost of substitution is usually the major hurdle. In addition, the cost effectiveness must be clearly established in the total product life cycle [51]. Excluding the pressed and sintered parts, the cost of wrought powder metallurgy (PM) aluminum alloys are substantially greater than the cost of alloys produced by conventional ingot metallurgy (IM) techniques because of the additional processing steps involved (Figure 9.12). The powder metallurgy (PM) processed alloys cost 1.5–5 times as much as the conventional ingot metallurgy processed counterpart. Therefore, the early applications of the PM alloys were limited to components where superior performance and effective material utilization justified their use [11,50,51]. However, with time and concurrent advances in technology the overall cost of the PM alloys was reduced as the volume of powder metallurgy components increased.

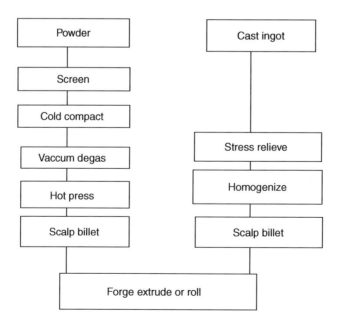

FIGURE 9.12 Comparison of the steps involved in powder metallurgy and ingot metallurgy processing in producing the end product from starting stock. (From Hildeman, G. J. and Koczak, M. J., *Treatise on Materials Science and Technology*, Academic Press, New York City, USA, Vol. 31, pp. 323–360, 1989.)

9.12   TEMPER DESIGNATION SYSTEM

Heat treatment or temper designation system was developed and put forth by the Aluminum Association [56]. It has also become part of the International Alloy Designation System (IADS) and is adopted by most countries. It is used for all the wrought alloys and takes the form of letters added as suffixes to the alloy number. One or more digits following the letter indicate subdivisions of the tempers, which have a significant influence on characteristics of the alloy. The temper designation system that is accepted and applicable to the high strength alloys used in aerospace-related applications is given in Table 9.9.

Designations for the common tempers and a description of the sequence of operations used to produce these tempers are given:

1. F: As-Fabricated
   Aluminum alloy shaped by cold working and hot working
2. O: Annealed
   Wrought aluminum alloy products annealed to obtain lowest strength temper and cast aluminum alloy products annealed to improve ductility and dimensional stability.
3. H: Strain-hardened (wrought products only).
   Products strengthened by strain hardening, with or without supplementary thermal treatment to produce some reduction in strength.
4. W: Solution heat-treated
   Unstable temper applicable to aluminum alloys only, whose strength tends to change at room temperature, over the time duration of months or years, following solution heat treatment.

TABLE 9.9
Temper Nomenclature: Symbols Added as Suffix Letters or Digits to the Alloy Number

Suffix Letter F,O,H,T, or W Indicates Basic Treatment Condition	First Suffix Digit Indicates Secondary Treatment Used to Influence Properties	Second Suffix Digit for Condition H Only Indicates Residual Hardening
F—As-fabricated		
O—Annealed-wrought products only		
H—Cold worked strain hardened	1. Cold worked only	2. 1/4 hard
	2. Cold worked and partially annealed	4. 1/2 hard
		6. 3/4 hard
	3. Cold worked and stabilized	8. Hard
		9. Extra hard
W—Solution heat treated	1. Partial solution + natural aging	
T—Heat-treated stable	2. Annealed cast products only	
	3. Solution + cold worked	
	4. Solution + natural aging	
	5. Artificially aged only	
	6. Solution + artificial aging	
	7. Solution + stabilizing	
	8. Solution + cold work + artificial aging	
	9. Solution + artificial aging + old-work	

5. T: Solution heat-treated and aged

Alloys whose strength becomes stable within a few weeks of solution heat treatment. The digits following T designate the type of aging treatment.

9.12.1 SYSTEM FOR HEAT-TREATABLE ALUMINUM ALLOYS

The temper designation system for wrought alloy and cast alloy products that are strengthened by heat treatment employs the W and T designation. The "W" designation denotes an unstable temper, whereas the "T" designation denotes a stable temper other than F (as fabricated), O (annealed) or H (strain hardened). A number from 1 to 10 follows the letter T. Each number indicating a specific sequence of basic treatments.

(i) T1: Cooled from an elevated-temperature shaping process and naturally aged to a substantially stable condition

This designation applies to products that are not cold worked after an elevated-temperature shaping process such as casting or extrusion, for which the mechanical properties have been stabilized by room temperature aging. It also applies to products that have been flattened or straightened after cooling from the shaping process, for which the effects of cold working imparted by flattening or straightening are not accounted for in specified property limits.

(ii) T2: Cooled from an elevated-temperature shaping process, cold worked, and naturally aged to a substantially stable condition

This temper refers to products that have been cold worked to improve strength after cooling from a hot-working process such as rolling or extrusion for which the mechanical properties have been stabilized by room-temperature aging. It also applies to products wherein the effects of cold work, either imparted by a flattening or straightening operation, are accounted for in specified property limits.

(iii) T3: Solution heat-treated, cold worked, and naturally aged to a substantially stable condition

T3 applies to products that have been cold worked specifically to improve strength subsequent to solution heat treatment for which the mechanical properties have been stabilized by room temperature aging. It applies to products in which the effects of cold work, either imparted by a flattening or straightening operation, are accounted for in specified property limits.

(iv) T4: Solution heat-treated and naturally aged to a substantially stable condition

This signifies products that are not cold worked following solution heat treatment and for which the mechanical properties have been stabilized by room-temperature aging. If the products are either flattened or straightened, the effects of cold work imparted by the flattening or straightening operation are not accounted for in specified property limits.

(v) T5: Cooled from an elevated-temperature shaping process and artificially aged

T5 includes products that are not cold worked following an elevated-temperature shaping process such as casting or extrusion, for which the mechanical properties have been substantially improved by precipitation heat treatment.

(vi) T6: Solution heat-treated and artificially aged

This group encompasses products that are not cold worked following solution heat treatment and for which the mechanical properties or dimensional stability, or both, have been substantially improved by the precipitation heat treatment. If the products are either flattened or straightened, the effects of cold work imparted by flattening or straightening are not accounted for in specified property limits.

(vii) T7: Solution heat-treated and over-aged or stabilized

This temper applies to wrought products that have been precipitation heat treated beyond the point of maximum strength to provide some special characteristic, such as enhanced resistance to stress-corrosion cracking or exfoliation corrosion. It also applies to cast products that are artificially aged following solution heat treatment to provide dimensional stability and stability in strength.

(viii) T8: Solution heat-treated, cold worked, and artificially aged

This designation applies to products that have been cold worked specifically to improve strength following solution treatment and for which the mechanical properties, or dimensional stability, or both, have been substantially improved by precipitation heat treatment. The extrinsic influence of cold work to include work imparted by the flattening or straightening operation is accounted for in specified property limits.

(ix) T9: Solution heat-treated, artificially aged, and cold worked

This grouping is comprised of products that are cold worked specifically to improve strength after they are precipitation heat-treated.

(x) T10: Cooled from an elevated-temperature shaping process, cold worked, and artificially aged

T10 identifies products that are cold worked specifically to improve strength after cooling from a hot-working process such as rolling or extrusion and for which the mechanical properties have been substantially improved by precipitation heat treatment. The effects of cold work include any cold work imparted by the flattening or straightening operation and are accounted for in specified property limits.

9.13 MECHANICAL BEHAVIOR OF ALUMINUM ALLOYS: ROLE OF MICROSTRUCTURE

The key microstructural features that control the properties of aluminum alloys are [5,24,31,38]

1. Coarse intermetallic compounds, usually in the range of 0.5–10 μm in size, which form during ingot solidification or during subsequent processing. These compounds contain the impurity elements iron and silicon. A few of the insoluble compounds are Al_6 (Fe, Mn), Al_3Fe, Alpha-Al (Fe, Mn) Si, and Al_7Cu_2Fe while the soluble compounds are Al_2Cu, Mg_2Si, and Al_2CuMg. These particles are often aligned as stringers in fabricated products Figure 9.13.

2. The submicron size particle referred to as dispersoids, of size 0.05–0.5 μm, which are the intermetallic compounds containing the transition metals chromium, manganese, and zirconium or other high melting point elements. Examples of these particles are (i) $Al_{20}Cu_2Mn_3$ (Figure 9.14), (ii) $Al_{12}Mg_2Cr$, and (iii) Al_3Zr. To a certain extent, these particles serve to inhibit recrystallization and grain growth in the alloys.

3. Fine precipitates, up to 0.01 μm in size, which form during precipitation hardening or age hardening heat treatments and which promotes or enhance strengthening (Figure 9.15).

4. The size and shape of grains.

5. The dislocation substructure resulting from cold working. Each of these features is influenced by the various stages involved during ingot solidification and subsequent processing, referred to as thermomechanical processing (TMP) of both the wrought and cast aluminum alloys.

(a) 50 µm
Second-phase particles

(b) 20 µm
Second-phase particles

FIGURE 9.13 Bright field optical micrographs showing the distribution of coarse second-phase particles in microstructure of aluminum alloy 2024-T8.

9.13.1 TENSILE BEHAVIOR (PROPERTIES)

Essentially aluminum alloys are divided into two distinct groups depending upon whether or not they respond to precipitation hardening or age hardening [23,31,38]. For alloys that do respond to age hardening treatments, the finely dispersed precipitates have a dominant effect in raising both yield and tensile strengths. For the alloys that do not respond to age hardening, the dislocation substructure produced by cold working in the case of wrought alloys and grain size in the case of cast alloys are important [5,16,31].

The presence of coarse intermetallic compounds has relatively little positive influence on both yield and tensile strengths but can cause a notable loss of ductility and toughness of both the cast and wrought products [16,57]. These particles tend to either crack or decohere at small values of plastic strain forming internal voids, which under the action of continued plastic strain, tend to coalesce leading to premature failure (Figure 9.16). The fabrication of wrought products often tends to cause directional grain flow. Elongated grain structure is retained by the addition of elements such as manganese and chromium that form submicron particles that prevent the

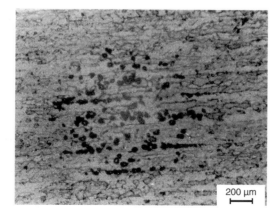

200 µm

FIGURE 9.14 Bright field transmission electron micrograph showing the distribution of $Al_{20}Cu_2Mn_3$ dispersoids in the microstructure of an Al–Cu–Li–Mn alloy 2020-T651. (From Srivatsan, T. S., Yamaguchi, K., and Starke, E. A. Jr., *Materials Science and Engineering*, 83(1), 87–107, 1986; Srivatsan, T. S. and Coyne, E. J. Jr., *Aluminium, an International Journal*, 62(6), 437–442, 1986.)

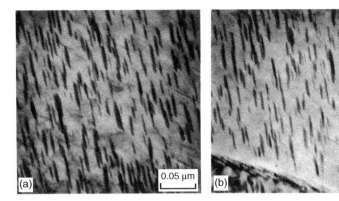

FIGURE 9.15 Bright field transmission electron micrograph showing the matrix strengthening precipitates [θ′ (Al$_2$Cu) and T$_1$(Al$_2$CuLi)] in aluminum alloy 2020-T651. (From Srivatsan, T. S., Coyne, E. J. Jr., and Starke, E. A. Jr., *Journal of Materials Science*, 21, 1553–1560, 1986.)

occurrence of recrystallization during mechanical deformation and subsequent heat treatment [5,31]. The coarse intermetallic particles also become aligned to form stringers in the direction of metal flow (Figure 9.13). These microstructural features are known as mechanical fibering, and in conjunction with crystallographic texture, they promote anisotropy in tensile and other properties. Accordingly, the tensile properties are greatest in the longitudinal direction and least in the short transverse direction in which stressing is normal to the stringers of intermetallics. The influence of test specimen orientation (longitudinal (L) versus long transverse (LT) versus short transverse (ST)) on tensile properties of two alloys, i.e., 7075 and 2014, is summarized in Table 9.10.

The governing criteria for design often involve the use of either yield strength or ultimate tensile strength, two key parameters obtained from the tensile test. Quantitative relationships do exist between strength and (i) precipitate particles, (ii) sub-grain and grain size, and (iii) texture effects [16,57]. For the age hardenable aluminum alloys, strength is primarily controlled by the volume fraction, size, and spacing of the matrix strengthening precipitates. A high volume fraction of small hard particles is desired but this is often difficult to obtain in precipitation hardening

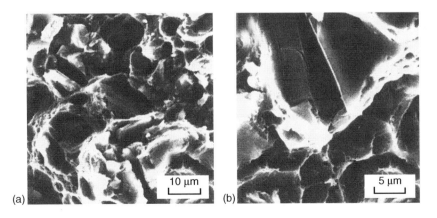

FIGURE 9.16 Scanning electron micrograph of a precipitation hardened aluminum alloy 2014 reinforced with 15 vol.% alumina particulates and deformed in cyclic loading showing. (a) Particulate failure by cracking (b) Decohesion or separation at the matrix-particulate interfaces.

TABLE 9.10
Effect of Purity on the Fracture Toughness of Some High-Strength Aluminum Alloys

Alloy and Temper	% Fe Maximum	% Si Maximum	0.2% Proof Stress (MPa)	Tensile Strength (MPa)	Fracture Toughness (MPa m$^{1/2}$)	
					Longitudinal	Short Transverse
2024-T8	0.5	0.5	450	480	22–27	18–22
2124-T8	0.3	0.2	440	490	31	25
2048-T8	0.2	0.15	420	460	37	28
7075-T6	0.5	0.4	500	570	26–29	17–22
7075-T73	0.5	0.4	430	500	31–33	20–23
7175-T736	0.2	0.15	470	540	33–38	21–29
7050-T736	0.15	0.12	510	550	33–39	21–29

Source: From Speidel, M. O., *Proceedings of the Sixth International Conference on Light Metals*, Leoben, Austria, Aluminum-Verlag, 1975.

aluminum alloy systems. In an ideal precipitation hardenable aluminum alloy, the precipitation and presence of the equilibrium phase is preceded by the formation of Guinier–Preston (GP) zones and intermediate precipitates that are easily sheared by the moving dislocations. During artificial aging the Guinier–Preston zones and intermediate precipitates grow in size with time and the strength progressively increases. Thus, the strength of a particular precipitation hardenable aluminum alloy can be optimized by maximizing those elements that participate in the aging sequence and can be put in solid solution at the solution heat treatment temperature [16,31].

Distribution and size of the strengthening precipitates significantly affects the attainable strength and can be controlled by aging temperature and time. For some alloys, the precipitation kinetics and resultant strength are controlled by the amount of deformation prior to aging. When the precipitates have large coherency strains, or interfacial energies, defects such as dislocations, sub grain boundaries and grain boundaries are the preferential sites for the nucleation of precipitates in the high strength 2XXX and 8XXX series alloys. For most of the age hardenable aluminum alloys, yield strength anisotropy is observed in thick plates, forgings, and extrusions and in thin sheet stock. For the thick products, the anisotropy is normally observed through the thickness while in thin products the anisotropy occurs in the plane of the sheet [57].

Early research work on high strength aluminum alloys was directed primarily at maximizing tensile properties in materials chosen and used for aircraft construction. With time and the gradual evolution and emergence of a newer generation of civilian and military aircraft, the emphasis on alloy development shifted away from tensile strength as an overriding consideration and much more attention was given to the behavior or mechanical response of alloys under the actual conditions experienced during service [55]. Tensile strength controls the resistance to failure by mechanical overload.

9.13.2 FRACTURE TOUGHNESS

In the presence of macroscopic fine microscopic cracks and other flaws in the microstructure, the property of toughness (i.e., fracture toughness) of the alloy becomes the most important parameter. Similar to other structural metallic materials, the toughness of aluminum alloys decreases as the level of strength is raised by either the independent or the conjoint influence of (a) alloying, (b) heat treatment, and (c) cold working (strain hardening). Minimum fracture toughness requirements become more stringent and for the high strength aluminum alloys it is necessary to place a ceiling on the level of yield strength that can be safely chosen and used by the design engineer [5,11,31].

Since crack extension in commercial aluminum alloys often proceeds by the ductile, fibrous mode involving the growth and coalescence of the fine microscopic voids that nucleate either by cracking of the coarse and intermediate-size second-phase particles (Figure 9.16a) or by decohesion at interfaces between the second-phase particle and the aluminum alloy metal matrix (Figure 9.16b). A major advancement in the development of new and improved alloys having greatly improved fracture toughness resulted from a careful control of the level of impurity elements, namely: iron and silicon. This effect is shown in Figure 9.17 for the Al–Cu–Mg alloy system. The plane strain fracture toughness values may be doubled by maintaining the combined levels of these elements below 0.5% when compared to similar alloys in which this value exceeds 1.0%. This resulted in a range of high toughness versions of the older alloy compositions that are in commercial use in which the levels of impurities have been noticeably reduced (Table 9.10).

The "toughness tree" shown in Figure 9.18 is an effective tool for classifying the contribution of microstructural features to toughness [57]. The metallic and nonmetallic inclusions on the right-hand side of the "toughness tree" can be visualized as extrinsic effects primarily because they are controlled by defects due to processing than by alloy composition. Metallic and nonmetallic inclusion content is minimized by controlling composition, melt temperature, filtering or direct-chill processing [57]. The hydrogen level is kept low by minimizing exposure to moisture while mechanical deformation processing aids in minimizing porosity. The features on left of the "toughness tree" are termed intrinsic because they are characteristic of composition and microstructure of the alloy. These features depend on the response of the material to deformation processing and heat treatment [57].

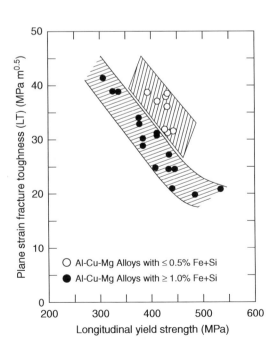

FIGURE 9.17 Plane strain fracture toughness of commercial Al–Cu–Mg sheet with differing levels of iron and silicon content. (From Srivatsan, T. S. and Coyne, E. J. Jr., *Aluminium, an International Journal*, 62(6), 437–442, 1986.)

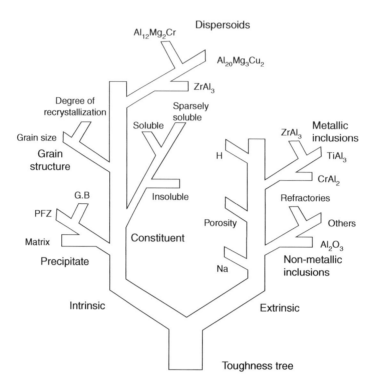

FIGURE 9.18 Toughness tree representation of microstructural features that affect toughness in age hardenable aluminum alloy. (From Starke, E. A. Jr., and Staley, J. T., *Progress in Aeronautical Science*, 32, 131–172, 1996.)

9.13.3 Fatigue Response

Contrary to the iron base alloys (i.e., steels), the increase that has been achieved in the tensile strength of most nonferrous alloys have not be accompanied by proportionate improvements in fatigue properties. This is shown in Figure 9.19, which reveals the relationship between fatigue endurance limit (5×10^8 cycles) and tensile strength for the different alloys. The fatigue ratios are the lowest for the precipitation-hardened alloys. As a rule, the more an alloy is dependent on precipitation hardening for its total strength, the lower this ratio becomes [5,59].

Disappointing fatigue properties of the precipitation or age-hardened aluminum alloys are also attributed to an additional factor, which is the metastable nature of the metallurgical structure under conditions of cyclic stressing. The localization of strain is particularly harmful because the matrix strengthening precipitates may be removed by repeated shearing caused by the motion of dislocations. This causes localized softening, which leads to a concentration of stress and a concomitant localization of strain at intrinsic microstructural features so that the process of cracking is accelerated. This effect is shown for the recrystallized aluminum alloy in Figure 9.20 [60,61].

Fatigue behavior of precipitation hardened aluminum alloys can be significantly improved if the fatigue deformation could be dispersed more uniformly. Factors that prevent the formation of coarse slip bands and localized heterogeneous deformation should assist in this regard. Thus, it is expected that the commercial purity alloys will perform much better than equivalent high purity compositions because the presence of second-phase inclusions and intermetallic compounds would tend to disperse slip. This effect is shown for an Al–Cu–Mg alloy 2024 for fatigue curves of the commercial purity and high purity versions of the alloy (Figure 9.21) [63]. The superior fatigue property of the commercial purity 2024 alloy arises because slip is more uniformly dispersed by the

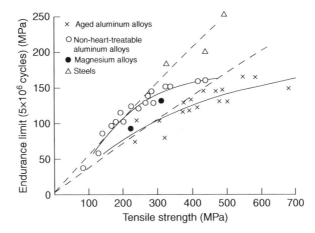

FIGURE 9.19 Comparison of the fatigue ratio (endurance limit to tensile strength) for the aluminum alloys with other materials. (From Speidel, M. O., *Proceedings of the Sixth International Conference on Light Metals*, Leoben, Austria, Aluminum-Verlag, 1975.)

submicron size dispersoid Al_6Mn particles. In subsequent research studies, thermomechanical processing (TMP) whereby plastic deformation, either before or during the aging sequence (treatment), increases the dislocation density was found to be beneficial for improved fatigue performance of aluminum alloy 7075 (Figure 9.22) [64]. This effect arises in part from an increase in tensile properties caused by such a thermal-mechanical treatment. The promising results obtained were based on tests conducted on smooth specimens of alloy 7075. However, the improved fatigue behavior was not sustained for the severely notched test specimens since local stress concentration and resultant strain localization overrides the more subtle microstructural effects on mechanical response.

The microstructure does exert a much greater influence on fatigue properties of aluminum alloys than on tensile properties has been demonstrated for an Al–Mg alloy containing trace amounts of silver. The binary Al–Mg alloy such as the Al–5Mg, the magnesium is essentially present in solid solution and the alloy develops a relatively high level of strength [65]. The same applies for an Al–5Mg–0.5Ag alloy in the as-quenched condition. For this alloy, the endurance limit after 10^8 cycles is 87 MPa, which equals 0.2% the proof stress (Figure 9.23). This was attributed to the interaction of magnesium atoms with moving dislocations, which minimizes the formation of coarse slip bands during fatigue deformation. The silver containing alloy responds to age hardening at elevated temperatures due to the formation of finely dispersed precipitates [65]. The endurance limit for 10^8 cycles decreases to 48 MPa due essentially to the localization of strain in a limited number of coarse slip bands.

9.13.4 FATIGUE CRACK INITIATION

Fatigue cracks often initiate either on the surface or at internal sites. The initiation event is associated with surface flaws, defects, and slip steps, which tend to concentrate stress. Other intrinsic microstructural features such as constituent particles, precipitate-free zones (PFZs), and micro-porosity that either concentrate the stress or localize the strain exert an adverse effect on fatigue crack initiation resistance. Thus, a microstructure that is strengthened by hard, non-shearable particles, free from constituent phases and porosity, and having a small recrystallized or unrecrystallized grain structure is preferred for enhanced fatigue crack initiation resistance [66].

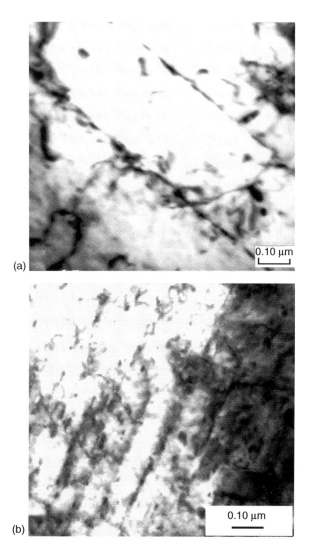

FIGURE 9.20 Bright field transmission electron micrograph showing planar slip deformation for alloy 2020-T651 deformed in cyclic tension compression loading leading to localization of stress and concomitant strain concentration at the microscopic level. (a) Partially recrystallized microstructure deformed at plastic strain amplitude of 0.22%. (b) Unrecrystallized microstructure deformed at plastic strain amplitude of 0.05%. (From Srivatsan, T. S., Yamaguchi, K., and Starke, E. A. Jr., *Materials Science and Engineering*, 83(1), 87–107, 1986.)

9.13.5 FATIGUE CRACK GROWTH

While fatigue crack growth is of interest the performance of aluminum alloy is measured by recording the crack growth rate (da/dN) as a function of the stress intensity range (ΔK). This type of fatigue crack growth test is of prime importance for alloys used in aerospace applications [57,66]. Crack growth is influenced by the following factors: (i) alloy composition and microstructure, (ii) the presence of oxygen and other deleterious species in the environment, (iii) temperature, (iv) load ratio (R), (v) material thickness (i.e., thickness relative to size of the plastic zone), (vi) stress intensity range, and (vii) the processes used in preparing the alloy [66]. It has been fairly well recognized and documented in the published literature that a mutual

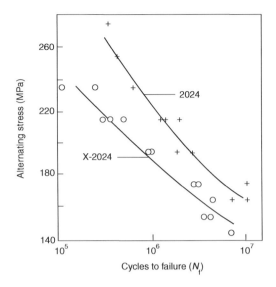

FIGURE 9.21 Effect of reducing the concentration of submicron particles in an Al–Cu–Mg alloy. X2024 is high purity version of aluminum alloy 2024. (From Lutjering, G., *Proceedings of Third International Conference on Strength of Metals and Alloys*, Institute of Metals, London, 1980.)

interaction among many of these variables does complicate the proper interpretation and extrapolation of the data while introducing significant uncertainties with respect to damage tolerant design and failure analysis [67]. For the 7XXX alloys, no consistent or appreciable differences were observed for the low- and high-purity versions at low and intermediate levels of stress intensity ranges (ΔK). However, at the high stress intensity ranges, the fatigue crack growth rates were

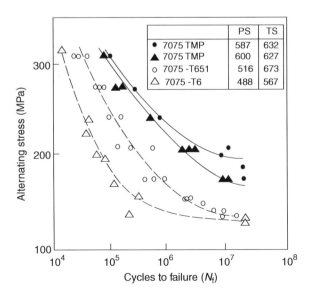

FIGURE 9.22 Influence of thermo mechanical processing and heat treatment on high cycle fatigue response of unnotched specimen of commercial Al–Zn–Mg–Cu alloy 7075. (From Ostermann, F. G., *Metallurgical Transactions*, 2A, 2897–2907, 1971.)

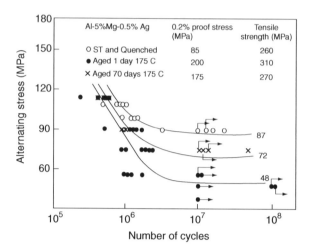

FIGURE 9.23 Stress-fatigue life curves for the alloy Al-5Mg-0.5Ag in different conditions. (From Boyapati, K. and Polmear, I. J., *International Journal of Fatigue of Engineering Materials and Structures*, 2, 23, 1979.)

noticeably reduced for the iron and silicon containing alloys. The reason for the observed improvement was related to the higher fracture toughness of the high purity metals. At high stress intensity levels or ranges, where the values of crack growth rate per cycle (d*a*/d*N*) are large, localized fracture and void nucleation at the constituent particles become the dominant mechanism controlling fatigue crack growth [66]. A comparison of crack growth rates (d*a*/d*N*) of aluminum alloy 2024 and 7075 is shown in Figure 9.24 [68].

For the precipitation hardenable aluminum alloys, microstructure is the principal independent variable, which can be used to control fatigue crack growth (FCG) once the loading conditions and environment are established. Key microstructural factors that exert an influence on fatigue crack growth are (i) slip character, (ii) slip length, and (iii) slip reversibility [69]. Each of these factors is affected by both alloy composition and microstructure [70]. Localization of strain within the

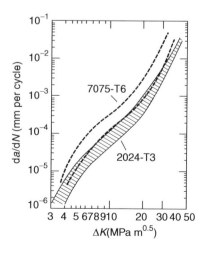

FIGURE 9.24 Comparison of fatigue crack growth rates of aluminum alloy 2024-T3 and 7075-T6 in laboratory air environment at varying humidity. (From Hahn, G. T. and Simon, R., *Engineering Fracture Mechanics*, 5, 523–533, 1973.)

microstructure tends to reduce fatigue crack growth while concurrently increasing the threshold stress intensity factor (ΔK_{TH}).

9.13.6 CORROSION FATIGUE

All structural materials are susceptible to degradation upon exposure to the environment, and aluminum alloys are no exception. Under the conjoint influence of conditions of simultaneous corrosion and cyclic stressing, known as corrosion fatigue, the reduction in strength is greater than the cumulative effects if each is considered separately. While it is possible to provide adequate protection for the metallic parts, which are stressed under static conditions, most surface films (including the protective oxides) are easily broken during cyclic loading. The observed reduction in fatigue strength of a material in a particular corrosive medium can be related to corrosion resistance of the material in that medium. Under conditions of corrosion fatigue, all types of aluminum alloys exhibit about the same percentage reduction in strength when compared with their fatigue strength in the room temperature laboratory air environment. Under fresh water conditions, the fatigue strength at 10^6 cycles is about 60% that in laboratory air and in the sodium chloride solution it is between 25 and 35% that in laboratory air. Further, the corrosion fatigue strength of an aluminum alloy is independent of its metallurgical condition [5,16,31].

Overall, aluminum alloy products corrode in a localized manner either by pitting, intergranular attack, or exfoliation corrosion. Pitting is essentially the removal of metal at localized sites resulting in the formation of cavities. Pits normally initiate at the coarse and intermediate size intermetallic particles, so their size, quantity, location, continuity, and corrosion potential relative to the adjacent aluminum alloy matrix does exert an influence on pitting corrosion behavior [71]. Intergranular corrosion is selective attack of the grain boundary regions with no appreciable attack of the metal matrix. Electrochemical cells are formed between the precipitates located on the boundaries and the depleted solid solution. Exfoliation corrosion initiates at the surface but is primarily a subsurface attack that proceeds along narrow paths parallel to the fabricated surface of the wrought or cast alloy product [71,72].

9.13.7 CREEP CHARACTERISTICS

Fracture due to creep, even in pure metals, normally occurs by the initiation of cracking in grain boundary regions. The susceptibility of grain boundaries to cracking in the precipitation-hardened aluminum alloys is enhanced because the grains are harder and less willing to accommodate plastic deformation than the relatively softer precipitate-free zones immediately adjacent to the boundaries. Further, the strength of grain boundaries is modified by the presence of precipitate particles. The precipitation-hardened alloys are normally aged at either one or two temperatures, which facilitate the peak strength to be realized in a relatively short amount of time. Continued exposure to elevated temperature results in over-aging and concomitant softening. Hence, the service temperature must be well below the final aging temperature if loss of strength due to over-aging is to be either avoided or minimized. Further, resistance to creep is enhanced by the presence of submicron iron-rich and silicon-rich intermetallic compounds and other fine particles that are stable at the required service temperature (normally below 200°C). The presence of fine alumina (Al_2O_3) particles in an aluminum alloy matrix also serves the same purpose [5,16]. While affording dispersion strengthening the alumina (Al_2O_3) particles aid in suppressing grain boundary migration.

9.13.8 PROPERTIES AT ELEVATED-TEMPERATURES

The 7XXX series of precipitation-hardened alloys that are based on the Al–Zn–Mg–Cu system develop the highest room-temperature tensile properties of any of the aluminum alloys produced from conventionally cast ingots. However, the strength of the 7XXX series alloys declines rapidly

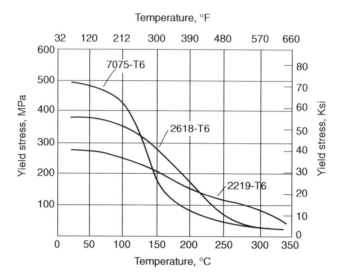

FIGURE 9.25 Values of 0.2% yield stress of aluminum alloy after exposure for 1000 h at temperatures between and 350°C. (From Bray, J. W., *ASM Metals Handbook*, Vol.2, 10th ed., pp. 31–60, 2000.)

when they are exposed to elevated temperatures (Figure 9.25) due mainly to coarsening of the fine precipitates, which contribute to strength of the alloys [67]. Alloys of the 2XXX series such as 2014 and 2024 perform better above these temperatures but are normally not used for the elevated-temperature applications [67]. Strength at temperatures above 100–200°C is improved mainly by solid solution strengthening or second-phase hardening. An alternative approach to improving the elevated temperature performance of aluminum alloys has been the use of rapid solidification processing to produce powders or foils containing a high super-saturation of the elements iron, silicon, and chromium, which diffuse slowly through solid aluminum [11]. An experimental Al–Cu–Mg alloy containing silver resulted in improved creep properties as shown in Figure 9.26 [67].

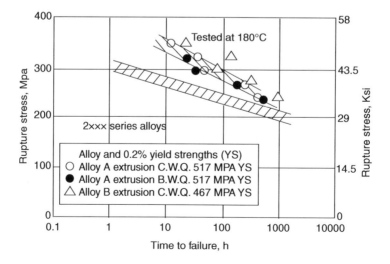

FIGURE 9.26 Stress rupture results for creep tests conducted at 180°C on aluminum alloys containing silver and compared with traditional 2XXX series alloys. Alloy A: Al-6.3Cu-0.5Mg-0.5Ag-0.5Mn, Alloy B: Al-6Cu-0.45Mg-0.5Ag-0.5Mn-0.14Zr. (From Rooy, E. L., *ASM Metals Handbook*, Vol.2, 10th ed., pp. 46–483.)

9.13.9 Low-Temperature Properties

Aluminum alloys represent a very important class of structural metals for subzero-temperature applications and find use for structural parts for operation at temperatures as low as -270°C (-450°F). Below zero, most aluminum alloys show marginal change in properties, i.e., the yield and tensile strengths may increase resulting in a slight decrease in elongation. Impact strength remains approximately constant. Consequently, aluminum is a useful material for many low-temperature applications. The chief deterrent to its extensive use is its relatively low elongation compared with the austenitic ferrous alloys [67].

9.14 CORROSION BEHAVIOR OF WROUGHT ALUMINUM ALLOYS

Aluminum is an active metal that tends to readily oxidize under the influence of high free energy of the reaction whenever the necessary conditions for oxidation are conducive and/or prevailing. However, overall the alloys of aluminum are stable in most environments due to the rapid formation of a natural oxide of alumina (Al_2O_3) on the surface. Oxide inhibits bulk reaction predicted from thermodynamic data. When the surfaces of aluminum are scratched sufficiently to remove the oxide film, a new film will quickly re-form in most environments. As a rule, the protective film is stable in aqueous solutions of the pH range 4.5–8.5 whereas it is easily soluble in both strong acids and alkalis leading to rapid attack of the aluminum alloy surface. Exceptions are concentrated nitric acid, glacial acetic acid, and ammonium hydroxide.

The oxide film that forms on a freshly rolled aluminum alloy exposed to ambient air is very thin and has been measured to be around 2.5 nm. It tends to grow at a decreasing rate for several years reaching a thickness of some tens of nanometers. The rate of growth of the film becomes rapid at the higher temperatures and at the higher humidities.

The presence of a thicker oxide film gives enhanced corrosion resistance to aluminum and its alloys. Various chemical and electrochemical reagents can produce a thicker oxide film. Natural film can be thickened about 500 times, say 1–2 µm, by immersion of components in certain hot acid and/or alkaline solutions. Although the films produced are mainly Al_2O_3, they also contain chemical such as chromates, which are collected from the bath to render them corrosion resistant.

9.15 RECENT DEVELOPMENTS ON ALUMINUM ALLOYS: ALUMINUM–LITHIUM ALLOYS

With the evolution of technology the conventional aluminum alloys face stiff competition from emerging composite material technologies, particularly in the structural aerospace market. Hybrid materials based on organic and metal matrices with whisker, fiber, or particle ceramic reinforcements offer impressive combinations of strength, stiffness and high temperature resistance [73–78]. Besides, the aramid polymer-reinforced aluminum alloy (ARALL) laminates [79–81] fabricated by resin bonding aramid fibers sandwiched between thin aluminum alloy sheets show exceptional promise as fatigue resistant materials. In light of these advances, the aluminum industry introduced a new generation of aluminum alloys, i.e., the aluminum–lithium alloys obtained by incorporating ultra-low density lithium into traditional aluminum alloys. These alloys were representative of a new class of lightweight, high modulus, high strength, monolithic structural materials, which are cost effective compared to the more expensive composite counterpart [82–92]. Despite the limitations posed by specific-stiffness and high temperature stability, aluminum–lithium alloys enjoy several advantages over the composite materials. Economically, the aluminum–lithium alloys were three times as expensive as conventional high strength aluminum alloys, whereas the competing hybrid materials can be up to 10 to 30 times more expensive [90]. Secondly, the fabrication technology for the lithium-containing aluminum alloys is quite compatible with existing manufacturing methods such as extrusion, sheet forming, and forging to obtain finished products for the conventional aluminum alloys. Finally, the lithium-containing aluminum alloys offer considerably higher ductility and fracture toughness properties compared to most metal-matrix composites [62,74–77].

Ingot metallurgy aluminum–lithium alloys have greater structural efficiency than the conventional high strength aluminum alloys because of reduced density, increased strength, and enhanced elastic modulus [82]. The use of the aluminum–lithium alloys in lieu of the currently used high strength aluminum alloys can reduce structural weight by 7–15% and increase the elastic modulus by 15–20% [82,85,88]. In addition, the fatigue crack propagation rates and spectrum fatigue crack growth resistance of the aluminum–lithium alloys are better than those of the 7XXX alloys [93]. The superior fatigue crack growth resistance of ingot aluminum–lithium alloy was rationalized to microstructural influences and related deformation characteristics. An un-recrystallized microstructure, strong texture, and sharable precipitates promote planar slip deformation [94]. Since planar slip increases slip reversibility, less damage tends to accumulate at the tip of a propagating crack. Planar slip deformation (Figure 9.27) also promotes the development of a tortuous crack path. This lowers the effective stress intensity by promoting roughness-induced crack closure and the tortuous nature of the crack path [95]. In fact, the observed variations in the fatigue resistance of aluminum–lithium alloys with respect to microstructure, crack size, loading sequence, and product form are largely a manifestation of the degree to which shielding is promoted or restricted in the different microstructures under various loading conditions [96–100]. The constant amplitude fatigue crack growth rate of peak aged commercial Al–Li alloys and traditional aluminum alloys is shown in Figure 9.28.

Another benefit of using aluminum–lithium alloys is their high temperature capability compared with the 7XXX alloys. This is due to the stable precipitates of Al–Li alloys at the high temperatures. Lithium-containing aluminum alloys exhibit approximately 40°C increases in use temperature over the 7XXX alloys for the same heat treatment condition. It was found that aluminum–lithium alloys exhibit an excellent combination of strength and ductility at the liquid nitrogen temperatures. Yield strength, ultimate tensile strength, and elongation to fracture of alloy 2090 at 77 K was 620 MPa, 745 MPa and 12% [101]. The high temperature capability of these alloys is beneficial to the high performance aircraft, and the excellent cryogenic mechanical properties make these alloys attractive for rocket applications [93]. With progressive evolution in processing science and technology, newer alloys were developed having lower lithium

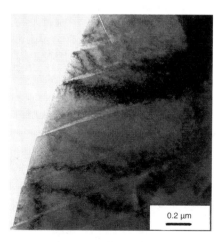

FIGURE 9.27 Cyclic deformation structures in an Al–Cu–Li–Mn alloy deformed at 433 K showing inhomogeneous deformation (a) Partially recrystallized microstructure (plastic strain amplitude 0.02%). (b) Unrecrystallized microstructure (plastic strain amplitude of 0.045%). (From Srivatsan, T. S., Yamaguchi, K., and Starke, E. A. Jr., *Materials Science and Engineering*, 83(1), 87–107, 1986.)

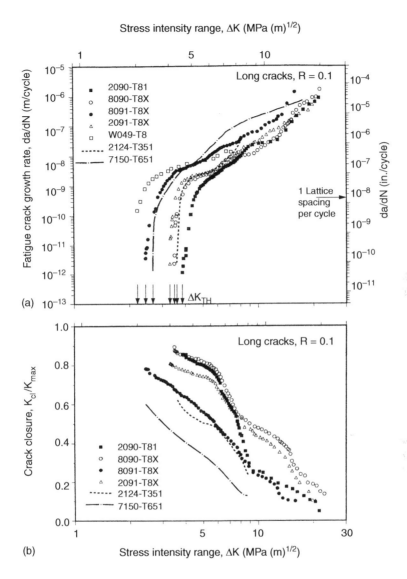

FIGURE 9.28 Constant amplitude (a) fatigue crack growth rates, and (b) corresponding crack closure levels for the long cracks in peak aged commercial Al–Li and traditional aluminum alloys in plate form (L–T orientation, $R = 0.1$, moist ambient air environment, using 6.4 mm thick specimen). The growth rate behavior of the Al–Li alloys is superior to traditional aluminum alloys consistent with the higher crack-closure levels. (From Venkateshwara Rao, K. T. and Ritchie, R. O., *Materials Science and Technology*, 5, 896–905, 1989.)

concentration than the initially developed and commercialized alloys 2090, 2091, and 8090. The first of the newer generation was Weldalite 049, [102] which can attain yield strength as high as 700 MPa and an associated elongation of 10%. Refinement of the original alloy, 2195 is being considered for cryogenic tanks for the U.S. Space Shuttle. Alloy 2195 offers several additional advantages over alloy 2219 for use in cryogenic tanks. Its higher strength coupled with higher modulus and lower density can lead to significant savings in weight. Further, alloy 2195 has good corrosion resistance, excellent fatigue properties, a significantly higher strength, and fracture toughness at cryogenic temperatures than at room temperature, can be near net shape formed, and can be welded provided adequate precautions are taken [103]. Over time, several other aluminum alloys containing less that 2% lithium have been developed and emerged. Research studies on these alloys have indicated that they provide a superior combination of properties for the bulkheads of high-performance aircraft [104,105]. The flat rolled products and extrusions of the new alloys would be competitive with polymer matrix composites for the horizontal stabilizer of commercial jetliners [57].

Several disadvantages exist, which are detrimental to implementation and prevent widespread use of these alloys on aircraft structures. Low ductility and fracture toughness in the short transverse direction, easy availability, loss of lithium, microstructural and mechanical anisotropy, and problems associated with delamination are the limiting factors [55,62,82,85]. Low ductility and poor fracture toughness in the short transverse direction are typical problems for high strength aluminum–lithium alloys for plate thickness exceeding one-inch. In addition, anisotropy in mechanical properties is the primary problem affecting the application of sheet metal stock [84,85].

9.16 ALUMINUM MACRO-LAMINATES

A relatively new class of materials showing promise for applications related to aerospace is the aluminum-based metal polymer laminates (referred to as macro-laminates). This technology was first developed at Delft University in the Netherlands during the 1970s. The first generation of these laminates was called ARALL, a succinct abbreviation for aramid aluminum laminates. These laminates are essentially composed of high strength aluminum sheets and unidirectional aramid fibers impregnated with a metal adhesive [93]. The layers are stacked, cured, and pre-strained subsequent to curing. In collaboration with the Delft University, Fokker Aircraft, and 3M Company, the Aluminum Company of America (ALCOA, USA) began evaluation and manufacturing studies of ARALL in 1983. The currently available ARALL laminates include ARALL 1 through 4. ARALL 1 is made up of 7475-T61 or 7075-T6 aluminum sheet and 3M Company's SP-366 pre-preg. The pre-preg is 50 wt.% unidirectional aramid fibers in an AF-163-2 adhesive. ARALL 2 and ARALL 3 consist of 2024-T8 and 7475-T651, respectively, and the sane pre-preg system. ARALL 4 is composed of aluminum alloy 2024-T8 and a high temperature pre-preg SP-376, which is made from AF-191 adhesive [93].

An attractive property of these macro-laminates is excellent fatigue resistance compared with the conventional aluminum alloys (Figure 9.29). Other advantageous properties macro-laminates offer include a 15–20% lower density and a 60% higher longitudinal strength than aluminum alloy 7075 and 2024 (Table 9.11). ARALL laminates are also resistant to fire and lightning strike damage and exhibit good mechanical damping characteristics [106]. The resistance to penetration by a lightning bolt determines the thickness of areas on the aircraft, which are prone to be hit by lightning. ARALL laminates provide the same protection as monolithic aluminum at significantly thinner gauges, so weight is saved. ARALL laminates can be fabricated using conventional aluminum machining, joining, and forming techniques [106,107]. It has been estimated that these laminates would tend to significantly increase durability in fatigue-prone areas with a potential weight savings of 30% with respect to aluminum because of higher design stress and lower specific weight [108]. Unfortunately, there are some drawbacks. The mechanical properties of the laminates are very anisotropic due to the unidirectional nature of fiber reinforcement. ARALL laminates are very sensitive to blunt notch stress and exhibit very low strain to failure and poor impact resistance [93].

9.17 THE ENGINEERED PRODUCTS FROM ALUMINUM ALLOYS

9.17.1 Castings

Pressure-die, permanent mold, green-sand mold, dry-sand mold, investment casting, and plaster casting [5,15,16] routinely produce castings of aluminum alloys. The key process variations include

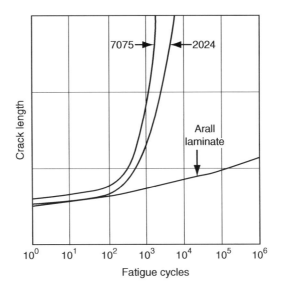

FIGURE 9.29 Comparison of the fatigue behavior of ARALL with conventional aluminum alloys. (From Frazier, W. E., Lee, E. W., Donnellen, M. E., and Thomoson, J. J., *Journal of Metals*, 22–25, May, 1989.)

(a) vacuum, (b) low-pressure, (c) centrifugal, and (d) pattern-related processes such as lost foam. Castings are produced by filling the mold with molten aluminum and are used for products having (a) intrinsic contours, (b) hollow or cored areas, and (c) combinations of the two. Choice of castings over other product forms is often based on net shape consideration. Reinforcing ribs, internal passageways, and complex design features, which are expensive to machine in a component made from the wrought alloy, can often be cast by appropriate pattern and mold or die design. Premium engineered castings display the characteristics of extreme integrity, close dimensional

TABLE 9.11
Typical Properties of ARALL Laminates and 2024-T3 Sheet

Material	Test Direction	UTS (MPa)	0.2% Offset YS (MPa)	Modulus (GPa)	Elongation (%)
ARALL1[a]	L	773	656	69	0.6
	LT	380	324	48	6.9
ARALL2[a]	L	704	352	66	1.3
	LT	317	235	45	12.4
ARALL3[b]	L	828	587	—	—
	LT	373	317	—	—
ARALL4[c]	L	745	380	61	2.6
	LT	338	311	49	4.6
2024-T3[d]	L	442	324	72	15
	LT	435	290	72	15

ARALL samples were 3/2 laminates, 1.3 mm thick. The 2024-T3 sample was 1.6 mm thick aluminum sheet.

[a] Alcoa technical fact sheet.

[b] Alcoa preliminary technical information.

[c] Alcoa preliminary technical information and NADC internal data.

[d] Military Standardization Handbook: Metallic Materials and Elements for Aerospace Vehicle Structures V.

tolerances, and consistently controlled mechanical properties in the upper range of existing high strength capabilities of selected alloys and tempers.

9.17.2 Extrusions

Aluminum alloy extrusions are produced by forcing solid metal through aperture dies. Designs that are symmetrical around the axis are especially adaptable for production in the extruded form. With incremental advances in technology, it is possible to extrude complex mandrel cored and even asymmetrical configurations. Major dimensions usually require no machining. Close tolerance of the as-extruded product often permits completion of part manufacture using simple cutoff, drilling, broaching, or other minor machining operations.

9.17.3 Forgings

Forgings are produced by inducing plastic flow through the application of kinetic, mechanical, or hydraulic forces in either a closed-die or an open die. A large number of aluminum alloy forgings are produced in closed dies to produce parts having good surface finish, dimensional control, exceptional soundness, and properties. Forgings are available both as rolled or mandrel-forged rings.

9.17.4 Powder Metallurgy

Parts made by the technique of powder metallurgy (PM) are formed by a variety of processes. For less demanding applications, the metal powder is compressed in a shaped die to produce green compacts. The diffusion bonded compacts are then sintered at an elevated temperature under a protective atmosphere. During sintering, the compacts consolidate and strengthen. The density of the sintered compact is increased by re-pressing. The use of re-pressing for improving dimensional accuracy is termed as "sizing," and when performed to alter configuration, it is termed as "coining." Powder metallurgy parts are very competitive with castings, forgings, stampings, machined components, and fabricated assemblies. Certain metal products can be produced only by powder metallurgy. Among these are the oxide dispersion strengthened alloys and materials whose porosity (number, distribution, and size of pores) is controlled.

9.18 CLASSIFICATION OF ALUMINUM ALLOY PRODUCTS

The aluminum industry has identified the major markets to be (a) building and construction, (b) transportation, (c) consumer durables, (d) electrical, machinery and equipment, (e) containers and packaging, (f) exports, and (g) other useful products that demand lightweight. Each major market comprises a wide range of end uses.

9.18.1 Building and Construction Application

In more recent years, alloys of aluminum have found extensive use in building bridges, towers, and storage tanks [5,16]. Since the initial cost of structural steel and plate is usually lower, alloys of aluminum find use when engineering advantages, construction features, architectural designs, lightweight and corrosion resistance are key considerations.

In the area of truck trailers, which are designed for maximum payload and operating economy from the standpoint of legal weight requirement, aluminum is chosen for use in frames, floors, roofs, cross-sills, and shelving. The wheels are made from forged aluminum, while the tanker and dump bodies are made from both sheet and plate forms, which are either riveted or welded together.

Aluminum alloy sheets have also found extensive use for the construction of mobile homes and travel trailers. Manufacturers of buses are concerned with minimizing dead weight. Sheet, plate, and extrusions of aluminum alloys have found use for body components and bumpers. The engine and structural components of the bus are made from cast, forged, and extruded stock. Over time, alloys of aluminum have been used in the construction of railroad hopper cars, boxcars, refrigerator cars, and tank cars [1,15,23]. Aluminum is also being used in passenger rail cars, particularly those for mass transit system.

Alloys of aluminum have grown in stature for use for a large variety of marine applications, including the main strength members such as hulls, deckhouses, and numerous other applications. The relatively low modulus of elasticity of aluminum alloys (10 million psi) offers advantages in structures erected on a steel hull. Flexure of the steel hull results in low stresses in an aluminum superstructure.

9.18.2 ELECTRICAL APPLICATIONS: CONDUCTOR ALLOYS

Use of aluminum predominates in most conductor applications. Aluminum of controlled composition is treated with trace additions of boron to remove titanium, vanadium, and zirconium, each of which increases resistivity [23]. The use of aluminum rather than competing materials is based on a combination of low cost, high electrical conductivity, adequate mechanical strength, low specific gravity, and excellent resistance to corrosion [5,15,16,23]. The most common conductor alloy (Alloy 1350) offers a minimum conductivity of 61.8% of the International Annealed Copper Standard (IACS) and a minimum tensile strength from 55 to 124 MPa depending on size. The minimum conductivity of hand drawn Aluminum 1750 is 204%. A limited number of alloys find use in bus bar for service at elevated temperatures in the installation of cable television [23].

9.18.3 HOUSEHOLD APPLIANCES

Lightweight and excellent appearance, coupled with adaptability to all forms of fabrication, and overall low cost of fabrication are the appealing reasons for the broad usage of aluminum in household electrical appliances. Lightweight is an important characteristic for selection and use in vacuum cleaners, electric irons, portable dishwashers, food processors, and even blenders. Low fabricating costs depend on several properties ranging from adaptability to die casting and the ease of finishing. Based on appearance and good corrosion resistance, expensive finishing of the alloy is often not required or necessary. Further, the ease of brazeability of aluminum makes it a useful choice for refrigerator and freezer evaporators. The tubing is placed on embossed sheet strips of a brazing alloy with a suitable flux. The assembly is then brazed in a furnace and the residual flux removed by washing first in boiling water and subsequently in nitric acid and cold water.

Wrought forms fabricated from sheet, tube, and wire are used in approximately the same quantities as die-castings. The wrought alloys are selected based on corrosion resistance, anodizing characteristics, formability, and other engineering properties. The lightweight, low maintenance costs coupled with corrosion resistance, durability, and attractive appearance make alloys of aluminum a prime candidate for portable furniture.

Since reflectivity of light is as high as 95% on specially prepared surfaces of high purity aluminum, it finds use as reflectors. Alloys of aluminum are often superior to other metals in its ability to reflect infrared or heat rays. It is capable of resisting tarnish from sulphides, oxides, and atmospheric contaminants [5,15,23] and has about three to ten times the useful life of that of silver, for use in mirrors, search lights, telescopes, and reflectors.

The addition of aluminum flakes to paint pigments explores the intrinsic advantages of high reflectance, durability, low emissivity, and minimum moisture penetration. Other applications for powders and pastes include printing inks, pyrotechnics, floating soap, aerated concrete, thermit welding, and energy enhancing fuel additives [1].

9.19 CONCLUDING REMARKS

The literature presented in this paper on the science, technology, and applications of aluminum alloys presents a cross section of results obtained and documented in the open literature. Results, documented by researchers, from both academia and industry, have certainly enriched our knowledge and understanding of the interrelationships between composition, processing, microstructural characteristics, and macroscopic behavior. Development of aluminum alloys began when the properties of unalloyed aluminum were insufficient to meet the needs of potential customers. A realization that properties were controlled by microstructure, which in turn is controlled by alloy composition, casting, fabrication, and heat treatment practices came considerably before the ability to identify the particular microstructural features that controlled properties. This led the alloy developer to create a host of heuristic rules relating composition and process to properties. The relationship between structure and properties is better known today than it has ever been. The alloy developer determines which existing alloy-process combination develops the property closest to the target. This provides valuable information on microstructural features controlling the property of interest. As a sound understanding exists on the relationship between structure and properties, alloy developers are now concentrating on science and less on heuristics. It has been demonstrated that the processing of novel aluminum alloys and the composite counterparts can be carried out using existing commercial presses and rolling mills, thus, obviating the need for investment in special purpose equipment.

Aerospace demands the development of high strength aluminum alloys. Newer developed alloys have the potential to provide higher margins of safety for the design of critical structures. Newer alloys and process modifications dominate the aluminum applications arena primarily because of cost effectiveness. Other technologies such as aluminum–lithium alloys, the rapidly solidified aluminum alloys, macro-laminates, and aluminum alloy-based metal matrix composites are commercially available but suffer from high cost and limited applications. Aluminum–lithium alloys are getting added attention, but they are two or three times as expensive as baseline aluminum alloy. These alloys are being used where high costs are not a significant factor, such as in military aircraft. Rapid solidification processed aluminum alloys have a limited number of applications. Rapid solidification enables the production of finer grained and homogeneous crystalline material with resultant higher strength. The potential for tough discontinuously reinforced aluminum alloy products, which match the specific strength and moduli of the graphite-epoxy composite offers the design engineer a choice for applications in both the existing and newer generation aircraft. In fact, it is aluminum's wide range of strong physical characteristics that offers the incentive for the development of new alloys, manufacturing processes, tempers, and aluminum-based composites. This is well supported by its low cost and environment friendliness. For the subsonic aircraft, the high strength alloys of aluminum continue to offer many advantages over other competing materials. The combination of specific properties, cost, predictable behavior, recyclability, coupled with years of experience in manufacturing, recyclability, and impact, will ensure that the aluminum alloys will continue to be the primary material of choice for the emerging subsonic aircraft.

Through the years, the demands put forth by the high performance aerospace industry provided the impetus for the development of high strength aluminum alloys. However, during the same period the demands put forth by the automotive industry were driven by manufacturability, cost, low structural weight, and improved fuel economy. Many automotive manufacturers did renew their interest in aluminum alloy structures after evaluating many composite materials and plastic materials in the 1980s and 1990s. New developments in body frame structures have provided the aluminum companies the much-needed impetus for developing newer alloys. While saving weight is the primary objective for aluminum body frames, the alloys must also resist corrosion, be recyclable and energy absorbent. Even today, the newer alloys and process modification dominate the arena for aluminum applications primarily because of cost-effectiveness. However, the newer and emerging technologies of aluminum–lithium, macro-laminates, and aluminum alloy metal matrix composites, while being commercially available, suffer from high cost and limited applications.

REFERENCES

1. Rooy, E. L., Introduction to aluminum and aluminum alloys, *ASM Metals Handbook*, Vol. 2, 10th ed., 46, 48, Materials Park, Ohio, USA.
2. Aluminum Alloys in the 1970s, *Metal Progress*, September, 68, 1970.
3. Speidel, M. O., Stress corrosion cracking of aluminum alloys, In *Metallurgical Transactions*, 6A, 631, 1975.
4. Studt, T., *R&D Magazine*, June, 48–51, 1992.
5. Polmear, I. J., *Light Alloys: Metallurgy of the Light Metals*, Edward Arnold Publishers, London, 1981.
6. Staley, J. T., Liu, J., and Hunt, W. H., *Advanced Materials Processes*, 10, 17–21, 1997.
7. Staley, J. T., *Canadian Aeronautics and Space Journal*, 31(1), 1–28, 1954.
8. ALCOA Aluminum and its Alloys: The Aluminum Company of America, Pittsburgh, PA, USA, 1950.
9. Ekvall, J. C., Rhodes, E. J., and Wald, G. G., Methodology for evaluating aircraft weight savings from basic material properties, *ASTM STP 761, Symposium on Design of Fatigue and Fracture Resistant Structures*, pp. 328–344, 1980.
10. Seventy-Five years of Alloys from ALCOA: A record of Alloy and Metallurgical Development at USA. Aluminum Company of America, Pittsburgh, PA, USA.
11. Lavernia, E. J., Ayers, J.D., Srivatsan, T. S. Rapid solidicication processing with specific aplication to aluminum alloys, *International Materials Reviews*, 37(1), 1–41, 1992.
12. Sanders, T. H., In *Fatigue and Microstructure*, Meshii, M., Ed., ASM, Metals Park, Ohio, 1979.
13. Peel, C. J., "Aluminum Alloys for Airframes: Limitations and Developments," presented at the Conference on Materials at their Limits, The Institute of Metals, Birmingham, U.K., 1985.
14. Epstein, S. G., *Technical Report on Aluminum and its Alloys*, The Aluminum Association, USA, 1978.
15. Hatch, J. E., *Aluminum: Properties and Physical Metallurgy*, ASM International, Materials Park, Ohio, 1984.
16. Varley, P. C., *The Technology of Aluminum and its Alloys*, Newnes-Butterworth, London, 1970.
17. Sanders, R. E., Jr., Baumann, S. F., and Stumpf, H. C., In *Aluminum Alloys: Their Physical and Mechanical Properties*, Sanders, T. H. and Starke, E. A., Eds., Vol. II, EMAS, United Kingdom, 1986.
18. Kelly, A. and Fine, M. E., *Acta Metallurgica*, 2, 9, 1954.
19. Hansen, M., *Constitution of Binary Alloys*McGraw Hill Publishers, New York, 1958, p. 107.
20. Cohen, J. B., Properties of ordered alloys, *Journal of Materials Science*, 4, 1012, 1969.
21. Buchner, A. R. and Pitsch, W., *Z. Metallkunde*, 76, 651, 1985.
22. Stoloff, N. S. and Davies, R. G., In *Progress in Materials Science*, Chalmers, B. and Hume-Rothery, W., Eds., Vol. 13, Pergamon Press, New York, p. 1, 1966.
23. Starke, E. A., Aluminum alloys of the 70's: Scientific solutions to engineering problems. An invited review, *Materials Science and Engineering*, 29, 99, 1977.
24. Starke, E. A., Kralik, G., and Gerold, V., *Materials Science and Engineering*, 11, 319, 1973.
25. Gragg, J. E. and Cohen, J. B., *Acta Metallurgica*, 19, 507, 1971.
26. Bauer, R. and Gerold, V., *Acta Metallurgica*, 10, 637, 1962.
27. Kelly, A. and Hirsch, P. B., *Philosophical Magazine*, 12, 881, 1965.
28. Kelly, A., *Philosophical Magazine*, 3, 1472, 1958.
29. Cottrell, A. H., *Dislocations and Plastic Flow in Crystals*, Oxford University Press, Oxford, 1953.
30. Fleischer, R. L., *Strengthening in Metals*, Van Nostrand, Princeton, New Jersey, p. 63.
31. Starke, E. A., Aluminum alloys of the 70's: Scientific solutions to engineering problems. An invited review, *Materials Science and Engineering*, 29, 40, 1977.
32. Wilm, A., *Metallurgie*, 8, 225, 1911.

33. Merica, P. D., Waltenberg, R. G., and Scott, H., *Transactions AIME*, 64, 41, 1920.
34. Mott, N. F. and Nabarro, F. R. N., *Proceedings Physics Society*, 52, 86, 1940.
35. Orowan, E., *Symposium on Internal Stress in Metals and Alloys*, The Institute of Metals, London, United Kingdom, p. 451, 1948.
36. Ardell, A. J., *Metallurgical Transactions*, 16A, 2131, 1985.
37. Blazer, J., Edgecombe, T. S., and Morris, J. W., In *Aluminum–Lithium III*, Baker, C., Harris, S. J., Gregson, P. J., and Peel, C. J., Eds., The Institute of Metals, London, United Kingdom, p. 369, 1986.
38. Starke, E. A., Causes and effects of "denuded" or "precipitate-free" zones at grain boundaries in aluminum-base alloys, *Journal of Metals*, 23(1), 5, 1970.
39. Peter Haasen, *Physical Metallurgy*Cambridge University Press, Cambridge, p. 200, 1978.
40. Srivatsan, T. S., Coyne, E. J., and Starke, E. A., Microstructural characterization of two lithium-containing aluminum alloys, *Journal of Materials Science*, 21, 1553–1560, 1986.
41. Guinier, A., *Nature*, 142, 569, 1938.
42. Preston, G. D., *Nature*, 142, 580, 1938.
43. Fisher, J. C., *Acta Metallurgica*, 2, 9, 1954.
44. Ashby, M. F. and Smith, G. C., *Philosophical Magazine*, 5, 298, 1960.
45. Thomas, G. and Nutting, J., The plastic deformation of aged aluminum alloys, *Journal Institute of Metals*, 86, 7, 1957.
46. Staley, J. T., History of wrought aluminum alloy development, In *Treatise on Materials Science and Technology*, Vol. 31, Academic Press, New York City, USA, 1989.
47. *Aluminum Alloys in Aircraft: Aluminum Company of America*, Pittsburgh, PA, USA, 1943.
48. *Age Hardening of Metals*, Tower Press, Cleveland, Ohio, 1950.
49. Anderson, W. S., Precipitation hardening aluminum-base alloys, In *Precipitation from the Solid Solution*, ASM Materials Park, Cleveland, 1957.
50. Sanders, R. E., Baumann, S. F., and Stumpf, H. C., Wrought non-heat treatable aluminum alloys, *Treatise on Materials Science and Technology*, Vol 31, Academic Press, New York, USA, pp. 65–104, 1989.
51. Hildeman, G. J. and Koczak, M. J., *Treatise on Materials Science and Technology*, Vol. 31, Academic Press, New York City, USA, 1989.
52. Irmann, R., *Metallurgie*, 46, 125–135, 1952.
53. Gregory, E. and Grant, N. J., *Transactions AIME*, 200, 247–257, 1954.
54. Lyle, J. P., *Metal Progress*, December, 110–115, 1952.
55. Grant, N. J., In *High Strength Powder Metallurgy Aluminum Alloys*, Koczak, M. J. and Hildeman, G. J., Eds., The Minerals–Metals and Materials Society, Warrendale, PA, pp. 3–30, 1982.
56. *Aluminum Standards Data Handbook*, The Aluminum Association, Washington DC, 1984.
57. Starke, E. A. and Staley, J. T., *Progress in Aeronautical Sciences*, 32, 131–172, 1996.
58. Srivatsan, T. S., Yamaguchi, K., and Starke, E. A., *Materials Science and Engineering*, 83(1), 87–107, 1986.
59. Speidel, M. O., *Proceedings of the Sixth International Conference on Light Metals*, Leoben, Austria, Aluminum-Verlag, 1975.
60. Srivatsan, T. S. and Coyne, E. J., *Aluminium, an International Journal*, 62(6), 437–442, 1986.
61. Srivatsan, T. S. and Coyne, E. J., Mechanisms governing cyclic fracture in an aluminum-copper-lithium alloy, *Materials Science and Technology*, 3, 130–138, 1987.
62. Lavernia, E. J., Srivatsan, T. S., and Mohamed, F. A., Review: strength, deformation, fracture behavior and ductility of aluminum–lithium alloys, *Journal of Materials Science*, 25, 1137–1158, 1990.
63. Lutjering, G., *Proceedings of Third International Conference on Strength of Metals and Alloys*, Institute of Metals, London, 1980.
64. Ostermann, F. G., *Metallurgical Transactions*, 2A, 2897–2907, 1971.
65. Boyapati, K. and Polmear, I. J., *International Journal of Fatigue of Engineering Materials and Structures*, 2, 23, 1979.
66. Starke, E. A. and Lutjering, G., In *Fatigue and Microstructure*, Meshii, M., Ed., American Society of Metals, Materials Park, Ohio, USA, pp. 205–239, 1979.
67. Bray, J. W., *ASM Metals Handbook*, Vol. 2, pp. 31–60, 2000.
68. Hahn, G. T. and Simon, R., *Engineering Fracture Mechanics*, 5, 523–533, 1973.

69. Starke, E. A. and Williams, J. C., In *Fracture Mechanics: Perspectives and Directions, STP 1020*, Wei, R. P. and Gangloff, R. P., Eds., ASTM, Philadelphia, pp. 184–205, 1989.

70. Starke, E. A., In *Aluminum Alloys: Contemporary Research and Applications*, Vasudevan, A. K. and Doherty, R. D., Eds., Academic Press, San Diego, CA, pp. 33–63, 1989.

71. Piascik, R. S. and Willard, S. A., *Fatigue and Fracture of Engineering Materials*, 17, 1247–1259, 1994.

72. Hornbogen, E. and Starke, E. A., *Acta Metallurgica Materialia*, 41, 1–16, 1993.

73. Steinberg, M. A., *Scientific American*, 55(4), 66, 1986.

74. Nair, S. V., Tien, J. K., and Bates, R. C., *International Metals Reviews*, 30, 275–290, 1985.

75. Shang, J. K., Wu, W., and Ritchie, R. O., *Materials Science and Engineering*, 102A, 181, 1988.

76. Chou, T. W., McCullough, R. L., and Pipes, R. B., *Scientific American*, 255(4), 192, 1986.

77. Geiger, A. L. and Jackson, M., *Advanced Materials and Processes*, 137(7), 23–33, 1989.

78. Kamat, S. V., Hirth, J. P., and Mehrabian, R., *Acta Metallurgica*, 37, 2395–2400, 1989.

79. Marissen, R., Ph.D. Thesis, Delft University of Technology, 1988.

80. Ritchie, R. O., Yu, W., and Bucci, R. J., *Engineering Fracture Mechanics*, 32, 361, 2000.

81. Bucci, R. J., Mueller, L. N., Vogelsang, L. B., and Gunnink, J. W., In *Aluminum Alloys: Contemporary Research and Applications*, Vasudevan, A. K. and Doherty, R. J. Eds., Eds. Treatise on Materials Science and Technology, Vol. 31, Academic Press, New York, USA, pp. 295–322, 1990.

82. Sanders, T. H. and Starke, E. A., Eds., *Aluminum–Lithium Alloys I, Proceedings of the First International Conference on Aluminum Lithium Alloys*, The Metallurgical Society of AIME, Warrendale, Pennsylvania, 1981.

83. Starke, E. A., Sanders, T. H., and Palmer, I. G., *Journal of Metals*, 33(8), 24–30, 1981.

84. Sanders, T. H. and Balmuth, E. S., *Metal Progress*, 121, 32–35, 1978.

85. Sanders, T. H. and Starke, E. A., Eds., *Aluminum–Lithium Alloys II, Proceedings of the Second International Conference on Aluminum–Lithium Alloys*, The Metallurgical Society of AIME, Warrendale, PA, 1984.

86. James, R. S., *Metals Handbook,* Vol. 2, 10th ed., ASM International, Metals Park, Ohio, 2000.

87. Grimes, R., Cornish, A. J., Miller, W. S., and Reynolds, M. A., *Metals Materials*, 4(7), 357, 1988.

88. Sanders, T. S. and Starke, E. A., Eds., *Aluminum–Lithium Alloys V, Proceedings of the Fifth International Conference on Aluminum–Lithium Alloys*, Materials and Component Engineering Publications, Edgbaston, United Kingdom, 1989.

89. Quist, W. E. and Narayanan, G. H., In *Aluminum Alloys: Contemporary Research and Applications*, Vasudevan, A. K. and Doherty, R. D. Eds., Eds. Treatise on Materials Science and Technology, Vol. 31, Academic Press, New York, USA, pp. 219–254, 1990.

90. Peel, C. J., *Materials Science and Technology*, 4, 1169–1179, 1987.

91. Dorward, R. C. and Tritchett, T. R., *Materials Design*, 9(2), 63, 1988.

92. Fielding, P. S. and Wolf, G. J., *Advanced Materials and Processes*, 150(4), 21–24, 1996.

93. Frazier, W. E., Lee, E. W., Donnellen, M. E., and Thomoson, J. J., *Journal of Metals*, 22–25, May, 1989.

94. Lee, E. W. and Waldman, J., *Aluminum–Lithium Symposium*, Los Angeles, CA, 1988.

95. Venkateshwara Rao, K. T. and Ritchie, R. O., *Materials Science and Engineering*, 100, 23–33, 1988.

96. Venkateshwara Rao, K. T. and Ritchie, R. O., *Materials Science and Technology*, 5, 896–905, 1989.

97. Venkateshwara Rao, K. T. and Ritchie, R. O., *Scripta Metallurgica*, 23, 1129–1135, 1989.

98. Venkateshwara Rao, K. T. and Ritchie, R. O., *Acta Metallurgica Materialia*, 38, 23–99, 1990.

99. Venkateshwara Rao, K. T. and Ritchie, R. O., *Metallurgical Transactions*, 22A, 191–200, 1991.

100. Venkateshwara Rao, K. T., McNulty, J. C., and Ritchie, R. O., *Metallurgical Transactions*, 22A, 191–202, 1991.

101. Lee, E. W. and Frazier, W. E., The effect of stretch on the microstructure and mechanical properties of 2090 aluminum-lithium alloy, *Scriuotas Metallurgucvsa*, 22, 53–57, 1988.

102. Pickens, J. R., Heubaum, F. H., Langan, T. J., and Kramer, L. S., Al-4.5-6.3 Cu-1.31 Li-0.4 Ag-0.4 Mg-0.14 Zr Alloy WELDALITE 049, In *Aluminum Lithium Alloys, Proceedings of the Fifth International Conference on Aluminum–Lithium Alloys,* Williamsburg, Virginia, Sanders, T. H. and Starke, E. A., Eds., MCEP, Birmingham, United Kingdom, pp. 1397–1414, 1989.

103. Starke, E. A. and Bhat, B. N., In *Aluminum–Lithium Alloys for Aerospace Applications Workshop: NASA Conference Publication 3287*, Bhat, B. N., Bales, T. T., and Vesely, E. J., Eds., Marshal Space Flight Center, Alabama, USA, pp. 3–5, 1994.

104. Lee, E. W. and Frazier, W. E., In *Aluminum–Lithium Alloys V, Proceedings of the Fifth International Conference on Aluminum–Lithium Alloys*, Sanders, T. H. and Starke, E. A., Eds., Williamsburg, PA., USA, pp. 100–110, 1989.
105. Donnellan, M. E. and Frazier, W. E., In *Aluminum–Lithium Alloys V, Proceedings of the Fifth International Conference on Aluminum–Lithium Alloys*, Sanders, T. H. and Starke, E. A., Eds., Williamsburg, VA, 1989.
106. Bucci, R. J. and Mueller, L. N., "ARALL Laminate Performance Characteristics," presented at the ARALL Laminates Technical Conference, Champion, PA, October, 1987.
107. Bentely, R. M. and Van Gent, P. H., "Formability in Bending of ARALL Laminates," special conference on ARALL Laminates, The Netherlands, Delft, October, 1988.
108. Fraterman, N. J., West, J., and cavan der Schjee, P. A., The Application of ARALL in the Wingbox of a Fokker 50, special conference on ARALL Laminates, The Netherlands, October, 1988.

10 Metal Matrix Composites: Types, Reinforcement, Processing, Properties, and Applications

T. S. Srivatsan
Department of Mechanical Engineering, University of Akron

John Lewandowski
Department of Materials Science and Engineering, Case Western University

CONTENTS

10.1   Introduction 277
10.2   Types of Metal Matrix Composites 278
      10.2.1   Fiber (Whisker)-Reinforced 279
      10.2.2   Particulate-Reinforced 279
      10.2.3   Dispersion Strengthened 279
10.3   Nature of the Reinforcement 280
      10.3.1   Whisker Reinforcement 281
      10.3.2   Particulate Reinforcement 282
      10.3.3   Particulate Versus Whisker 283
      10.3.4   Fiber Reinforcements 284
10.4   Matrix Material 285
10.5   Salient Characteristics of Metal Matrix Composites 286
10.6   Primary Processing of Discontinuously Reinforced
      Metal-Matrix Composites 289
      10.6.1   Primary Processing: Solid Phase Processes 290
            10.6.1.1   Powder Metallurgy 290
      10.6.2   The Technique of Spray Processing 294
            10.6.2.1   Spray Atomization and Deposition Processing 295
            10.6.2.2   Spray Atomization and Co-Deposition 296
            10.6.2.3   Spray Deposition Processing of Premixed MMCs 298
                  10.6.2.3.1   Method 1 298
                  10.6.2.3.2   Method 2 299
      10.6.3   Low Pressure Plasma Deposition 300
      10.6.4   Modified Gas Welding Technique 300
      10.6.5   The High Velocity Oxy-Fuel Spraying 301
10.7   Secondary Processing 302
      10.7.1   Forging of Discontinuously Reinforced Composites 302
      10.7.2   Hydrostatic Extrusion of Composite Materials 305
10.8   Properties of Discontinuously Reinforced Metal Matrix Composites 307

    10.8.1   Physical Properties 307
             10.8.1.1   Density, Heat Capacity, and Conductivity 307
    10.8.2   Mechanical Properties 309
             10.8.2.1   Modulus of Elasticity 309
             10.8.2.2   Strength 310
             10.8.2.3   Ductility 312
             10.8.2.4   Fracture Toughness 318
                        10.8.2.4.1   Effects of Changes in Notch Root
                                     Radium on Toughness 320
                        10.8.2.4.2   Effects of Changes in Volume Fraction
                                     on Toughness 320
                        10.8.2.4.3   Effects of Matrix Microstructure and
                                     Heat Treatment 323
                        10.8.2.4.4   Effects of Changes in Reinforcement Size 325
                        10.8.2.4.5   Effects of Changes in Reinforcement
                                     Clustering 325
                        10.8.2.4.6   Mixed Mode Toughness Studies 327
                        10.8.2.4.7   Models of Fracture Toughness in
                                     Particle Hardened Materials 328
                        10.8.2.4.8   Toughened MMCs 329
             10.8.2.5   Fatigue Behavior 332
                        10.8.2.5.1   Stress Amplitude and Strain Amplitude Controlled
                                     Fatigue Behavior 332
                        10.8.2.5.2   Fatigue Crack Growth Behavior 338
10.9   Applications 340
    10.9.1   Composites for Space Applications 341
    10.9.2   Composites for Automotive Applications 342
    10.9.3   Electronic Packaging 345
    10.9.4   Commercial Products 346
Acknowledgments 347
References 347

Abstract

There exists considerable scientific and technological interest in the development and use of metal matrix composites for a variety of performance critical and even nonperformance critical applications. In this paper, we discuss the techniques related to the processing of discontinuously reinforced metal matrix composites. The attractive combinations of microstructure and properties achievable through rapid solidification processing have prompted use of this technique as a means to synthesize discontinuously reinforced composites based on metal matrices. These hybrid materials have combined the standard alloys of metals with a wide variety of discontinuous reinforcements such as particulates, whiskers, and short fibers of ceramic and metallic materials. The spray processing techniques have engineered considerable technological interest for synthesizing discontinuously reinforced metal matrices and few of the spray base methods are examined. These include spray atomization and deposition processing, low-pressure plasma deposition, modified gas welding and high velocity oxyfuel spraying. The salient aspects pertaining to the secondary processing techniques of forging and extrusion are presented. An overview of the physical and mechanical properties of discontinuously reinforced metal matrix composites is presented and discussed with appropriate examples. A few practical applications that have reached commercial success are highlighted.

10.1 INTRODUCTION

The established need for lightweight and high-performance materials to satisfy the stringent demands of the burgeoning industries of air and ground transportation and several commercial products has provided the much-needed impetus for a plethora of research studies. This has culminated in the development and emergence of new and improved materials resulting from novel and innovative combinations of conventional materials. During the time period spanning the early 1970s to the present, the preponderance of research efforts have essentially focused on the areas of alloy design and use of novel processing techniques to synthesize high-performance, hybrid materials, termed as composites, as economically attractive and potentially viable alternatives to the traditional and newer generation of unreinforced metallic alloys [1–5].

A large majority of these composite materials are metallic matrices reinforced with high strength, high modulus and often-brittle second-phases, which can be either continuous in the form of fibers or discontinuous in the form of whisker, short fiber, platelet and particulate reinforcements embedded in a ductile metal matrix. Incorporation of hard, brittle, and essentially elastically deforming ceramic reinforcements in a soft, ductile and plastically deforming metal matrix has been shown, in a number of cases, to offer improvements in elastic modulus, wear resistance, strength, structural efficiency, reliability, and the control of physical properties such as density and coefficient of thermal expansion. A combination of these properties has enabled improved mechanical performance in direct comparison with the unreinforced metal matrix [6–14].

A composite, sometimes referred to as a hybrid material by the scientific and technical communities, can be defined as a heterogeneous structural material consisting of two or more distinct conventional materials, which are either mechanically or metallurgically bonded together to achieve a definite goal for a specific purpose [13,14]. The purpose behind composite material development and commercialization is to

1. Synergize the desirable properties of its constituents, that is, the (a) reinforcement or second phase, (b) bulk or parent material called the matrix (primary phase), and
2. Suppress the shortcomings of each of the constituents.

This combination can result in a newly synthesized material containing unique properties for a spectrum of structural and even nonstructural applications [14,15]. Interest in the use of reinforcements in a continuous matrix phase, dating back to the early 1960s, was due to the limitation associated with conventional ingot metallurgy (I/M) processing techniques. At the same time, there was a growing need for newer generation materials to meet the ever-increasing demands of industries such as the space, ground transportation, and commercial products industries. Thus, a radical new approach to material synthesis and preparation was required to meet the demand for newer generation materials [15–17].

Reinforcement can be either a continuous phase or a discontinuous phase. Continuous reinforcements are typically fibers and occasionally laminates. The discontinuous reinforcements include particulates, platelets, chopped fibers, and whiskers. Typical reinforcement geometries are exemplified in Figure 10.1 [12,13,15]. Discontinuous reinforcements are advantageous in that the composites are relatively easy to synthesize by the techniques of powder metallurgy (P/M), ingot metallurgy (IM), and mechanical alloying (MA).

In the ingot metallurgy (I/M) processing technique, the ceramic reinforcement is mixed with molten metal and the resulting mixture is subsequently cast and then subjected to mechanical deformation or working. In the powder metallurgy (P/M) processing technique, one or more reinforcements are thoroughly blended with pre-alloyed powder. The resulting mixture is canned, vacuum degassed, and consolidated into a billet using the technique of hot pressing or hot isostatic pressing. The consolidated composite specimen is subsequently forged

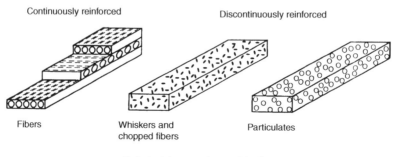

FIGURE 10.1 Schematic showing typical reinforcement geometries for metal matrix composites.

or extruded into the desired shape [12]. In the technique of mechanical alloying, the pure metal powder and the alloying ingredients are mechanically alloyed using high-energy ball mills while concurrently incorporating submicron oxide or carbide filler to form the discontinuously reinforced metal matrix composite (MMC). During this process, heavy working of the powder particles results in alloying by the process of repeated welding, fracturing, and re-welding. The mechanically alloyed product is subsequently consolidated into a suitable shape.

In this paper, we present and discuss the types of metal matrix composites, the most popular techniques used for the synthesis of discontinuously reinforced metal matrix composites, secondary processing of the composite using the techniques of forging and extrusion, an overview of the salient properties of composite material, and briefly highlight the potential applications. In the first part of the paper, we provide an overview of the types of metal matrix composites and the specific roles of the metal matrix and the reinforcement. The important and most widely used techniques for the synthesis or processing of discontinuously reinforced composites are discussed in the second part of the paper. In the third part, we provide an overview of the key properties of discontinuously reinforced metal matrix composites (DRMMCs), concluding with an overview of potential applications of this material in existing technology.

10.2 TYPES OF METAL MATRIX COMPOSITES

Metal matrix composites (denoted henceforth through the text of this chapter as MMCs) are materials consisting of metal alloys reinforced with fibers, whiskers, particulates, platelets, wires, or filaments [12,15]. Because of their superior mechanical properties and unique physical characteristics, such as low coefficient of thermal expansion, they have become attractive for use in a wide range of structural and nonstructural applications. A few notable production applications include (a) space shuttle struts, (b) NASA space telescope boom/waveguides, (c) Toyota and Mitsubishi diesel engine pistons, and (d) Pratt & Whitney exit fan guide vanes on Boeing 777 commercial aircraft. The space shuttle utilizes about 240 struts made from aluminum reinforced with boron fibers. The NASA space telescope has two rectangular antenna support booms that also serve as waveguides. The antenna support boom is fabricated from aluminum reinforced with graphite fibers. Toyota, the automobile manufacturer, has announced that it is currently mass-producing aluminum diesel engine pistons, which are locally reinforced with ceramic fibers, eliminating the need for nickel and cast-iron inserts. Currently, the annual production rate is estimated to be around 500,000. Mitsubishi is also using aluminum diesel engine pistons reinforced with silicon carbide whiskers.

The unique physical characteristics of MMCs have been exploited to a certain extent. The most noteworthy example is Hitachi, which has developed an experimental low thermal expansion heat sink of copper reinforced with graphite fibers [19]. Although many different MMCs comprised of widely different properties exist, these materials have proven to provide some general advantages over the (i) monolithic (unreinforced) counterpart and (ii) polymer matrix composites (PMCs). In direct comparison with the monolithic metals, reinforced metal matrices offer a combination of (a) high strength-to-density ratio (σ/ρ), (b) high stiffness-to-density (E/ρ) ratio, (c) better fatigue resistance, (d) better elevated temperature properties to include higher strength and lower creep rate, (e) a lower coefficient of thermal expansion, and (f) improved wear resistance [18].

The specific advantages of reinforced metal matrices over polymer matrix composites (PMCs) are (i) high temperature capability, (ii) fire resistance, (iii) high transverse stiffness and strength, (iv) no moisture absorption, (v) high electrical and thermal conductivities, (vi) better radiation resistance, (vii) no out-gassing, and (viii) the capability to fabricate the whisker-reinforced and particulate reinforced metal matrices using conventional metal working equipment. Some of the distinctly glaring disadvantages of the reinforced metal matrices when compared to the monolithic metals and polymer matrix composites are the (1) higher cost of some material systems, (2) relatively new/immature, growing and evolving technology, (3) complex fabrication methods for the fiber reinforced systems (except for casting), and (4) limited service experience.

Numerous combinations of matrices and reinforcements have been tried since work on metal matrix composites began in the late 1950s. The distinction of metal matrix composites, i.e., MMCs, from the other two or more phase alloys comes about from processing of the composite. In the production of a metal matrix composite, the matrix and the reinforcing phases are mixed together. This is to help distinguish a composite material from a two or more phase alloy, where the second-phase forms as a particular eutectic or eutectoid reaction [19]. In all cases, the matrix is essentially a metal. However, a pure metal is seldom used as the matrix. The matrix is generally an alloy. Subsequently, and until now, the matrices chosen have been alloys of aluminum, magnesium, copper, titanium, titanium-aluminides, nickel, nickel-aluminides, nickel-base superalloys, and various types of ferrous-based (i.e., iron) systems. Of those developed and effectively utilized, the aluminum alloy-based matrix composites are those that have been the subject of numerous development studies with the purpose of promoting their use in commercial products. Depending on the nature of the reinforcing phase, the metal matrix composite is categorized as follows:

10.2.1 Fiber (Whisker)-Reinforced

The reinforcing phase in a fiber composite material (i) spans the entire size range, from a fraction of a micron to several units in diameter, (ii) spans the length from mils to continuous fibers, and (iii) spans an entire range of volume concentration, from a few percent to greater than 80%. The distinguishing microstructural feature of fiber-reinforced materials is that the reinforcing fiber has one long dimension whereas the reinforcing phase of the other two types does not.

10.2.2 Particulate-Reinforced

The composite is characterized by dispersed particles of greater than 1.0 micrometer (i.e., μm) in diameter with a dispersed concentration in excess of 25%.

10.2.3 Dispersion Strengthened

The composite is characterized by a microstructure consisting of an elemental matrix within which fine particles are uniformly dispersed. The particle diameter ranges from about 0.01 to 0.1 micrometers (μm) and the volume concentration of particles ranges from 0 to 15%.

FIGURE 10.2 Tennis racket frame and golf club made from aluminum-SiC composites. (Courtesy of Alcan International, Taya, M. and Arsenault, R. J., Ed., *Metal Matrix Composites: Thermomechanical Behavior,* Pergamon Press, New York, 1989.)

Metal matrix composite (MMC) materials have been in development since the early 1970s. Initially, the emphasis was on continuous fiber-reinforced metal matrices, first developed for applications in the aerospace industry, followed by their application and use in other industries. Gradual expansion and consolidation into non-aerospace and nonmilitary applications came about slowly as the price of MMC gradually lowered, primarily as the result of the development and emergence of low cost fibers [20]. In the period following the 1990s, the preponderance of research activity has focused on the synthesis, characterization, and quantification of discontinuously reinforced metal matrices.

Discontinuously reinforced metal matrices are a class of materials that exhibit a blending of properties of the reinforcing phase and metal matrix. A resurgence of interest in discontinuously reinforced metal matrices has arisen due to advancements in the economical production of silicon carbide fibers and silicon carbide whiskers [SiC_w], which has concurrently led to the economic development and use of platelets and particulates of silicon carbide (SiC_p) in a metal matrix [21]. Each of the reinforcements, be it whisker, short fiber, chopped fiber, platelets or particulates, has property and cost attributes, which dictate their selection and use in a given situation. The distinct advantage of the discontinuously reinforced metal matrix composites (DRMMCs) is that they can be shaped by standard metallurgical processes, such as (a) forging, (b) rolling, and (c) extrusion. Relative ease of formability, existing metal forming infrastructure, and relatively low material cost were largely responsible for the discontinuously reinforced metal matrices being considered for a spectrum of applications. Examples of such applications are tennis rackets and heads of golf clubs (Figure 10.2), and automobile engine components, piston rod and connecting rod, which all have been made from Al/SiC_p composite (Figure 10.3) [22]. The many positive qualities of these low-cost components coupled with the use of existing metal working equipment have contributed to the resurgence of interest and growth in the use of discontinuously reinforced metal matrix composites.

10.3   NATURE OF THE REINFORCEMENT

Composite material owes its unique balance of properties to the appropriate combination of metal matrix and the reinforcing phase or phases, if more than one is present. However, it is the specific attribute of the reinforcing phase that is responsible for structural properties such as strength and stiffness. Whether particulate, filler, microsphere, whisker, or long fiber, the reinforcement is the key to optimizing cost-performance for a given application. Minerals and conductive fibers,

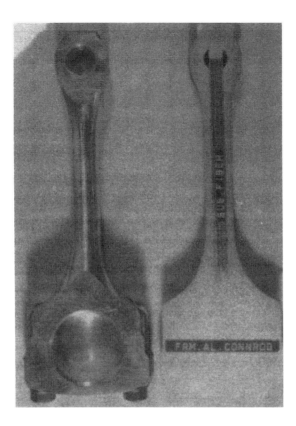

FIGURE 10.3 Connecting rod made of aluminum reinforced with continuous stainless steel fibers. (From Taya, M. and Arsenault, R. J., Ed., *Metal Matrix Composites: Thermomechanical Behavior,* Pergamon Press, New York, 1989.)

whiskers, and microspheres have become important constituents in many composite systems. However, during the early stages of material development, fiber reinforcement dominated the field in terms of volume, properties, and design versatility. It is the glass, carbon/graphite, ceramic, aramid, and a wide range of other organic fibers that have become established depending on specific performance requirements.

10.3.1 WHISKER REINFORCEMENT

Early work in this area was done by Brenner [23,24] and Sutton [25,26] using alpha alumina (Al_2O_3) whiskers. The cost of whiskers was relatively high and the resultant strength of the metal matrix was lower than expected. This was ascribed to the bonding difficulties of the metal matrix with the reinforcing alumina whiskers. Despite research efforts, these difficulties were never overcome and the composite system never matured. An independent research study by Divecha and co-workers [27] using beta SiC whiskers in aluminum demonstrated very good strength, modulus, fatigue resistance, and elevated temperature properties. However, the high cost of these whiskers was an impediment to continued development of this system.

In the late 1970s, researchers at the University of Utah developed a process for making SiC whiskers using rice hulls as the starting ingredient. This made the SiC whiskers potentially low cost and prompted a resurgence of research into this form of composite [4,28–31]. The SiC_w was approximately 0.1 micrometer (μm) in diameter and with a length (l)-to-diameter (d) ratio of up to 100 to 1. A high percentage of the original whisker product in 1980 was fine powder rather than whisker. During subsequent years, considerable improvement occurred in the processing of reinforcing whiskers. This resulted in the present product being a high aspect ratio (length/diameter) whisker. The most common techniques used to produce the whisker-reinforced aluminum alloy-based composite in the United States is powder metallurgy (PM) processing. The blending of high aspect ratio whiskers with metal powders is limited to a volume loading of approximately 25% due to the geometry of the whiskers. Whisker breakage during consolidation and secondary processing is also a potential problem. Health hazards with whiskers have also been cautioned.

10.3.2 PARTICULATE REINFORCEMENT

Particulates of silicon carbide (SiC_p) emerged in the late 1970s as a potentially attractive reinforcement for ductile metal matrices. Examples of silicon carbide particulates are shown in Figure 10.4. These particulates are commercially available in sizes from approximately 0.5 micrometers (μm) to greater than 100 μm. They are also available in narrow size ranges, as shown in the AISI size distribution specification in Figure 10.5. The particulates more closely match both the particle size and size distribution for commercially available aluminum powders. The particulates can be blended more efficiently at the higher volume percent than can the whisker products. Less breakage of the reinforcement is obtained and health risks of micrometer particulate have not been documented. Composites resulting from the addition of 40 vol.% particulates are common, and composites containing in excess of 55 vol.% particulates in a metal matrix are in the early stages of development. The particulates are being produced in large quantities for the abrasive industry and are relatively low in cost. The SiC particulates (SiC_p) have a high modulus ($E = 380$ GPa) and a low density ($\rho = 3.21$ g/cc).

FIGURE 10.4 Scanning electron micrograph showing the nature and morphology of the silicon carbide particulates. (From Harrigan, W. C. Jr., Metal Matrix Composites, In *Metal Matrix Composites, Mechanisms and Properties*, Everett, R. K. and Arsenault, R. J., Eds., Academic Press, London, pp. 1–15, 1991.)

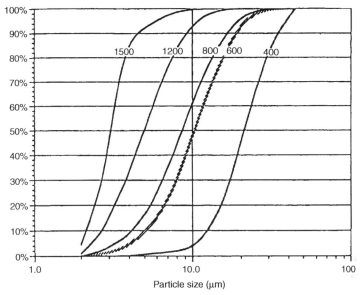

Particle size distribution for silicon carbide.

FIGURE 10.5 Size distribution for the silicon carbide particles. (From Harrigan, W. C. Jr., Metal Matrix Composites, In *Metal Matrix Composites, Mechanisms and Properties*, Everett, R. K. and Arsenault, R. J., Eds., Academic Press, London, pp. 1–15, 1991.)

10.3.3 PARTICULATE VERSUS WHISKER

A large concentration of research work during the late 1970s and early 1980s involved the testing of metal matrices that contained either particulate or whisker reinforcement. These composites were made by blending atomized powders with the SiC reinforcing phase. The studies attempted to demonstrate that whisker or particulate reinforcements produced a superior product. However, several of these studies were flawed because the composites that were compared had reinforcement levels that were different or the amount and type of secondary processing was different for each system. When using particulate reinforcements, extrusion of the composite material through a conical or a shear face die was found to be acceptable. In the case of whisker-reinforced metal matrices, attempts to preserve the whiskers resulted in extrusion of the reinforced metal or composite material to be restricted to conical or streamline dies [32]. Even with adequate precautions taken during metalworking, substantial breakage of the reinforcing whiskers occurred. Extrusion of the metal matrix composites produced alignment of the whiskers and resultant anisotropy in mechanical properties.

For a given volume fraction of the reinforcement, an appropriate matrix alloy, and comparable metal working and heat treatment prior to testing, the whisker composites tend to have higher elastic modulus in the extrusion (alignment) direction and lower values than the particulate reinforced composites in the perpendicular, i.e., transverse direction. Whisker composites have approximately the same yield strength (σ_{YS}), a higher ultimate tensile strength (σ_{UTS}), and a significantly lower strain to failure (ε_f) than the particulate reinforced composites in the extrusion direction. In the direction perpendicular, i.e., transverse direction, to the extrusion direction, whisker composites have significantly lower properties. Further, based on several studies, the key finding is that they are limited to a maximum of 25 vol.% due to blending-related difficulties. In addition, their cost is substantially higher than the particulates used for reinforcing aluminum alloy metal matrices, and is a major constraint to the development, commercialization, and use of whisker-reinforced metal matrix composites [28].

10.3.4 FIBER REINFORCEMENTS

The most widely used reinforcement is fiberglass, which accounts for some 90% of the reinforced metal market. First commercialized in 1939 by Owens Corning, glass fiber for composite material use gained acceptance during World War II, because of its lightweight, high strength, and nonmetallic characteristics. Since that time, both its volume and sophistication have continued to grow because of the versatility of glass fiber in its many forms coupled with a high level of performance at a relatively low cost [32]. The two most common reinforcement grades of glass fiber are "E" (for electrical) and "S" (for high strength) grades. The E-glass provides a combination of (a) high strength-to-weight (σ/ρ) ratio (roughly twice that of patented and drawn steel wire), (b) good fatigue resistance, (c) outstanding dielectric properties, (d) retention of strength up to 50% at temperatures up to 650°F, and (e) excellent resistance to chemical, corrosion and environmental attack. E-glass is available as continuous filament, chopped fiber, random fiber mats, and many other forms suitable for composite material fabrication [32]. The glass fibers retain much of their strength and stiffness at high temperatures. However, considering that the matrix is a metal and the temperatures associated with component fabrication and eventual use can be extremely high, the use of ceramic fibers is often more appropriate. The alumina-silica fibers, marketed under the trade name of *Fiberfrax*, offer an attractive balance of properties at about 20% below the price of chopped fiberglass [32]. Discontinuous Al_2O_3–SiO_2 fibers have found use for their application in (i) friction products (example: brake pads and clutch facings), (ii) metal matrix composites, and (iii) high-temperature thermoset composites. Because of its relatively low cost, the alumina-silica (Al_2O_3–SiO_2) fiber is a viable replacement for the more expensive reinforcement fibers in components requiring thermal stability, resistance to oxidation/reduction, and an inertness to moisture [32]. Silicon carbide (SiC) and silicon nitride (Si_3N_4), high purity alumina (Al_2O_3), metal oxides, and zirconia are other ceramic fibers commercially available in one form or the other. Silicon carbide (SiC) and alumina (Al_2O_3) fibers have exceptional compressive strength, as do the metal oxides. Whisker forms of these ceramics are used in extremely demanding applications because they offer the highest specific structural properties available from any existing material. The physical properties of selected inorganic reinforcements are summarized in Table 10.1 [32].

Other continuous fibers chosen and used for metal matrices are boron, graphite (carbon), alumina, and silicon carbide. The representative fiber properties of these are summarized in Table 10.2. Boron fibers are made by chemical vapor deposition (CVD) of the material on a

TABLE 10.1
Physical Properties of Inorganic Reinforcements

	E-Glass	S-Glass	Fused Silica	Alumina	Boron	Silicon Carbide	Alumina Silica
Specific gravity	2.54	2.49	2.2	3.9	2.58	—	2.73
Tensile strength (10^3 psi)	500	665	500	200	510	470	250
Tensile modulus (10^6 Psi)	10.5	12.5	10	35	58	29	15
Elongation (%)	48	54	—	—	0.9	1.5	—
Coef.of Therm.Exp. (10^{-6}/°F)	2.8	1.6	0.3	—	205	1.7	—

Source: Kreider, K. G. *Composite Materials*, Vol. 4, Academy Press, New York, p, 1, 1974.

TABLE 10.2
Representative Fiber Properties

Fiber	Density (lb/in.^{3})	Modulus (10^6 psi)	Tensile Strength (10^3 psi)
Alumina	0.14	50–55	200
Boron	0.09	58	400
Graphite-PAN high strength	0.065	34	360
Graphite-pitch, ultra high modulus	0.077	100	250
Silicon carbide-monofilament	0.11	62	400
Silicon carbide-multifilament	0.093	26–29	350

Source: Piggot, M. R. *Load Bearing Fiber Composites*, Pergamon Press, New York, 1980.

tough tungsten core. Carbon cores have also been used. To retard the chemical reactions that can occur between boron and the metals at elevated temperatures, fiber coatings of materials such as silicon carbide or boron carbide are sometimes used. The CVD process, using a tungsten or carbon core is also used to make the monofilaments of silicon carbide. Continuous alumina fibers are available from several suppliers. However, the chemical composition and properties of the various fibers are significantly different. Graphite fibers as a continuous reinforcement for metal matrices are made from two precursor materials: polyacrilonitrile (PAN) and petroleum pitch [18]. Graphite fibers have a wide range of strengths and moduli. Alumina and alumina-silica are the leading discontinuous fiber reinforcements for metal matrices and initially developed as insulating materials [18]. A number of metal wires including tungsten, beryllium, titanium, and molybdenum have been used to reinforce metal matrices. Currently, the most important wire reinforcements chosen and used are tungsten wire for the superalloys and superconducting materials incorporating niobium-titanium or niobium-tin in a copper matrix.

10.4 MATRIX MATERIAL

The matrix is responsible for overall integrity of the composite component. It binds the reinforcements together to (a) allow for effective distribution of the load and (b) protect the flaw sensitive reinforcement from self-abrasion and externally induced scratches and defects. The matrix also protects the reinforcement from moisture in the environment and chemical corrosion or oxidation, which is susceptible to embrittlement and premature failure [33]. Although the reinforcing phase provides much of the tensile strength and stiffness, the shear strength, compression strength, and transverse tensile strength of the composite material are usually dominated by the matrix [33]. Further, since the matrix aids in holding the composite component together, heat resistance of the metal matrix dictates the overall thermomechanical behavior of the composite material.

Numerous metals have been used as matrices and are summarized in Table 10.3. The most widely prepared and studied are the nonferrous metals, which include aluminum, titanium, magnesium, copper, and the superalloys. The resultant MMC systems have industrial applications ranging from performance-critical to commercial products. The important MMCs developed are listed as follows:

a. Aluminum matrix
  • Continuous fiber: boron, silicon carbide, alumina and graphite
  • Discontinuous fibers: alumina, alumina–silica

TABLE 10.3
Typical Matrix Alloys

- Aluminum
- Titanium
- Magnesium
- Copper
- Bronze
- Nickel
- Lead
- Silver
- Super alloys (Nickel and Iron based)
- Niobium (Columbium)
- Intermetallics

Source: Srivatsan, T. S., Sudarshan, T. S., In *Rapid Solidification Technology: An Engineering Guide*, Technomic Publishing Inc., PA, pp. 603–700, 1993.

           • Whiskers: silicon carbide
           • Particulates: silicon carbide and boron carbide
   b. Magnesium matrix
           • Continuous fibers: graphite and alumina
           • Whiskers: silicon carbide
           • Particulates: silicon carbide and boron carbide
   c. Titanium matrix
           • Continuous fibers: silicon carbide, coated boron
           • Particulates: titanium carbide
   d. Copper matrix
           • Continuous fibers: graphite and silicon carbide
           • Wires: niobium–titanium, niobium–tin
           • Particulates: silicon carbide, boron carbide and titanium carbide
   e. Superalloy matrices
           • Wires: tungsten

Representative properties of some of these metal matrix composites (MMCs) are summarized in Table 10.4 [18].

10.5 SALIENT CHARACTERISTICS OF METAL MATRIX COMPOSITES

The superior mechanical properties of MMCs are the major driving force behind their selection and use in both existing applications and a wide variety of emerging engineering applications. An important characteristic MMCs share with other composites is that by the appropriate selection of matrix material, reinforcement(s), and orientation of the reinforcement with respect to the metal matrix layered or dispersed, it is possible to tailor the properties of a component to meet the needs of a specific design. For example, within broad limits, it is possible to specify strength and stiffness in one direction, coefficient of expansion in another, and so forth [18]. This is not possible with monolithic metals. As a point of departure for consideration of the characteristics of MMC, a cursory overview of those of monolithic metals and polymer matrix composites (PMCs) is essential.

The monolithic metals tend to be isotropic, that is, to have the same properties in all directions. Some mechanical deformation processes such as rolling, tend to impart anisotropy, such that

TABLE 10.4
Representative Properties of Metal Matrix Composites

Matrix	Reinforcement	Reinforcement (%vol)	Modulus (10^6 psi)		Tensile Strength (10^3 psi)	
			Longitudinal	Transverse	Longitudinal	Transverse
Aluminum	None	0	10	10	40–70	40–70
Epoxy	High-strength graphite fibers	60	21	1.5	180	6
Aluminum	Alumina fibers	50	29	22	150	25
Aluminum	Boron fibers	50	29	18	190	15
Aluminum	Ultrahigh modulus graphite fibers	45	50	5	90	5
Aluminum	Silicon carbide particles	40	21	21	80	80
Titanium	Silicon carbide monofilament fibers	35	31	24	250	60

Source: Metal Matrix Composites Overview: Paper Number MMCIAC No. 253, Santa Barbara, California, U.S.A.

properties of the rolled-end product vary with direction. The stress versus strain behavior of monolithic metals is typically elastic–plastic. Most commercially used structural metals have adequate ductility and acceptable fracture toughness [18].

The most important structural polymer matrix composites (PMCs) are reinforced with straight, parallel, continuous fibers or with fabrics [18]. The key PMC reinforcing fibers are E-glass, S-glass, graphite (carbon), and aramid (Kevlar 49). Polymer matrix composites are highly anisotropic. They are strong and stiff parallel to the fiber direction, but their stiffness and strength are low perpendicular to it. The PMCs do not yield elastically, and their stress versus strain curves are generally linear to failure [14,16,18]. An exception is when transverse cracking in some plies causes a "knee" in the curve with a concurrent reduction in modulus [18].

The wide varieties of MMCs that have been engineered and put forth for commercial use have properties that differ dramatically. Key factors influencing their characteristics include the following [20,22,25]:

1. Reinforcement properties, form, and geometric shape
2. Reinforcement volume fraction
3. Matrix properties to include the intrinsic influence of porosity
4. Properties of the reinforcement-matrix interface
5. Residual stresses arising from the thermal and mechanical history of the composite
6. Possible degradation of the reinforcement resulting from chemical reactions at high temperatures and mechanical damage from processing and impact

Variation of specific moduli with specific strengths of aluminum and magnesium matrices reinforced with various fibers and particulate ceramics are shown in Figure 10.6. These values are compared to monolithic metallic materials such as aluminum, magnesium, steel, and titanium.

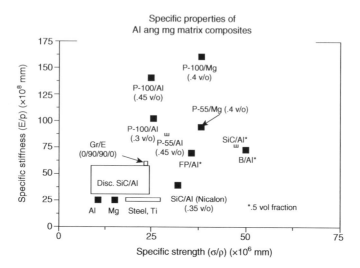

FIGURE 10.6 Specific properties of aluminum matrix composites and magnesium matrix composites compared with the unreinforced alloys. Properties of the continuous fiber reinforced materials are calculated parallel to the fibers. (From Mortensen, A., Cornie, J. A. and Flemings, M. C., *Journal of Metals,* 12, 1988.)

They are based on rule-of-mixtures (ROM) calculations, for the continuously reinforced composites, which are representative of the actual measured properties of the fiber [34,35]. In applications where such extreme properties are not a requirement, the family of discontinuously reinforced metal matrix composites based on particulates, whiskers, nodules, platelets, and short fiber reinforcements have emerged as attractive and economically viable commercial materials for the automotive, aerospace, and several other high performance markets [36–42]. These composites are preferred primarily because they offer a number of advantages, such as a (a) 15–40% increase in strength and (b) 30–100% increase in stiffness compared with unreinforced alloy [43–45], while generally maintaining receptiveness to processing and characterization techniques used for the conventional unreinforced counterpart.

Particulate reinforced metal matrix composites, like monolithic metals, tend to be isotropic. The presence of brittle reinforcements along with traces of metal oxides tends to impair their ductility and fracture toughness. Ongoing research and development efforts have attempted to reduce some of these deficiencies. The properties of metal matrices reinforced with whiskers tend to depend strongly on the orientation of the whisker reinforcement. Randomly oriented whiskers produce a material that is essentially isotropic in behavior. Processes such as extrusion can orient the whiskers resulting in anisotropic properties. Whiskers are also detrimental to ductility and fracture toughness [18].

Metal matrices reinforced with aligned fibers have anisotropic properties. They are stronger and stiffer in the direction of the fibers than perpendicular to them. The transverse strength and stiffness of unidirectional MMCs (materials having the fibers oriented parallel to one axis) frequently suffice for use in components such as stiffeners and struts [12,13,15,18]. This is one of the major advantages of metal matrix composites (MMCs) over polymer matrix composites (PMCs), which can rarely be used without transverse reinforcement. Further, since the modulus and strength of metal matrices are significantly higher with respect to those of the reinforcing fibers, their contribution to composite behavior is important. Stress versus strain curves of the MMC often exhibit significant nonlinearity resulting from the yielding of the metal matrix [12,18].

Another factor that exerts a significant influence on the behavior of fiber-reinforced metals is the large difference in coefficient of expansion between the two constituents [8,9,15]. This tends to cause large residual stresses in the composite microstructure when they are subjected to significant changes in temperature. In some cases, during the cool down from the processing temperature, the thermal stresses induced in the metal matrix are severe enough to cause yielding at the local (microscopic) level. The local stresses are often exacerbated by the large residual stresses induced by mechanical loading.

10.6 PRIMARY PROCESSING OF DISCONTINUOUSLY REINFORCED METAL-MATRIX COMPOSITES

A preponderance of well-focused research efforts undertaken during the last two decades have led to the evolution of two major methods for the preparation of particulate reinforced metal matrix composites [46,47]. In the first method, the reinforcing particles are added to the melt while it is held in the crucible, where they are distributed either by the stirring action of the induction heater or by a stirring apparatus, prior to final ejection onto the chill roller. In the alternative method, the reinforcing particles, entrained in an inert carrier gas, are injected onto the melt puddle while the metal is still liquid on the chill roller. In this case, the particles are distributed in the melt by the concurrent and competing influences of (i) the flow of liquid matrix material in the puddle and (ii) by the continuous addition of particles [46]. Attempts have also been made wherein the particles are injected into the molten metal, held in the crucible, through a vortex introduced by mechanical agitation prior to its ejection onto the chill roller. This can be viewed as a modification of the above two methods, depending on whether the concern is either with the precise location where the reinforcing particles are injected into the melt or when the reinforcing particles are added [5,9,46,47].

Rapid solidification (RS), defined as a "rapid" quenching from the liquid state, covers a broad range of material processes [48]. For the process to be considered in the RS regime its cooling rate would have to be in the order of $>10^4$ K/s. The time of contact at high temperatures is usually limited to milliseconds followed by rapid quenching to room temperature. Choice of the quenching medium (be it water, brine solution, or liquid nitrogen) has a profound influence on (a) the resultant microstructure, and (b) distribution of the phases in the final product. The high cooling rate results in significant under-cooling of the metal and leads to several metastable effects, which can be categorized as being either constitutional or microstructural. Segregation of the phase or phases, if more than one is present, results in better matrix properties, or alternatively, improves the quality of the bond at the reinforcement-matrix interfaces [49–52].

Rapidly solidified composite materials can be either an in situ composite (i.e., formed during processing) or a composite consisting of a rapidly solidified matrix and a reinforcement that is subsequently formed into a composite material using several of the available traditional processes [53,54]. In principle, the rapid solidification processing methods tried and used for the manufacture of discontinuous particulate reinforced metal matrices can be grouped according to the temperature of the metal matrix during processing. Accordingly, the processes are classified into the following categories [15,46–48]:

1. Solid phase processes
2. Liquid phase processes
3. Two-phase (solid-liquid) processes

In the following section, we provide a limited discussion of the salient features of the solid phase process. Details of the liquid phase processes and two-phase processes can be found elsewhere [15,55–62].

10.6.1 PRIMARY PROCESSING: SOLID PHASE PROCESSES

Fabrication of particulate reinforced MMCs from blended elemental powders involves a number of stages prior to final consolidation. The salient features of the key processes, namely, (i) powder metallurgy, (ii) high energy rate processing, and (iii) dynamic powder compaction are presented and discussed.

10.6.1.1 Powder Metallurgy

Solid phase processes involve the blending of rapidly solidified alloy matrix powders with reinforcing particulates, platelets, or whiskers through a series of steps summarized in Figure 10.7.

The sequence of steps summarized in this figure includes the following:

1. Rapidly solidified particles are sieved.
2. Precision blending of the reinforcing phase, i.e., whiskers or particulates, with the powdered alloy matrix is performed.
3. Reinforcements and metal powder mixture are compressed to approximately 75% of theoretical density.
4. Metal powder-reinforcement mixture is then loaded into a die, heated, and subsequently, degassed to remove absorbed and adsorbed gases, and ultimately consolidated to get a billet.
5. Billet is cooled, homogenized, and scalped prior to the secondary processing by extrusion, forging, rolling, and drawing.

Secondary fabrication of the billets can involve all normally applied metal working practices [63–66]. A flow chart depicting the process sequence, used at the Advanced Composite Materials Corporation (ACMC: Greer, South Carolina, USA) for the fabrication of powder metallurgy discontinuously reinforced metal matrix (aluminum alloy) DRA composites is shown in Figure 10.8 [67]. This technology has also been extended and modified, with various degrees of success by several other commercial manufacturers.

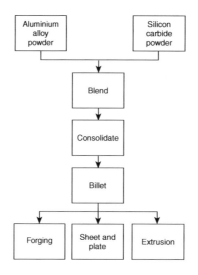

FIGURE 10.7 Flow chart showing the key fabrication steps for powder metallurgy aluminum alloy metal matrix composite.

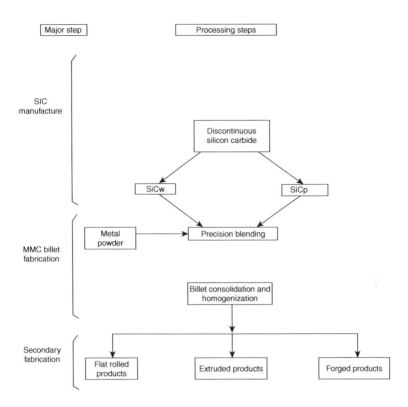

FIGURE 10.8 Flow chart showing fabrication steps for a discontinuously reinforced silicon carbide metal matrix composite at ACM Corporation. (From Geiger, A. L. and Walker, J. A., *Journal of Metals*, August, pp. 8–28, 1991.)

The most important steps in the manufacturing of discontinuously reinforced MMCs involve the prior selection of the discontinuous reinforcement phase and matrix alloy. The key criteria for the selection of a reinforcement phase include the [68,69] (a) elastic modulus, (b) tensile strength, (c) density, (d) melting temperature, (e) thermal stability, (d) compatibility with the matrix, (g) thermal coefficient if expansion, (h) size and shape, and (i) cost. The various possible discontinuous reinforcements and their properties are summarized in Table 10.5.

A spectrum of silicon carbide particulate (SiC_p) sizes and shapes are available and have been tried as reinforcements for the metal matrix. Typically, the alpha and beta silicon carbide crystal structures have been examined as possible reinforcements [68]. A typical particle size distribution for Grade 3 silicon carbide particulates is shown in Figure 10.9. Selection of the most appropriate silicon carbide particulate (SiC_p) size is determined by the ratio of the appropriate size of silicon carbide particulate to the size of the metal powder [70,71,72]. Several aluminum alloys belonging to the low, medium, and high strength categories have been examined as possible matrix alloys [68]. Initially, the most popular choices were based on the age hardenable alloys belonging to the 2XXX (2124) and 6XXX (6061 and 6092) series [4,47,50,53,54]. These alloys were prepared as elemental or pre-alloyed helium inert gas atomized powders. The powders utilized in the 2124 aluminum alloy metal matrix composites had a mean size of 15 micrometers (μm) and a mesh of 325 (Figure 10.10) [68]. Other matrix compositions investigated include the high strength P/M alloys: 7091 and 7090, the experimental high temperature Al–Ce–Fe alloys, and the lightweight Al–Li–Cu–Mg alloys.

In the powder metallurgy (P/M) approach developed by the Aluminum Company of America (ALCOA), the reinforcement material is bended with rapidly solidified powders. The blend is then

TABLE 10.5
Properties of Selected Matrix and Reinforcement Materials

	Young's Modulus		Poisson's Ratio	Bulk Modulus		Thermal Conductivity		Coefficient of Thermal Expansion		Density	
	GPa	$(10^6\ psi)$		GPa	$(10^6\ psi)$	W/mK	BTU/ft hr°F	$10^{-6}/K$	$(10^{-6}/°F)$	g/cm^3	(lb/in^3)
Matrix Alloys											
6061 Al (T6)	70.3	10.2	0.34	75.2	10.9	171	99	23.4	13.0	2.68	0.097
2124 Al (T6)	72.3	10.5	0.34	77.7	11.3	152	88	23.0	12.8	2.75	0.099
ZK60A Mg	44.8	6.5	0.29	35.9	5.2	117	68	26.0	14.4	1.83	0.066
Reinforcements											
Carbides											
SiC	400	58	0.2	221	32	32	18.5	3.4	1.9	3.21	0.116
SiC_p (Grade 3)	400	58	0.2	221	32	120	69	3.4	1.9	3.21	0.116
B_4C	448	65	0.21	255	37	39	22.5	3.5	1.9	2.52	0.091
Nitrides											
AlN	345	50	0.25	228	33	150	87	3.3	1.8	3.26	0.118
Si_3N_4	207	30	0.27	152	32	28	16	1.5	0.8	3.18	0.115
Oxides											
Al_2O_3	379	55	0.25	255	37	30	17	7.0	3.9	3.98	0.144
SiO_2 (fused quartz)	73.1	10.6	0.17	36.6	5.3	1.4	0.8	<1	<0.6	2.66	0.096
$LiO_2\ Al_2O_3\ 4SiO_2$	67.6	9.8	0.19	36.6	5.3	1.3	0.8	<1	<0.6	2.38	0.086
$LiO_2\ Al_2O_3\ 8SiO_2$	69	10	0.18	35.9	5.2	1.3	0.8	<1	<0.6	2.39	0.086
$Al_2\ TiO_5$	30.3	4.4	0.2	16.6	2.4	2	1.2	1.0	0.6	3.68	0.133
Others											
Si	112.4	16.3	0.42	2.35	34	100	58	3.0	1.7	2.33	0.084
C fiber (P100)(L)	690	100	—	—	—	400	231	-1.5	-0.8	2.18	0.079
C fiber (P100)(T)	—	—	—	—	—	—	—	30	16.7	2.18	0.079

Source: Schoutens, J.E., Introduction to Metal Matrix Composite Materials, MMCIAC Tutorial Series No. 272, 1982.

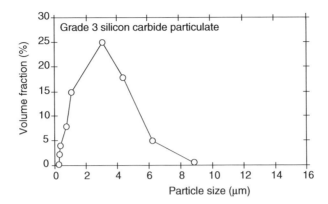

FIGURE 10.9 Size distribution for Grade 3 silicon carbide particles. (From Rack, H. J., *Advance Materials and Manufacturing Processes*, 3(3), 327, 1988.)

consolidated into billets by cold compaction, subsequently outgassed, and followed by hot isostatic, or vacuum hot pressed (Figure 10.11) to form starting products (billets) for secondary processing and subsequent fabrication. Cold compaction density is controlled in order to maintain an open, interconnecting porosity. This is highly essential during the subsequent stage of degassing or outgassing during the pressing operation. The process of outgassing involves the removal of adsorbed gases, chemically combined water, and other volatile species in the consolidated billet. This is achieved through the synergistic action of heat, vacuum, and inert gas flushing. For the SiC_p reinforced aluminum alloy-based metal matrix composites, the outgassing aids in the removal of (a) water adsorbed from both the reinforcing silicon carbide particles (SiC_p) and the aluminum alloy matrix, and (b) chemically combined water from the metal matrix [68]. An advantage of the powder metallurgy technique is an ability to make use of the improved properties of advanced rapidly solidified powder technology for the synthesis of metal matrix composites [68,69]. In addition, the technique has the added advantage of offering near isotropic properties to the composite body [70].

In the process developed by Advanced Composite Materials Corporation (ACMC) to make discontinuously reinforced MMCs, the atomized metal powders and particulate reinforcements are

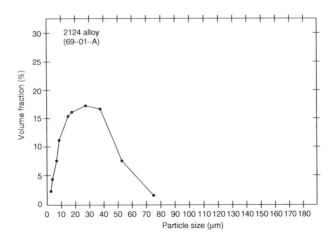

FIGURE 10.10 Powder size distribution for 2124 helium inert gas atomized aluminum. (From Rack, H. J., *Advance Materials and Manufacturing Processes*, 3(3), 327, 1988.)

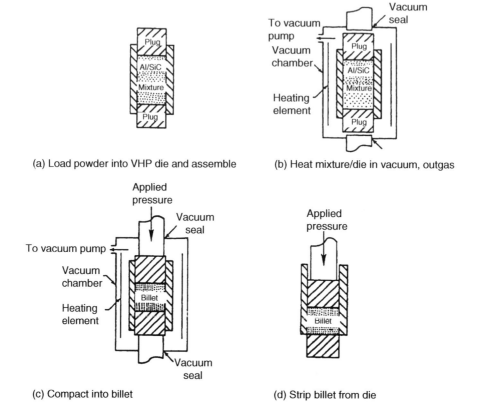

(a) Load powder into VHP die and assemble					(b) Heat mixture/die in vacuum, outgas

(c) Compact into billet					(d) Strip billet from die

FIGURE 10.11 Schematic representation of the vacuum hot pressing technique. (From Rack, H. J., *Advanced Materials and Manufacturing Processes*, 3(3), 327, 1988., Hunt, W. H., Jr., Cook, C. R., Armanie, K. P. and Gareganus, T. B., In *Powder Metallurgy Composites*, Kumar, P., Ritter, A., and Vedula, K., Eds., Metallurgical Society of AIME, Warrendale, PA, USA, 1987.)

mixed using proprietary processes. The processes are critical in the production of homogeneous, high-performance composite materials. Mixtures are loaded into a die, degassed, and vacuum hot pressed at a temperature slightly above the solidus temperature of the matrix alloy. The presence of molten metal during hot pressing allows for the occurrence of full densification of the billet at moderate pressures, even when very small (less than five μm) high surface-area reinforcements are used [67,68].

10.6.2 THE TECHNIQUE OF SPRAY PROCESSING

Use of the principles of rapid solidification has resulted in the development of a variety of processing techniques that have emerged as potentially attractive, technically viable, and economically affordable alternatives to the optimization of the microstructure and properties of particulate reinforced metal matrix composites. Of the several techniques that have been developed and put to use since 1990, the technique of spray processing offers an opportunity to utilize the benefits associated with fine particulate technology. The benefits include the following: (i) an overall refinement in intrinsic microstructural features, (ii) modifications in alloy chemistry, (iii) improved microstructural homogeneity, (iv) in situ processing, and, in some cases, (v) near net shape manufacturing [73–75]. Since the 1990s, the technique based on the principles of spray technology has rapidly evolved to gain the attention and interest of even the commercial producers of material and manufacturers of the product. Now, a variety of spray-based methods exists as listed below:

  a. Spray atomization and deposition processing [73–86]
  b. Low-pressure plasma deposition [87–89]
  c. Modified gas welding technique [90]
  d. High velocity oxyfuel thermal spraying [91,92]

We provide a discussion of the technique of spray atomization and deposition processing, followed by briefly highlighting the key features of the techniques of low-pressure plasma deposition (LPPD), modified gas welding (MGW), and high velocity oxy-fuel thermal spraying (HVOF).

10.6.2.1 Spray Atomization and Deposition Processing

The technique of spray atomization and deposition processing of discontinuously reinforced metal matrices is essentially a semisolid synthesis method in which the reinforcing phase, i.e., reinforcement or reinforcements, if more than one, and the metal matrix are thoroughly blended, while the latter is in the semi-liquid state. The processing technique offers several unique characteristics that differentiate it from the liquid, solid, and two-phase processing approaches. Most importantly, mixing and consolidation are completed in a single operation thereby reducing the number of processing steps. The technique has been used to produce preforms with the as-spray deposited density approaching 97% of the theoretical density. This makes the spray deposited MMC conducive for densification with limited post deposition thermomechanical processing (TMP). The technique has also been used for the near net shape forming of composites in the geometrical configurations of tubes and discs. In direct comparison with the techniques of liquid phase processes and rheo-casting, the contact time between the metal matrix and reinforcement is significantly reduced.

Development of spray deposition processing for the manufacturing of composite materials has engendered considerable scientific and technological interest. The preponderance of research activity has focused on examining several salient aspects of this processing technique. These include the following: (a) methods for incorporation of the reinforcing phase or phases, if more than one is present; (b) an interaction behavior between the metal matrix and the reinforcing phase; and (c) microstructural evolution. In addition to this processing technique, other related processing techniques have been developed for the synthesis of composite materials. These include thermal spray deposition and plasma spray deposition [84,86]. Despite intrinsic differences in the methods used for this generation of semiliquid/semisolid droplets, the deposition and consolidation stages involved in these two techniques are essentially similar.

Early studies on spray deposition processing of MMCs, first reported during the late 1980s and early 1990s, involved simultaneous spraying of the metal and reinforcing ceramic particulates, a process referred to as co-deposition [93–103]. These early studies made use of pure aluminum and commercial aluminum alloys as metal matrices. Since then, a wide variety of matrix-reinforcement combinations have been successfully processed using the technique of spray deposition. Matrix materials that have been used include the aluminum-base alloys such as aluminum–copper [104–106], aluminum–iron [107], aluminum–lithium [108–110], Al–Mg–Si [111–113], aluminum–silicon [114–117], aluminum–titanium [118] and Al–Zn–Mg–Cu [119,120], copper alloys [121–123], iron-based alloys [124], titanium-based alloys [125,126], and even intermetallic compounds such as molybdenum disilicide ($MoSi_2$) [127,128] and nickel aluminide [129–131]. The reinforcements frequently tried and used in spray deposition include the following: (a) aluminum oxide (Al_2O_3), (b) graphite, (c) silicon carbide (SiC), (d) titanium diboride (TiB_2), (e) titanium carbide (TiC), and (f) steel in both powder and fiber forms [132,133]. A number of spray deposition techniques have been developed for the synthesis of discontinuously reinforced composite materials. These include spray atomization and co-deposition [102,104,106,108–110,118], spray deposition of premixed MMCs [105,117], modified Osprey method [94,96,98], reactive spray deposition [129,133], and plasma spray deposition [125,126].

10.6.2.2 Spray Atomization and Co-Deposition

This is the first and most frequently used spray-processing technique that has been applied for the synthesis of composite materials. The term co-deposition is used to describe simultaneous deposition of both the metal matrix and reinforcing phase. Experimentally, co-deposition is accomplished by entraining the reinforcements within a flowing gas and aiming the gas-reinforcement mixture at a fine dispersion of atomized droplets. This method is suitable for the synthesis of discontinuously reinforced metal matrix composites. The technique of co-deposition was initially used for the fabrication of discontinuous particle-reinforced metal matrix composites (DRMMCs) due to the relative ease of availability of reinforcements coupled with the low costs associated with this technique. A schematic of the equipment used for spray atomization and co-deposition of composites is shown in Figure 10.12. The technique involves three consecutive steps: (a) atomization, (b) ceramic particle co-injection, and (c) deposition. During spray deposition, the matrix material is atomized into a fine dispersion of droplets using a high velocity gas. A simplified representation of the technique is shown in Figure 10.13. Following atomization, the droplets travel toward a deposition surface under the combined influence of forces of fluid, drag, and gravity. Prior to arrival of the droplets at the deposition surface, the ceramic particulates are injected into the atomized spray using one or more gas injectors. The injection allows the ceramic particulates to interact with the atomized droplets. The spray of droplets interspersed with the reinforcing ceramic particulates is collected on either a substrate or a shaped container. The distance between the atomizer and the injection position is defined as injection distance and that between the atomizer and the deposition surface is termed as deposition distance. It is possible to direct the flow of the reinforcing particulates directly onto the deposition surface. This makes the injection distance identical to the deposition distance. The injection distance must be selected in order to either minimize

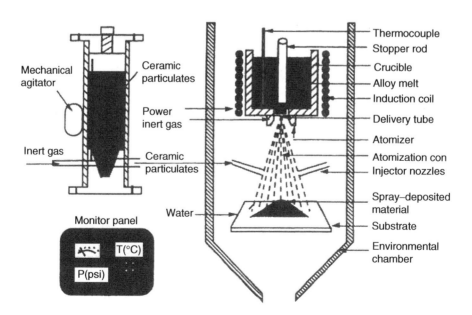

FIGURE 10.12 Schematic diagram showing spray atomization and deposition processing of particulate reinforced metal matrix composites. (From Srivatsan, T. S., and Sudarshan, T. S. and Lavernia, E. J., *An International Review of Journal*, 39, (4/5), 317–409, 1995; Lavernia E. J., *SAMPLE Quarterly*, 22, 2, 1991.)

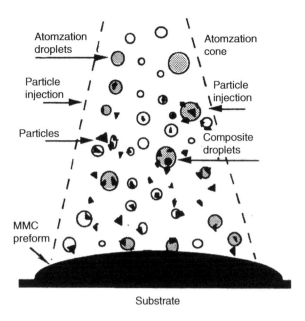

Substrate

FIGURE 10.13 A schematic illustrating the spray atomization and co-deposition process. (From Zeng, X. and Lavernia, E. J., *An International Journal of Rapid Solidification*, 7, 219, 1992.)

or avoid interfacial interactions that can occur as the particles are exposed to the liquid and partially liquid droplets.

In spray atomization and co-injection, the ceramic particulates are accelerated and directed by the droplets using gas injectors. Generally, two or more injectors may be used in a single experiment depending on the desired volume fraction of reinforcements. Although a variety of designs have been used to entrain the ceramic particulates with appropriate gas flow control, there are two basic types of injectors that have been used for this purpose, namely (a) a fluidized bed injector, and (b) a coaxial injector. The injectors are generally located outside an environmental chamber at a location where gas injection is facilitated directly into the chamber. A typical fluidized bed injector is shown in Figure 10.14 [118]. In a fluid bed injector, the ceramic particulates are placed inside of a tube containing

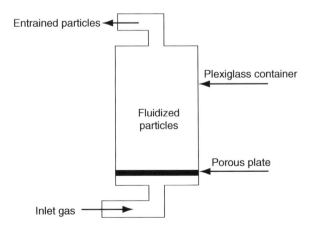

FIGURE 10.14 Schematic showing a fluidized bed injector. (From Gupta, M., Juarez-Islas, J., Frazier, W. E., Mohamed, F. A., and Lavernia, E. J., *Metallurgical Transaction B*, 23B, 719, 1992.)

a porous material. During injection, a pressurized gas (usually inert) passes through the fluidized bed and entrains the ceramic particulates. As the gas pressure increases, the fluidized bed begins to expand. This is accompanied by a concurrent decrease in density. When a critical pressurizing condition is achieved, the drag force exerted by the fluid on the particulates equals the gravitational force that holds these particulates inside the injector container. In spray atomization and co-injection experiments the reinforcing ceramic particulates are incorporated into the metal matrix while the latter is in a semiliquid state. A few of the distinct advantages with this particular technique are as follows [101,107,118,134,135]:

1. Reduced contact temperature between the ceramic reinforcements and the metal matrix
2. Minimized interfacial interactions
3. Rapidly solidified microstructures
4. Limited reinforcement segregation

In addition, when conditions for incorporation of the reinforcing phase(s) are properly controlled, spray atomization and co-deposition can be used for the synthesis of hybrid or layered MMCs, as well as functionally gradient materials (FGMs) [136].

10.6.2.3 Spray Deposition Processing of Premixed MMCs

In spray deposition processing of premixed MMCs the starting materials for atomization and deposition experiments are composites [105,108,117]. The reinforcement phases are introduced by using any one of the two methods shown in Figure 10.15.

10.6.2.3.1 Method 1

In situ formation of the reinforcing phase(s). In this method, the matrix material is initially melted and the reinforcements are formed in a crucible by chemical reactions between the molten matrix and the gaseous or solid phases. The resultant composite slurry is spray atomized and deposited.

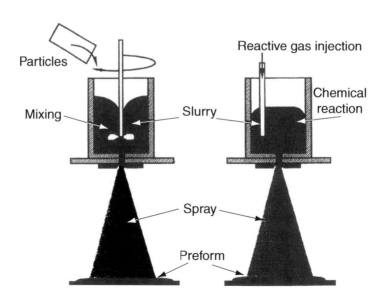

FIGURE 10.15 Schematic diagram showing spray deposition synthesis of premixed metal matrix composites.

10.6.2.3.2   Method 2

A solid MMC is used as the starting material for melting and atomization. Alternatively, particles and the molten matrix are mixed to form a slurry. Subsequently, the slurry is atomized into droplets, which are deposited onto a three-dimensional preform. This method successfully synthesized aluminum–lithium based MMCs containing 10 vol.% SiC and 5 vol.% graphite [108].

The primary advantage with the technique of spray atomization and deposition of premixed composites is an ability to obtain a near uniform distribution of the reinforcement at both the macroscopic and microscopic levels. This method was used to synthesize and characterize a spray deposited cast aluminum alloy 356 reinforced with SiC_p [116]. In this study, an $AA356/SiC_p$ MMC produced by the DURAL method was used as the starting material. During spray deposition, the composite material was re-melted by induction heating and deposited as a preform. Microstructures of $AA356/SiC_p$ MMCs produced by ingot casting and spray deposition are shown in Figure 10.16 [116]. In the ingot cast material, an agglomeration of the reinforcing SiC particulates was evident resulting in particulate-rich and particulate-depleted regions. In the cast material, the reinforcing SiC_p were present at the dendritic boundaries forming an inter-dendritic network. However, in the spray deposited material, the reinforcing SiC_p were distributed uniformly through the aluminum alloy metal matrix. Disadvantages associated with spray deposition of premixed MMCs are as follows:

1. When the reinforcing phases are introduced by re-melting premixed MMCs, they are exposed to liquid matrix material for a prolonged period. Thus, the thermal exposure promotes interfacial reactions, which cannot be avoided. Interfacial reactions place a limit on the combination of metal matrix and reinforcement(s) that can be successfully synthesized using this approach.
2. Fluidity of the composite slurry decreases with an increase in particle volume fraction. Consequently, increasing the reinforcement volume fraction in the metal matrix may make it difficult to deliver the composite slurry to the gas atomizer. Therefore, viscosity places a limit on the volume fraction of particulate reinforcement that can be successfully incorporated into the metal matrix using this method.

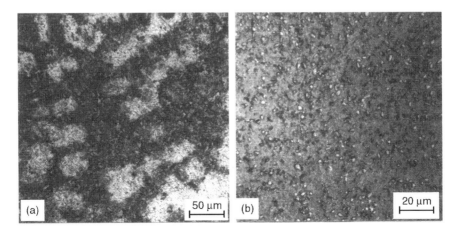

FIGURE 10.16 Optical micrographs showing the distribution of reinforcing SiC particulates in cast aluminum alloy A356. (a) Ingot cast microstructure and (b) Spray deposited microstructure.

10.6.3 Low Pressure Plasma Deposition

A technique was developed by General Electric Company for the net shape processing by plasma spraying in a reduced pressure environment and was termed as Low Pressure Plasma Deposition (denoted henceforth as LPPD). Unlike the technique of spray deposition, the unique feature of the LPPD process is that it makes use of pre-alloyed powders as the feedstock coupled with an intrinsic difference in the manner in which the sprayed deposit is built up into a monolithic shape.

This LPPD technique combines particle melting, quenching, and consolidation in a single operation. The process involves an injection of a variety of powder particles ranging from metallic to ceramic, and including mixtures, into a plasma jet stream created by heating an inert gas in an electric arc confined within a water-cooled nozzle. The particles injected into the plasma jet, at temperatures of 10,000 K and higher, undergo rapid melting, and are accelerated toward the work-piece surface. Rapid quenching of the molten particles occurs when the droplets impact the substrate. The cooling rates are very high, of the order of 10–10 K/s, and the resulting microstructure is fine grained and homogeneous. Compared to other rapid solidification processing approaches, plasma-spraying offers the advantages of large throughput (Kg/h), high density, and an ability to deposit objects of complex shape and to produce near net shape bulk forms.

The technique of conventional plasma spraying is normally carried out at atmospheric pressure. The resultant deposit contains oxidation products together with traces of porosity due to incomplete melting, wetting, or fusing together of the deposited particles. The problem of oxidation is minimized either by shielding the plasma arc in an inert gas atmosphere, or by enclosing the entire plasma spraying unit in an evacuated chamber, which is maintained at about 30–60 torr inert gas pressure by high speed pumping. Under the latter operating conditions of LPPD, the gas velocities are high (typically in the Mach 2–3 range) due to the higher permissible pressure ratios. When the inert environment chosen is vacuum, the technique is referred to as vacuum plasma spraying [86–88,137,138]. In vacuum plasma spraying, although the occurrence of solidification is rapid, the deposit undergoes continuous annealing due to exposure to elevated temperatures (>800°C). However, annealing was found to be beneficial since it provided stress relief and recrystallization while concurrently facilitating improved inter-particle bonding [139,140].

10.6.4 Modified Gas Welding Technique

In this technique, the gas metal arc (GMA) welding torch is modified, wherein, pure aluminum or aluminum alloy wire feedstock is melted and combined with silicon carbide particulates (SiC_p) or silicon carbide whiskers (SiC_w) entrained in an inert gas. Upon striking a substrate or mold, the mixture of aluminum alloy and silicon carbide particulates solidifies into a composite structure [90]. A schematic of the deposition process is shown in Figure 10.17. The thermal history of the droplets is controlled to some extent by the interactive influences of the [90] (a) electrical parameters of the melting process, (b) shielding gas used, and (c) distance from the orifice to the mold or substrate. Thermal control of solidification is affected by the competing influences of the thermal properties of the mold and substrate. At any point in time, the small amount of liquid present is inadequate to bond successive droplets and the SiC reinforcement present. In this technique, the reinforcement material is not substantially affected by fluid flow and macro-segregation during solidification. Using appropriate molds and substrates, complex parts and near net shapes of uniform composition and structure are produced. The process has been used as a welding technique for MMCs that provides a weld metal of approximately the same composition as the base material. This enables in reducing the need for mechanical fasteners and adhesives in the fabrication of large structures.

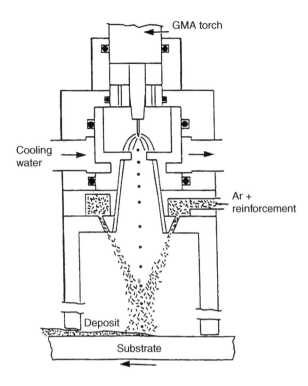

FIGURE 10.17 Schematic of the modified gas metal arc welding deposition process. (From Buhrmaster, C. L., Clark, D. E., and Smart, B. B., *Journal of Metals*, November, 44, 1988.)

A few of the notable advantages of this technique include the following [90,91]:

a. An ability to melt by the arc, any metal, including the very high-temperature alloys
b. An ability to entrain reinforcement in the shielding gas thus providing a wide variety of reinforced metallic matrices or composite materials
c. Melting is more convenient than conventional melting methods, and off with relative ease.
d. An ability to produce large droplets, thereby, minimizing explosion with finely divided aluminum powder.
e. The modified gas metal arc torch is compact, inexpensive, and controllable.

Overall, the modified gas welding technique has been found to be stable and can be operated for several minutes, with controllable termination as the end of the substrate is approached. Potential problems associated with operation of this process are (a) clogging of the orifice, (b) arc instabilities, and (c) arc extinction due to gas flow imbalances. For the process to be successful, proper shielding of the droplet stream and solidifying metal are both important and essential in order to produce a good surface finish while keeping internal porosity at a low level [90].

10.6.5 The High Velocity Oxy-Fuel Spraying

The technique of high-velocity oxyfuel thermal spraying (HVOF) is the most notable development in the thermal spray industry since the development of plasma spraying. Rapid growth of this technology can be attributed, to a large degree, to an extensive development of materials and equipment following the introduction of the Jet Kote HVOF spray process in 1982 [91,92]. During the early years, this technique was used exclusively for aircraft and aerospace-related applications and part

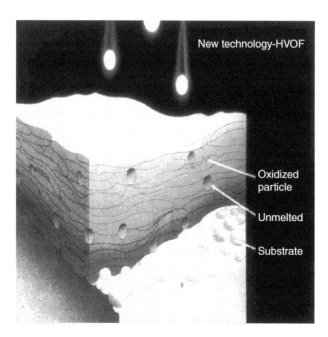

FIGURE 10.18 Micrograph showing high velocity oxyfuel thermal spraying propelling costing powders onto the substrate at very high velocities, which flattens the power particles creating a dense coating than is possible using other spraying processes. (From Parker, D. W. and Kutner, G. L., *Advanced Materials and Processes*, April, 68, 1991.)

conditioning. However, with time and advances in technology, the applications of HVOF did expand from an initial use of tungsten carbide coatings to include hundreds of different coatings that provided a combination of wear, erosion resistance, corrosion resistance, restoration, and clearance control of components [91].

In this technique, the flow of powder is electronically controlled and feed rates are monitored automatically. Powders deposited using the HVOF thermal spraying include pure metals, metal alloy, carbides, certain ceramics, and even plastics. Preliminary use of the HVOF thermal spraying technology was for carbide coatings on aircraft gas turbine engine components, which experienced observable wear from a combination of abrasion, adhesion, erosion, fretting, and corrosion. The key parameters influencing wear resistance of the carbide coatings is flame temperature and particle velocity. Other noteworthy advantages of this spray technique is that the tungsten-carbide coatings produced are better than those obtained using plasma synthesis from the standpoint of (a) enhanced hardness, (b) higher coating bond strength (c), lower oxide content and porosity, and (d) improved wear resistance combined with low residual stress in the coating. Overall, the HVOF coatings show no cracking, spalling, or delamination upon heating to temperatures as high as 1095°C (2000°F) in laboratory air and quenching in liquid Freon. The microstructures resulting from HVOF spraying are equal to, or better than, those of the highest quality plasma sprayed coatings (Figure 10.18).

10.7 SECONDARY PROCESSING

10.7.1 FORGING OF DISCONTINUOUSLY REINFORCED COMPOSITES

Secondary processing such as forging and extrusion can improve the mechanical properties of MMC materials by breaking up the particle agglomeration, reducing or eliminating porosity, and improving particle-to-particle bonding [141,142]. A potential problem with open-air forging is the tendency for cracking to occur on the outer surface, possibly due to the secondary tensile stresses involved in forging that are imposed relatively quickly. This results in matrix-reinforcement debonding, cavitation, reinforcement fracture, and even macroscopic cracking [143–145]. Very high temperatures can also induce macroscopic defects such as hot tearing or hot shortness [12,142].

The enhancement of forgeability in discontinuous particulate reinforced metal matrices arises from two main factors: (a) matrix grain size, and (b) ductility. A finer grain size matrix material that is forged at elevated temperatures maintains a lower flow stress, thus reducing the tendency for cracking. The strength of the particle-matrix interface is not that critical since the fracture path typically traverses through the metal matrix. However, early fracture is possible when perturbation of flow around the reinforcing particles is so significant that both local shear strain and hydrostatic tension is generated between the particles. The fine SiC particles exhibit noticeably less damage than the polycrystalline microspheres during forging [142].

Noticeably less quantity work has been conducted on cold forming or cold forging. This is due to the limited ductility of such materials, although hydrostatic extrusion of these materials is possible at room temperature [146,147]. In all cases, damage in the form of (a) reinforcement cracking and (b) voiding at the reinforcement-matrix interfaces tends to occur, even in compression. Often there is a change in failure mechanism with increasing test temperature and changes in strain rate. A number of studies have investigated the compressive behavior of the discontinuously reinforced aluminum alloy metal matrices at both room and elevated temperatures under conditions of quasi-static and dynamic strain rates. The room temperature results are documented in references [143,148–152], while the high temperature results are documented in references [149,152,153–155].

Several investigators [148,152,154–156] have independently examined the forging behavior of aluminum–copper base metal matrix composites reinforced with ceramic particles such as alumina (Al_2O_3) and silicon carbide (SiC). The emphasis of these studies involved developing the compressive stress versus strain response of the material under conditions of open die forging. Typical findings of this study revealed the following:

a. An increase in offset yield strength with increasing strain rate
b. Extensive cracking of the reinforcing particles at elevated temperatures and high strain rates
c. Densification in the case of powder forging of composites containing different levels of starting porosity

Around the same time, a comprehensive study on fully dense powder metallurgy processed aluminum alloy 2080 reinforced with 15 vol.% SiC_p revealed no evidence of macro-cracking in the subscale forged billets. However, forging of the porous P/M 2080 reinforced with 20 vol.% SiC_p showed different amounts of surface cracking in the subscale billets forged to different levels of strain at 500°C (930°F) (Figure 10.19). The outer portions of the forged billet experienced significant tensile stresses depending on the level of barreling, thereby producing cracking. No cracking was observed at the lower strain rates (Figure 10.19), indicating that powder forging of some metal matrix composites may be a viable near-net-shape manufacturing route. Furthermore, enhanced densification and no cracking were observed in the central region of the billets regardless of strain rate. For the case of forging a fully dense discontinuously reinforced aluminum extensive particle cracking was observed for both aluminum alloy 2124 reinforced with 20 vol.% SiC and aluminum alloy 2618 reinforced with 20 vol.% of Al_2O_3 [154,155]. Extensive cracking of the reinforcing particulates in aluminum alloy 2124, as shown in Figure 10.20, follows the manifestation of instability in the form of flow localization [155].

Several studies have attempted to investigate the effects of different processing parameters (i.e., strain rate, test temperature, and stress state) on the flow stress response and damage development in subscale forgings of discontinuously reinforced aluminum alloys [141,143,148,151,152].

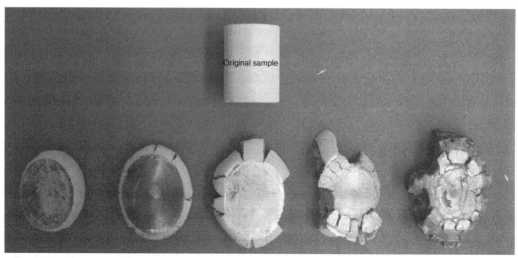

Forged at 0.001 /s to 0.4 ε Forged at 0.001 /s to 0.7 ε Forged at 0.005 /s to 0.7 ε Forged at 1 /s to 0.7 ε Forged at 10 /s to 0.7 ε

⊢———⊣
1 in.

FIGURE 10.19 Macroscopic appearance of powder metallurgy 2080/20 vol.% SiC powder compacts forged at different strain-rate/strain combinations at 500°C. (From Awadallah, A., Prabhu, N. S., and Lewandowski, J. J., Forging/forming simulation studies on a unique, high capacity deformation simulator apparatus, *Materials and Manufacturing Processes*, 17(6), 737–764, 2002.)

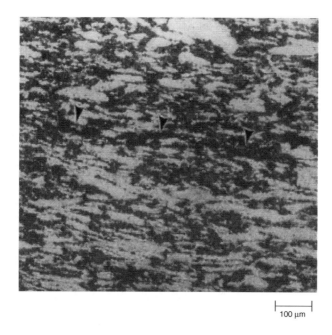

⊢——⊣
100 μm

FIGURE 10.20 Microstructure of aluminum alloy 2124 reinforced with 20 vol.% of SiC and mechanically deformed at 350°C at strain rate of 1 s, showing the manifestation of instability as flow localization and cracking. (From Radhakrishna Bhat, B. V., Mahajan, Y. R., Roshan, H. Md., and Prasad, Y. V. R. K., *Metallurgical Transactions* 23A, 2223–2230, 1992.)

Only a few studies have attempted to evaluate the mechanical behavior of subscale forged billets. This is because of the difficulty in testing adequately sized subscale billets due to equipment capacity limitations. However, the general effects of deformation processing on microstructure and mechanical properties have been determined and documented for both the P/M and cast composites on a limited number of systems [141,143,157]. The availability of high capacity forging simulation equipment provided an opportunity for studies of this type [143]. Preliminary studies have convincingly shown that the distribution of the reinforcing phase, that is, the particulates, is affected differently in different regions of the subscale billets forged under different conditions. This does influence the mechanical properties, as has been shown in an independent study [143].

Few recent studies have attempted to explore the performance of sinter-forged powder metallurgy composites [158–161]. The microstructure of the sinter-forged composites exhibited near uniform distribution of the SiC particles, which are aligned perpendicular to the direction of forging. The sinter-forged composite exhibited higher elastic modulus and ultimate tensile strength than the extruded material, but noticeably lower strain to failure (ε_f). Higher modulus and strength were attributed to an absence of any significant processing-induced particle fracture, while a poorer matrix-particle bonding compared with the extruded material was the cause for the lower strain to failure. It was also found that the fatigue behavior of the sinter-forged composites was similar to that of the extruded material [143].

10.7.2 Hydrostatic Extrusion of Composite Materials

Hydrostatic extrusion involves extruding the billet through a die using the action of fluid pressure medium instead of direct application of the load through a ram, which is used in conventional extrusion. In hydrostatic extrusion, a fluid surrounds the billet. The fluid is then pressurized and this provides the means to extrude the billet through the die. The advantages of this technique are three fold [162]:

a. Extrusion pressure is independent of the length of the billet because friction at the billet-container interface is eliminated.
b. Combined friction of billet/container and billet/die contact reduces to billet/die friction only.
c. Pressurized fluid gives lateral support to the billet and is hydrostatic in nature outside the deformation zone. This aids in preventing the billet from buckling. Skewed billets have been successful extruded under hydrostatic pressure [163].

The limitations inherent with hydrostatic extrusion are:

i. Use of repeated high pressure makes container vessel design crucial for safe operation.
ii. Presence of fluid and high-pressure seals complicates loading, with fluid compression reducing the overall efficiency of the process.

A typical ram versus displacement curve for hydrostatic extrusion is compared with the curve for conventional extrusion in Figure 10.21 [164]: A limited number of investigations have been conducted on discontinuously reinforced composites [165–167], where there is potential interest in the cold extrusion of such systems [168–170]. A potential problem in such systems during deformation processing relates to damage to the reinforcing phase as well as fracture of the billet because of limited ductility of the material, particularly at room temperature. Potential advantages of low-temperature processing include the ability to significantly strengthen the composite and inhibit the formation of any reaction products at the particle-matrix interfaces, since deformation processing is conducted at temperatures lower than that where significant diffusion, recovery, and recrystallization occur [165–167].

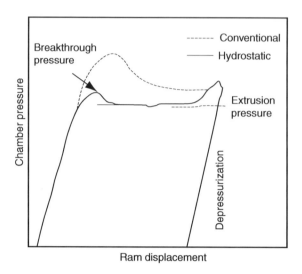

FIGURE 10.21 A typical ram-displacement curve for hydrostatic extrusion. (From Lewandowski, J. J. and Awadallah, A., *ASM Metals Handbook: Metal Working: Bulk Forming*, Vol. 15A, Materials Park, Ohio, pp. 440–445, 2005.)

The experimentally obtained extrusion pressures [165,166] are compared with those predicted by the models of Pugh [171] and Avitzur [172] in Table 10.6. It appears that the high level of work hardening experienced by such metal-based composites [165,166,173] provides a considerable divergence from the values for extrusion pressure predicted by the models based on non-work hardening materials. Monolithic aluminum alloy X2080, which exhibits lower work hardening, extrudes at pressures more closely estimated by the models for a non-work hardening material. Additional work is needed, and is in progress, to establish a wider range of matrix alloy compositions, reinforcement size, shape, and volume fraction to support this observation [162].

TABLE 10.6

Comparison of Hydrostatic Extrusion Pressures Obtained for Monolithic 2080 and 2080 Composites Containing Different Size SiCp to Model Predictions

Material	Extrusion Pressure (MPa)	Work Hardening Model of Pugh		Non Work Hardening Model of Pugh		Non Work Hardening Model of Avitzur	
		$\mu = 0.2$	$\mu = 0.3$	$\mu = 0.2$	$\mu = 0.3$	$\mu = 0.2$	$\mu = 0.3$
Monolithic X2080X2080-15SiCp (SiC$_P$ size)	476	654	771	557	663	559	656
4 μm	648–662	698	824	608	724	611	717
9 μm	648–676	695	820	607	723	610	682
12 μm	572	661	780	579	689	581	682
17 μm	552–559	653	771	579	689	581	682
37 μm	552–579	615	725	558	665	561	658

Source: Lewandowski, J. J., Lowhaphandu, P. *Effects of Hydrostatic Pressure on Mechanical Behavior and Deformation Processing of Metals*, International Materials Reviews, Vol. 43(No 4), pp. 145–188, 1998,

10.8 PROPERTIES OF DISCONTINUOUSLY REINFORCED METAL MATRIX COMPOSITES

An extensive amount of published work on the properties of discontinuously reinforced metal matrix composites (DRMMCs) is available in the open literature. A brief overview, covering the key highlights of the influence of discontinuous reinforcement on properties of the metal matrix is presented in this section.

10.8.1 PHYSICAL PROPERTIES

10.8.1.1 Density, Heat Capacity, and Conductivity

Many physical properties of discontinuously reinforced metal matrices can be described by mathematical expressions that are functions of the properties of the independent matrix and the reinforcing phases. Density and heat capacity are accurately predicted by the simple rule-of-mixtures (ROM) theory. The density of the composite is described by the equation

$$\rho_{DRA} = \rho_m V_{m,} + \rho_r V_r \qquad (10.1)$$

where the subscripts m and r denote the density (ρ) and volume fraction (V) of the matrix and reinforcing phases, respectively. From this expression, it is evident that the addition of moderate volume fractions of silicon carbide to an aluminum alloy results in only a slight increase in density because the two components have similar densities. Thus, the DRMMCs based on aluminum alloy metal matrices can directly replace an aluminum alloy without a significant increase in the weight of the structural component.

Mathematical models also exist for predicting the thermal and electrical conductivity of composites. For the discontinuously reinforced metal matrix composites based on an aluminum alloy metal matrix, i.e., DRAMMC, and containing small volume fraction of the spherical particulate reinforcement, the Rayleigh-Maxwell equation is a good measure to predict conductivity:

$$K_{DRA} = K_M \frac{1 + 2V_r[(1 - K_m/K_r)/(2K_M/K_r + 1)]}{1 - V_r[(1 - K_M/K_r)/(2K_M/K_r + 1)]} \qquad (10.2)$$

For composites containing high volume fraction of the reinforcement, or a range of reinforcement size, the equations put forth by Bruggeman [174] are applicable and accurate. In the case of non-spherical reinforcement particles in a metal matrix, the models derived and presented by Niesel [175] and Fricke [176] are usable. In general, the conductivity of the reinforcement phase is not well characterized, so composite material conductivity cannot be accurately predicted a priori. A measurement of matrix and composite conductivity allows for $K_{reinforcement}$ to be calculated using the existing models in the open literature, allowing for the prediction of conductivity for the other composites having the same matrix and reinforcement phases. Thermal conductivity of 2009/SiC/xxp and 2009/SiC/xxw composites as a function of volume fraction of the reinforcement is shown in Figure 10.22. Electrical conductivity of the 6061 and 2009 matrix alloys discontinuously reinforced with SiC is shown in Figure 10.23.

Thermal expansion of composites has also been extensively modeled. An equation for spherical reinforcement particles, which considers only isostatic stress, was derived by Turner (α is the coefficient of thermal expansion) [177].

$$\alpha_{DRA} = [\alpha_M V_M K_M + \alpha_r V_r K_r]/[V_M K_M + V_r K_r] \qquad (10.3)$$

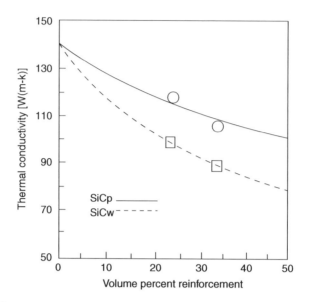

FIGURE 10.22 Thermal conductivity versus volume percent reinforcement for 2009/SiC/xx-T6 DRA composites. (From Geiger, A. L. and Walker, J. A., *Journal of Metals*, August, 8–28, 1991.)

A more sophisticated model considering shear as well as isostatic stress was provided by Kerner [178]. Measured values of the thermal expansion coefficient for a DRA billet for an aluminum alloy MMC, i.e., 6061/SiC/xxp, were between the values predicted by the existing models and is shown in Figure 10.24 [179]. For metal matrices strengthened with non-spherical reinforcements, thermal expansion is anisotropic when processing introduces a preferred orientation of the reinforcing phase.

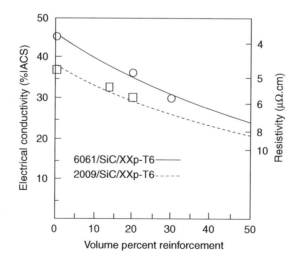

FIGURE 10.23 Electrical conductivity versus volume percent reinforcement for 2009/SiC/XXp-T6 and 6061/SiC/XXp-T6 DRA composites. (From Geiger, A. L. and Walker, J. A., *Journal of Metals*, August, 8–28, 1991.)

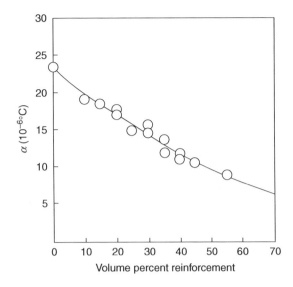

FIGURE 10.24 Coefficient of thermal expansion versus volume percent reinforcement for 6061/SiC/xxp-T6 composites (valid for 0–150°C). (From: Geiger, A. L. and Jackson, M., *Technology Materials and Processes*, Shieh, W. T., Ed., ASM International, Metals Park, Ohio, p. 93, 1989.)

10.8.2 MECHANICAL PROPERTIES

10.8.2.1 Modulus of Elasticity

Young's modulus is an elastic property, which is well bracketed by two models. The linear upper bound is defined by the rule-of-mixtures (ROM):

$$E_{DRA} = [E_M V_M + E_r V_r]/\{V_M + V_r\} \tag{10.4}$$

While the nonlinear lower bound is defined by a more complex expression for materials containing spherical particles, which is simplified by assuming Poisson's ratio is a universal constant with a value of 0.2 [180].

$$E_{DRA} = [E_M V_M + E_r(V_r + 1)]/\{E_r V_M + E_M(V_r + 1)\} \tag{10.5}$$

In both the models, represented by Figure 10.4 and Figure 10.5, the Young's Modulus of the composite material increases monotonically as the volume fraction of SiC increases. The shape of the reinforcing phase, which affects the load carrying capability, has a direct influence on magnitude of the increase. The average Young's moduli of a discontinuously reinforced aluminum alloy composite reinforced with SiC_P and SiC_W is shown in Figure 10.25. Each point on the plot represents an average from extruded and rolled products made using the 2XXX, 6XXX and 7XXX series matrix alloys. The moduli of the DRA composites made using the SiC_W aligned in the longitudinal direction falls midway between the two bounds, while moduli in the transverse direction falls near the lower bound. Although the reinforcing SiC particles (SiC_P) have more load carrying capability than predicted for the spherical particles due to their irregular shape, the Young's modulus of the composite is considered isotropic.

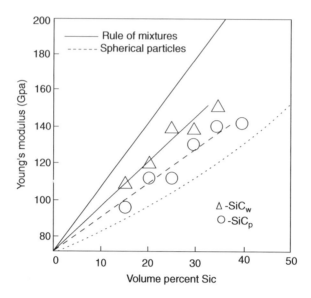

FIGURE 10.25 Variation of Young's Modulus with volume percent of SiC_p and SiC_w. (From Geiger, A. L. and Walker, J. A., *Journal of Metals*, August, 8–28, 1991.)

10.8.2.2 Strength

A relationship between particle volume fraction and strength is shown in Figure 10.26. An increase in the volume fraction of the SiC from 15 to 25 vol.% produces an increase in the proportional limit, tensile yield strength, and ultimate tensile strength. The parabolic relationship between size of the reinforcing particle ($d^{0.5}$) and strength of aluminum alloy 6013 is shown in Figure 10.27. A reduction in particle size from 28 to 2 μm increases the (a) proportional limit, (b) tensile yield strength, and (c) ultimate tensile strength. A further reduction in particle size to 0.7 μm increases the tensile yield strength and ultimate tensile strength but decreases the proportional limit [67].

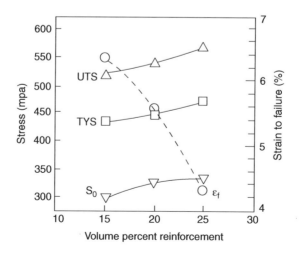

FIGURE 10.26 Influence of volume percent reinforcement on ultimate tensile strength, tensile yield strength, and proportional limit for 6013/SiC/xxp-T6 DRA composites. (From Geiger, A. L. and Walker, J. A., *Journal of Metals*, August, 8–28, 1991.)

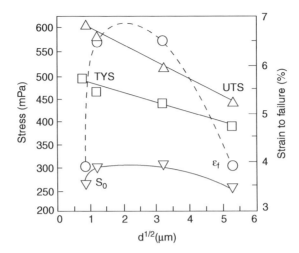

FIGURE 10.27 Influence of square root of reinforcement size on ultimate tensile strength, tensile yield strength, proportional limit, and strain to failure of 6013/SiC/xxp-T6 composites. (From Geiger, A. L. and Walker, J. A., *Journal of Metals*, August, 8–28, 1991.)

Considering the size and spacing of the reinforcing SiC particles to be large, a dislocation looping mechanism alone cannot account for the differences measured in proportional limit and tensile yield strength [181]. Assuming that the matrix precipitates and crystallographic texture do not vary significantly with volume fraction or size of the reinforcing phase, it is reasonable to postulate that the reinforcing SiC particles affect strength via thermally induced plastic strains and resulting residual stresses.

Plastic strains are introduced in the matrix during quenching from the solution heat treatment as a direct consequence of the thermal expansion mismatch between the metal matrix and the reinforcing phase [182–184]. The thermal expansion mismatch results in an increase in residual back stress (compressive) in the matrix, although some tensile residual stresses exist between the closely spaced particles [182] or around the particles having a high aspect ratio [185]. An increase in volume fraction of the reinforcing phase from 10 to 20 vol.% SiC_p was observed to increase the proportional limit via the back stress. Tensile yield strength increased with an increase in back stress and aided by an increase in the average dislocation density [180]. Smaller particulate reinforcements rather than the larger particulates, tend to increase the average dislocation density, and accordingly, the magnitude of the residual stress state. This prediction cannot account for the observed decrease in proportional limit, which occurs when the SiC particle size is decreased to less than 2 μm.

The exact manner in which particle size affects ultimate tensile strength can be best described in terms of work hardening. The most common expression of work hardening is empirical and relates the true flow stress to the true plastic strain (ε_P)

$$\sigma = K(\varepsilon_P)^n \tag{10.6}$$

where n is an index and represents the work hardening rate and K is the strength coefficient. The conjoint influence of reinforcement volume fraction and particle size on flow stress can be best rationalized through the spacing parameter, which represents the physical constraint exerted by the intrinsic microstructural features on the motion of dislocations [186]. This relationship suggests that work hardening becomes stronger as the particles are made either smaller or become more numerous (both of which decreases particle spacing). If the reinforcing particles influence flow stress by generating geometrically necessary dislocations, the appearance of the fracture surfaces is indicative of this mechanism of strengthening [187].

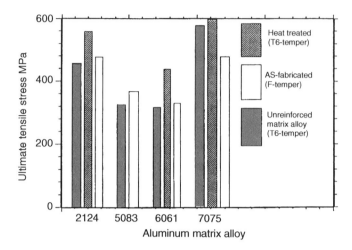

FIGURE 10.28 Effect of matrix alloy and heat treatment on ultimate tensile strength of 20 vol.% SiC/Al composites. (From McDanels, D. L., *Metallurgical Transactions*, 16A, 1105–1115, 1985.)

The increase in strength of a discontinuously reinforced metal matrix as compared with the ultimate tensile strength of the matrix is shown in Figure 10.28. From this data, the observed increase in strength decreases as ultimate tensile strength of the metal matrix increases. Arsenault and co-workers [188] attributed this to the fact that higher yield strength would result in the generation of fewer dislocations due to differences in coefficient of thermal expansion and a higher tensile thermal residual stress. A combination of these two factors results in a smaller increase in strength of the metal matrix composite. Around the same time, a study by Taya and Arsenault [189] revealed that the increase in composite yield strength (σ_{yc}) over the matrix yield strength (σ_{ym}) is relatively independent of the yield strength of the metal matrix.

Another very important factor that affects both the level of work hardening, UTS, and fracture strain/ductility, relates to the evolution of damage in the composite material [190–210]. Damage may evolve in a number of different ways. These include (i) decohesion of reinforcement/matrix interface, (ii) cracking of the reinforcement, and (iii) failure of the matrix. In particular, the cracking of reinforcement particles during straining has been shown to reduce the instantaneous modulus of the composite since the particles are not very effective in the stiffening of the composite [190,191]. This is reflected in a loss of work hardening capacity while concurrently providing sites for the accumulation of damage with increasing stress and strain. Damage can also evolve during deformation processing (e.g., extrusion, drawing, forging) with a similar reduction of the modulus and other related mechanical properties.

The effects of suppressing damage on the mechanical properties of the component is easily demonstrated by conducting compression tests and comparing the results with those obtained in tension on the same material. Presence of residual stresses and evolving damage contributes to the observed difference in yield strength obtained both in tension and in compression. Another viable way for suppressing damage is by the superposition of hydrostatic pressure during tension or compression experiments. Dramatic effects of superimposed pressure on both yield and tensile strength have been reported, arising as a consequence of suppressing damage through changes in stress state (147).

10.8.2.3 Ductility

With the onset of plastic deformation, the particle-matrix interface is the most highly stressed part in the composite microstructure and tends to fail either through the formation of a cavity or fracture of the reinforcing particles [211]. Fracture surfaces reveal the reinforcing SiC particulates to be

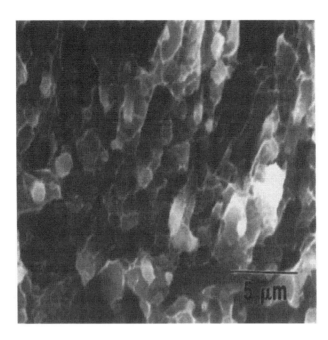

FIGURE 10.29 Scanning electron micrograph showing tear ridges on fracture surface of 6061 aluminum alloy sample reinforced with SiC_p and cyclically deformed under strain amplitude control.

surrounded by ductile regions described as tear ridges because of their shape [212] (Figure 10.29). Submicron-dimpled surfaces observed in the tear ridges of aluminum alloy 7034 reinforced with SiC particulates are indicative of extensive ductility or localized micro-plastic deformation (Figure 10.30) [213]. Larger reinforcing particles tend to cavitate and fracture more easily and readily than the smaller ones. However, a metal matrix composite containing smaller reinforcing particles can accommodate greater strains without appreciable cavitation than the composite microstructure containing larger reinforcing particles [211]. The improvement in ductility that results in aluminum alloy 6013 reinforced with SiC particulates, i.e., 6013/SiC/xxp, when the particle size is reduced from 28 μm to 10 μm was attributed to improved cavitation resistance [67]. A clustering of the reinforcing particles tends to cause the microscopic cracks to initiate at relatively low values of strain. The strength of a discontinuously reinforced aluminum alloy metal matrix can be increased more effectively by reducing the reinforcement particle size than by increasing its volume fraction [211,213–215]. This facilitates an improvement in tensile ductility rather than a detrimental influence of ductility, if segregation of the reinforcing particles can be either minimized or prevented.

A key factor affecting the improvement in ductility of an aluminum alloy reinforced with SiCp is the degree of metalworking from billet to final product. A higher degree of reduction by mechanical working helps to increase composite ductility by the conjoint and mutually interactive influences of the following:

a. Reducing matrix porosity while generating fresh surfaces to improve aluminum powder binding
b. Breaking up the second-phase inclusions and more effectively stringering them
c. Making the dispersion of reinforced particles finer and increasing uniformity

In general, the ductility decreases to a low level upon increasing the volume fraction of reinforcing particles in the metal matrix [212,216].

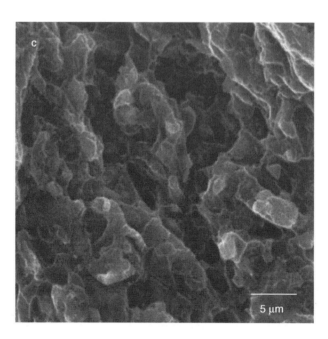

FIGURE 10.30 Scanning electron micrograph showing submicron size shallow dimples on the fracture surface of 7034 aluminum alloy reinforced with SiC_p and cyclically deformed under total strain amplitude control, indicative of localized micro-plastic deformation.

Over the past decade, extensive experimental studies have been conducted by a variety of researchers in order to document the effects of changes in various microstructural parameters on the evolution of damage in discontinuously reinforced MMCs [192–210], as reviewed recently [190,191]. It has been experimentally demonstrated that damage evolution in such materials is affected by changes associated with the matrix as well as the reinforcement. Various studies have demonstrated damage in the form of (i) fractured particulate; (ii) debonding/cracking of reinforcement/matrix interfaces; and (iii) failure of the matrix through microvoid coalescence or shear processes. Figure 10.31, redrawn from earlier work [216] schematically summarizes the various damage processes, which tend to operate in such materials. While one mode of damage is often dominant in a particular heat treatment/reinforcement size/matrix alloy combination, all of the damage modes listed has been observed (to varying degrees) in a single specimen. A number of studies have quantified the extent of reinforcement cracking during tension testing of a variety of reinforcement/matrix/heat treatment combinations. A general summary of the effects of matrix and reinforcement type on the extent and/or severity of damage evolution is provided in Figure 10.32 [217], while Figure 10.33 [218] illustrates the general effects of strength level on ductility of such composites. The percentage of cracked reinforcement observed on the polished surfaces of uniaxial tension specimens deformed to sequentially higher strains is plotted in Figure 10.32 and Figure 10.34 and reveals significant differences between various material/reinforcement combinations. This occurs in part due to (a) differences in the matrix strength levels (illustrated in Figure 10.33) and heat treatment, and (b) differences in the reinforcement size and distribution. Figure 10.34 illustrates more detail on the evolution of cracked reinforcement in a 7XXX powder metallurgy alloy where the effects of heat treatment and reinforcement size on such events were quantified [190,191,202]. The evolution of damage across the specimen gage length throughout the loading history is very inhomogeneous, as noted in other studies on similar topic [193,196,200,202], while final fracture does not typically occur where the level of damage is highest on the surface of the specimen. This latter point indicates that the spatial distribution of damage in three dimensions may be important for both

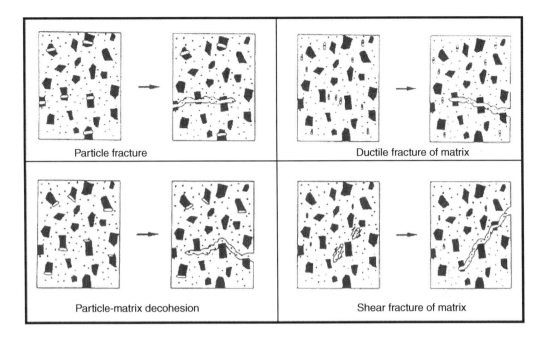

FIGURE 10.31 Sequence of damage events leading to fracture in particulate composites. (From Zok, F., Embury, J. D., Ashby, M. F., and Richmond, O., In *9th Riso Int'l Symp. On Metallurgy and Materials Science*, Anderson, S. I., Lilholt, H., and Pederson, O. B., Eds., Riso National Lab, Roskilde, Denmark, pp. 517–527, 1987.)

nucleation and coalescence of damage in such materials. Consistent with such arguments, early work [193,195,196,218] illustrated that fracture nucleation in such materials typically occurs in regions of clustered reinforcement, while crack propagation aids in linking the damage contained in clustered regions [193].

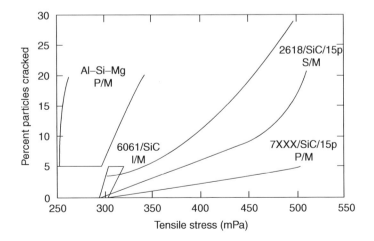

FIGURE 10.32 Stress of damage of particle cracking for DRA produced by different routes (From Lloyd, D. J., *Acta Metallurgica*, 39, No. 1, 59–71, 1991, Singh, P. M. and Lewandowski, J. J., *Metallurgical Transactions*, 24A, 2531–2543, 1993, Llorca, J., Martin, A., Riuz, J., and Elices, M., *Metallurgical Transactions* 24A, 1575–88, 1993.) compared with an Al-Si-Mg model system. (From: Hunt, W. H., Jr., Brockenbrough, J. R., and Magnusen, P. E., *Scripta Metallurgica et Materialia*, 25, 15–20, 1991.)

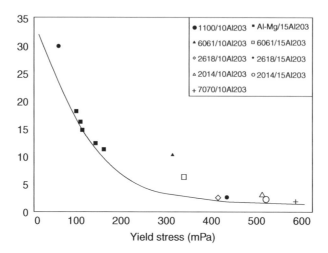

FIGURE 10.33 Variation of elongation with yield strength in Al_2O_3 composites. (From Hahn, G. T. and Rosenfield, A. R., *Metallurgical Transactions*, 6A, 653, 1975.)

The evolution of damage in the form of cracked reinforcement has been shown to reduce the work hardening rate [204,207,209] and Poisson's ratio [190] of composites, in addition to reducing the instantaneous modulus obtained upon unloading [195,200,202,205]. The reduction in instantaneous modulus compared to the starting modulus of the composite, (E/E_o), is summarized for a variety of composites and is found to be nonidentical, consistent with the observation of different levels of damage evolution (shown in Figure 10.32), coupled with the generally unquantifiable effects of other types of damage, such as debonding or cracking at the reinforcement-matrix interfaces, on both work hardening and modulus reduction. In comparison, the stiffness ratio (E/E_o) for the unreinforced matrices over the same strain range would be 1.0 [190,202]. Only a few studies have convincingly identified damage modes in the matrix and at/near the interfaces [202,219]. This is because of the associated higher difficulty in specimen preparation and damage detection. In one case on a 7XXX matrix heat treated to the solution annealed, underaged, and overaged conditions, it was clearly demonstrated that damage in the matrix and near the reinforcement/matrix interfaces evolved at a lower stress and strain in the overaged material than that observed in the solution annealed and underaged conditions [202]. The observed accelerated rate of damage evolution in the overaged condition resulted in significantly lower ductility and toughness compared with that of the underaged condition [192,193,202].

Considerable research on monolithic metals has demonstrated that the imposed or residual stress state significantly affects the ductile fracture process [147,220]. Discontinuously reinforced aluminum alloys fail by the microvoid coalescence (MVC) process whereby the microscopic voids nucleate, grow, and eventually coalesce to produce final failure. In monolithic metals, the presence of triaxial tensile stresses through the introduction or presence of a notch, or crack, has been shown to accelerate the stages of microvoid coalescence by promoting void nucleation and exacerbating void growth and coalescence near the notch/crack [220]. The ductility of notched and precracked specimens is typically lower due to the negative effects of local stress concentration on the ductile fracture process. Few studies have been conducted to determine the effects of machined notches on the fracture behavior [221] and ductility [222,223] of a metal matrix composite, primarily because of the low ductility they possess, as shown in Figure 10.33. However, it should be noted that ductility of the composite in simple tension loading is dependent on strength of the metal matrix and the microstructure, as indicated in Figure 10.33. A recent study on 2XXX aluminum composite [222] indicated that the introduction of a notch decreased the ductility of the composite to nearly zero, and that changing the severity of the notch did not appreciably change the notch ductility of the 2XXX

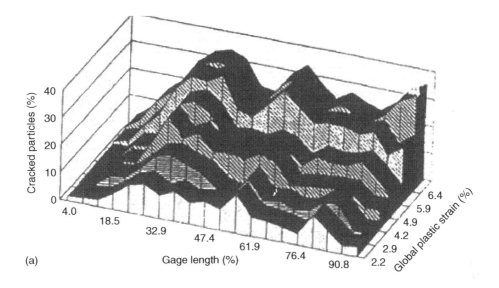

(a)

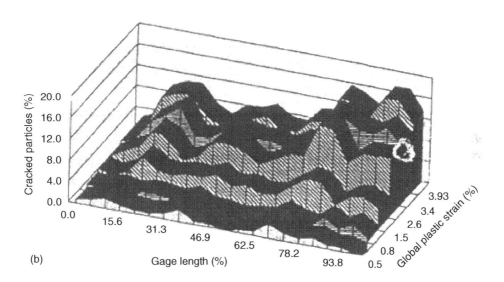

(b)

FIGURE 10.34 Distribution of the local percentage of cracked SiC$_p$ along the gage length of an (a) under-aged and (b) overaged MB78 + 15% SiC$_p$/13 μm tensile specimen. (From Singh P. M. and Lewandowski, J. J., *Intrinsic and Extrinsic Fracture Mechanisms in Inorganic Composites,* Lewandowski, J. J. and Hunt, W. Jr. Eds., TMS, Warrendale, PA. pp. 57–68, 1995; Lewandowski, J. J. and Singh, P. M., *ASM Metals Handbook,* Volume 19-*Fracture and Fatigue,* ASM International, Materials Park, Ohio, pp. 895–904, 1966, Singh, P. M. and Lewandowski, J. J., *Metallurgical Trans,* 24A, 2531–2543, 1993.)

alloy examined. Such an observation might not be general to all composite systems due to the large differences in tensile ductility as shown in Figure 10.33, in addition to the following observations.

While the introduction of triaxial tensile stresses has been shown [222,223] to be detrimental to the ductility of composites, the superposition of triaxial compressive stresses during tension testing was observed to be beneficial as reviewed recently [147]. In Figure 10.35 [147] summarizes the results of tension testing various composites [201,202,224–226] in a pressure vessel maintained at different levels of confining pressure. In general, the tensile ductility of all of the composites increased

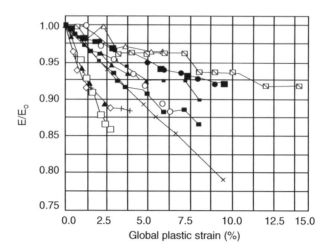

FIGURE 10.35 Effect of global plastic strain on reduction in elastic modulus in various DRA. (From Singh, P. M. and Lewandowski, J. J., In *Intrinsic and Extrinsic Fracture Mechanisms in Inorganic Composites*, Lewandowski, J. J. and Hunt, W. Jr., Eds., TMS, Warrendale, PA. pp. 57–68, 1995; Lewandowski, J. J. and Singh, P. M., *ASM Metals Handbook*, Volume 19-*Fracture and Fatigue,* ASM International, Materials Park, Ohio, pp. 895–904, 1966.)

with an increasing level of the confining pressure (i.e., negative hydrostatic stress). Further, the magnitude of pressure-induced ductility improvement is clearly dependent on the following: (i) matrix strength, (ii) heat treatment/microstructure, and (iii) reinforcement volume fraction. This apparently arises through their effects on the type and pressure-dependence of the damage that evolves. These studies have clearly revealed that damage evolution in particulate MMCs is stress state dependent, while complementary research studies [210,227,228] have shown how such experiments are relevant to the deformation processing of these materials.

10.8.2.4 Fracture Toughness

Analytical models documented in the open literature are conducive for predicting that large reinforcement particles [184] and low work hardening rates [185] would result in improved fracture toughness. However, large particles and the associated low work hardening rate arising from damage are detrimental to both strength and ductility. Since tensile ductility of discontinuously reinforced metal matrices is indicative of cavitation (initiation) and particle fracture resistance, no exact correlation with toughness, which is an index of the coalescence (propagation) resistance, is expected unless a weak interfacial bond exists between the metal matrix and the reinforcing particle. Lower rates of work hardening tend to reduce the flow stress and facilitate the accumulation of more strain during a simple tension test. However, this does not facilitate an improvement in fracture toughness (Figure 10.36) [212].

Fracture toughness is one of the original factors used in the damage tolerant design of structures. Materials engineering has through the years attempted to increase the fracture toughness of metal matrix composites to satisfy the design criteria, which has led to a large number of investigations aimed at understanding the origins of fracture toughness [229–236]. Rapid fracture resulting from lack of toughness in a flawed body occurs when the driving force imposed on the crack exceeds the availability of material to resist rapid crack extension. Fracture toughness is characterized as a material property, and the American Society for Testing and Materials (ASTM) has expended considerable effort in standardizing tests for this property [237]. The fracture toughness of a material is often given the symbol K_C or J_C, although other terms have been used to specify the fracture toughness determined under special conditions, such as (i) the fracture toughness of thin sheets, K_Q,

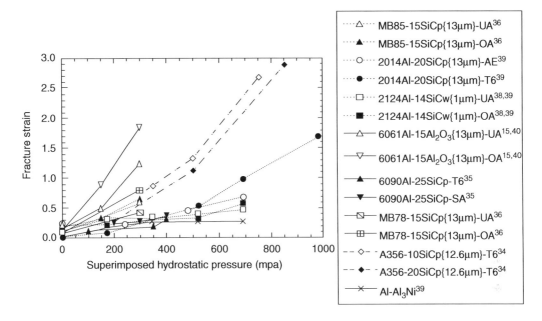

FIGURE 10.36 Effect of superimposed pressure on fracture strain of DRA. (From Lewandowski, J. J. and Lowhaphandu, P., *International Materials Review*, 43(No 4), 145–188, 1998.)

and (ii) that of thick sections K_{IC} or J_{IC}. The loading rates used for these tests are quasi-static. If rapid loading is used, the dynamic fracture toughness is determined. Under conditions of dynamic loading fracture, toughness is defined as the condition required for the initiation of material tearing (stable crack extension) (J_C) if the material exhibits this behavior. However, it is defined as the stress intensity factor required for rapid, sustained (unstable) crack growth (K_{IC}) if the material does not tear in a stable manner. In understanding the fracture toughness of metal matrix composites the matrix characteristics of interest are typical of the usual metallurgical description of microstructure and includes such factors as (i) composition, (ii) grain size, (iii) precipitate and dispersoid content, (iv) volume fraction of dislocation density, and (v) sub grain size.

There are a number of factors contributing to the fracture characteristics of particulate reinforced composites that are the subject of several intensive investigations [238–243]. The addition of particles to ductile metallic matrices resulted in the following changes in material behavior:

a. Slip characteristics of the matrix are altered. The particles exert the primary effect of blocking the slip lines, but there also exists secondary effects, such as limiting the operation of secondary dislocation sites and limiting the extent of cross-slip. These effects will have such consequences as limiting the size of the plastic zone surrounding a crack tip, thereby decreasing the volume of deforming material within it. The volume of deforming material is responsible for magnitude of fracture toughness.

b. Presence of particles near the crack tip limits the strain to fracture at the crack tip. This, combined with limitations on size of the plastic zone, alters the distribution of strain within the plastic zone. Strain at the crack tip is limited in direct comparison with the unreinforced material [232]. The brittle nature of the metal matrix when reinforced with particles, as compared with the unreinforced metal matrix, is direct evidence of the slip-limiting characteristics of the reinforcing particles. Fatigue crack threshold stress intensity factor (ΔK_{TH}) can be computed from the mean free slip of dislocations through the metal matrix and the yield stress [236].

c. The effect of adding stiff particles to the metal matrix is an increase in yield stress of the metal matrix composite. This effect has been found and documented in a few particle-reinforced aluminum alloy metal matrices, whereas, for many others it has not. A number of factors tend to influence a change in yield stress besides the presence of reinforcing particles. These include the following: (i) matrix residual stresses and an increase in dislocation density in the metal matrix arising from a thermal expansion mismatch between the reinforcing particulate and the matrix, (ii) an alteration of the recrystallization characteristics of the reinforced metal matrix during thermal treatment, which tends to control the size of grains and formation of sub-grains, (iii) an alteration of precipitation kinetics and the distribution of precipitate particles. The independent or conjoint influence of these factors does exert an influence on deformation of the metal matrix during final fracture.

d. Debonding of reinforcing particles near the crack tip to form microscopic cracks not directly connected to the main crack is another mechanism that has the potential to influence the fracture toughness. The magnitude of this effect is controlled by the size of the particles and their bonding with the metal matrix. The presence and location of fine microscopic cracks is important. Those present at the particle-matrix interfaces facilitate the promotion of microvoid initiation, which has a detrimental influence on fracture toughness.

e. The influence of particulate fracture on composite material fracture toughness has been the subject of considerable research. Hard and brittle particles within the plastic zone of a growing crack suppress the regions in a material that could plastically deform. The work-to-fracture ratio is decreased in direct proportion to the volume fraction of the reinforcing particulates. The elastic distortional energy of the reinforcing particles is released once the crack tip has passed. When the particle breaks, then the energy of new surface formation is absorbed. Because this energy is low, the effect on fracture toughness is negligible.

The remainder of this section reviews the fracture toughness characteristics of discontinuously reinforced aluminum (DRA) in terms of the variables and its influence, i.e., (i) particle spacing, (ii) particle size, (iii) volume fraction, (iv) matrix alloy, and (v) matrix microstructure. Where available, information on J_{IC} and tearing modulus data for alumina and SiC reinforced aluminum alloys are presented and reviewed in the following section in light of existing models in the published literature. Recent research result on the fracture behavior of toughened structures such as DRA/Aluminum alloy laminates is also presented in order to demonstrate the potential beneficial effects of these and similar approaches.

10.8.2.4.1 Effects of Changes in Notch Root Radium on Toughness

There are significant effects of changes in notch root radius on the fracture toughness of DRA as shown in Figure 10.37 [221]. These observations are consistent with the earlier work on steels [244,245] and other monolithic metals [246], although only two specific references exist to document such effects in DRA [221,247].

10.8.2.4.2 Effects of Changes in Volume Fraction on Toughness

In compiling the results on toughness of DRA, the authors have located and summarized the published data, which investigated the effects of systematic changes in reinforcement volume fraction, reinforcement size, and resultant distribution on the resulting toughness as done in two earlier publications [190,191]. Figure 10.38 summarizes the influence of reinforcement volume fraction on fracture toughness of various DRA systems. The J_{IC} data presented in Figure 10.39 were collected from research works where (a) J_{IC} testing was independently conducted and (b) by converting some of the K_{IC}, K_Q and K_{EE} data to J_{IC} data using the relationship

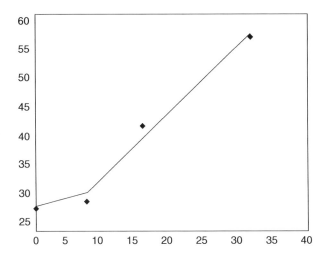

FIGURE 10.37 Effect of changes in Notch Root Radius on Fracture Toughness of a DRA. (From Manoharan, M. and Lewandowski, J. J., *International Journal of Fracture*, 40–2, 31–34, 1989.)

$$J = K^2(1 - v^2)/E \qquad (10.7)$$

The data demonstrates the general loss in toughness with an increase in volume fraction of the reinforcement up to 40 vol.%, coupled with the significant effects of both matrix selection and matrix temper. The toughness decreases rapidly with the addition of reinforcement for low volume fraction (e.g., 0–10%), with relatively smaller losses in toughness accompanying an increase in reinforcement content, i.e., between 10 and 40%. Interestingly, notched toughness data obtained on very high volume fraction DRA (i.e., 55%) was observed to remain in excess of 20 MPa.m $^{1/2}$.[248] The high value of toughness results from the use of a notch to determine the fracture toughness instead of a

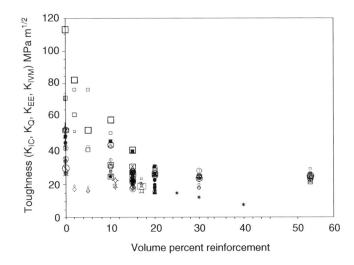

FIGURE 10.38 Compilation of Fracture Toughness data. (From Lewandowski, J. J. and Singh, P. M., *ASM Metals Handbook*, Volume 19-*Fracture and Fatigue*, ASM International, Materials Park, Ohio, pp. 895–904, 1966, Osman, T. M., Singh, P. M., and Lewandowski, J. J., *Scripta Metallurgica et Materialia*, 607–612, 1994.)

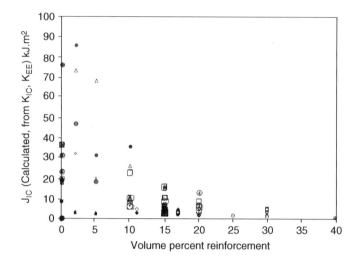

FIGURE 10.39 Compilation of J_{IC} data. (From Lewandowski, J. J. and Singh, P. M., *ASM Metals Handbook*, Volume 19-*Fracture and Fatigue,* ASM International, Materials Park, Ohio, pp. 895–904, 1966, Osman, T. M., Singh, P. M. and Lewandowski, J. J., *Scripta Metallurgica et Materialia*, 607–612, 1994.)

fatigue precrack, as shown in Figure 10.39. Nonetheless, such values are encouraging since they indicate the existence of some level of damage tolerance even for high reinforcement contents. Figure 10.38 and Figure 10.39 include data for DRA based on various aluminum alloys including 1100, 2XXX, 6XXX, 7XXX, 8XXX and the model Al–Si alloys containing brittle Si particles. The processing techniques utilized for the composites include the following: (i) casting, (ii) powder metallurgy, (iii) in situ techniques, and (iv) spray deposition.

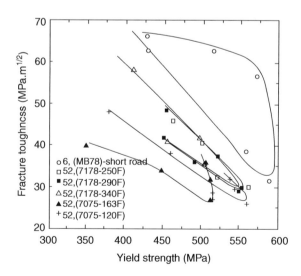

FIGURE 10.40 Fracture toughness versus yield strength for monolithic aluminum alloys. (From Lewandowski, J. J. and Singh, P. M., *ASM Metals Handbook*, Volume 19-*Fracture and Fatigue,* ASM International, Materials Park, Ohio, pp. 895–904, 1966; Osman, T. M., Singh, P. M., and Lewandowski, J. J., *Scripta Metallurgica et Materialia*, 607–612, 1994.)

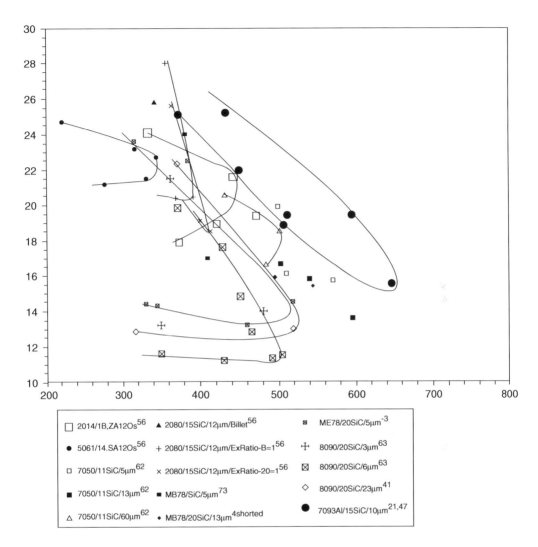

FIGURE 10.41 Fracture toughness versus yield strength for DRA.

10.8.2.4.3 Effects of Matrix Microstructure and Heat Treatment

The effects of matrix strength and heat treatment on fracture toughness of both monolithic and DRA materials are shown in Figure 10.40 through Figure 10.42, which shows the variation of fracture toughness with yield strength and tensile strength. For a given alloy system, the changes in strength level were obtained through heat treatment. For the unreinforced alloys, it has been observed that the toughness decreases with an increase in strength up to peak strength, and the toughness recovers upon continued aging [192,249,250], as shown in Figure 10.40. This has been demonstrated for a variety of 7XXX monolithic materials [192,249–251], 6XXX [252] and 2XXX [252] alloy systems. In contrast, none of the early studies on toughness of composite systems appeared to exhibit a recovery of the toughness upon overaging, as shown in Figure 10.41 and Figure 10.42 for DRAs based on 2XXX [253–255], 6XXX [253,256], 7XXX [192,193,257,258], or 8XXX [259] systems. The lack of toughness recovery upon overaging has been attributed to a change in the mechanisms of fracture nucleation and growth (i.e., damage evolution) in such systems [192,193,253,257]. For the 7XXX DRA system, analysis of sequentially strained tensile specimens revealed that the mechanisms of

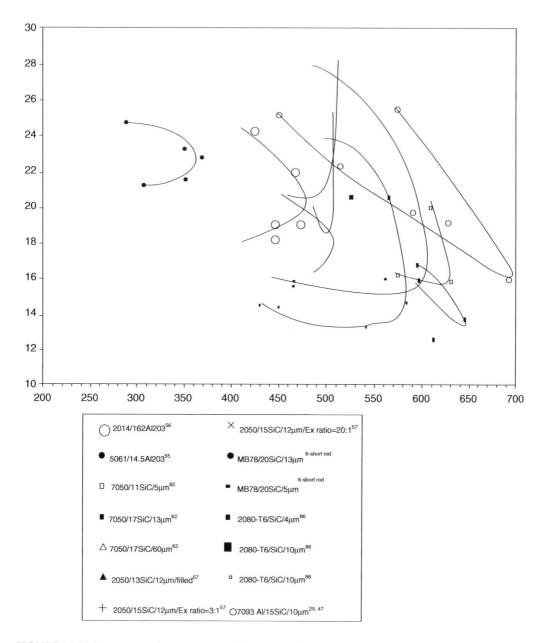

FIGURE 10.42 Fracture toughness versus tensile strength for DRA.

damage accumulation to be significantly affected by the aging treatment [202]. The underaged specimens exhibited a predominance of reinforcement fracture at all values of strain, while the overaged specimens exhibited (i) failure of the matrix and (ii) failure near the reinforcement/matrix interfaces. The rate of damage accumulation, as monitored by the decay in elastic modulus, was shown to occur at a faster rate in the overaged materials [202] resulting in much lower ductility than the underaged specimens. An analysis of the fracture surface features of the underaged and overaged DRA specimens revealed differences in fracture surface morphologies between the underaged and overaged specimens, indicative of the intrinsic influence of microstructure on the operative damage mechanisms occurring ahead of a crack tip in such materials [192,193,253,257]. In situ straining coupled with observations of fracture evolution ahead of the crack tips revealed an influence of matrix microstructure on the nature of crack propagation [193,257].

While Figure 10.41 and Figure 10.42 reveal that most recent investigations have found a lack of recovery of the toughness upon overaging of the DRA. However, recent study on a 7XXX powder metallurgy produced DRA has revealed that certain overaging heat treatments do provide for recovery of the fracture toughness [219,260]. The mechanism(s) contributing to recovery of the fracture toughness is still under investigation.

10.8.2.4.4 Effects of Changes in Reinforcement Size

Reinforcement size and aspect ratio have been considered. One of the more dramatic differences in behavior exists between whisker-reinforced and particle reinforced materials. The whisker-reinforced materials show substantially different strength and toughness behavior and exhibit greater strengthening, greater anisotropy, and lower toughness values [67,191]. On a more limited scale, the intrinsic influence of size of the reinforcement particle has been examined for the powder metallurgy DRA materials. Recent experiments on X2090/SiC extrusions, which examined reinforcement particles of (a) 4 μm average size (Federation of European Producers and Abrasives [FEPA] grade F-1000, (b) 10 μm average size (FEPA grade F-600), and (c) 36 μm average size (FEPA grade F0280) at particle loadings of 10, 20, and 30 vol.%, have convincingly shown that the toughness/yield strength relationship is not greatly affected by particle size, Figure 10.43 [261], while the toughness/tensile strength relationship shown in Figure 10.44 favors the finer-particle materials [261]. Such behavior is consistent with the higher work hardening and lower damage development rate (e.g., via reinforcement cracking) in the fine-particle materials [190,191,195,202,207,228].

10.8.2.4.5 Effects of Changes in Reinforcement Clustering

It is generally agreed that changes in the level of reinforcement clustering may exert significant effects on the local stresses present in DRA and thereby affect various mechanical properties as well as the accumulation of damage in the composite [190–192,218]. The techniques used to characterize clustering as well as some of their implications on mechanical properties have been covered in various journal publications. Part of the difficulty in comparing data between different investigations on DRA lies in the general lack of information on the level of reinforcement clustering between similar materials processed using different techniques. In addition, it is not yet clear which of the various

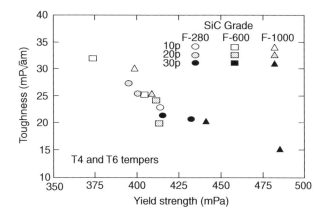

FIGURE 10.43 Toughness-yield strength relationship for X2080/SiC/X$_p$ DRA where both particle size and particle volume are varied. (From Seleznev, M. L., Argon, A. S., Seleznev, I. L., Cornie, J. A., and Mason, R. P., SAE Trans. Paper No. 980700, 1998.)

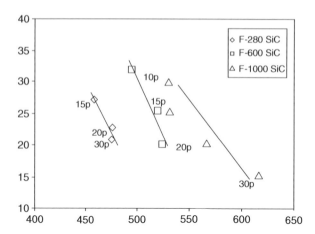

FIGURE 10.44 Toughness-yield strength relationship for X2080/SiC/X_p DRA demonstrating the effect of particle size. (From Seleznev, M. L., Argon, A. S., Seleznev, I. L., Cornie, J. A., and Mason, R. P., SAE Trans. Paper No. 980700, 1998.)

means of measuring clustering is most meaningful in establishing a correlation with various mechanical properties.

Despite the uncertainties in the quantification of clustering, it has been clearly shown by examining the techniques used for the processing of various DRA that the extreme examples of clustered particle distributions occur in molten metal processed DRA [262]. If the choice to process DRA is by molten metal processing, it has been shown that squeeze casting promotes a more homogeneous particulate distribution and a more refined matrix microstructure having better tensile properties than that obtained by casting into sand molds where the solidification rate is significantly slower [263]. It has further been shown that subsequent hot working (i.e., via swaging or extrusion) of the squeeze case and sand-cast composites changes the reinforcement size and distribution in both cases, with accompanying improvements in strength and ductility. Interestingly, hot working the squeeze cast composite containing a relatively homogeneous distribution of the reinforcement does not change the reinforcement distribution as significantly as it does for the clustered sand cast composite. Further, for the sand cast composite segregation of the reinforcement into bands was observed after hot working [263]. Although hot working of both the sand cast and squeeze cast composites improved the mechanical properties in comparison to the as-cast counterpart, the squeeze cast composites exhibited better combinations of strength and ductility after hot working [263]. These differences were attributed to changes in the level of reinforcement clustering as well as due to differences in the matrix microstructure produced through sand casting and squeeze casting.

Published work on cast DRA has indicated that clustering of reinforcements is detrimental to the fracture toughness [262]. In that study, the minimum reinforcement spacing was measured for composites cast under different conditions. For a constant global volume fraction of the reinforcement, an increase in the minimum reinforcement spacing (i.e., more clustering) reduced the fracture toughness as is shown in Figure 10.45.

In powder metallurgy processed materials, earlier studies have examined the effects of changes to the ratio of reinforcement particle size with aluminum powder particle size on the level of clustering [192,193,218]. Clustering in those studies was quantified by first creating a two-dimensional computer representation of the reinforcement present in a metallographic sample of the DRA. The reinforcement were first represented as circles [192,218] and the center of each circle (i.e., particle) was used to construct cell boundaries around each particle in order to represent the material closest to each particle. The regions of locally high volume fraction of reinforcement (i.e., clustered regions)

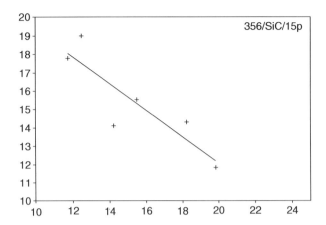

FIGURE 10.45 Effect of clustering on fracture toughness of cast DRA. (From Hahn, G. T. and Rosenfield, A. R., *Metallurgical Transactions* 6A, 653, 1975.)

could be quantified for different ratios of the reinforcement/matrix powder particle size. Subsequent revisions to this technique have enabled the representation of the particles as ellipses [193], with the cell boundaries constructed around each particle using the surface of each particle. Three-dimensional versions of such approaches are also being developed.

In one study, it was shown that the level of clustering in a powder metallurgy processed DRA containing small (e.g., 5 μm average size) reinforcements was noticeably greater than that containing larger (e.g., 13 μm average size) reinforcements [192,193]. While it has been demonstrated that the level of reinforcement cracking is generally lower in finer-particle material due to the higher strength of such small particles, the toughness of the finer-particle material was also somewhat lower than that of the coarser-particle material over a range of heat treatments [192,193]. The researchers observed that the higher level of, or enhanced, clustering in the finer-particle material was a least partly responsible for the lower toughness. In addition, in situ fracture toughness experiments were conducted to determine the precise location of damage nucleation and growth during loading [193,257]. It was shown that fracture nucleation ahead of a crack tip occurred preferentially in clustered regions. Furthermore, macroscopic crack extension occurred by linkage of the crack tip with the damage present in clustered regions ahead of the crack tip. While such results suggest the importance of clustered reinforcement to both damage initiation and crack extension, the need to extend such studies to 3-D was emphasized in the earlier study since crack extension also depends on distribution of the reinforcement through the thickness of the specimen.

10.8.2.4.6 Mixed Mode Toughness Studies

Relatively few studies have attempted to investigate the effects of mixed mode loading on fracture toughness of DRA [264,265], although a number of studies on the monolithic metals have revealed significant effects of changes in loading mode on the magnitude of toughness. While there is no ASTM standard for mixed mode toughness testing, recent investigations have utilized compact tension specimens [264] as well as bend bar specimens [265]. Notches/precracks oriented at different angles to the loading direction were utilized to vary the amount of mixed mode loading. The general observations made in the two earlier studies were that an introduction of the non-Mode I loading (i.e., increased mixed mode loading) produced an increase in fracture toughness, consistent with similar results obtained on monolithic metals. The magnitude of increase in fracture toughness due to mixed mode loading and its dependence on (i) matrix type, (ii) temper, (iii) volume fraction, and (iv) size of reinforcement, have not yet been investigated to any extent.

10.8.2.4.7 Models of Fracture Toughness in Particle Hardened Materials

The DRA systems summarized general fail by the nucleation, growth, and coalescence of voids. A possible approach to the prediction of fracture toughness of materials failing by the microvoid coalescence type of fracture is to make assumptions about conditions ahead of the crack tip. Rice and Johnson [266] considered the true strain ahead of a blunt crack and assumed that metals containing a small volume fraction of void nucleating particles (e.g., inclusions) would fracture at a strain between 0.2 and 1. According to their model, fracture occurs when the high strain region attains a size comparable with the microstructural dimension characteristic of the fracture process. With regard to DRA materials, Crowe et al. [247] and Kamat et al. [267] utilized the approach similar to that proposed by Rice and Johnson [253], assuming that such an argument could be extended to materials containing a high volume fraction of reinforcement particles. The region of plastic flow is limited to a volume of width δ, a value which corresponds to the interparticle spacing, λ. If such a relation were to hold, the following equation applies:

$$\delta = \alpha J_{\text{IC}}/\sigma_{\text{f}} = \lambda \tag{10.8}$$

It should be emphasized that this model was developed for metals containing a small volume fraction of void nucleating particles. In addition, the model assumes a near uniform distribution of the equiaxed reinforcement and microscopic cracks associated with each particle. Although this is not generally observed in DRA, experimental work continues on determining the effects of various microstructural features on toughness and appropriate measures of the interparticle spacing. In Figure 10.46 is plotted the value of δ versus the interparticle spacing (λ), for all of the data presented in Figure 10.38 and Figure 10.39. The interparticle spacing was determined using the relationship [254,255]

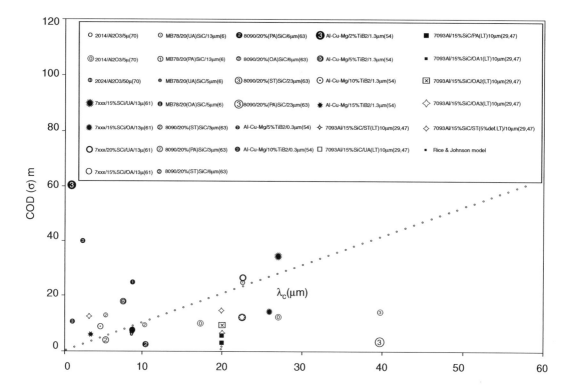

FIGURE 10.46 Plot of COD (δ) versus calculated center-to-centre interparticle spacing, λ_{c}.

$$\lambda = 0.77 N_S^{1/2} \tag{10.9}$$

In Figure 10.46, this represents the center-to-center spacing of the reinforcing particles. Included in Figure 10.46 is a line corresponding to the Rice and Johnson analysis where the critical condition is assumed to occur when $\delta = \lambda$, although the exact relationship according to the analysis will depend on the work hardening rate.

The data summarized in Figure 10.46 reveals a general trend of increasing δ with increasing interparticle spacing (λ). However, the exact relationship between δ and λ is not clear at this time. The uncertainty relates to the nature of the reinforcement spatial distribution and reinforcement size distribution, which does affect the calculated λ. In addition, other particles (e.g., precipitates, grain boundary regions) tend to affect both the location of damage sites and the types of damage mechanisms, which occurs between the reinforcement particles and thereby reduces the value of δ and λ. This emphasizes the need for continued studies on the nature of damage, which evolves ahead of a crack tip in such systems, as well as the effects of reinforcement clustering and matrix microstructure on the evolution of damage. Such observations also question the extension of the Rice and Johnson model to systems having such high volume fractions of reinforcement where the stress fields around each particle tend to interact, as the particles get sufficiently close.

It is clear that the flaw tolerance of both monolithic materials as well as DRA is of critical importance in design. A number of investigators have utilized the approach of Rice and Johnson [266] to rationalize the toughness of DRA. It has also been observed in experimental fracture toughness tests on such materials that the fracture often propagates catastrophically. Thus, the issue of crack growth resistance of such materials also becomes relevant. The data presented for fracture toughness generally characterizes the crack initiation resistance of such materials, it is equally important to characterize their crack growth resistance. Of the various methods proposed to evaluate crack growth toughness of materials, the tearing modulus concept introduced by Paris et al. [268] has found wide acceptance. The tearing modulus parameter is a measure of the material's resistance to tearing and is an indication of the stability of crack growth. It is defined as

$$T_{r} = \frac{E}{\sigma_0^2} \left(\frac{dJ}{da} \right) \tag{10.10}$$

In the expression E is the Young's modulus, δ_0 is the flow stress, and dJ/da is the slope of the J-R curve. The tearing modulus (T_r), proportional as it is to dJ/da, is a measure of the strain energy that must be provided to the crack tip to enable it to advance by a unit crack length. Few investigations have measured both the crack initiation and crack growth resistance of both the monolithic and DRA materials. The available data on 6XXX and 7XXX aluminum alloys as well as model Al-Si alloy containing Si particles is summarized in Figure 10.47. The data suggests a large effect of matrix alloy composition and heat treatment on the tearing modulus of the materials systems chosen. It was found that addition of reinforcements significantly reduces the crack growth resistance, while aging the matrix to the overaged condition in the 7XXX alloy further reduces the crack growth resistance. The effects of matrix strength and ductility on crack growth resistance are shown in the model Al-Si materials where aging to the T6 condition eliminates any tearing resistance compared with the T4 condition [269]. Few other recent studies have shown that the crack growth resistance is affected by changes in specimen thickness [219,254,260].

10.8.2.4.8 Toughened MMCs

Increasing the strength/toughness combinations of DRA materials may be possible through improvements in both primary and secondary processing routes coupled with a judicious selection of the matrix, reinforcement, and heat treatment. While improvements in the spatial distribution of the

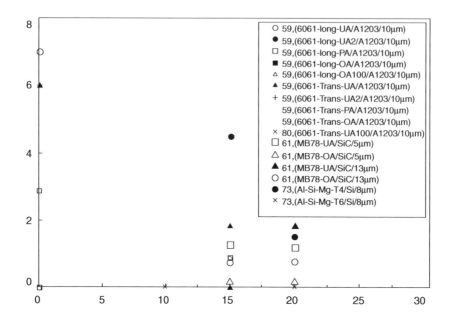

FIGURE 10.47 Tearing modulus data for various DRA. (From Lewandowski, J. J. and Singh, P. M., *ASM Metals Handbook*, Volume 19-*Fracture and Fatigue,* ASM International, Materials Park, Ohio, pp. 895–904, 1966; Osman, T. M., Singh, P. M., and Lewandowski, J. J., *Scripta Metallurgica et Materialia*, 607–612, 1994.)

reinforcement results in an improvement in mechanical properties, these approaches have focused on either preventing or minimizing the evolution of damage in DRA systems. More significant enhancements to properties can be realized via other extrinsic toughening approaches, which are designed to contain any evolving damage.

Several recent studies [191,193,261,270–272,273–293] have convincingly demonstrated that extrinsic toughening approaches may offer more dramatic increases in the fracture-critical properties than those obtained by manipulation of the microstructural variables. Such extrinsic approaches inherently accept that damage to the DRA will evolve during mechanical loading and are designed to contain progression of the damage. In such cases, it is beneficial to create inhomogeneous structures of various size and length scales where the clustered reinforcement is surrounded by ductile/tough ligaments of either the matrix material or another material. Although a limited amount of research of this nature has been conducted in comparison to the work outlined above, preliminary results on a number of DRA systems have shown a significant increase in the fracture-critical properties while concurrently retaining the other benefits, such as, enhanced stiffness, of the composite.

Inhomogeneous DRA structures have been created using a number of different processing routes. These have produced structures, which are either 2-D or 3-D in nature, and have been critically reviewed in two recent publications [191,294]. Examples of 2-D structures include various layered and/or laminated DRA, which have been produced by the techniques of (a) spray deposition [283,289], (b) roll bonding [291], (c) extrusion [270–272,273], and (d) vacuum hot pressing [274–276]. The 3-D structures include (i) microstructurally toughened (MT) [277–279] DRA and (ii) ductile phase toughened DRA [285]. Examples of such 2-D and 3-D structures are shown in Figure 10.48 and Figure 10.49.

A recent study on the effects of lamination on the toughness of DRA/Aluminum alloy laminates has shown that dramatic improvements in both the static and impact toughness can be obtained using such approaches [261,270–276]. The two most commonly used orientations for such laminates, denoted as the crack arrestor and crack divider, are shown in Figure 10.48. A careful selection

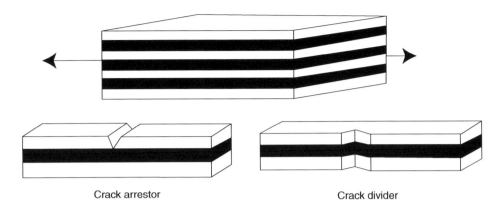

Crack arrestor   Crack divider

FIGURE 10.48 Crack arrester and divider orientations. (From Osman, T. M., Singh, P. M. and Lewandowski, J. J., *Scripta Metallurgica et Materialia*, 607–612, 1994.)

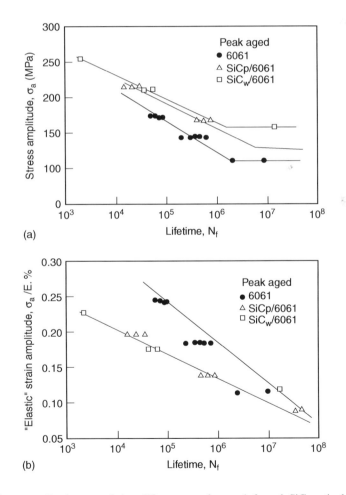

FIGURE 10.49 Stress amplitude versus fatigue life response for unreinforced, SiC particulate reinforced, and SiC whisker reinforced aluminum alloy 6061 in the peak-aged condition. (From Hassen, D. F., Crowe, C. R., Ahearn, J. S., and Cooke, D. C., In *Failure Mechanisms in High Performance Materials*, Early, J. G., Robert Shives, T., and Smith, J. H., Eds., p. 187, 1984.) (a) Variation of stress amplitude with fatigue life and (b) Variation of elastic strain amplitude with fatigue life.

of lamina and interface characteristics has revealed that such laminated systems exhibit a combination of properties approaching those of the constituents [191,290,294]. A variant of the laminate approach uses microstructurally toughened (MT) regions, which have been shown to impart incredible improvements in toughness while simultaneously retaining the beneficial properties of the constituents [277–279]. In order to compare the properties of such toughened structures to conventional DRA, it is necessary to compare the properties at an equivalent stiffness, which is controlled by the global volume fraction of reinforcement.

10.8.2.5 Fatigue Behavior

Fatigue is a phenomenon of mechanical property degradation leading to failure of a material or a component under cyclic loading. Understanding the fatigue behavior of a metal matrix composite is of vital importance because without such an understanding it would be virtually impossible to gain acceptance of the design, aerospace, mechanical, and structural engineers. Many high volume applications of the DRMMCs involve cyclic loading, such as components used in an automobile. An understanding of the fatigue behavior of DRMMCs has lagged behind the other important and relevant aspects, such as elastic stiffness and strength. The major difficulty in this regard is an application of conventional approaches to studying and understanding the fatigue behavior of MMCs. Although improved processing techniques are still required to overcome the poor ductility and fracture toughness properties displayed by many of these composites [295,296], studies on the initiation and subcritical crack growth of incipient cracks in both the particulate reinforced and whisker-reinforced aluminum alloys have shown noticeably improved endurance strengths and high cycle fatigue (HCF) resistance coupled with a potential for superior fatigue crack growth (FCG) properties, compared with the unreinforced counterpart [297–306].

10.8.2.5.1 Stress Amplitude and Strain Amplitude Controlled Fatigue Behavior

The stress (S) versus fatigue life (N_f) behavior for several powder metallurgy processed particulate reinforced and whisker-reinforced 6061 aluminum alloy in various environments has been studied by Hassen and co-workers [299]. Their results, shown in Figure 10.49, as variation in stress amplitude (σ_{amp}) as a function of number of cycles-to-failure (N_f) in the room temperature laboratory air environment reveals that the two forms of reinforcement (particulate and whisker) have a similar effect. The composites being superior in high cycle fatigue resistance (HCF) but offer no particular advantage over the unreinforced counterpart in low cycle fatigue (LCF). The superior HCF properties were attributed to result from the higher stiffness of the composites, as differences between the MMC and the unreinforced alloy are diminished by replotting the test data in terms of elastic strain amplitude ($\sigma_{amplitude}/E$).

In a study of cast saffil reinforced Al–Mg–Si and Al-Si composites under cyclic strain control, Hurd and co-workers [298] found that in the binary Al–Si alloys, which are weakened by the addition of saffil fibers, the low cycle fatigue properties of the composite were inferior to the unreinforced alloy. However, for the ternary Al–Mg–Si alloy, which was strengthened by the addition of fiber reinforcement, the LCF properties were comparable (Figure 10.50). Microscopically, the initiation cracks during low cycle fatigue were found to occur much later in the fatigue life of many composites compared with their constituent matrices [307]. It was observed that the SiC_W–matrix (AA 2124) interface was a preferential site for crack initiation in the $2124/SiC_W$ composites, and subsequent small-crack growth was generally inhibited [307]. Such beneficial properties were not found in the cast saffil reinforced Al–Si alloys since the weakly bonded saffil fibers readily promote interfacial crack initiation resulting in a large fraction of cracks being present throughout the life [307].

Srivatsan and co-workers [308–310] evaluated the influence of carbide particle content on both the high cycle fatigue and low cycle fatigue behavior of aluminum alloy 6061. Increasing the volume fraction of SiC_p in the aluminum alloy (6061) metal matrix resulted in higher fatigue strength (taken

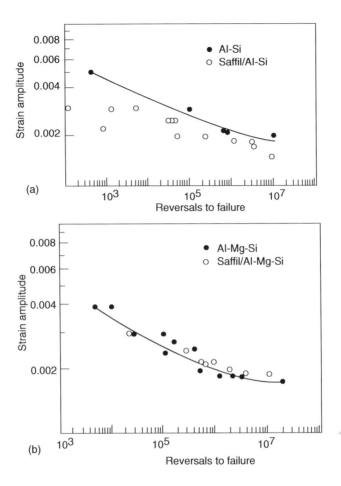

FIGURE 10.50 Strain amplitude versus fatigue live curves for cast unreinforced and saffil fiber reinforced aluminum alloy composites (a) Al-Si alloy and (b) Al-Mg-Si alloy. (From Hurd, N. J., *Materials Science and Technology*, 4, 513, 1988.)

as the highest stress at which the specimen survived one million cycles). In particular, the highest volume fraction (15%: 6061/SiC/15p-T6) composite showed noticeable improvement in fatigue strength compared to the unreinforced metal matrix (6061-T6) (Figure 10.32) [309]. The high strength of the 6061/SiC/xxp-T6 MMCs compared with the unreinforced 6061-T6 metal matrix was responsible for the superior high cycle fatigue resistance of the composites. Further, for a given volume fraction of the SiC particulate reinforcement phase in the 6061 aluminum alloy metal matrix, the cyclic fracture morphology was observed to be essentially similar at the different cyclic stress amplitudes. Macroscopically, the fracture was reminiscent of brittle failure, that is: low cyclic ductility, but microscopically the fracture surfaces revealed features reminiscent of locally ductile (voids and dimples) and brittle (microscopic and macroscopic cracks) mechanisms (Figure 10.51).

Under total strain amplitude control, at equivalent plastic strain amplitudes an increase in reinforcement content resulted in degradation in cyclic plasticity and resultant low-cycle fatigue life [310]. The degradation in cyclic fatigue life was far more noticeable at the lower cyclic strain amplitudes (Figure 10.52). Cyclic response of the discontinuous particulate reinforced 6061/SiC composites revealed a combination of initial hardening followed by gradual softening to failure at

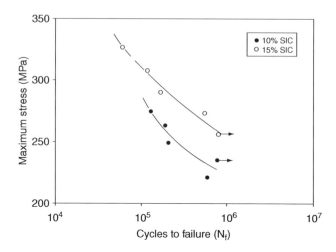

FIGURE 10.51 Influence of reinforcement volume fraction on maximum stress versus fatigue life response of 6061/SiC/xxp-T6 composites. (From Srivatsan, T. S., Meslet Al-Hajri, Petraroli, M. Hotton, B., and Lam, P. C., *Materials Science Engineering*, A325, 202–214, 2002.)

all cyclic strain amplitudes, and at both volume fractions of the particulate reinforcement (10 and 15 vol.%) of the particulate reinforcement in the aluminum alloy metal matrix. These researchers observed that for a given volume fraction of the particulate reinforcement in the metal matrix the degree of cyclic hardening was greater at the higher cyclic strain amplitudes and resultant higher response stress (Figure 10.53). The mechanisms responsible for the rapid softening prior to catastrophic failure were rationalized to be a combination of

1. separation or decohesion at the matrix-particulate interfaces,
2. formation of a number of microscopic cracks through cracked particulates,
3. the growth and coalescence of microscopic cracks to form macroscopic cracks, and

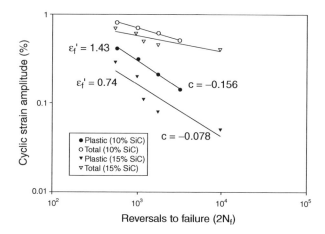

FIGURE 10.52 Influence of reinforcement volume fraction on plastic strain versus fatigue life response of aluminum alloy 6061 reinforced with SiC particulates. (From Srivatsan, T. S., Lam, P. C., Hotton, B., and Meslet, Al Hajri, *Applied Composite Materials* 9, 131–153, 2002.)

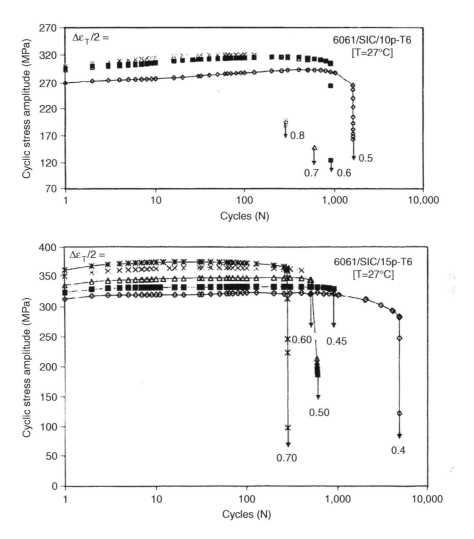

FIGURE 10.53 Cyclic stress response curves for aluminum alloy 6061-T6 reinforced with SiC particulates and deformed at room temperature (27°C), showing the variation of cyclic stress amplitude with cycles (a) 10 vol.% SiC_p and (b) 15 vol.% SiC_p. (From Srivatsan, T. S., Lam, P. C., Hotton, B., and Meslet Al Hajri, *Applied Composite Materials* 9, 131–153, 2002.)

4. concurrent growth of both the microscopic and macroscopic cracks through the composite microstructure.

The composite with a higher volume fraction of the SiC reinforcement exhibited higher cyclic strength and stability when compared to the lower volume fraction counterpart. This was rationalized to be due to reinforcement influences on composite microstructure and strengthening of the metal matrix [310].

In a study on understanding the influence of volume fraction of Al_2O_3 particulate reinforcement on cyclic strain resistance of aluminum alloy 2014 at two elevated temperatures of 100°C and 180°C [311,312] it was observed that the $2014/Al_2O_3$ composites exhibited a linear trend for the variation of elastic strain amplitude with reversals-to-failure and plastic strain amplitude with reversals-to-failure (Figure 10.54 and Figure 10.55). At equivalent plastic strain amplitudes, for a given composite, an

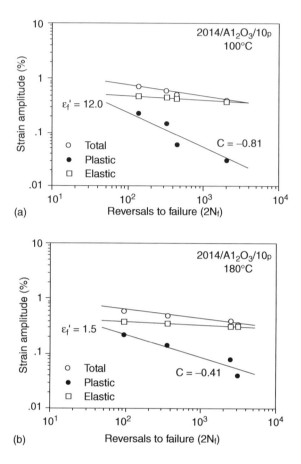

FIGURE 10.54 Cyclic strain amplitude versus fatigue life response for 2014/Al₂O₃/10p-T6 composite at (a) 100°C and (b) 180°C. (From Srivatsan, T. S. and Annigeri, R., *Metallurgical Materials Transactions*, 31A, 959–974, 2000.)

increase in test temperature from 100 to 180°C, enhanced cyclic plasticity and improved the low cycle fatigue (LCF) life. The improvement in cyclic strain resistance and resultant fatigue life was more noticeable at the lower cyclic strain amplitudes (Figure 10.56) [312]. Cyclic stress response of the 2014/Al₂O₃ composites revealed softening to failure from the onset of fully reversed cyclic deformation at all cyclic strain amplitudes, and at the two test temperatures (Figure 10.57 and Figure 10.58) [311]. The softening, or decrease in cyclic stress with cycles, was more pronounced at the higher test temperature (Figure 10.57b and Figure 10.58b). At both test temperatures, the researchers found the degree of softening to be higher at the higher cyclic strain amplitudes and resultant higher response stress [311].

A study was conducted to understand the influence of aging condition (peak aged versus underaged) on stress amplitude controlled high cyclic fatigue resistance of aluminum alloy 7034 reinforced with silicon carbide particulates (SiC$_p$) [313–315]. Because of the limited ductility of the composite at ambient temperature (28°C), the tests were conducted at an elevated temperature (120°C) corresponding to the aging temperature of the alloy. It was observed by the researchers that at equivalent values of maximum stress the degradation in high cycle fatigue life was in the range 200–300%. The degradation in fatigue life was more pronounced for the underaged (UA) microstructure and the peak aged (PA) microstructure (Figure 10.59). Also, for a given aging condition, increasing the load ratio from −1.0 to 0.1 resulted in higher fatigue strength (taken as the highest

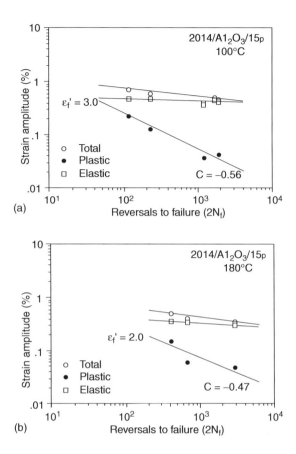

FIGURE 10.55 Cyclic strain amplitude versus fatigue life response for 2014/Al$_2$O$_3$/15p-T6 composite at (a) 100°C and (b) 180°C. (From Srivatsan, T. S. and Annigeri, R., *Metallurgical Materials Transactions*, 31A, 959–974, 2000.)

stress at which the composite specimen survived one-million cycles) [313]. At a given load ratio, variation of maximum stress with fatigue life revealed minimal to no influence of aging condition (i.e., underaged versus peak aged) for the 7034/SiC/15p composite (Figure 10.60). This observation was rationalized to be due to the constant nature of yield strength and ultimate tensile strength of the two microstructures [313].

Cyclic stress response of the ceramic particle (SiC$_p$) reinforced 7034 aluminum alloy metal matrix revealed rapid initial hardening in the first few cycles of cyclic straining followed by stability for a fraction of the fatigue life and culminating in rapid softening to failure at the high cyclic strain amplitudes. At the lower cyclic strain amplitudes, the initial hardening during straining was low, followed by stabilization for most of the fatigue life and culminating in gradual softening to failure (Figure 10.61). The possible mechanisms contributing to hardening of the composite matrix during fully reversed cyclic straining were ascribed to the conjoint and interactive influences of (a) an increase in dislocation density during fully reversed cyclic straining, (b) mutual interaction of dislocations with other dislocations (Figure 10.62), (c) dislocation multiplication, and (d) dislocation-silicon carbide particulate interaction (Figure 10.63). Growth and coalescence of the microscopic cracks to form one or more macroscopic cracks causes the load carrying, and thus the stress-carrying capability of the structure to decrease and the material gradually softens to failure. Essentially, the 7034/SiC composites followed the Basquin and Coffin-Manson relationships for strain-reversals to

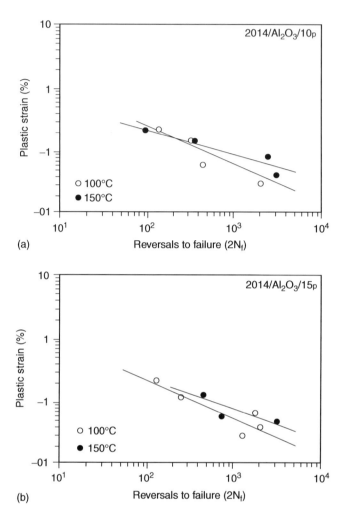

FIGURE 10.56 A comparison of the influence of test temperature on plastic strain versus fatigue life response of the composites (a) 2014/Al$_2$O$_3$/10p-T6 and (b) 2014/Al$_2$O$_3$/15p-T6. (From Srivatsan, T. S. In *Proceedings of the International Conference and Exhibition on Advances in Production and Processing of Aluminum* [APPA 2001] Suliaman, S., Ed., Bahrain, 2001.)

fatigue life response. The cyclic ductility of the composite microstructure was found to be noticeably inferior to the true monotonic fracture ductility [314].

10.8.2.5.2 Fatigue Crack Growth Behavior

The fatigue crack growth rate test data, shown in Figure 10.64, for particulate reinforced composites often displays several distinct regimes of behavior, each characterized by a definite crack-extension mechanism, as shown in Figure 10.65. At near threshold levels (below 10^{-9} m/cycle) the crack follows a path that tends to avoid the reinforcement particles. These result in a superior crack growth resistance in many composites compared with the constituent matrices, due primarily to crack closure [303] and crack trapping mechanisms [305]. The magnitude of the fatigue threshold was found to be dependent upon mean particle size, an observation, which is shown to be consistent with the crack trapping mechanism [305]. At the higher growth rates, generally between 10^{-9} m/cycle and 10^{-6} m/cycle,

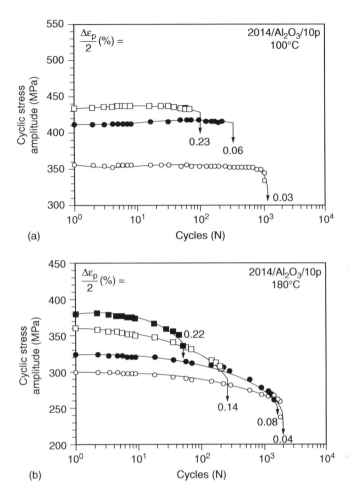

FIGURE 10.57 Cyclic stress response curves for the 2014/Al$_2$O$_3$/10p-T6 composite at (a) 100°C, and (b) 180°C. (From Srivatsan, T. S., In *Proceedings of the International Conference and Exhibition on Advances in Production and Processing of Aluminum* [APPA 2001] Suliaman, S., Ed., Bahrain, 2001.)

thecrack growth resistance is somewhat improved in the composite alloys. In this regime, the occurrence of limited fracture of the reinforcing particles located ahead of the main crack tip leads to the development of un-cracked ligaments, which act as crack bridges (Figure 10.66) [304].

Quite similar to the monolithic alloys, the near threshold fatigue crack growth in metal matrix composites shows a strong dependence on load ratio, principally due to the development of significant levels of crack closure, i.e., premature contact of the crack surfaces during the loading cycle [316–318]. At very low growth rates, or ultra-low growth rates, the dominant mechanism impeding crack advance is the promotion of crack deflection around the reinforcing particulates to induce crack closure levels from asperity wedging and crack trapping by the reinforcing particles [303–305].

Although at low R values (e.g., less than 0.3), reinforcement additions have generally been observed to produce higher fatigue thresholds, a steeper Paris law regime, and a transition to unstable fracture at lower ΔK levels than their monolithic matrices [191,319]. The latter behavior is characteristic of all DRA and is a direct consequence of the lower fracture toughness of the composites in comparison to their monolithic matrices. In the higher ΔK portions of the Paris law regime it has been

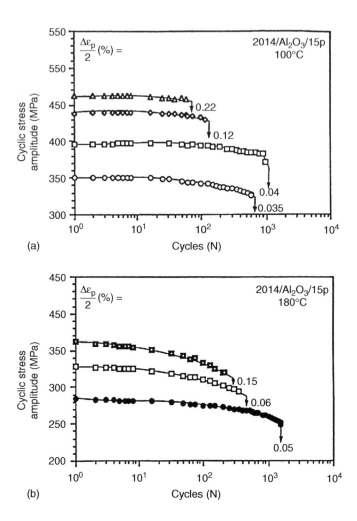

FIGURE 10.58 Cyclic stress response curves for the 2014/Al$_2$O$_3$/15p-T6 composite at (a) 100°C and (b) 180°C. (From Srivatsan, T. S., In *Proceedings of the International Conference and Exhibition on Advances in Production and Processing of Aluminum* [APPA 2001] Suliaman, S., Ed., Bahrain, 2001.)

observed that significant levels of reinforcement cracking may occur, and this may also contribute to the higher crack growth rates in this region [319–322].

Of the numerous studies that have been conducted to understand the key aspects relevant to the fatigue behavior of DRMMCs, the essential finding is that whether their primary use be aerospace or automotive applications, the resistance offered by the metal matrix composites to failure under cyclic or otherwise varying load remains paramount. The addition of particulate or whisker reinforcements to a metal matrix does result in low ductility and fracture toughness properties. Overall, the fatigue properties of metal matrix composites are comparable and in many cases exceed that of their unreinforced matrices.

10.9　APPLICATIONS

Of the numerous applications now in actual use, only a few key ones are highlighted.

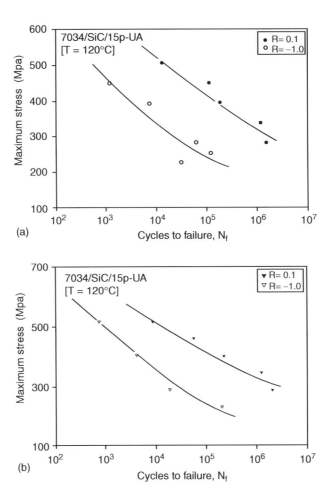

FIGURE 10.59 Variation of maximum stress with cyclic fatigue life for the 7034/SiC/15p composite at $R =$ 0.1 and $R = -1.0$. (a) Under-aged (UA) and (b) peak-aged (PA). (From Srivatsan, T. S. and Meslet Al Hajri, *Composites Engineering*, Part B, 33. (5), 391–404, 2002.)

10.9.1 COMPOSITES FOR SPACE APPLICATIONS

The silicon carbide particulate reinforced aluminum alloy based MMCs have been used in the U.S. Air Force F-16 aircraft. An application being the ventral fin for the aircraft, which provided lateral stability during high angle of attach maneuvers. A 6092/SiC MMC sheet material replaced the unreinforced aluminum skins in the honeycomb structure of the fin. Due to the increased specific stiffness, erosion resistance, and high cycle fatigue behavior, of the DRMMC the service life increased significantly. The same material was also used for the fuel access door covers of the F-16. The increased specific stiffness property of the DRMMC allowed for its substitution for the unreinforced aluminum alloy.

The rotor sleeve in a Eurocopter is made from a forged SiC particulate reinforced aluminum alloy 2009 having good stiffness, high cycle fatigue resistance, and damage tolerance. The aluminum MMC replaced titanium, reducing both weight and production cost [28].

Graphite fiber and whisker reinforced metal matrix composites represent the next generation of high stiffness, low thermal expansion materials for applications on dimensionally critical spacecraft structures. These structures require high specific stiffness. One such structure is the wrap rib concept

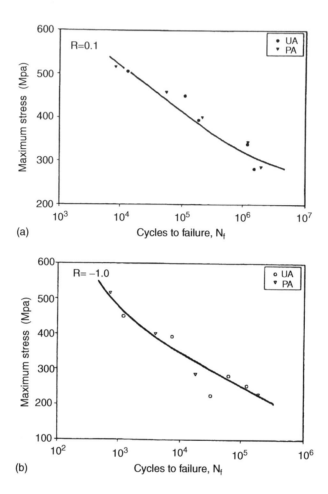

FIGURE 10.60 Influence of aging condition (under-aged versus peak aged) on high cycle fatigue response for (a) $R = 0.10$ and (b) $R = -1.0$. (From Srivatsan, T. S. and Meslet Al Hajri, *Composites Eng*, Part B, 33 (5), 391–404, 2002.)

space antenna. MMCs offer many advantages over existing resin matrix composites, including higher electrical and thermal conductivity, better radiation resistance, coupled with the need for no outgassing. Combinations of commercial high-strength aluminum alloy matrices and post fabrication processes developed at the National Aeronautics and Space Administration (Langley Research Center) have resulted in MMCs that do not exhibit residual thermal strain or strain hysteresis during thermal cycling. This enabled the MMCs to be excellent candidate materials for dimensionally critical space structures.

10.9.2 Composites for Automotive Applications

Properties of interest to the automotive engineer include specific stiffness, wear resistance, and cyclic fatigue resistance. While weight savings are equally important in automotive applications, the need for achieving performance improvements with much lower cost premiums than tolerated by aerospace applications drives the attention toward low-cost materials and processes. Several MMCs of the discontinuous type have been introduced only when the combination of properties and cost satisfied a

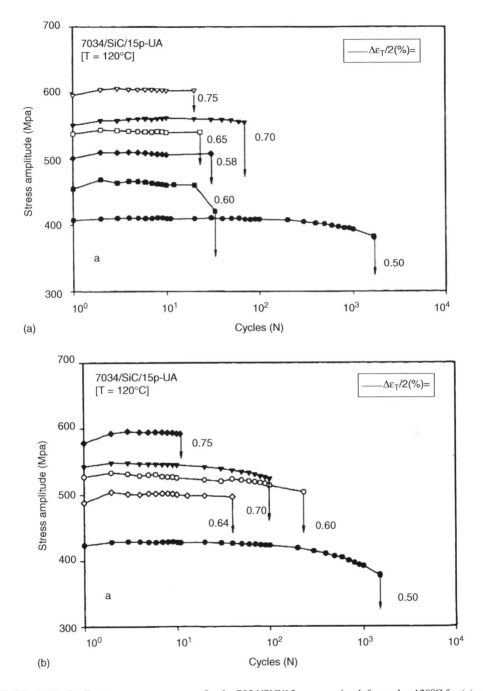

FIGURE 10.61 Cyclic stress response curves for the 7034/SiC/15p composite deformed at 120°C for (a) under aged (UA) and (b) peak aged (PA). (From Srivatsan, T. S., Meslet Al Hajri, Hannon, W., and Vasudevan, V. K., *Materials Science and Engineering*, 79, 181–196, 2004.)

particular need. Replacement of the steel and cast iron in engine applications relies on a combination of increased specific stiffness, improved wear resistance, and increased cyclic fatigue resistance provided by the discontinuously reinforced MMCs. An early application of the aluminum MMC was the Toyota piston for diesel engines. Reinforcing an aluminum alloy with chopped fibers

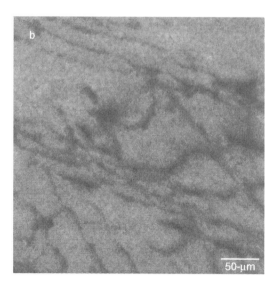

FIGURE 10.62 Bright field transmission electron micrograph showing distribution and interaction of dislocations in cyclically deformed specimen of 7034/SiC/15p PA. (From Meslet Al-Hajri, PhD, The University of Akron, Ohio, 2004.)

provided improved wear and thermal fatigue resistance. These pistons are now being commercially produced in Japan. Honda Motor Company is also using MMCs it its engines for the Prelude model. In this case, the hybrid preforms consisting of carbon and alumina fibers are infiltrated by molten aluminum to form the cylinder liners during medium pressure squeeze casting of the engine block.

Aluminum alloy based MMCs offer a useful combination of properties for brake system applications and is a viable alternative to traditionally used cast iron. The high wear resistance and high thermal conductivity of aluminum MMCs enables substitution in disk brake rotors and brake drums

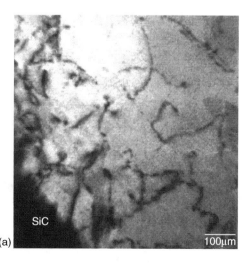
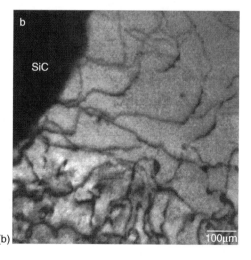

FIGURE 10.63 Bright field transmission electron micrograph showing interaction of dislocations with the reinforcing SiC particulate in cyclically deformed specimen of 7034/SiC/15p-UA. (From Meslet Al-Hajri, PhD, The University of Akron, Ohio, 2004.)

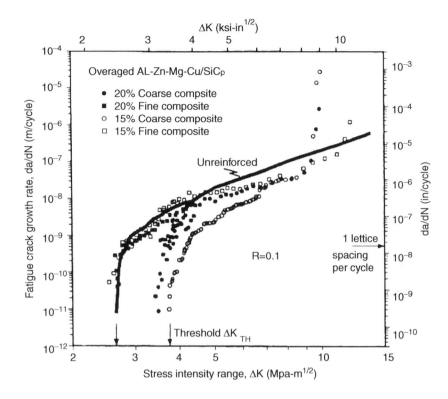

FIGURE 10.64 Variation of fatigue crack growth rates (da/dN) with stress intensity range for overaged SiC reinforced Al–Zn–Mg–Cu alloy at room temperature, showing behavior at load ratio of 0.1. (From Shang, J. K. and Ritchie, R. O., In *Metal Matrix Composites: Mechanisms and Propertie*, Everett, R. K. and Arsenault, R. J., Eds., Academic Press, New York, p. 255, 1991.)

with an attendant weight savings of about 50–60%. MMC rotors have been found to reduce brake noise and wear, besides ensuring uniform friction over the entire testing and use sequence when compared with cast iron rotors.

Aluminum MMCs, when used for drive shafts take advantage of the increased specific stiffness of these materials. Currently used drive shafts are constrained by the speed at which the shaft becomes dynamically unstable. The critical speed of a drive shaft is a function of its length, radius and specific stiffness. In vehicles with space and packaging constraints that do not allow for an increased drive shaft diameter, the MMCs offer a promising solution. The aluminum alloy-based MMCs enable longer drive shafts for a given diameter, or a smaller-diameter shaft of given length. This has enabled the manufacture of drive shafts from $6061/Al_2O_3$ by conventional die-casting with subsequent extrusion into tube.

10.9.3 ELECTRONIC PACKAGING

Packaging material used in Microwave Integrated Multifunction Assemblies (MIMAs) must exhibit good thermal conductivity for efficient heat removal from power dissipating systems. The design often requires lightweight materials for purposes of reduced package density in aerospace applications. High flexural and shear strengths are also desired for reduced cross section of the structural elements. Aluminum alloy metal matrix composites, consisting of an aluminum alloy reinforced with 40–55 vol.% silicon carbide particulates, have been developed and used in electronic and microwave packaging applications. The electronic grade composite material has density similar

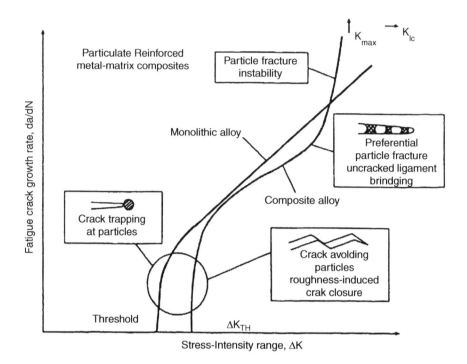

FIGURE 10.65 A schematic illustration showing a comparison between the variation of fatigue crack propagation with stress intensity for unreinforced and particulate reinforced metal matrix composites. The salient micro-mechanisms associated with crack growth in each regime are shown. (From Shang, J. K. and Ritchie, R. O., In *Metal Matrix Composites: Mechanisms and Properties*, Everett, R. K. and Arsenault, R. J., Eds., Academic Press, New York, 1991, p. 255.)

to aluminum, while retaining good thermal conductivity and in some cases electrical conductivity along with superior flexural and shear strength properties. Thus, these components can be considered as high value added. Increased application of Al/SiC microprocessor lids in packaging has been driven by the need for a lightweight and potentially lower cost replacement to pure copper. The coefficient of thermal expansion of the DRAMMC can be tailored to match the adjoining packaging material by the careful control of the reinforcement volume fraction. In recent years, printed wiring board cores are being made from Al–SiC material, as a viable replacement for the conventional copper and aluminum core. Besides the compatible CTE property, the higher specific stiffness reduces thermal cycling and vibration-induced fatigue.

10.9.4 COMMERCIAL PRODUCTS

A few interesting applications have been prototyped or commercially produced in the general sector of commercial and industrially oriented products. Aluminum reinforced with boron carbide (B_4C) has shown great promise for nuclear shielding applications because the isotope present in boron carbide can naturally absorb neutron radiation. As a result, this type of MMC is being considered for storage casks that contain spent fuel rods from nuclear reactors. Aluminum alloy based MMCs have made much headway as an industrial material for applications requiring a lower weight with improved precision in performance. This led to the development and emergence of a six-meter long MMC needle as a replacement for the conventionally used steel needle in carpet weaving applications.

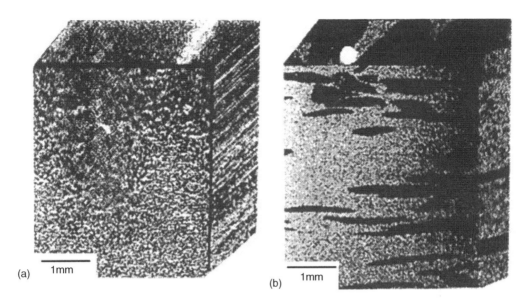

FIGURE 10.66 The ductile phase toughened discontinuously reinforced aluminum composite. (From Lewandowski, J. J., Liu, C., and Hunt, W. H. Jr., In *Processes and Properties for Powder Metallurgy Composites*, Kumar, P., Vedula, K. and Ritter, A., Eds., The Metallurgical Society, Warrendale, PA, pp. 117–131,1988.)

ACKNOWLEDGMENTS

Dr. T. S. Srivatsan acknowledges the Aluminum Company of America, DuralCan U.S.A. and U.S. Air Force Materials Laboratory (Wright Patterson Air Force Base, Dayton, Ohio, USA) and Ohio Aerospace Institute (Cleveland, Ohio) for providing partial support of some of the research presented in this paper. Most sincere thanks, appreciation and gratitude is extended to Mr. Satish Vasudevan for his understanding and assistance with the figures. Dr. J.J. Lewandowski acknowledges D. Padhi and P.M. Singh for assistance with data gathering and plotting while ARO-DAALO3-89-K-0068, NSF-PYI-DMR-8958326, ONR-N00014-91-J-1370, ONR-N00014-99-1-0327, ALCOA, ALCAN, and Ohio Aerospace Institute are acknowledged for partial support of some of the work reviewed presently, as well as supply of materials and interaction on various aspects.

REFERENCES

1. Koss, D. A. and Copley, S. M., *Metallurgical Transactions*, 2A, 1557–1560, 1971.
2. Brown, L. M. and Stobbs, W. M., *Philosophical Magazine*, 23, 1185–1189, 1971.
3. Divecha A. P. Crowe C. R., Fishman S. G. *Failure Modes in Composites IV*, Metallurgical Society of AIME, Warrendale, PA, pp. 406–411, 1977.
4. Divecha, A. P., Fishman, S. G., and Karmarkar, S. G., *Journal of Metals*, 33, 12–16, 1981.
5. Nair, S. V., Tien, J. K., and Bates, R. C., *International Metals Reviews*, 30(6), 285–296, 1985.
6. McDanels, D. L., *Metallurgical Transactions*, 16A, 1105–1115, 1985.
7. Dermarkar, S., *Metals and Materials*, 2, 144–147, 1986.
8. Chawla, K. K., *Composite Materials*, Springer Verlag, New York, USA, 1987.
9. Srivatsan, T. S. and Sudarshan, T. S., *Rapid Solidification Technology: An Engineering Guide*, Technomic Publishing Inc., Lancaster, PA, pp. 603–700, 1993.
10. Hunt, W. H. Jr., Cook, C. R., and Sawtell, R. R., Cost effective high performance powder metallurgy aluminum matrix composites for automotive applications, *SAE Technical Paper Series 91-0834*, Warrendale, PA, February 1991.

11. Hunt, W. H. Jr., Cost effective high performance aluminum matrix composites for aerospace applications, *International Conference on PM Aerospace Materials*, Lausanne, Switzerland, November 1991.
12. Clyne, T. W. and Withers, P. G., *An Introduction to Metal Matrix Composites Cambridge Solid State Science Series*, Cambridge University Press, Cambridge, 1993.
13. Kreider, K. G., *Composite Materials*, Vol. 4, Academy Press, New York, p. 1, 1974.
14. Kelley, P., *Composites*, 10, 2, 1979.
15. Srivatsan, T. S., Sudarshan, T. S., and Lavernia, E. J., Progress in materials science, *An International Review Journal*, 39(4/5), 317–409, 1995.
16. Piggot, M. R., *Load Bearing Fiber Composites*, Pergamon Press, New York, 1980.
17. Broutman, L. J. and Krock, R. H., *Modern Composite Materials*, Addison Wesley Publishing Company, New York, 1969.
18. Metal Matrix Composites Overview: Paper Number MMCIAC No. 253, Santa Barbara California, U.S.A.
19. Schoutens, J. E., Introduction to Metal Matrix Composite Materials, MMCIAC Tutorial Series No. 272, 1982.
20. Fishman S. G., *A Metal Matrix Composite Requirement: More Reliable Mechanical Properties Data*, MMCIAC, Current Highlights, Vol. 1, May 1981.
21. Cutler, B., *American Ceramic Society Bulletin*, 52, 425, 1973.
22. Taya, M. and Arsenault, R. J., Eds., *Metal Matrix Composites: Thermomechanical Behavior* New York, 1989, Pergamon Press, 1989.
23. Brenner, S. S., *Journal of Applied Physics*, 33, 33, 1962.
24. Brenner, S. S., *Journal of Metals*, 14(11), 808, 1962.
25. Sutton, W. H., *Journal of Aeronautics and Astronautics*, August 46, 1966.
26. Sutton, W. H. and Chorn, J., *Metals Engineering Quarterly*, 3(1), 44, 1963.
27. Divecha, A. P., Lare, P., and Hahn, H., *Silicon Carbides Whisker Metal Matrix Composites*, Technical Report, Air Force Materials Laboratory, AFML-TR-69-7, May 1969.
28. Harrigan, W. C., In *Metal Matrix Composites*, Everett, R. K. and Arsenault, R. J., Eds., Academic Press, London, pp. 1–15, 1991.
29. Arsenault, R. J., *Materials Science and Engineering*, 64, 171, 1984.
30. McDaniels, D. L., *Metallurgical Transactions*, 16A, 1105, 1985.
31. Crow, C. R., Gray, R. A., and Hasson, D. F., In *Proceedings of ICCM V*, Harrigan, W. C, Jr., Strife, J., and Dhingra, A. K., Eds., AIME, Wavendale, PA, p. 843, 1985.
32. Lawrence, K., English: Reinforcements: Part II, *Materials Engineering*, September, 25–27, 1987.
33. Lawrence, K., English: matrix: Part III, *Materials Engineering*, September, 33–36, 1987.
34. Arsenault, R. J., In *Composite Structures*, Marshall, I. H., Ed., Elsevier Applied Science, London, 1987.
35. Mortensen, A., Cornie, J. A., and Flemings, M. C., *Journal of Metals*, 12, 1988.
36. Fishman, S. G., *Role of Interfaces on Material Damping*, ASM International, Material Park, Ohio, USA, pp. 35–41, 1985.
37. Hasson, D. F. and Crowe, C. R., *Strength of Metals and Alloys, ICSMA VII*, Oxford Press, United Kingdom, pp. 1515–1521, 1985.
38. Mandell, J. F., Kong, K. C., and Grande, D. H., *Ceramic Engineering Science Proceedings*, 7(8), 937–940, 1987.
39. Liaw, P. K., Greggi, H. G., and Logsdon, W. A., *Journal of Materials Science Letters*, 22(5), 1613–1617, 1987.
40. Niskanen, P. and Mohn, W. R., *Advanced Materials and Processes*, 133(3), 39–42, 1988.
41. Grande, D. H., Mandell, J. F., and Hong, K. C., *Journal of Materials Science*, 23, 311–320, 1988.
42. Zedalis, M. S., Bryan, J. D., Gilman, P. S., and Das, S. K., *Journal of Metals*, 29–32, 1991.
43. Willis, T. C., *Metals and Materials*, August, 485–489, 1988.
44. East, W. R., *Materials Engineering*, March, 33–36, 1988.
45. DeMeis, R., *Aerospace America*, March, 26–29, 1989.
46. Ibrahim, I. A., Mohamed, F. A., and Lavernia, E. J., *Journal of Materials Science*, 26, 1137, 1991.
47. Srivatsan, T. S., Ibrahim, I. A., Mohamed, F. A., and Lavernia, E. J., *Journal of Materials Science*, 27, 1991.

48. Grant, N. J., In *Rapid Solidification Processing: Principles and Technologies*, Mehrabian, Kear, B.H. and Cohen, M., Eds., Claitors Publications, Baton Rouge, LA, pp. 230–245, 1978.

49. Lawley, A., Atomization of specialty alloy powders, *Journal of Metals*, 33, 13–20, 1981.

50. Lawley, A., Modern powder metallurgy science and technology, *Journal of Metals*, 38, 15, 1986.

51. Suryanarayana, C., Froes, F. H., and Rowe, R. G., *International Materials Reviews*, 36(3), 85, 1991.

52. Galbraith, J. M., Tosten, M. H., and Howell, P. R., *Journal of Material Science*, 22, 27, 1987.

53. Eylon, D., Cooke, C. M., and Froes, F. H., In *Titanium: Rapid Solidification Technology*, Froes, F. H. and Eylon, D., Eds., AIME, Warrendale, PA, p. 311, 1986.

54. Cooke, C. M., Eylon, D., and Froes, F. H., *Proceedings of the Conference on Titanium Science: Technology and Applications* by Les Editions de Physique, France, June 1988.

55. Spencer, D. B., Mehrabian, R., and Flemings, M. C., *Metallurgical Transactions*, 3A, 1925, 1972.

56. Mehrabian, R., Riek, R. G., and Flemings, M. C., *Metallurgical Transactions*, 5A, 1899, 1974.

57. Jolly, P. S. and Mehrabian, R., *Journal of Materials Science*, 121, 1393, 1976.

58. Jung Jen Allen Cheng D., Apelian, and Doherty, R. D., *Metallurgical Transactions*, 17A, 2049, 1986.

59. Vogel, A., Doherty, R. D., and Cantor, B., In *Solidification and Casting of Metals*, Hunt, J., Ed. University of Sheffield, pp. 518–525, 1977.

60. Doherty, R. D., Lee, Ho-In, and Feest, E. A., *Materials Science and Engineering*, 65, 181, 1984.

61. Flemings, M. C., *Metallurgical Transactions*, 221, 957, 1991.

62. Rohatgi, P. K., Asthana, R., and Das, S., *International Metals Reviews*, 31, 115–139, 1986.

63. Rack, H. J., Baruch, T. R., and Cook, J. L., In *Proceedings of Science and Engineering of Composites*, Hasyashi, T., Kawata, K., and Umekawa, S., Eds., Japan Society for Ceramic Materials, Japan, p. 1465, 1982.

64. Rack, H. J. and Niskanen, P., *Light Metals Age*, February, 9, 1984.

65. Rack, H. J. and Ratnaparkhi, P., *Journal of Metals*, November, 55–57, 1988.

66. Chawla, K. K., *Journal of Metals*, December, 25, 1985.

67. Geiger, A. L. and Walker, J. A., *Journal of Metals*, August, 8–28, 1991.

68. Rack, H. J., *Advanced Materials and Manufacturing Processes*, 3(3), 327, 1988.

69. Rack, H. J., In *Processing and Properties of Powder Metallurgy Composites*, Vedula, K., Kumar, P., and Ritter, A., Eds., The Minerals, Metals and Materials Society, Warrendale, PA, USA, p. 155, 1987.

70. Hunt, W. H. Jr., Cook, C. R., Armanie, K. P., and Gareganus, T. B., In *Powder Metallurgy Composites*, Kumar, P., Ritter, A., and Vedula, K., Eds., Metallurgical Society of AIME, Warrendale, PA, USA, 1987.

71. Hunt, W. H., Richmond, O., and Young, R. D., In *Sixth International Conference on Composite Materials*, Mathew, F. L., Buskell, N. C. R., Hodgkinson, J. M., and Moron, J., Eds., Elsevier Applied Science Publishers, London, p. 2209, 1987.

72. Cebulak W. S., *Advanced Metallic Structures: A Review*, Air Force Wright Aeronautical Laboratories, Technical Report, TR-87-3042, October 1986.

73. Gupta, M., Mohamed, F. A., and Lavernia, E. J., In *Proceedings of the International Symposium on Advances in Processing and Characterization of Ceramic Matrix Composites, Vol. 17, CIM/ICM*, Mostaghci, H., Ed., Pergamon Press, NY, USA, p. 236, 1989.

74. Gupta, M., Mohamed, F. A., and Lavernia, E. J., *Materials and Manufacturing Processes*, 5(2), 165, 1990.

75. Ibrahmin, I. A., Mohamed, F. A., and Lavernia, E. J., In *Internation Conference on Advanced Aluminum and Magnesium Alloys*, Kahn, T. and Effenberg, G., Eds., ASM International, Amsterdam, p. 745, 1990.

76. Bricknell, R. H., *Metallurgical Transactions*, 17A, 583, 1986.

77. Fiedler, H. C., Sawyer, T. D., Koop, R. W., and Leatham, A. G., *Journal of Metals*, 39(8), 28, 1987.

78. Willis, T. C., *Metals Materials*, 4, 485, 1988.

79. Lavernia, E. J., *International Journal of Rapid Solidification*, 5, 47, 1989.

80. Singer, A. R. E. and Ozbek, S., *Powder Metallurgy*, 28(2), 72, 1985.

81. Grant, P. S., Kim, W. T., Bewlay, B. P., and Cantor, B., *Scripta Metallurgica*, 23, 1651, 1989.

82. Gutierrez, E., Lavernia, E. J., Trapaga, G., Szekely, J., and Grant, N. J., *Metallurgical Transactions*, 20A, 71, 1989.

83. Srivatsan, T. S. and Lavernia, E. J., In *Processing and Fabrication of Advanced Materials for High Temperature Applications*, Ravi, V. A. and Srivatsan, T. S., Eds., The Minerals, Metals and Materials Society, Warrendale, PA, p. 141, 1993.

84. Gupta, M., Mohamed, F. A., Lavernia, E. J., and Srivatsan, T. S., *Journal of Materials Science*, 28, 2245, 1993.

85. Zeng, X., Nutt, S. R., and Lavernia, E. J., In *Processing and Fabrication of Advanced Materials III*, Ravi, V. A., Srivatsan, T. S., and Moore, J. J., Eds., The Minerals, Metals and Materials Society, Warrendale, PA, 1993.

86. Yue, Wu and Lavernia, E. J., In *Processing and Fabrication of Advanced Materials III*, Ravi, V. A., Srivatsan, T. S., and Moore, J. J., Eds., The Minerals, Metals and Materials Society, Warrendale, PA, 1993.

87. Tiwari, R., Herman, H., and Sampath, S., *Spray Forming of MoSi2 and MoSi2-Based Composites*, Presented at Materials Research Society Meeting, Boston, MA, 1990.

88. Sampath, S., Tiwari, R., Gudmundsson, B., and Herman, H., *Scripta Metallurgica*, 25, 1425, 1991.

89. Tiwari, R., Herman, H., Sampath, S., and Gudmundsson, B., In *Innovative Inorganic Composites*, Herman, H., Ed., The Minerals, Metals and Materials Society, Warrendale, PA, 1991.

90. Buhrmaster, C. L., Clark, D. E., and Smart, B. B., *Journal of Metals*, November, 44, 1988.

91. Parker, D. W. and Kutner, G. L., *Advanced Materials and Processes*, April, 68, 1991.

92. Busby, P. O. and Nikitich, J., *Advanced Materials and Processes*, December 35, 1991.

93. Singer, A. R. E., *Annals CIRP*, 32, 145, 1983.

94. Buhrmaster, C. L., Clark, D. E., and Smart, H. O., *Journal of Metals*, 40, 44, 1988.

95. Willis, J., *Metals and Materials*, 4, 485, 1988.

96. White, J., Palmer, I. G., Hughes, I. R., and Court, S. A., In *Aluminum-Lithium Alloys V*, Sanders, T. H. and Starke, E. A., Eds., MCE Publications, Birmingham, UK, p. 1635, 1989.

97. Gupta, M., Mohamed, F. A., and Lavernia, E. J., In *Proceedings of 17th International Symposium on Advances in Processing and Characterization of Ceramic Matrix Composites*, Mostaphaci, H., Ed., Pergamon Press, NY, U.S.A., p. 236, 1989.

98. Kojima, K. A., Lewis, R. E., and Kaufman, M. J., In *Aluminum-Lithium Alloys V*, Sanders, T. H. and Starke, E. A., Eds., MCE Publications, Birmingham, UK, p. 85, 1989.

99. Gupta, M., Mohamed, F. A., and Lavernia, E. J., In *Proceedings 17th International Symposium On Advances in Processing and Characterization of Ceramic Metal Matrix Composites*, Mostaphaci, H., Ed., Pergamon Press, Oxford, p. 236, 1989.

100. Gupta, M., Mohamed, F. A., and Lavernia, E. J., *Materials and Manufacturing Processes*, 5, 165, 1990.

101. Ibrahim, I. A., Mohamed, F. A., and Lavernia, E. J., *Journal of Material Science*, 26, 1137, 1991.

102. Wu, Y. and Lavernia, E. J., *Journal of Metals*, 43(8), 16, 1991.

103. Wu, Y., *Master of Science Thesis*, University of California at Irvine, Irvine, CA, USA, 1994.

104. Zhang, J., Perez, R. J., Gupta, M., and Lavernia, E. J., *Scripta Metallurgica Materialia*, 28, 91, 1993.

105. Wu, M., Srivatsan, T. S., Pickens, J. R., and Lavernia, E. J., *Scripta Metallurgica Materialia*, 27, 761, 1992.

106. Gupta, M., Srivatsan, T. S., Mohamed, F. A., and Lavernia, E. J., *Journal of Material Science*, 28, 2245, 1993.

107. Kim, N. J., Park, W. J., Ahn, S., and Elias, L., In *High Performance Metal and Ceramic Composites*, Upadhya, K., Ed., TMS, Warrendale, PA, p. 137, 1994.

108. Gupta, M., Bowo, K., Lavernia, E. J., and Earthman, J. C., *Scripta Metallurgica Materialia*, 28, 1053, 1993.

109. Gupta, M., Ibrahim, I. A., Mohamed, F. A., and Lavernia, E. J., *Journal of Material Science*, 26, 6673, 1991.

110. Gupta, M., Mohamed, F. A., and Lavernia, E. J., *International Journal of Rapid Solidification*, 6, 247, 1991.

111. Gupta, M., Mohamed, F. A., and Lavernia, E. J., *Metallurgical Transactions A*, 23A, 831, 1992.

112. Perez, R. J., Zhang, J., and Lavernia, E. J., *Scripta Metallurgica Materialia*, 27, 1111, 1992.

113. Zeng, X. and Lavernia, E. J., *International Journal of Rapid Solidification*, 7, 219, 1992.

114. Wu, Y. and Lavernia, E. J., *Metallurgical Transactions A*, 32A, 2923, 1992.

115. Wu, Y. and Lavernia, E. J., In *Processing and Fabrication of Advanced Materials III*, Ravi, V. A., Srivatsan, T. S., and Moore, J. J., Eds., TMS, Warrendale, PA, p. 501, 1994.

116. Srivatsan, T. S. and Lavernia, E. J., *Composites Engineering*, 4(4), 459, 1994.
117. Gupta, M., Lane, C., and Lavernia, E. J., *Scripta Metallurgica Materialia*, 26, 825, 1992.
118. Gupta, M., Juarez-Islas, J., Frazier, W. E., Mohamed, F. A., and Lavernia, E. J., *Metallurgical Transactions B*, 23B, 719, 1992.
119. Armanie, K. P., and Zaidi, M. A., Metallographic examination of osprey MMC sample, Technical Report, Alcoa Technical Center, Pittsburgh, PA, p.1, 1987.
120. Maher, P. P., Cantor, B., and Katgerman, L., In *International Conference of Advanced Aluminum and Magnesium Alloys*, Khan, T. and Effenberg, G., Eds., ASM International, Materials Park, OH, p. 659, 1990.
121. Mathur, P., Kim, M. H., Lawley, A., and Apelian, D., *Powder Metallurgy: Key to Advanced Materials Technology*, ASM International, Materials Park, OH, p. 55, 1990.
122. Majagi, S. I., Ranganathan, K., Lawley, A., and Apelian, D., In *Microstructural Design by Solidification Processing*, Lavernia, E. J. and Gungor, M. N., Eds., TMS, Warrendale, PA, p. 139, 1992.
123. Perez, J. F. and Morris, D. G., *Scripta Metallurgica Materialia*, 31, 231, 1994.
124. Ranganathan, K., Lawley, A., and Apelian, D., In-Situ *Spray Casting of Dispersion Strengthened Alloys-II: Experimental Studies*, Paper No. 14E-T6-6, 1993 Powder Metallurgy World Congress, Kyoto, Japan Powder Metallurgy Association, Japan, 12–15, July 1993.
125. Zhao, Y. Y., Grant, P. S., and Cantor, B., *Journal of Microscopy*, 169(2), 263, 1993.
126. Zhao, Y. Y., Grant, P. S., and Cantor, B., *Journal of Physique*, 3, 1685, 1993.
127. Jeng, Y. L., Wolfenstine, J., and Lavernia, E. J., *Scripta Metallurgica Materialia*, 28, 453, 1993.
128. Jeng, Y. L., Lavernia, E. J., and Wolfenstine, J., *Scripta Metallurgica Materialia*, 29, 107, 1993.
129. Zeng, X., Liu, H., Chu, M. G., and Lavernia, E. J., *Metallurgical Transactions*, 23A, 3394, 1992.
130. Liang, X. and Lavernia, E. J., *Materials Science and Engineering*, A153, 654, 1992.
131. Lawley, A. and Apelian, D., *Powder Metallurgy*, 37, 123, 1994.
132. Wu, Y., Zhang, J., and Lavernia, E. J., *Metallurgical and Materials Transactions B*, 25B, 135, 1994.
133. Lawrynowicz, D. E. and Lavernia, E. J., *Scripta Metallurgica Materialia*, 31, 1277, 1994.
134. Ibrahim, I. A., Mohamed, F. A., and Lavernia, E. J., In *International Conference on Advanced Aluminum and Magnesium Alloys*, Khan, T. and Effenberg, G., Eds., ASM International, Materials Park, OH, p. 745, 1990.
135. Lavernia, E. J., *SAMPLE Quarterly*, 22, 2, 1991.
136. Wu, M., Hunt, W. H. Jr., Lewandowski, J. J., Zhang, J., and Lavernia, E. J., *Materials Science and Engineering*, 1995.
137. Apelian, D., Wei, D., and Farouk, B., *Metallurgical Transactions*, 20B, 251, 1989.
138. Chang, K. M., Taub, A. I., and Huang, S. C., *Materials Research Society Symposium Proceedings*, 39, 335, 1985.
139. Enami, K. and Nenno, S., *Transactions Japan Institute of Metals*, 19, 571, 1978.
140. Taub, A. I. and Jackson, A. R., *Materials Research Society Symposium Proceedings*, 58, 389, 1986.
141. Rozak, G. A., Lewandowski, J. J., Wallace, J. F., and Altmisoglu, A., Effects of casting conditions and deformation processing on A356 Aluminum and A356-20 vol% SiC composites, *Journal of Composite Materials*, 26(14), 2076–2106, 1992.
142. Ghosh, A. K., Solid state processing, In *Fundamentals of Metal Matrix Composites*, Suresh, S., Mortensen, A., and Needleman, A., Eds., Butterworth-Heinemann, Oxford, pp. 23–41, 1993.
143. Awadallah, A., Prabhu, N. S., and Lewandowski, J. J., Forging/forming simulation studies on a unique, high capacity deformation simulator apparatus, *Materials Manufacturing Processes*, 17(6), 737–764, 2002.
144. Hilinski, E. J., Lewandowski, J. J., Rodjom, T. J., and Wang, P.T., *Flow Behaviour and Stress Evolution Modelling for Discontinuously Reinforced Composites*, Vol. 7, 1994 World P/M Congress, Lall, C. Neupaver, A., Eds., Metal Powder Industries Federation, pp.119–131, 1994.
145. Hilinski, J., Lewandowski, J. J., Rodjom, T. J., and Wang, P. T., *Development of Densification Model for DRA Composites*, Vol.7, 1994 World P/M Congress, Lall, C. Neupaver, A., Eds., Metal Powder Industries Federation, pp. 83–93, 1994.
146. Grow, A. L., Lewandowski, J. J., *Effects of Reinforcement Size on Hydrostatic Extrusion on MMC's*, Paper No. 950260, *SAE Transactions*, 1995.
147. Lewandowski, J. J. and Lowhaphandu, P., Effects of hydrostatic pressure on mechanical behavior and deformation processing of metals, *International Materials Reviews*, 43(4), 145–188, 1998.

148. Xu, H. and Palmiere, E. J., Particulate refinement and redistribution during the axisymmetric compression of an Al/SiCp metal matrix composite, *Composites Part A, Applied Science Manufacturing*, 30, 203–211, 1999.

149. Ganguly, P., Poole, W. J., and Lloyd, D. J., Deformation and fracture characteristics of AA6061 particle reinforced metal matrix composite at elevated temperatures, *Scripta Materialia*, 44, 1099–1105, 2000.

150. San Marchi, C., Cao, F., Kouzeli, M., and Mortensen, A., Quasistatic and dynamic compression of aluminum oxide particle reinforced pure aluminum, *Materials Science and Engineering*, A337, 202–211, 2002.

151. Kouzeli, M. and Dunand, D., Effect of reinforcement connectivity on the elasto-plastic behavior of aluminum composites containing sub-micron aluminum particles, *Acta Material*, 51, 5121–6105, 2003.

152. Cavaliere, P., Cerri, E., and Evangelista, E., Isothermal forging modeling of $2618 + 20\%$ Al_2O_{3P} metal matrix composite, *Journal of Alloys and Compounds*, 378, 117–122, 2004.

153. Wang, G. S., Geng, L., Zheng, Z. Z., Wang, D. Z., and Yao, C. K., Investigation of compression of SiCw/6061 Al composites around the solidus of the matrix alloy, *Materials Chemistry and Physics*, 70, 164–167, 2001.

154. Badini, C., La Vecchia, G. M., Fino, P., and Valente, T., Forging of 2124/SiCp Composite: Preliminary studies of the effects on microstructure and strength, *Journal Materials Processing Technology*, 116, 289–297, 2001.

155. Radhakrishna Bhat, B. V., Mahajan, Y. R., Roshan, H. Md., and Prasad, Y. V. R. K., Processing map for hot working of powder 2124 Al-20 vol pct SiCp metal matrix composite, *Metallurgical Transactions A*, 23A, 2223–2230, 1992.

156. Syu, D. G. C. and Ghosh, A. K., Forging limits for an Aluminum matrix composite: Part 1. experimental results, *Metallurgical Transactions A*, 25A, 2027–2038, 1994.

157. Osman, T. M., Lewandowski, J. J., and Hunt, W. H., In *Microstructure-Property Relationships for an Al/SiC Composite with Different Deformation Histories, Fabrication of Particles Reinforced Metal Composites*, Masonnave, J. and Hamel, F. G., Eds., ASM International, Materials Park, Ohio, USA, pp. 209–216, 1990.

158. Hunt, W. H., New directions in Aluminum-based P/M materials for automotive applications, *International Journal of Powder Metallurgy*, 36(6), 51–60, 2000.

159. Zhao, N., and Nash, P., The Processing of 6061 Aluminum Alloy Reinforced SiC Composites by Cold Isostatic Pressing, Sintering and Sinter-Forging, *Proceedings of the Second International Powder Metallurgy Aluminum & Light Alloys for Automotive Applications Conference*, Metal Powder Industries Federation, pp.145–152, 2000.

160. Chawla, N., Williams, J. J., and Saha, R., Mechanical behavior and microstructure characterization of sinter-forged SiC particle reinforced aluminum matrix composites, *Journal of Light Metals*, 2, 215–227, 2002.

161. Awadallah, A., and Lewandowski, J. J., *ASM Metals Handbook*, Vol. 14A, Materials Park, Ohio, USA, pp. 440–445.

162. Lewandowski, J. J. and Awadallah, A., *ASM Metals Handbook: Metal Working: Bulk Forming*, Materials Park, Ohio, pp. 440–445, 2005.

163. Avitzur, B., *Metal Forming: Process and Analysis*, McGraw-Hill, NY, USA, 1968.

164. Pugh, H. Ll. D., In *The Mechanical Behavior of Materials under Pressure*, Pugh, H. Ll. D., Ed., Elsevier Publishing, Amsterdam, p. 391, 1970.

165. Patankar, S. N., Grow, A. L., Margevicius, R. W., and Lewandowski, J. J., Hydrostatic extrusion of 2014 and 6061 composites, In *Processing and Fabrication of Advanced Materials III*, Ravi, V. A. and Srivatsan, T. S., Eds., TMS, Warrendaler, PA, pp. 733–745, 1994.

166. Grow, A. L. and Lewandowski, J. J., SAE Trans, *Materials Science Technology*, 8, 611, 1992.

167. Patankar, S. N., Grow, A. L., and Margevicius, W., In *Processing and Fabrication of Advanced Materials III*, Ravi, V. A. and Srivatsan, T. S., Eds., TMS, Warrendale, PA, p. 733, 1994.

168. Wagener, H. W. and Wolf, J., Mater, J., *Processing Technology: First Asia-Pacific Conference on Materials Processing*, Vol. 37, p. 253, 1993.

169. Wagener, H. W. and Wolf, J., *Key Engineering Materials*, 104–107, 99, 1995.

170. Fuchs, F. J., *Engineering Solids under Pressure*, Pugh, H. Ll. D. Ed., Butterworth Publishers, London, p.145, 1970.
171. Pugh, H. Ll. D., *Journal of Mechanics and Engineering Society*, 6, 362, 1994.
172. Avitzur, B., Journal Eng. Ind. Transactions, ASME, Ser.B, 87, 487, 1965.
173. Lewandowski, J. J., Liu, D. S., and Liu, C., *Scripta Metalurgica*, 25, 21, 1991.
174. Bruggeman, D. A. G., *Annalen Physik*, 24, 636–676, 1935.
175. Niesel, W., *Annalen Physik*, 10(1), 336–348, 1952.
176. Fricke, H., *Physics Reviews*, 24, 575, 1924.
177. Turner, P. S., *Journal of Research NBS*, 37, 239, 1946.
178. Kerner, E. H., *Proceedings of the Physics Society*, 68B, 808, 1956.
179. Geiger, A. L. and Jackson, M., In *Technology Materials and Processes*, Shieh, W. T., Ed., ASM International, Metals Park, Ohio, p. 93, 1989.
180. Taya, M., Lulay, K., and Lloyd, D., *Acta Metallurgica*, 39(1), 73, 1991.
181. Kelly, P., *International Metallurgical Reviews*, 18, 31, 1973.
182. Kim, C., Lee, L., and Plichta, M., *Metallurgical Transactions*, 21A, 1837, 1990.
183. Lee, J., *Metallurgical Transactions*, 17A, 379, 1986.
184. Vogelsang, M., Arsenault, R., and Fisher, R., *Metallurgical Transactions*, 17A, 379, 1986.
185. Poviek, G., Needleman, A., and Nutt, S. R., *Materials Science and Engineering*, A132, 31, 1991.
186. McNelly, T., Edwards, B., and Sherby, O., *Acta Metallurgica*, 25, 117, 1977.
187. McDanels, D. L., *Metallurgical Transactions*, 16A, 1105–1115, 1985.
188. Arsenault, R. J., *Journal of Composite Technology Research, ASTM*, 10, 140, 1988.
189. Taya, M. and Arsenault, R. J., *Scripta Metallurgica*, 21, 349, 1987.
190. Singh, P. M. and Lewandowski, J. J., In *Intrinsic and Extrinsic Fracture Mechanisms in Inorganic Composites*, Lewandowski, J. J. and Hunt, W., Eds., TMS, Warrendale, PA, pp. 57–68, 1995.
191. Lewandowski, J. J., and Singh, P. M., *ASM Metals Handbook*, Volume 19-*Fracture and Fatigue*, ASM International, Materials Park, Ohio, pp. 895–904, 1966.
192. Lewandowski, J. J., Liu, C., and Hunt, W. H., In *Processes and Properties for Powder Metallurgy Composites*, Kumar, P., Vedula, K., and Ritter, A., Eds., The Metallurgical Society, Warrendale, PA, pp. 117–131, 1988.
193. Lewandowski, J. J., Liu, C., and Hunt, W. H., *Materials Science & Engineering*, A107, 142–155, 1989.
194. You, C. P., Thompson, A. W., and Bernstein, I. M., *Scripta Metallurgica*, 21, 181–187, 1987.
195. Hunt, W. H. Jr., Brockenbrough, J. R., and Magnusen, P. E., *Scripta Metallurgica et Materialia*, 25, 15–20, 1991.
196. Lewandowski, J. J. and Liu, C., Proc. Int'l Symp. On *Adv. Structural Materials*, In *Proc. Me. Soc. Of Canadian Inst. Mining and Metallurgy*, Wilkinson, D., Ed., Pergamon Press, Oxford, pp. 22–33, 1988.
197. Brechet, Y., Embury, J. D., Tao, S., and Luo, L., *Acta. Metallurgica*, 39, 1781–1786, 1991.
198. Mummery, P. and Derby, B., *Materials Science & Engineering*, A135, 221–224, 1991.
199. Arsenault, R. J., Si, N., Feng, C. R., and Wang, L., *Materials Science & Engineering*, A131, 55–68, 1991.
200. Lloyd, D. J., *Acta Metallurgica*, 39(1), 59–71, 1991.
201. Liu, D. S., Lewandowski, J. J., *Metallurgical Transactions A*, 24A, 609.
202. Singh, P. M. and Lewandowski, J. J., *Metallurgical Transactions A*, 24A, 2531–2543, 1993.
203. Llorca, J., Martin, A., Riuz, J., and Elices, M., *Metallurgical Transactions A*, 24A, 1575–1588, 1993.
204. Corbin, S. F. and Wilkinson, D. S., *Acta Metallurgica Materialia*, 42(4), 1329–1335, 1994.
205. Mochida, T., Taya, M., and Lloyd, D. J., *Materials Transactions JIM*, 32(10), 931–942, 1991.
206. Mummery, P. M., Anderson, P., Davis, G. R., Derby, B., and Elliot, J. L., *Scripta Metallurgica*, 29, 1457–1462, 1993.
207. Lewandowski, J. J., Kiu, D. S., and Liu, C., *Scripta Metallurgica*, 25, 21–26, 1991.
208. Hong, S. I. and Gray, G. T., *Acta. Metallurgica*, 38, 1581, 1990.
209. Vasudevan, A. K., Richmond, O., Zok, F., and Embury, J. D., *Materials Science & Engineering*, A107, 63–69, 1989.
210. Embury, J. D., Zok, F., Lahie, D. J., and Poole, W., In *Intrinsic and Extrinsic Fracture, Mechanisms in Inorganic Composite Systems*, Lewandowski, J. J. and Hunt, W. H., Eds., TMS-AIME, Warrendale, PA, pp. 1–7, 1995.
211. Ashby, M., *Philosophical Magazine*, 14, 1157, 1966.

212. Davidson, D. L., *Metallurgical Transactions*, 22A, 113, 1991.
213. Srivatsan, T. S., Al-Hajri, M., Hannon, W., and Vasudevan, V. K., *Materials Science and Engineering*, 379, 181–196, 2004.
214. Srivatsan, T. S., *Journal of Materials Science*, 31, 1375–1388, 1996.
215. Srivatsan, T. S., *International Journal of Fatigue*, 17(3), 183–199, 1995.
216. Zok, F., Embury, J. D., Ashby, M. F., and Richmond, O., In *9th Riso Int'l Symposium On Metallurgy and Materials Science*, Anderson, S. I., Lilholt, H., and Pederson, O. B., Eds., Riso National Lab, Roskilde, Denmark, pp. 517–527, 1987.
217. Hunt, W. H. Jr., In *Intrinsic and Extrinsic Fracture Mechanisms in Inorganic Composite Systems*, Lewandowski, J. J. and Hunt, W. H., Eds., TMS-AIME, Warrendale, PA, pp. 31–39, 1995.
218. Hunt, W. H. Jr., Richmond, O., and Young, R. D., *ICCM-7 Elsevier*, 2, 189–198, 1987.
219. Pandy, A. B., Majumdar, B. S., and Miracle, D. B. *Metallurgical & Materials Transactions A*, In press, 1999.
220. Thomason, P. F., *Ductile Fracture of Metal*, Pergamon Press, Oxford, 1990.
221. Manoharan, M. and Lewandowski, J. J., *International Journal of Fracture*, 40–2, $31–$34, 1989.
222. Somerday, B. P. and Gangloff, R. P., *Metallurgical & Materials Transactions A*, 25A, 1471–1479, 1994.
223. Lloyd D. J., and Morris, P. L., *Science & Engineering of Light Metals* (Japan Inst. Light Metals, Tokyo), 465–469.
224. Embury, J. D., Newell, J., and Tao, S., *Proceedings of the 12th Riso Symposium on Mats. Sci.*, Fiso Nat'l Lab, Rosklide, Denmark, p.317, 1991.
225. Liu, D. S., Manoharan, M., and Lewandowski, J. J., *Journal of Materials Science Letters*, 8, 1447, 1989.
226. Vasudevan, A. K., Richmond, O., Zok, F., and Embury, J. D., *Materials Science & Engineering*, A107, 63, 1989.
227. Grow, A. L., and Lewandowski, J. J., SAE Trans., Paper No. 950260, 1993.
228. Grow, A. L., M.S. Thesis, Case Western Reserve University, Cleveland, OH, 1994.
229. Friend, C. M., *Materials Science and Technology*, 5, 1–7, 1989.
230. Lawn, B. R. and Wilshaw, T. R.,*Fracture of Birttle Solids*, Cambridge University Press, Cambridge, UK, pp. 46–76, 1975.
231. Davidson, D. L., *Metallurgical Transactions A*, 18A, 2115–2128, 1987.
232. Hutchinson, J. W., Rice, J. R., and Rosengren, G. F., *Journal of Mechanics Physics of Solids*, 16, 1–31, 1968.
233. Hutchinson, J. W., *Acta Metallurgica*, 35, 1605, 1987.
234. Rice, J. R., *Fracture*, Vol. 2, Academic Press, London, p. 287, 1975.
235. Davidson, D. L., *Journal of Materials Science*, 24, 681–687, 1989.
236. Davidson, D. L., *Engineering Fracture Mechanics*, 33, 965–977, 1989.
237. ASTM Standard Test Method E-399-83. Plane Strain Fracture Toughness of Metallic Materials, *Annual Book of Standards*, Vol. 02.02, American Society for Testing and Materials, Philadelphia, PA, USA.
238. Crowe, C. R., Gray R. A. and Hasson, D. F., *Proceedings of Fifth International Conference on Composite Materials*, ICCM V, TMS, AIME, Warrendale, PA, p. 843, 1985.
239. Hutchinson, J. W., *Acta Metallurgica*, 35, 1605, 1987.
240. Ruse, L. R. F., *International Journal of Fracture*, 31, 233, 1986.
241. Kachanov, M., *International Journal of Fracture*, 30, R-65, 1986.
242. Davidson, D. L., In *Metal Matrix Composite: Mechanisms and Properties*, Everett, R. K. and Arsenault, R. J., Eds., Academic Press, London, p. 217, 1991.
243. Clyne, T. S., Ed, In *Metal Matrix Composites*, Elsevier Applied Science Publishers, NY, USA, 2000.
244. Knott, J. F., *Fundamentals of Fracture Mechanics*, Butterworthhs, London, UK, 1973.
245. Irwin, G. R., *Applied Materials Research*, 3, 65, 1964.
246. Lowhaphandu, P. and Lewandowski, J. J., *Scripta Metallurgica et Materialia*, 38(12), 1811–1817, 1998.
247. Crowe, C. R. and Hassen, D. F., In *Strength of Metals and Alloys, ICSMA*, Vol. 2.6, Giffkins, R. C., Ed., Pergamon Press, Oxford, UK, pp. 859–865, 1982.

248. Seleznev, M. L., Argon, A. S., Seleznev, I. L., Cornie, J. A., and Mason, R. P., SAE Trans. Paper No. 980700, 1998.
249. Hahn, G. T. and Rosenfield, A. R., *Metallurgical Transactions A*, 6A, 653, 1975.
250. Garrett, G. G. and Knott, J. F., *Metallurgical Transactions A*, 9A, 1187, 1975.
251. Aiken, R. M., Jr., NASA Contractor Report 365, May 1991.
252. Develay, R., *Metals and Materials*, 6, 404, 1972.
253. Klimowicz, T. F. and Vecchio, K. S., In *Fundamental Relationship between Mircostructure and Mechanical Properties in Metal Matrix Composites*, Gungor, M. and Liaw, P. K., Eds., TMS, Warrendale, PA, p. 255, 1990.
254. Osman, T. M. and Lewandowski, J. J., *Scripta Metallurgica et Materialia*, 31, 191–195, 1994.
255. Lewandowski, J. J., Metal matrix composites, In *Comprehensive Composite Materials*, Vol. 3, Kelly, A. and Zweben, C., Eds., Elsevier, Amsterdam, pp. 151–187, 2000.
256. Manoharan, M. and Lewandowski, J. J., Unpublished Research, Case Western Reserve University, Cleveland, OH, 1988.
257. Manoharan, M. and Lewandowski, J. J., *Acta Metallurgica Materialia*, 389, 489–496, 1990.
258. Doel, T. J. A., Loretto, M., and Bowen, P., *Composites*, 24, 270–276, 1993.
259. Downes, T. J. and King, J. E., *Composites*, 24, 276–284, 1993.
260. Pandey, A. B., Majumdar, B., and Miracle, D. B., *Metallurgical and Materials Transactions A*, 29A, 1237–1243, 1998.
261. Hunt, W. H., Osman, T. M., and Lewandowski, J. J., *Journal of Metals*, 45, 30–35, 1993.
262. Lloyd, D. J., In *Intrinsic and Extrinsic Fracture Mechanisms in Inorganic Composite Systems*, Lewandowski, J. J. and Hunt, W. H., Eds., TMS-AIME, Warrendale, PA, pp. 39–49, 1995.
263. Rozak, G. A., Altmisolgu, A. A., Lewandowski, J. J., and Wallace, J. F., *Journal of Composites Materials*, 26, 2076–2106, 1992.
264. Kamat, S. V., Hirth, J. P., and Mehrabian, R., *Scripta Metallurgica*, 23, 523–528, 1989.
265. Manoharan, M. and Lewandowski, J. J., *Journal of Composite Materials*, 25, 831–841, 1991.
266. Rice, J. R. and Johnson, M. A., In *Inelastic Behavior of Solids*, Kanninen, M. F., Lilholt, H., and Pederson, O. B., Eds., McGraw-Hill, NY, USA., p. 641, 1969.
267. Kamat, S. V., Hirth, J. P., and Mehrabian, R., *Acta Metallurgica*, 37, 2395, 1989.
268. Paris, P. C., Tada, H., Zahoor, A., and Ernst, H., ASTM-STP, 668, 5, 1979.
269. Manoharan, M., Lewandowski, J. J., and Hunt, W. H., *Materials Science & Engineering*, A172, 63–69, 1993.
270. Manoharan, M., Ellis, L. Y., and Lewandowski, J. J., *Scripta Metallurgica*, 24, 1515, 1990.
271. Ellis, L. Y. and Lewandowski, J. J., *Journal of Materials Science Letters*, 10, 461, 1991.
272. Ellis, L. Y. and Lewandowski, J. J., *Materials. Science & Engineering*, A183, 59, 1994.
273. Ellis, L. Y., M.S. Thesis, Case Western Reserve University, Cleveland, OH, USA, 1992
274. Syn, C. K., Lesuer, D. R., and Sherby, O. D., *High Performance Ceramic and Metal Matrix Composites*, TMS, Warrendale, PA, p. 125, 1994.
275. Syn, C. K., Leuser, D. R., Cadwell, K. L., Sherby, O. D., and Brown, K. R., In *Developments in Ceramic and Metal Composites*, Upadhya, K., Ed., TMS, Warrendale PA, p. 311, 1992.
276. Osman, T. M., Lewandowski, J. J., Leuser, D. R., Syn, C. K., and Hunt, W. H. Jr., In *Aluminum Alloys: Their Physical and Mechanical Properties (Proc. ICAA-4)*, Sanders, T. H. and Starke, E. A., Eds., Georgia Institute of Technology, GA, p. 706, 1994.
277. Nardone, V. C., Strife, J. R., and Prewo, K., *Metallurgical Transactions A*, 22A, 171, 1991.
278. Nardone, V. C. and Strife, J. R., *Metallurgical Transactions A*, 22A, 183, 1991.
279. Nardone, V. C., In *Intrinsic and Extrinsic Fracture Mechanisms in Particulate Reinforced Materials*, Lewandowski, J. J. and Hunt, W. H., Eds., TMS, Warrendale, PA, 1995.
280. Osman, T. M., Lewandowski, J. J., and Hunt, W. H. Jr., In *Advances in P/M and Particulate Materials—1994*, Lall, C. and Neupaver, A., Eds., MPIF, Princeton, NJ, p. 351, 1994.
281. Osman, T. M., Singh, P. M., and Lewandowski, J. J., *Scripta Metallurgica et Materialia*, 607–612, 1994.
282. Osman, T. M., Lewandowski, J. J., Lesuer, D. R., and Hunt, W. H., In *Micromechanics of Advanced Materials*, Chu, S. M. et al., Eds., TMS-AIME, Warrendale, PA, pp. 135–144, 1995.

283. Wu, M., Zhang, J. J., Lewandowski, J. J., Hunt, W. H., and Lavernia, E. J., In *Processing and Fabrication of Advanced Materials IV*, Srivatsan, T. S. and Moore, J. J., Eds., TMS-AIME, Warrendale, PA, pp. 155–165, 1996.

284. Osman, T. M., Lewandowski, J. J., Hunt, W. H., and Leuser, D. R., In *International Conference on Inorganic Matrix Composites*, Surappa, M. K., Ed., TMS-AIME, Warrendale, PA, pp. 155–165, 1996.

285. Ellis, L. Y., Lewandowski, J. J., and Hunt, W. H., Jr., In *Layered Materials for Structural Applications*, *MRS Proceedings*, Vol. 434, Lewandowski, J. J., Ward, C. H., Jackson, M. R., and Hunt, W. H. Jr., Eds., MRS, Pittsburgh, PA, pp. 213–219, 1996.

286. Osman, T. M. and Lewandowski, J. J., *Metallurgical Transactions A*, 27A, 3937–3947, 1996.

287. Leuser, D. R., Wadsworth, J., Riddle, C. K., Syn, J. J., Lewandowski, and Hunt, W.H., In *Layered Materials for Structural Applications, MRS Proceedings*, Vol. 434, Lewandowski, J. J., Ward, C. H., Jackson, M. R., and Hunt, W. H. Jr., Eds., MRS, Pittsburgh, PA, pp.205–213, 1996.

288. Osman, T. M. and Lewandowski, J. J., *Materials Science and Technology*, 1001–1006, 1996.

289. Wu, M., Zhang, J., Hunt, W. H., Lewandowski, J. J., and Laverina, E. J., *Journal of Materials Synthesis and Processing*, 4, 127–134, 1996.

290. Lesuer, D. R., Syn, C. K., Sherby, O. D., Wadsworth, J., Lewandowski, J. J., and Hunt, W. H., *International Materials Reviews*, 41, 169–197, 1996.

291. Zhang, J. J. and Lewandowski, J. J., *Journal Materials Science*, 32, 3851–3856, 1997.

292. Osman, T. M., Lewandowski, J. J., and Lesuer, D. R., *Materials Science & Engineering*, A229, 1–9, 1997.

293. Pandey, A. B., Majumdar, B., and Miracle, D. B., *Materials Science and Engineering.*, A259, 296–307, 1999.

294. Lewandowski, J. J. and Singh, P. M., In *Intrinsic and Extrinsic Fracture Mechanisms in Inorganic Composites*, Lewandowski, J. J. and Hunt, W. Jr., Eds., TMS, Warrendale, PA, pp. 129–147, 1995.

295. You, C. P., Thompson, A. W., and Bernstein, I. M., *Scripta Metallurgica*, 21, 181, 1987.

296. Mohn, W. R., *Research and Development*, 54, 1987.

297. Williams, D. R., and Fine, M. E., *Proceedings of the Fifth International Conference on Composite Materials*, Harrigan, W. C., Strife, J., and Dhingra, A. K., Eds., p. 639, 1985.

298. Hurd, N. J., *Materials Science and Technology*, 4, 513, 1988.

299. Hassen, D. F., Crowe, C. R., Ahearn, J. S. and Cooke, D. C., In *Failure Mechanisms in High Performance Materials* Early, J. G., Robert Shives, T., and Smith, J. H., Eds., p. 187, 1984.

300. Harris, S. J., *Materials Science and Technology*, 4, 231, 1988.

301. Logsdon, W. A. and Liaw, P. K., *Engineering Fracture Mechanics*, 24, 737, 1986.

302. Davidson, D. L., *Metallurgical Transactions*, 18A, 2115, 1987.

303. Shang, J. K., Yu, W., and Ritchie, R. O., *Materials Science and Engineering*, 102, 181, 1988.

304. Shang, J. K. and Ritchie, R. O., *Metallurgical Transactions*, 20A, 897, 1989.

305. Shang, J. K. and Ritchie, R. O., *Acta Metallurgica*, 37, 2267, 1989.

306. Christman, T. and Suresh, S., *Materials Science and Engineering*, 102, 211, 1988.

307. Shang, J. K. and Ritchie, R. O., In *Metal Matrix Composites: Mechanisms and Properties*, Everett, R. K. and Arsenault, R. J., Eds., Academic Press, New York, p. 255, 1991.

308. Srivatsan, T. S. and Lam, P. C., Aluminum Transactions, *An International Journal*, 2(1), 19–26, 2000.

309. Srivatsan, T. S., Meslet Al-Hajri, Petraroli, M., Hotton, B., and Lam, P. C., *Materials Science and Engineering*, A325, 202–214, 2002.

310. Srivatsan, T. S., Lam, P. C., Hotton, B., and Meslet Al Hajri, *Applied Composite Materials*, 9, 131–153, 2002.

311. Srivatsan, T. S. and Annigeri, R., *Metallurgical and Materials Transactions*, 31A, 959–974, 2000.

312. Srivatsan, T. S., In *Proceedings of the International Conference and Exhibition on Advances in Production and Processing of Aluminum* [APPA 2001], Suliaman, S., Ed., Bahrain, 2001.

313. Srivatsan, T. S. and Meslet Al Hajri, *Composites Engineering, Part B*, 33(5), 391–404, 2002.

314. Srivatsan, T. S., Meslet Al Hajri, Hannon, W., and Vasudevan, V. K., *Materials Science and Engineering*, 79, 181–196, 2004.

315. Meslet Al-Hajri, Doctor of Philosophy Thesis, The University of Akron, Akron, Ohio, USA, 2004.

316. Elber, W., *Engineering Fracture Mechanics*, 2, 37, 1970.

317. Suresh, S. and Ritchie, R. O., In *Fatigue Crack Growth Threshold Concepts*, Suresh, S. and Davidson, D. L., Eds., TMS-AIME, Warrendale, PA, USA, p. 227, 1984.

318. Hunt, W. H. Jr. and Herling, D. R., *Advanced Materials and Processes*, February, 39–41, 2004.
319. Allison, J. and Jones, J. W., In *Fundamental Mechanisms of Metal Matrix Composites*, Suresh, S., Mortensen, A., and Needleman, A., Eds., Butterworth-Heinemann, Oxford, pp. 269–297, 1993.
320. Lewandowski, J. J., *SAMPE Quarterly*, 20(12), 33–37, January 1989.
321. Shang, J. K. and Ritchie, R. O., *Metallurgical Transactions A*, 20A, 897–908, 1989.
322. Li, M., Ghosth, S., Richmond, O., Weiland, H., and Rouns, T. N., *Materials Science & Engineering*, A265, 1531–1537, 1999.

11 Titanium Alloys: Structure, Properties, and Applications

Fred McBagonluri
Siemens Hearing

W. O. Soboyejo
Department of Mechanical and Aerospace Engineering, Princeton Institute
for the Science and Technology of Materials (PRISM), Princeton University

CONTENTS

11.1   Introduction 360
11.2   Physical Metallurgy of Titanium Alloys 361
  11.2.1   Basic Properties and Engineering Applications 362
  11.2.2   Common Alloying Elements and Classifications 363
  11.2.3   Titanium Alloy/Intermetallic Phase Diagrams 363
11.3   Basic Mechanical Properties 365
  11.3.1   Creep Resistance 366
  11.3.2   Fracture Toughness 368
  11.3.3   Fatigue Behavior of $(\alpha+\beta)$ Titanium Alloys 368
    11.3.3.1   Effects of Microstructure and Micro-Texture 368
    11.3.3.2   Effects of Environment 370
    11.3.3.3   Effects of *R*-ratio 370
    11.3.3.4   Effects of Frequency 371
    11.3.3.5   Effects of Temperature 371
    11.3.3.6   Effects of Crystallography 372
    11.3.3.7   Effects of Crack Closure and Crack Deflection 373
    11.3.3.8   Mechanisms of Fatigue Crack Growth in Titanium Alloys 373
  11.3.4   Short Crack Growth in $(\alpha+\beta)$ Titanium Alloys 373
  11.3.5   Dwell Behavior of $(\alpha+\beta)$ Titanium Alloys 374
    11.3.5.1   Effects of Microstructural and Micro-Textural Influences 374
    11.3.5.2   Effects of Time Dependent Strain Accumulation 376
    11.3.5.3   Internal Hydrogen and Environmental Effects 378
    11.3.5.4   Effects of Crystallography 378
    11.3.5.5   Fatigue-Creep Synergism 379
    11.3.5.6   Effects of Temperature 380
    11.3.5.7   Effects of Stress Ratio 380
    11.3.5.8   Dwell Hold Time 381
  11.3.6   Proposed Mechanisms of Dwell Fatigue Crack Growth 381
    11.3.6.1   Evans and Gostelow 382
    11.3.6.2   Wojcik, Chan, and Koss...382

　　　　　11.3.6.3　Evans and Bache 382
　　　　　11.3.6.4　Hack and Leverant 383
　　11.3.7　Modeling of Fatigue and Dwell Fatigue...384
　　11.3.8　Titanium Matrix Composites 385
　　　　　11.3.8.1　Particulate-Reinforced TMCs 385
　　　　　11.3.8.2　Whisker-Reinforced TMCs 385
　　　　　11.3.8.3　Fiber-Reinforced TMCs 390
　　11.3.9　Biomedical Applications 391
　　11.3.10　Cylotoxicity in Biomedical Ti Alloys 393
　　11.3.11　Stress Shielding in Biomedical Titanium Alloys 395
　　11.3.12　Bone Detachment from Ti Implants by Osteolysis 395
11.4　Conclusions 395
References 396

11.1　INTRODUCTION

Titanium and its alloys have been used in numerous applications for more than forty years. The primary beneficiaries of titanium alloys applications have been the aerospace, biomedical, petrochemical, and jewelry industries. In each area of its application, the salient properties of titanium are tailored. For instance, in the aerospace industry resistance to crack initiation and growth, tailorability of microstructure, fatigue behavior under cryogenic conditions, and resilience to creep makes the use of titanium alloys ideal. A typical application of titanium alloys in aero-engines is shown in Figure 11.1 [1]. Figure 11.2 also shows the recreational use of titanium in high-end golf clubs [1]. In the biomedical industry, corrosion resistance, malleability, and biocompatibility are the desired properties of titanium alloys' use. In the petrochemical industries where corrosion of storage facilities may have dire environmental implications, the resistivity of titanium to corrosion makes it an ideal choice. Titanium research saw an increase in activity in the late 1960s and 1970s [2,3]. Preliminary work during this period focused on microstructural processing and phase transformations [4], the role of microstructure in fatigue crack growth behavior [5–10], and tailorability of microstructure to specific applications [4].

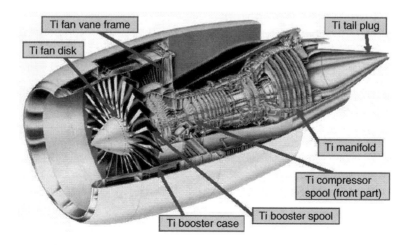

FIGURE 11.1 Titanium applications in aero-engines (Williams 2003).

FIGURE 11.2 Application of titanium in recreational golf clubs (Williams 2003).

Further interest in titanium research occurred in 1992, following a report of the Scientific Advisory Board of the United States Air Force. This report recommended that a unique consortium of government and engine manufacturers be set up to investigate uncontained failures in military aircrafts resulting from high cycle fatigue (HCF) [11]. The recommendations of the board emphasized the investigation of unpredictability in engine components' failures on military and aging aircraft engines and served as the basis for the Advanced High Cycle Fatigue Life Assurance Methodologies Program.

Major issues addressed in the HCF program included foreign object damage (FOD); fretting fatigue in Ti–6Al–4V, and mode mixity. In the civilian aircraft arena, where failures have been attributed to low cycle fatigue failures, significant titanium-based projects have addressed issues such as the roles of microstructures in the fatigue behavior of titanium alloys [5–20]. Other equally significant contributions have occurred in the following areas: crack closure and nonclosure mechanisms [21,22], short crack growth [23,24], and probabilistic fracture mechanics [25].

Dwell sensitivity in titanium alloys is another area of titanium alloy fatigue research that has attracted significant focus mechanisms [14,15,26–37]. This chapter begins with a brief review of the physical metallurgy of titanium alloys and an overview of the mechanical properties of titanium alloys with a special focus on fatigue and dwell fatigue. An overview of the structure, physical/mechanical properties, and applications of titanium alloys and composites follows. The chapter then examines some of the novel composites that are being developed for emerging and future applications. Finally, the biomedical applications of titanium are highlighted at the end of the chapter.

11.2 PHYSICAL METALLURGY OF TITANIUM ALLOYS

Recent advances in the development of aero-engines with increased specific thrust and efficiencies can be partly attributed to the use of titanium-based alloys in components such as fan blades, compressor blades, discs, hubs, and in numerous rotating parts [4]. Key advantages of titanium-based alloys harnessed in these applications include high strength/weight ratio, strength at moderate temperatures, and good resistance to creep and fatigue [3,4,38–40].

11.2.1 BASIC PROPERTIES AND ENGINEERING APPLICATIONS

Titanium is the fourth most abundant element in the earth's crust [4]. Titanium alloys have found widespread use in the biomedical industry as implants, in the aerospace industry for airframe and engine components, and in the petrochemical industry for storage [4]. It is also worth noting that super-conducting materials such as titanium–niobium alloys and titanium intermetallics are widely used as matrix in metallic matrix composite material [41–43].

These widespread uses of titanium alloys are due to their excellent strength/weight ratio at relatively high temperatures, biocompatibility, corrosion resistance, high fracture toughness, and crack growth retardation. Application of titanium in industry has been substantially limited, however, by machining challenges and limited reactivity with other elements, which results in protracted extraction processes [44,45].

In sheet form, titanium and its alloys are known to exhibit crystallographic textures [44]. These textures are pronounced in certain planes and directions and are intrinsically aligned to prior working directions. Material anisotropy resulting in differences in the physical and mechanical properties of titanium is the result of this texturing phenomenon. Notable among these affected properties are Young's modulus and Poisson's ratio. Texturing effects are reduced in $\alpha + \beta$ alloys more than in α titanium alloys [44]. Figure 11.3 shows a comparative plot of yield strength-density ratio as a function of temperature for titanium alloys and other industrial structural materials [4]. The superiority of titanium alloys over these traditional materials is apparent from this plot. A summary of the typical mechanical properties of titanium alloys is presented in Table 11.1.

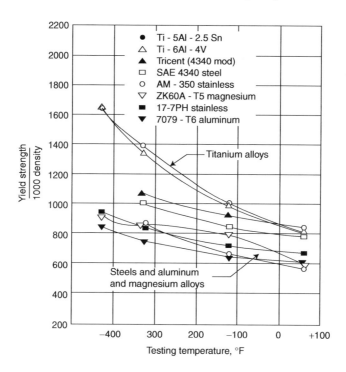

FIGURE 11.3 Yield strength–density ratio as a function of temperature for titanium alloys. (From Collings, E. W., *Physical Metallurgy of Titanium*, ASM, Materials Park, OH, 1994.)

TABLE 11.1
Examples of Ti Alloys in the Different Classes of Titanium Alloys

Alpha and Near-Alpha	Alpha + Beta	Beta, Near-Beta and Metastable
		Ti–13V–11Cr–3Al
Ti–5Al–2.5Sn (α)		Ti–15V–3Cr–3Sn–3Al
Ti–8Al–1Mo–1V (near-α)	Ti–6Al–4V	Ti–10V–2Fe–3Al
Ti–6Al–2Sn–4Zr–2Mo (near-α)	Ti–6Al–6V–2Sn	

Source: From Collings, E. W., *Physical Metallurgy of Titanium*, ASM, Materials Park, OH, 1994. With Permission.

11.2.2 COMMON ALLOYING ELEMENTS AND CLASSIFICATIONS

Titanium alloys phase chemistry and microstructure development is an allotropic transformation from a low temperature hexagonal close pack (HCP) alpha-phase to a high temperature body centered cubic (BCC) beta-phase. The lattice parameters in the BCC phase at 927°C are $a = b = 0.332$ nm. The BCC/beta phase has a density of 4.35 g/cm^3. In the lower temperature HCP phase, the lattice parameters are, $a = 0.295$ nm, $c = 0.468$ nm, and $c/a = 1.587$. The HCP phase has a room temperature density of 4.51 g/cm^3 [4].

The $\alpha + \beta$ transformation in pure titanium occurs at 883°C. Certain alloying elements, particularly transition metals such as Mo, Ta, Co, V, or Nb, stabilize the cubic beta-phase and decrease the alpha–beta transformation temperature. These transition metals are collectively referred to as beta stabilizers. Other metals, such as Al and interstitial solutes such as O, N, and C are called alpha-stabilizers. These elements increase the temperature of the alpha–beta transformation. Other elements such as Zr, Cu, Si, Au, and Sn have little effect on beta eutectoid transformation and are considered neutral. Titanium alloys are technically classified as alpha-alloys, beta-alloys, or alpha + beta alloys based on metallurgical differences [4,39].

Alpha (α) alloys are titanium alloys that typically contain aluminum and tin. Titanium forms an interstitial solid solution with these stabilizers. Al, in particular, is known to have a strong solution hardening effect on titanium but substantially reduces alloy ductility [4,40]. Beta (β) alloys contain one or more β stabilizers such as V, Nb, Mn, Co, Ta, Cr, or Cu. There are two principal stabilizers associated with the β alloys namely (a) isomorphous, e.g., V, Co, Ta, Rh, and Mn and (b) beta eutectoids e.g., Cu, Sn, Fe, Si, and Au [4,39].

In $\alpha + \beta$ alloys, both alpha and beta phases are present at room temperature. The alpha phase, in general, is similar to unalloyed titanium, but is strengthened by alpha stabilizers such as Al. The high temperature phase of titanium alloys is the beta phase, which is stabilized to room temperature by beta stabilizers such as vanadium, molybdenum, iron, or chromium [4]. Alpha–beta ($\alpha + \beta$) titanium alloys are also strengthened by heat treatment. With the exception of Ti–6Al–4V, they are not recommended in cryogenic applications. They have limited creep strength, especially in elevated temperature environments [4,39,41]. Figure 11.4 shows optical micrographs of Ti–6Al–2Sn–4Zr–2M0–0.1Si in two characteristic metallurgical conditions, (a) transformed β (Widmanstatten $\alpha + \beta$) after heat treatment 2 h/1024°C/air cool and (b) $\alpha + \beta$ after 2 h/968°C/air cool. Table 11.1 shows typical classes of titanium alloys in applications.

11.2.3 TITANIUM ALLOY/INTERMETALLIC PHASE DIAGRAMS

Characteristic equilibrium phase diagrams for four alloy classes of titanium are shown in Figure 11.5. It shows an equilibrium phase diagram of Ti–Al, where Al is an α-stabilizer. The most informative regions enclose the ordered intermetallics Ti$_3$Al and Ti–Al. Figure 11.5a shows a binary phase diagram for Ti–Mo, showing the temperature of the alpha and beta fields as a function of the Mo content [4]. Ti–Al intermetallics offer significant advantages in density (half the density

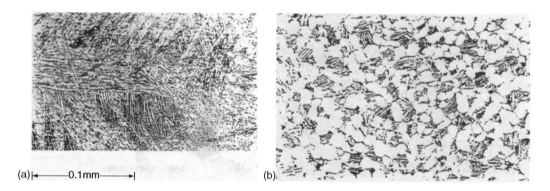

(a)|←——0.1mm——→| (b)

FIGURE 11.4 Optical micrographs of Ti–6Al–2Sn–4Zr–2M0–0.1Si in two characteristic metallurgical conditions. (a) Transformed β (Widmanstatten $\alpha + \beta$) after heat treatment 2 h/1024°C/air cool and (b) $\alpha + \beta$ after 2 h/968 °C/air cool. (From Collings, E. W., *Physical Metallurgy of Titanium*, ASM, Materials Park, OH, 1994.)

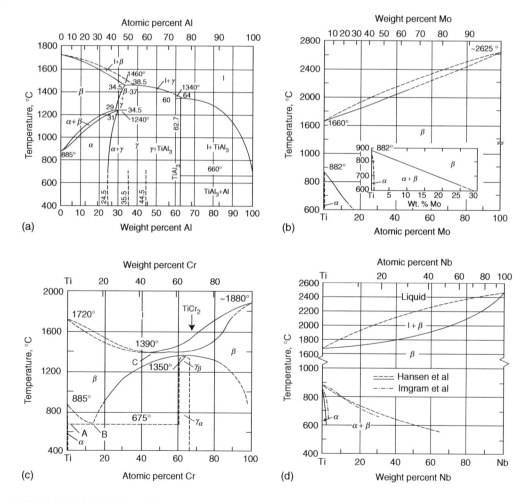

FIGURE 11.5 (a) Ti–Al equilibrium phase diagram; (b) Ti–Mo equilibrium phase diagram; (c) Ti–Cr equilibrium phase diagram (points indicated by A, B, and C are concentrations of 0.5, 14, ~45 at.% Cr, respectively); and (d) Ti–Nb equilibrium phase diagram. (From Collings, E. W., *Physical Metallurgy of Titanium*, ASM, Materials Park, OH, 1994.)

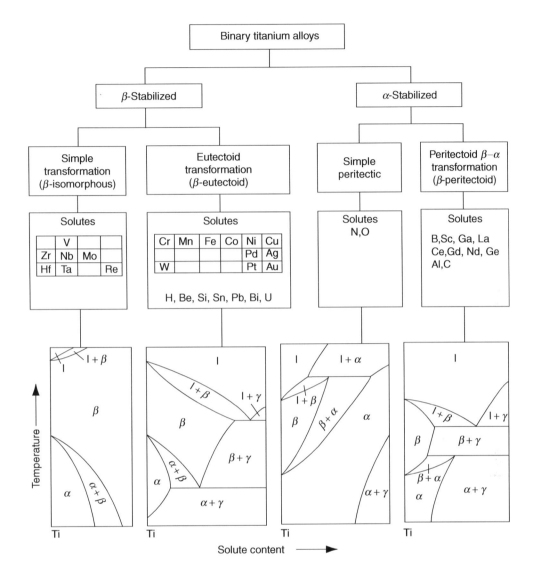

FIGURE 11.6 Classification scheme for binary titanium alloy phase diagrams. (From Collings, E. W., *Physical Metallurgy of Titanium*, ASM, Materials Park, OH, 1994.)

of Ni superalloys) over Ni superalloys. However, they have lower ductility and fracture toughness at room temperature [43]. Their creep resistance and ductility are also enhanced by the respective additions of Mo and Nb [43].

Extensive areas of the application of titanium alloys depend on these characteristic binary phase diagrams to study solubility of constituents at various temperatures and for materials' and microstructural enhancements' activities such as heat treatment. Detailed schematics of various forms of binary titanium and constituent alloying elements and stabilizers are presented in Figure 11.6 [4].

11.3   BASIC MECHANICAL PROPERTIES

Basic mechanical properties of α/β titanium alloys are reviewed in this section. These include creep resistance, fracture toughness, strength and ductility, and the effects of microstructure on these

TABLE 11.2
Mechanical/Physical Properties of Titanium Alloys, Titanium Aluminides, and Superalloys

Property	Ti-Base	Ti3Al-base	TiAl-base	Superalloys (Ni)
Structure	HCP/BCC	DO19	L10	fcc/L12
Density (g/cm^3)	4.5	4.1–4.7	3.7–3.9	7.9–8.5
Modulus (GPA)	95–115	110–145	160–180	206
Yield strength (MPa)	380–1,150	700–990	350–600	800–1,200
Tensile strength (MPa)	480–1,200	800–1,140	440–700	1,250–1,450
Room-temperature ductility (%)	10–25	2–10	1–4	3–25
High-temparature ductility (%/°C)	12–50	10–20/660	10–60/870	20–80/870
Room temperature fracture toughness (MPa m$^{1/2}$)	12–50	13–30	12–35	30–100
Creep limit (°C)	600	750	750–950	800–1,090
Oxidation (°C)	600	650	800–950	870–1,090

properties. The mechanical properties of titanium alloy systems are completely intertwined with alloying and heat treatment protocols. Hence, a wide variation in mechanical properties can be achieved by means of prudent selection of alloying elements and heat treatment methodologies. Improvements in mechanical properties, such as tensile strength, can be achieved without adversely impacting ductility [40]. Table 11.1 shows typical mechanical properties of titanium alloys systems [43].

Titanium alloys are significantly affected by the presence of interstitial elements, mainly carbon, and the reactive gases of the atmosphere. The presences of traces of interstitial elements in titanium alloys systems add strength to the metal, but at the expense of ductility (Table 11.2 and Table 11.3). In general, interstitial elements carbon, nitrogen, oxygen, boron, and hydrogen are referred to as contaminants, while substitutional elements additives, added for microstructural improvements, are referred to as alloying elements [40].

11.3.1 CREEP RESISTANCE

In spite of the obvious historical advantages offered by titanium alloys in terms of high strength, excellent corrosion resistance, and relatively good biocompatibility, it has been established that Ti alloys tend to exhibit a susceptibility to room-temperature creep. This phenomenon has been observed even at stresses well below the macroscopic yield strength of such alloys [46–48]. Figure 11.7 shows a creep plot for titanium systems. Figure 11.7 shows aging as having deleterious effects on titanium.

TABLE 11.3
Mechanical/Physical Properties of Ti Alloys, Titanium Aluminides, and Ni-base Alloys

	Ti	Ti3Al	TiAl	Ni-base
Density (g/cm^3)	4.54	4.28	3.83	8.45
Stiffness (GPa)	110	145	175	206
Max temperature/creep °C	535	815	900	1095
Ductility @ 25°C (%)	20	2–4	1–3	3–10

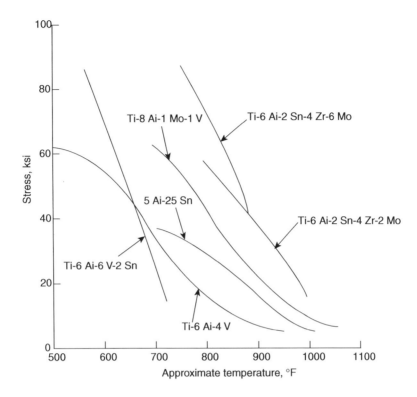

FIGURE 11.7 Comparison of creep strengths selected titanium alloys systems (0.1% creep strain at 150 h) (Lampman 2000).

In recent times, phenomenological and mechanistic investigations have been undertaken to investigate room temperature creep in titanium-based alloy systems [45]. Primary creep has been established as a dominant mode of deformation for titanium alloys at lower temperatures under most service conditions. Currently, there is no clear consensus on the underlying damage mechanisms associated with room-temperature creep. Some researchers have actually attributed dwell fatigue to a creep mechanism occurring at room temperature conditions [45]. Over the years, the requirement to design for intrinsic creep in titanium applications has led to modification in titanium alloys system [46–48].

In high temperature applications, creep is rarely a factor between 200 and 315°C, even at stresses approaching the yield strengths in this regime. However, above 315°C creep is a critical materials selection criterion. In alpha–beta alloys, small additions of beta stabilizing elements ensure retention of some beta phase at room temperature. Alpha–beta alloys are usually heat treated in the alpha–beta phase, and the resulting microstructure consists primarily of primary alpha and transformed beta. The maximum operating temperature under creep conditions for these alloy systems are normally between 300 and 450°C.

In the case of near-alpha alloys, however, the heat treatment occurs in the alpha–beta phase field. By prudently controlling the alpha and beta stabilizing elements, titanium alloys systems have been developed with improved creep resistance in the range 450–500°C. With further heat treatment and compositional modification, creep resistance can be engineered up to 600°C (TIMET). The drawback in this case is that the heat treatment usually occurs above the beta transus, and upon cooling, acicular alpha structure results. The acicular microstructure has reduced fatigue performance in spite of its superior creep performance [40].

Maximum creep performance can be achieved by using a duplex annealing method. In this approach, the alloy is solution annealed at a temperature in the alpha–beta range (28–56°C below the beta transus). The forgings are then held for about 1 h and fan or air cooled. Further stabilization is achieved by annealing for 8 h at 595°C [40]. Such heat treatment protocols have been applied on Ti–6Al–2Sn–4Zr–2Mo and Ti–8Al–1Mo–1V [40].

11.3.2 FRACTURE TOUGHNESS

Fracture toughness in titanium alloys is a function of strength, composition, microstructure and texture, and the interrelationships among these properties. Secondly, titanium alloys are strain-rate sensitive at elevated temperatures. Furthermore, titanium alloys are known to vary inversely with strength in the same way as other metallic structures such as steels or aluminum alloys.

For instance, the plane strain fracture toughness of the alpha–beta alloys is known to reduce from values between 60 and 100 MPa m$^{1/2}$ at proof stress levels of 800 MPa, to values between 20 and 60 MPa m$^{1/2}$ at proof stress levels of about 1200 MPa. It has been established within the titanium body of knowledge that for certain alpha–beta alloys, an increase in fracture toughness can be accomplished by modification using heat treatment procedures or through minor variation in alloy chemistry.

Also, extra low interstitial titanium alloys (ELI) grade can be realized by the reduction of oxygen levels in Ti–6Al–4V. Such modification in alloy chemistry enhances fracture toughness with marginal reduction in alloy tensile and fatigue strength. In general, beta alloys and heat-treated near-alpha alloys have better fracture toughness levels than the alpha–beta alloys. High toughness titanium alloys systems considered for Mach 2.4 applications include near-beta Beta-CEZ (Ti–5Al–4Mo–4Zr–2Sn–2Cr) and an alpha beta Ti-622-22 (Ti–6Al–2Sn–2Mo–2Zr–2Cr).

11.3.3 FATIGUE BEHAVIOR OF $(\alpha+\beta)$ TITANIUM ALLOYS

Fatigue crack growth behavior in titanium alloys has been studied extensively over the past three decades by many investigators [5–20,49–60]. These investigations of the fatigue crack nucleation and propagation mechanism have broadened our understanding on the behavior of titanium alloys under service conditions. For instance, these studies have shown that surface or subsurface crack nucleation is most likely to occur in α/β titanium alloys deformed under pure fatigue loading conditions [20,33]. In many cases, surface or subsurface nucleation are known to occur from inclusions such as hard α phase, rogue particles, or interfaces/grain boundaries [55]. Fatigue crack nucleation is also shown to occur by a Stroh mechanism, i.e., due to dislocation pileup [27,33,61,62].

11.3.3.1 Effects of Microstructure and Micro-Texture

Fatigue crack growth in low cycle and high cycle fatigue depends strongly on microstructural and micro-textural effects [14,15,19,23,49–51]. A number of microstructures of titanium alloys have been studied, and the most widely reported microstructures are β-annealed—Widmanstatten α [6,7]; mill annealed (MA)—elongated primary α [7,8,13]; recrystallized annealed (RA)—equiaxed $\alpha+$ transformed β microstructure [7,51]; and solution treated and over-aged—STOA primary α in martensitic matrix [51].

Numerous studies in the literature (listed in the preceding section) have highlighted the effects of microstructure on the fatigue crack growth behavior of titanium alloys. For instance, micro-structural sensitivity has been identified in the transition from near-threshold fatigue regime to intermediate fatigue regime. This transition has been attributed to metallurgical variables such as grain size and slip characteristics [5,9].

Yoder, Cooley, and Crooker [7–10] also reported that in beta-annealed alloys such as Ti–6Al–4V and Ti–8Al–1Mo–1V with Widmanstatten colonies, there is an apparent transition from microstructural-sensitive to microstructural-insensitive growth characteristics at ΔK_T. They observed that the transition from a facetted fracture mode to a striated mode was dependent on the microstructural dimension, l^*, given by Equation 11.1.

$$\Delta K_T = \frac{1}{4\pi} \sigma'_y \sqrt{l^*} \tag{11.1}$$

where ΔK_T is transition stress intensity factor range, σ'_y cyclic yield strength, and l^* is microstructural dimension.

Saxena and Radhakrishnan [38] investigated the effect of phase morphology on fatigue crack growth behavior of $\alpha + \beta$ titanium alloys using Ti–6Al–4Mo–2Zr–0.2Si. They observed that microstructures comprising metastable β matrix yielded higher FCG resistance or rates than those of transformed β matrix, independent of primary α phase morphology (equiaxed or elongated). They further noted, however, that the effect of primary α is dependent on the type of β phase present. They attributed the observed phase morphology effects on fatigue crack growth behavior in $\alpha + \beta$ titanium alloys to roughness-induced and plasticity-induced crack closure mechanisms (Figure 11.8).

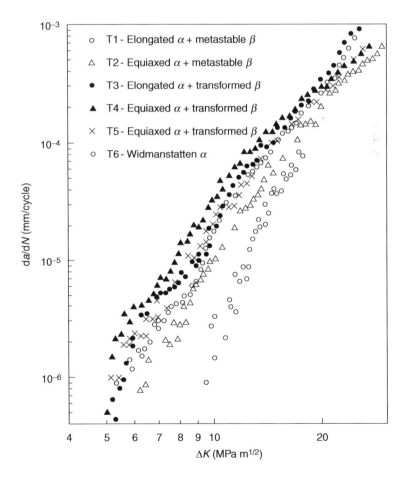

FIGURE 11.8 Fatigue crack growth rate as a function of effective SIF range, ΔK_{eff} for various microstructural/ heat treatment conditions.

In general, Widmanstatten colony microstructures have been observed to exhibit the highest FCG resistance [9,10,51,54] in the long crack regime, with the resistance to fatigue crack growth substantially improved with the coarsening of the colony structures, while STOA microstructure exhibits the lowest FCG resistance.

Recrystallized annealed microstructure represents an intermediate case [51]. In the case of Ti–6Al–4V, recrystallized annealed material is observed to exhibit marginal fatigue resistance over the mill-annealed microstructure. In the case of Ti–4.5Al–5Mo–1.5Cr, mill-annealed microstructure exhibited a higher FCG resistance than the recrystallized annealed microstructure [51].

Shen, Soboyejo, and Soboyejo [14,15] observed that in Ti–6Al–2Sn–4Zr–2Mo–1Si, the colony structure has the best fatigue resistance, followed by the elongated structure, and the equiaxed structure exhibited the lowest fatigue resistance. Similar observations were reported by Ravichandran et al. [54,55] for long cracks. However, Ritchie et al. [63] have also suggested that the Widmanstatten microstructures have limited resistance to short crack growth.

11.3.3.2 Effects of Environment

Environmental effects, such as temperature and corrosion, can play a significant role in fatigue failure. Constrained parts that are subject to fluctuating temperatures and repeated thermal expansion-contraction cycles may fail from thermal fatigue, even without external applied mechanical loads [64].

FCG in titanium alloys may also be affected by the environment. Corrosive environments, such as those created by 3.5% NaCl solution may accelerate FCG rates significantly by an order of magnitude or more [40]. The presence of environmental factors serves to accelerate environmentally assisted fatigue crack growth rates.

Diffusion of hydrogen atoms at crack tips and the subsequent formation of hydrides have been documented as possible dwell fatigue crack initiation mechanism [27]. Wojcik, Chan, and Koss [30] observed significant irreversibility in crack growth with the diffusion of water vapor into the crack tip as a result of the reduction in local stress intensity factor range.

Titanium alloys are known to form protective layers on the surface of titanium alloys when exposed to environments containing oxygen. This protective film of TiO_2 has a re-healing effect. Thus, the resistance of titanium alloys is generally limited in strong, highly reducing acid concentrated solutions such as HCl, HBr, H_2SO_4, and H_3PO_4 and HF. The solutions are the most caustic media for titanium-based alloys [40].

11.3.3.3 Effects of *R*-ratio

Titanium fatigue crack growth dependence on stress ratio has been extensively investigated [17–19,54,65,66]. These studies have shown that fatigue crack growth rates increase with increasing stress ratio in titanium alloys with different microstructures. In mill-annealed Ti–6Al–4V, fatigue crack growth rates were observed to increase with increasing stress ratios. Such trends have been attributed largely to the effects of crack closure [17–19,49,54]. Detailed studies of crack profiles have also shown that roughness-induced crack closure is the dominant type of crack closure mechanism in titanium alloys [67].

Ritchie et al. [68] investigated the effects of constant-*R* on fatigue crack propagation of Ti–6Al–4V at four load ratios (50 Hz) and compared the results with constant-K_{max} data at four K_{max} values: $K_{max} = 26.5, 36.5, 46.5,$ and 56.5 MPa $m^{1/2}$ (1000 Hz) as shown in Figure 11.9. They observed that higher load ratios induce lower ΔK_{th} thresholds and faster growth rates at a given applied K. Furthermore, using compliance measurements, they observed no closure effects above $R = 0.5$; although, at $R = 0.1–0.3$, K_{cl} values were somewhat constant at ~ 2.0 MPa $m^{1/2}$. Other studies have also concluded similarly that trends in stress ratio and fatigue growth dependence investigations are

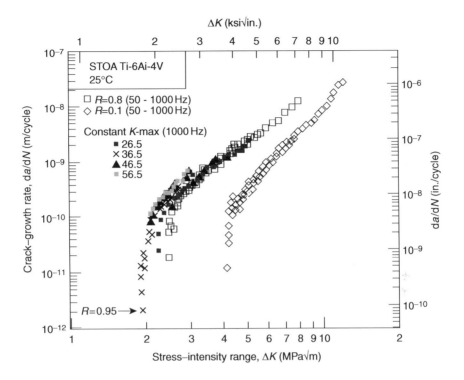

FIGURE 11.9 Constant-K_{max} fatigue crack propagation behavior at K_{max} values of 26.5, 36.5, 46.5, and 56.5 MPa m.

directly attributable to crack closure effects [17–19,49]. Thus, stress ratio effects in titanium alloys are essentially a crack closure phenomenon.

11.3.3.4 Effects of Frequency

There is no significant frequency effect (see Figure 11.10) between ~50 and 20,000 Hz on titanium alloys in ambient air, indicating the absence of frequency effects at near-threshold levels [68]. However, some frequency effects may be observed at lower frequency ($f < 1$ Hz) due to possible creep-fatigue and environmental interactions [69]. At increasing temperatures, however, the synergistic interaction with creep is further driven by temperature effects [40]. It is worth noting that fatigue crack growth rates in vacuo (10^{-6} torr) are ~2 orders of magnitude slower than in air at an equivalent ΔK, even though the non-propagation threshold has been observed to remain somewhat the same.

11.3.3.5 Effects of Temperature

Temperature does not have a significant effect on FCG in titanium alloys. According to the Metals Handbook, vol. 3, 9th ed., FCR for Ti–5Al–2.5Sn and Ti–6Al–4V (IN) showed no noticeable dependence on temperature [40], while Ti6Al–4V (ELI) showed higher FCG at cryogenic temperatures than at room temperatures under same ΔK loading [40]. At high temperatures, fatigue strength depends on total time stress applied rather than on the number of applied cycles, and hence, FCG is affected by deformation under creep loading conditions [40].

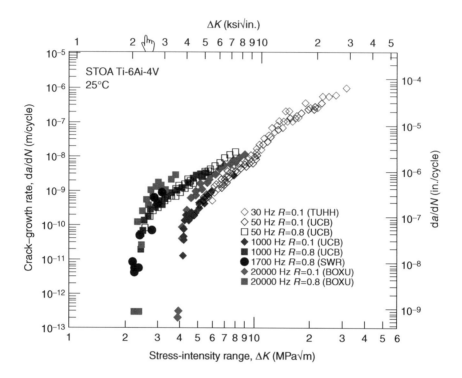

FIGURE 11.10 Effect of frequency on fatigue crack growth for bimodal Ti–6AL–4V in room air.

11.3.3.6 Effects of Crystallography

Wojcik, Chan, and Koss [30] investigated the effect of orientation on stage I fatigue crack of individual Widmanstatten colonies of $\alpha + \beta$ Ti alloy growth characteristics of Ti–8Al–1Mo–1V in air and in dry helium. The results of this investigation indicated that crack growth occurs along preferred crystallographic planes, especially near the basal $\{1\bar{1}00\}$ plane of the α-phase.

Furthermore, the fracture morphology is usually characterized by a cleavage-like facetted appearance, the absence of striations, and is characterized by sporadic micro-cracking. It was also observed that only crack tip shielding effects resulting from micro-cracking affected the sensitivity of colonies to orientation-dependent crack growth rates. A power law crack growth model related to all three modes of crack tip displacement range was proposed as

$$\frac{da}{dN} \propto (\Delta K_{\text{eq}})^2 \tag{11.2}$$

Wocjik et al. [30] also noted that individual Widmanstatten colonies exhibit anisotropic behavior, which is usually dependent on both α-phase orientation and β-phase platelets. Similar observations were reported by Shademan et al. [17,18].

Bache, Evans, et al. [70] characterized mechanical anisotropy in titanium alloys and its relevance to the application of LEFM in the stress analysis of titanium-based alloys, especially the assumption that mechanical behavior in titanium alloys is isotropic. They observed that slight variations in crystallography have significant effects on fatigue in titanium, which has the HCP crystallographic structure and limited slip systems. They observed further using the EBSD technique that interactions between fatigue cracks and microstructural features were related to crystallographic orientation.

These investigators concluded that (i) the features that affect crack propagation include colony/prior beta grain boundaries and secondary cracking; and (ii) that many fatigue crack propagation characteristics are a direct consequence of crystallographic orientation, especially the orientation of the alpha colony within which the crack initiates is a key factor in determining overall fatigue life.

11.3.3.7 Effects of Crack Closure and Crack Deflection

Shademan and Soboyejo [17], Shademan et al. [18], and Sinha and Soboyejo [19] studied the effects of positive stress ratio on the fatigue performance of Ti–6Al–4V. They concluded that the observed differences between short and long crack growth behavior at lower stress ratios are due to the absence of crack closure reduced levels of crack closure in the short crack growth regime.

Prior work on closure effects has shown that when the da/dN is plotted against the effective stress intensity factor range, ΔK_{eff}, the data collapses onto a straight line. This result suggests that the observed stress ratio effects in titanium alloys are essentially due to crack closure effects. Hence, by correcting for closure effects, the fatigue crack growth rate data appear to cluster within a narrow band of data scatter [49,65,66].

Crack closure mechanism(s) in $\alpha + \beta$ titanium alloys have been attributed to roughness induced crack closure [21,56,57,67,71]. Similar observations by Sehitoglu and Garcia [72] obtained using measurements techniques for crack closure/opening experiments on titanium alloys support the conclusion that asperity deformation due largely to contact forces may be the primary mechanism responsible for crack closure effect in $\alpha + \beta$ titanium alloys. Furthermore, plasticity induced closure mechanisms have also been proposed to account for the overall closure effects observed in experimental characterization of fatigue crack growth behavior in $\alpha + \beta$ titanium alloy systems [73–75].

11.3.3.8 Mechanisms of Fatigue Crack Growth in Titanium Alloys

Fatigue crack growth rates in titanium alloys are strongly dependent on microstructure [76,77]. Two principal reasons have been proposed to explain these effects: (i) crack path tortuosity and (ii) formation of secondary cracks. The former occurs when favorable microstructural features exist along which crack growth experiences little resistance. The latter case, results from the reduction of energy associated with the main crack. Both factors result in reduction in fatigue crack growth rates [76,77].

Eylon and Hall [26] and Yoder et al. [7–10] observed that fatigue crack growth in β-annealed Ti alloys occur in high dense coarse slip bands along individual Widmanstatten colonies. Hall and Hammond [78] observed that in β-annealed (Widmanstatten microstructure) Ti–6Al–4V alloy, fatigue crack occurred preferentially along the β-layers. In the equiaxed microstructure of Ti–6Al–4V, Dubey et al. [49] observed that fatigue crack growth occurs along intersecting slip bands [49].

11.3.4 SHORT CRACK GROWTH IN $(\alpha + \beta)$ TITANIUM ALLOYS

Short cracks in titanium alloys do not exhibit threshold values in the stress intensity factor range, and crack propagations rates are usually faster than those observed for long cracks. Small cracks growth in titanium alloys, in general, experienced little or no wake and hence experience no closure effects but the entire applied ΔK [54,55,79].

Small cracks are known to exhibit (i) faster propagation rates than long cracks at equivalent ΔK and (ii) significant crack-growth rates at applied ΔK below the long-crack ΔK_{th} [17–19,51,54,55,79]. McBagonluri et al. [60] using benchmarking, acoustic emission, and scanning electron microscopy (SEM) techniques, established crack evolution and growth for Ti-6242 under fatigue and dwell fatigue conditions. The underlying crack nucleation and fatigue fracture modes were elucidated for three microstructures: equiaxed, elongated, and colony of Ti-6242.

The fracture modes were found to be similar for both fatigue and dwell fatigue. The dominant crack nucleation mode is observed to involve a Stroh-type dislocation mechanism, where subsurface cracks are characterized by prominent facetted fracture modes in the near-threshold regime. Subsequent fatigue crack growth was observed to occur by fatigue striations and ductile dimples or cleavage-like static modes at higher stress intensity factor ranges.

The long crack growth data are similar for both dwell and pure fatigue. However, the dwell fatigue crack growth rates are shown to be much greater than those due to pure fatigue in the short crack growth regime. The differences between the dwell crack growth rates and the pure fatigue crack growth rates in the short regime are attributed to possible creep effects that give rise to a mean stress effect.

Sinha et al. [19] investigated short and long crack growth behavior of Ti–6Al–4V, which had a mill-annealed microstructure. Experiments conducted on positive stress ratio effects clearly indicated that microstructurally short cracks exhibit crack growth rates at significantly lower stress intensity factor ranges. This investigation also showed anomalous crack growth and retardation within the short crack regime. Observed differences in short and long crack growth behavior at lower stress ratios were attributed largely to intrinsic crack closure mechanisms occurring within short crack regime.

11.3.5 Dwell Behavior of $(\alpha+\beta)$ Titanium Alloys

Evans and Gostelow [27]; Evans and Bache [61]; Bache, Davies, and Evans [32]; and Spence, Evans, and Cope [33] observed that in structural components made from α and α/β titanium alloys, a substantial dwell life debit is realized between components differentially subjected to pure fatigue and dwell hold periods, with dwell showing marked reduction in life.

Prior research on α/β titanium alloys attributes the occurrence of dwell sensitivity to a range of possible deleterious mechanisms. These mechanisms include time-dependent strain accumulation [27,29,80]; microstructural and micro-textural influences [14,15,29,33,36,37]; stress ratio effects [81]; temperature effects [28]; internal hydrogen embrittlement and environmental effects [28,30,32]; crystallographic orientation dependence [30]; and synergistic interaction between fatigue and creep [34,80,82]. Furthermore, fatigue behavior in titanium alloys is known to occur as a result of subsurface crack nucleation from hard α phase and near-interstitials [11,19,52,55] or Stroh-type dislocation reactions [33,61]. Despite these investigations, there is still no consensus on the based cause of dwell fatigue in titanium-based alloys.

11.3.5.1 Effects of Microstructural and Micro-Textural Influences

The role of microstructural effects on dwell fatigue behavior in titanium-based alloys have been examined in numerous works [14,15,25,37,69]. Eylon and Hall [26] focused on the role of microstructure on the low-cycle fatigue (LCF) and low-cycle dwell fatigue (LCDF) behavior on IMI 685 (Ti–6Al–5Zr0.5Mo–0.25Si), and Goswami [37] collected extensive data on dwell related mechanisms in several microstructures and alloys, while Shen, Soboyejo, and Soboyejo [14,15] concentrated on long crack fatigue and dwell fatigue behavior of three microstructures: equiaxed, elongated, and colony of Ti-6242.

Eylon and Hall [26] investigated Widmanstatten and basket-weave microstructures obtained at different cooling rates. This investigation focused on fatigue crack initiation and propagation characteristics associated with dwell fatigue, the role of dwell cycles, and the effect of dwell hold time on dwell fatigue life. Materials with colony structures were observed to exhibit lower strengths, lower ductility, and shorter LCF lives than materials with basket-weave structures. Improvement in fatigue lives associated with LCF lives in basket-weave structures were attributed to slip characteristics.

These investigations first showed that fatigue crack initiation and propagation are strongly dependent on the occurrence of heterogeneous slip systems. In these systems, the propensity for crack nucleation is favored on, or parallel to, intense slip bands resulting from fatigue cycling. Secondly, the propagation of the slip bands was retarded by the colony boundaries, due to differences in the orientations between adjacent colonies. Finally, single crystal-like behavior was exhibited by colonies of similarly oriented α-platelets and deformation twins were observed to extend across entire colonies as shown in Figure 11.11 [26].

Shen, Soboyejo, and Soboyejo [69] examined long crack fatigue and dwell fatigue crack growth characteristics in three microstructures of Ti-6242 namely, equiaxed, elongated, and colony. They observed that basic material properties (yield, ultimate strength, and ductility) were statistically the same for all three microstructure types. Equiaxed microstructure had the fastest fatigue crack growth rate, followed by the colony microstructure. The elongated microstructure exhibited the best fatigue life. Crystallographic fracture modes were observed in the threshold and lower Paris regimes. In addition, transgranular fatigue crack growth occurred in all three microstructures.

This investigation showed that dwell and pure fatigue crack growth rates are generally similar in the long crack growth regime, and the underlying mechanisms of long crack growth are also generally similar under pure fatigue and dwell-crack growth conditions. Differences were, however, observed in dwell and fatigue sensitivities between the different microstructures in the mid- and high-ΔK regimes. These differences were attributed to possible interactions between fatigue and creep during dwell fatigue loading at room temperature.

Sansoz and Ghonem [69] investigated fatigue crack growth in Ti-6242 lamellar microstructures and studied the influence of loading frequency and temperature on its fatigue behavior. The Ti-6242 alloy was heat treated to three microstructures: fine, coarse, and extra coarse. Testing was performed at 520°C and 595°C using 10 and 0.05 Hz, respectively. There was no significant microstructural influence at 520°. Also, the addition of a 300-minute dwell hold time did not alter the crack growth rate. However, at 595°C the crack growth rate was higher than at 520°C. The influence of both loading frequency and hold time we observed to be similar at both

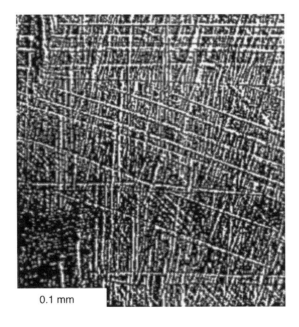

0.1 mm

FIGURE 11.11 Deformation twins across colony of large colony microstructures. (From Eylon, D. and Hall, J. A., *Met. Trans. A*, 8A, 981, 1977.)

temperatures. They concluded that a single mechanism is associated with trans-colony fracture and formation of quasi-cleavage facets characterizes fatigue crack growth in Ti-6242.

Spence, Evans, and Cope [33] investigated dwell behavior in Ti-6246 using a near-α microstructure consisting of small regions of α phase in acicular α matrix. In these investigations, Spence et al. [33] concluded that dwell sensitivity is driven by a host of synergistic interactions. They suggested that the susceptibility to dwell fatigue failures in titanium alloys is enhanced by (i) the propensity for stress redistribution between strong and weak microstructural sites, (ii) a preferred basal crystallographic orientation with a high propensity for crack nucleation, (iii) the interaction between shear and tensile stress components on this basal plane, (iv) the presence of α-phase, and finally, (v) the environment.

Bache, Cope, et al. [81] investigated dwell effects on two microstructures of titanium: a bimodal morphology with 15% equiaxed primary-α within a fine transformed β matrix and a microstructure with aggregation of elongated primary α grains. Differences in microstructural responses to dwell were observed for these microstructures. For instance, the reduction in life observed in the equiaxed primary-α within a fine transformed β matrix was marginal, when compared with the baseline data. On the other hand, the microstructure with the elongated α grains exhibited significant dwell effects. Differences in dwell effects were attributed to differences in grain sizes; the elongated primary α grains sizes were twice that of the equiaxed α grains, hence, longer crack growth segments resulted between deflections in this microstructure.

Kassner, Kosaka, and Hall [82] studied the effects of microstructure on LCF and LCDF behavior in Ti 6242 (Ti–6Al–2Sn–4Zr–2Mo–O.lSi). Dwell effects reduced fatigue life by a factor of 3–5 at lower stresses. They observed that a reduction in volume percent of α phase at a given stress level increased dwell fatigue life. There was, however, no noticeable effect of microstructure on low cycle fatigue life.

Woodfield [35] and Woodfield et al. [36] investigated the dwell fatigue sensitivity in Ti-6242. They observed, as did previous investigators [28,29], that dwell fatigue crack nucleation occurs in subsurface regimes. In their investigations, a detailed examination of the fracture morphology of a failed spool, revealed a subsurface initiation site with brittle cleavage-like, facetted features rather than the classical fatigue striations associated with pure fatigue [64].

11.3.5.2 Effects of Time Dependent Strain Accumulation

Evans and Gostelow [27] investigated the relationship between LCF and LCDF and the interaction between LCDF and creep in IMI 685 alloy. They observed that by imposing a 5-minute dwell hold, a fatigue life debit of factor 16 occurred (Figure 11.12). They also observed that the mode of fatigue crack nucleation was subsurface, and unlike in the LCD, nucleation sites in dwell were characterized by extensive facets. The observed facets sizes were comparable to the colony size in the IMI 685 alloy. There was also no indication of multiple crack initiation, which was indicative that a rapid transition from stage I to stage II fatigue crack growth might have occurred. Striations were observed in both LCD and LCDF specimens. Final ductile fracture occurred by micro-void formation and subsequent coalescence.

The results from this investigation provided evidence of a strong correlation between LCDF and creep and LCF and LCDF. For instance, strain-time data collected during dwell testing exhibited the three characteristic stages associated with creep phenomenon in engineering structures: primary, secondary, and tertiary stages (Figure 11.13). Secondly, there was strong correlation between the fatigue lives of the creep and dwell specimens that were tested at dwell stresses corresponding to the mean stresses in the creep tests (Figure 11.14).

By observing similarities between dwell and creep tests, Evans and Gostelow [27] suggested that dwell fatigue life is dominated principally by the accumulation of creep strain at peak stresses during dwell cycling. Hence, the ubiquity of facets and the similarity of observed fracture morphology on the creep and dwell fracture surfaces may be an indication of a common

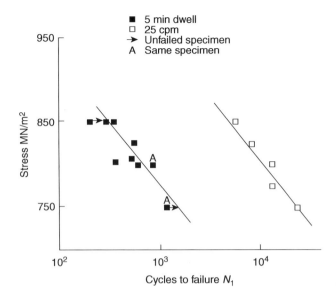

FIGURE 11.12 Effect of LCF and LCDF on fatigue life in IMI 685. (From Evans, W. J. and Gostelow, C. R., *Met. Trans. A*, 10A, 1837, 1979.)

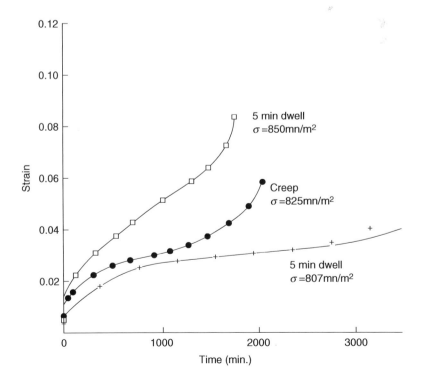

FIGURE 11.13 Strain Versus. time-to-failures curves for LCDF and creep.

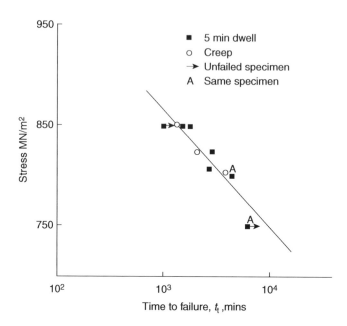

FIGURE 11.14 Stress Versus. time-to-failure behavior of LCDF and creep. (From Evans, W. J. and Gostelow, C. R., *Met. Trans. A*, 10A, 1837, 1979.)

mechanism. Furthermore, they asserted that facet formation in LCDF might be a function of the ratcheting effect of strain accumulated at peak stress rather than cyclic effect on material. Figure 11.15 shows observed differences in fatigue crack growth behavior in LCF and LCDF.

Hack and Leverant [28] observed that the overall strain to failure is similar for dwell fatigue, static loading, dwell fatigue with pre-strained, and fatigue with imposed dwell at minimum and maximum loads. These results indicate that failure mode is intrinsically dependent on strain accumulation. This observation may be an indication that failure is driven principally by the overall strain, not necessarily due to prior loading history, and consequently, LCDF may in essence be a creep problem rather than a cyclic one.

11.3.5.3   Internal Hydrogen and Environmental Effects

Hydrogen embrittlement during the forging of titanium alloys has been known to have a deleterious effect on titanium base structures. Hack and Leverant [28] observed that an increase in hydrogen content at room temperature dramatically resulted in life reduction by a factor of 10 in titanium alloys. Wojcik, Chan, and Koss [30] investigated the sensitivity of crack growth to the environment. They observed that crack growth reversibility is reduced with the diffusion of water vapor into the crack tip. The dwell effect has also been attributed to hydride formation at crack tips or at slip band [27–29]. It has been advanced that the resulting stress gradients enhance hydrogen atom diffusion. This local diffusion of hydrogen atoms results in hydride precipitation and consequently, increased strength, higher strain-hardening and reduced ductility.

11.3.5.4   Effects of Crystallography

Wojcik, Chan, and Koss [30] observed that crack growth rates in Widmanstatten single colonies of α–β Ti 811 are in general not sensitive to crystallographic orientation unless there are shielding effects at the crack tip emanating from micro-cracking at or near interfaces of α–β. This shielding

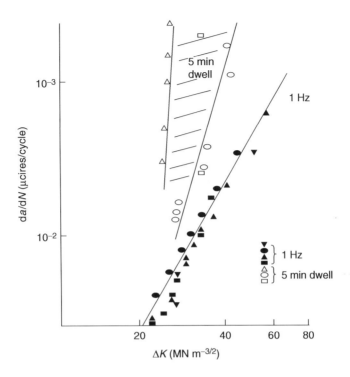

FIGURE 11.15 Crack growth data for LCDF and LCF showing the dwell effect. (From Evans, W. J. and Gostelow, C. R., *Met. Trans. A*, 10A, 1837, 1979.)

effect reduces the local critical stress intensity range and consequently, increasing cracks growth rates.

Davidson and Eylon [83] investigated the effect facet crystallographic plane on crack initiation and propagation in LCF and LCDF in titanium IMI 685 and Ti-11. They applied a technique called the selected area electron channeling. They observed that unlike in LCDF, basal plane facets were present in LCF specimens. Similar conclusions were drawn for the Ti-11 as well. The LCD specimens exhibited off-basal orientation between 6° and 10°, while the LCDF exhibited off-basal orientation of 4° and 10°. Calibrating the electron channeling and back scattered electron images, these investigators found that most of the β platelets were on or near α phase primary $\{1\,0\,\bar{1}\,0\}$ planes. Other significant findings included (i) a facet consisting of two α colonies transformed from prior β grain; (ii) an α colony was found to have β plates on $\{1\,0\,\bar{1}\,0\}$ planes, (iii) while another colony had its $\{1\,0\,\bar{1}\,0\}$ plane rotated through an angle of $\sim 70°$. (iv) Small plastic zones were correlated with crack nucleation sites and (v) most of the LCF specimens were found to exhibit surface nucleation, while the LCDF specimens exhibited subsurface nucleation.

Bache, Evans, et al. [31] used electron back scattered diffraction (EBSD) analysis to study quasi-cleavage and hydrogen induced fractures under cyclic and dwell loading in titanium alloys. They characterized the local grain orientation and microstructural conditions using micro-textural analysis technique based upon electron back scattered diffraction (EBSD) measurements. Their observations from this study supported previous account of facet formation based on the pileup of dislocations at grain boundaries [61].

11.3.5.5 Fatigue-Creep Synergism

Chu, MacDonald, and Arora [80] observed an increase in fatigue life on specimens subjected to prior creep deformation under strain controlled LCF conditions. This observation has been attributed to the emergence of complex dislocation networks resulting from prior creep deformation and the resulting interference of these dislocation networks to the onset of crack nucleation. For instance, microscopy of prior creep damage specimens showed, on the average, a limited number of micro-cracks as compared to fatigue only specimens. Hack and Leverant [28] observed, however, that under stress controlled LCF plus pre-strain, fatigue failures occurred in much shorter times than observed by Chu et al. [80]. While pre-deformation may create a network of dislocations, dwell fatigue may be driven entirely by the synergistic interactions between creep and fatigue, which may also be affected by preexisting dislocation structures. This observation is indicative of a time-dependent effect at play at low frequencies, and largely supports the strain accumulation postulate advanced by Hack and Leverant [28]. However, for Ti-alloys, it is not clear why creep occurs at room temperature, well below their recrystallization temperature.

11.3.5.6   Effects of Temperature

The effect of temperature on dwell was reported by Hack and Leverant [28]. In their investigation of temperature effects on dwell fatigue, they observed that at cryogenic and elevated temperature conditions the dwell effect is eliminated. Sansoz and Ghonem [69] studied the effects of temperature on fatigue crack growth rate in Ti-6242. They observed a slight increase in the fatigue crack growth rate at 595°C than that observed at 520°C.

11.3.5.7   Effects of Stress Ratio

Bache, Cope, et al. [81] studied the effect of dwell and R ratio on the fatigue life of Ti-6246 (Figure 11.16 and Figure 11.17). They observed that in general positive R ratios tend to have more severe effects on dwell fatigue, as compared with negative stress ratios. They suggested an inverse Stroh-type mechanism to explain the enhanced fatigue life (Figure 11.15) observed for

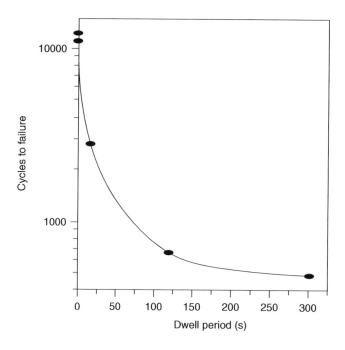

FIGURE 11.16 Effect of dwell hold period on fatigue life in disc material. (From Bache, M. R., Cope, M., Davies, H. M., Evans W. J., and Harrison, G., *Int. J. Fatigue*, 19, S1–S83, 1997.)

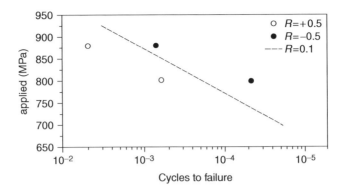

FIGURE 11.17 Effect of R ratio on dwell fatigue life in disc material. (From Bache, M. R., Cope, M., Davies, H. M., Evans W. J., and Harrison, G., *Int. J. Fatigue*, 19, S1–S83, 1997.)

negative stress ratios. They advanced that negative stress ratio may lead to slip reversal, which may cause the dislocation pileup stress field to relax (Figure 11.17). This plastic relaxation, results in increased dwell fatigue life. Shen, Soboyejo, and Soboyejo [14,15] also suggested that this apparent effect of stress ratio on dwell fatigue life could be explained by the lower crack-tip stress levels and stress triaxiality associated with lower stress ratio. They further suggested that this stress state might be partially responsible for the trends in cleavage facet formation.

11.3.5.8 Dwell Hold Time

Eylon and Hall [26] investigated the effect of dwell holding time on fatigue life reduction in dwell. They imposed a 5-minute dwell period at peak load and realized that a significant fatigue life debit occurs in colony structures. There was, however, no noticeable dwell effect on the long fatigue crack propagation rate. Thus, anomalous short crack fatigue effects emanating from dwell may be related to time dependent phenomena associated the nucleation, propagation, and retardation of slip bands. The interplay of these synergies, increase the propensity for stage I fatigue initiation and is further enhanced by localized deformation that may give rise to mean stress effects (Ritchie and Lankford 1986). Kassner, Kosaka, and Hall [82] observed that, when a dwell hold period of 2 minutes was imposed, specimens under dwell loading failed at significantly lower cycles. The factor decrease of LCF from dwell however, was nullified with decreasing stress levels. It was also observed from this investigation that increasing primary α phase associated with lower solution temperatures seems to increase the susceptibility to low-cycle dwell fatigue. They suggested that susceptibility to dwell fatigue might be associated with ambient-temperature time-dependent cyclic plasticity.

11.3.6 PROPOSED MECHANISMS OF DWELL FATIGUE CRACK GROWTH

Mechanisms of dwell fatigue crack nucleation in titanium base alloys have been attributed to a host of damage mechanisms including subsurface crack nucleation. Evans and Bache [61] proposed a model akin to Stroh-type dislocation pileup as giving rise to subsurface crack nucleation at stress levels below the yield stress. Ravichandran [54] and Ravichandran, Jha, and Shankar [55] have also observed surface and subsurface crack nucleation to be a characteristic crack nucleation mechanism associated with titanium-based alloys. Furthermore, the orientation of fatigue crack nucleation in titanium alloys has been observed to occur on the basal plane [30].

11.3.6.1 Evans and Gostelow

Evans and Gostelow [27] proposed a stress relief model to explain crack nucleation under dwell loading conditions. In this model, stress relief occurs when load is initially applied to a structure as a result of increased dislocation movement and increased strain-rate. In this scenario, the presence of hydrogen atoms does not significantly affect deformation. However, as creep progresses, strain rates and dislocation mobility effects are retarded due to strain hardening and hence stress relief mechanisms are also offset. Nucleation of cleavage cracks is enhanced subsequently by the increase in the interactions between hydrogen atoms and dislocations, which further increases strain-hardening resulting in crack nucleation.

11.3.6.2 Wojcik, Chan, and Koss

Wojcik, Chan, and Koss [30] suggested that crack occurs along slip bands as a result of the interaction between crack-tip stress field and slip. In this model, the interaction of the crack tip and slip are assumed to occur on a preferential coplanar crack plane, and is driven primarily by the inability of the coplanar slip to relax the normal elastic stress at the crack tip. Thus, facet formation appears to result from the interplay between shear stress and stresses normal to the slip plane.

11.3.6.3 Evans and Bache

Evans and Bache [61] suggested two stress states to explain dwell and cyclic failure namely, high and low stress states. In the high stress state, the facets are inclined at 45° to the tensile load axis and the fracture morphology is characterized by high levels of strain accumulation. In the high stress state, however, multiple grains undergo slip, even those not aligned with the plane of maximum shear. In the case of the lower stress, the facets occur perpendicular to the direction of tensile load and result in lower levels of strain accumulation. The authors used the Stroh model to explain the incidence of quasi-cleavage facetted morphologies in typical dwell failures. Figure 11.18 illustrates the Stroh model. This model proposes that crack growth is driven by slip and that decohesion of the slip band is dependent on large normal stresses. This indicates that facet formation dependents on critical combination of shear stress (strain) and tensile stress normal to the slip plane. Using Stroh's analysis, Evans and Bache [61] developed a shear

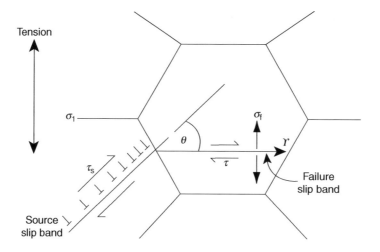

FIGURE 11.18 Schematic illustration of Stroh model.

stress long the slip band. This formulation is given by

$$\tau = \beta \tau_s \sqrt{\frac{L}{r}} \tag{11.3}$$

where τ_s is the average resolved shear stress on the source slip plane, θ, the angle of incidence between the source slip band and the failure slip band, and β is a function of θ, and L is the distance over which the lead dislocation in the pileup has traveled from the source.

11.3.6.4 Hack and Leverant

Hack and Leverant [28,29] suggested a slip band model to characterize damage evolution under dwell loading conditions. They proposed that slip bands emanating from creep deformation induced by dwell load extend across grain or colony boundaries. The progress of these slip bands across the microstructure is retarded at grain or colony boundaries. The pileup of these slip bands at colony boundaries induces shear bands, resulting in high local hydrostatic stresses.

The hydrostatic stress concentrates hydrogen atoms, which promote crack nucleation and/or hydride formation. The subsequent cracking of hydride results in crack nucleation. By idealizing the slip band as a single planar array of edges as shown in Figure 11.19, and using Irwin's formulation for a mode II fracture mode, the hydrostatic stress is defined as in Equation 11.4 by [28]

$$\sigma_H = \frac{1}{3}(\sigma_{xx} + \sigma_{yy} + \sigma_{zz}) \tag{11.4}$$

where

$$\sigma_{xx} = -\frac{K_{II}}{\sqrt{2\pi r}} \sin\frac{\theta}{2} \left[2 + \cos\frac{\theta}{2} \cos\frac{3\theta}{2} \right] \tag{11.5}$$

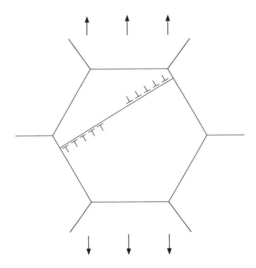

FIGURE 11.19 Schematic of double-ended dislocation pileup blocked at both ends by a grain or colony boundary. (From Hack, J. E. and Leverant, G. R., *Scripta Met.*, 14, 437–441, 1980.)

$$\sigma_{yy} = \frac{K_{II}}{\sqrt{2\pi r}} \sin\frac{\theta}{2} \left[\cos\frac{\theta}{2}\cos\frac{3\theta}{2}\right] \qquad (11.6)$$

$$\sigma_{zz} = -2\frac{K_{II}}{\sqrt{2\pi r}} \sin\frac{\theta}{2} \qquad (11.7)$$

$$K_{II} = \tau\sqrt{\pi a} \qquad (11.8)$$

Substituting Equation 11.5 through Equation 11.8 into Equation 11.4, we obtain

$$\sigma_H = -\frac{\sqrt{2}(1+v)}{3} \sqrt{\frac{a}{r}} \tau \sin\frac{\theta}{2} \qquad (11.9)$$

In the case of mixed mode I and II cracks (of equivalent length to a slip band) oriented at 45° to the tensile axis, the hydrostatic stress is given by

$$\sigma_H = \frac{(1+v)}{3} \sqrt{\frac{a}{2r}} \sigma_A \left[\cos\frac{\theta}{2} - \sin\frac{\theta}{2}\right] \qquad (11.10)$$

where K_{II} is mode II stress intensity factor, r is the distance of point of interest from the tip of the shear band and θ is the angle between the plane of the shear band and point of interest, τ is shear stress on the shear band and a is half shear band length. Thus, when the hydrostatic stress at the pileup compares in magnitude to the stress at the crack tip, the hydrostatic stress local concentration of hydrogen diffuses into the crack tip inducing crack initiation.

11.3.7 MODELING OF FATIGUE AND DWELL FATIGUE

This section presents some recently developed fracture mechanics concepts for the prediction of short and long fatigue crack growth in α/β titanium alloys. These are used to predict crack growth and crack shape evolution under fatigue and dwell fatigue loading. The modeling presumes that preexisting flaws exist due to processing/machining. Hence, the failure due to fatigue is due to the extension of short and long cracks. In any case, an initiation stage may be added simply by taking the total fatigue life, N_f to be the sum of the initiation fatigue life, N_i, and the propagation fatigue life, N_p. The resulting predictions protocols relied on fracture mechanics estimates of crack extension, Δa_i, that are given by Equation 11.11 [Dowling 1993; McDowell 1997; Anderson 1995].

$$\Delta a_i = \int_{a_i}^{a_{i+1}} da = \int_{N_i}^{N_{i+1}} f(\Delta K)dN \qquad (11.11)$$

where a is the current crack length; a_i is the initial crack length; a_{i+1} is the crack length after the crack extension, Δa_i; and the function $f(\Delta K)$ is a function that characterizes the dependence of da/dN on ΔK; and N is the number of fatigue cycles [64]. This model relies on three proposed phenomenological crack growth rates models that capture trends in the data investigated. These are given by Equation 11.12 through Equation 11.14.

These crack growth models were obtained by fitting experimental fatigue crack growth rates data with a series of fitting functions. The most appropriate functions, which captured the crack growth eccentricities of the three microstructures, were used in the life prediction protocols developed. Conservative estimates of stress intensity factor ranges were computed using elastic models proposed by Trantina et al. [84] Equation 11.15

$$\frac{da}{dN} = A(\Delta K)^n \tag{11.12}$$

$$\frac{da}{dN} = \frac{\eta}{\Delta Ka} + \beta[e^{\lambda \Delta Ka}] \tag{11.13}$$

$$\frac{da}{dN} = \alpha(\Delta Ka)^m e^{\text{Log}(\Delta Ka)} \tag{11.14}$$

$$\Delta K = F\left(\frac{2}{\pi}\right)\Delta\sigma\sqrt{\pi a} \tag{11.15}$$

where A, n, η, β, λ, α, m are fitting parameters obtained by fitting the three models to fatigue and dwell fatigue data, respectively. F is the crack shape parameter that accounts for the ellipticity of crack front evolution.

The life prediction result using this model captures the eccentricities in crack growth for Ti-6242. The results also indicate that with a representative crack growth rate model, dwell fatigue life can be predicted using existing fatigue crack growth data, hence, reducing laborious and complex experimental setups for dwell fatigue life testing [58,60].

11.3.8 Titanium Matrix Composites

Significant efforts have been made to develop titanium matrix composites (TMCs) over the past two decades [12,13,16,42,85–94]. These include composites with ceramic particulate reinforcements SiC and TiC [85,86]; whiskers of TiB [12,13,86–94,103]; and coated SiC fibers [16,42,85–89]. This section reviews the structure and properties of these composites. Some of the emerging/potential applications of titanium matrix composites are also highlighted.

11.3.8.1 Particulate-Reinforced TMCs

Particulate-reinforced TMCs are typically reinforced with SiC, TiC, or Ti_5S_3 particles [86]. Most of these are produced using low cost liquid infiltration techniques [86] or powder metallurgy methods [12]. Typical reinforcement volume fractions range between 5 and 20 vol.%, and the resulting composites have been tailored to have uniform dispersions of reinforcements (Figure 11.20). However, the basic mechanical properties of particulate-reinforced TMCs are somewhat incremental in nature (Figure 11.21). They suffer from significant loss of ductility [95], and, therefore, have limited potential for structural applications of particulate TMCs [96,97].

11.3.8.2 Whisker-Reinforced TMCs

In contrast to the particulate TMCs, whisker-reinforced titanium matrix composites have been particularly successful in emerging automotive applications [93]. There have also been efforts to develop them for potential applications in military and commercial aerospace vehicles [98,99]. Their relative success has been due largely to their attractive combinations of improved mechanical properties and low/moderate cost.

The most successful of the whisker-reinforced titanium matrix composites have been the in situ titanium matrix composites (in situ TMCs) produced via in situ reactions of Ti and B [12,93,103].

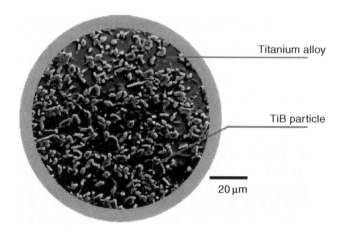

FIGURE 11.20 Microstructure of titanium matrix composite (Toyota Central R&D Lab).

These have been used successfully to fabricate ingot metallurgy (I/M), powder metallurgy (P/M) in situ TMCs, and self-propagating synthesis (SPS).

In the case of in situ I/M alloys, the in situ TMCs are produced simply by alloying with boron in the melt [13,99,100]. The solidification microstructures can also be broken down by wrought processing. However, these alloys suffer from "clustering" of reinforcements and variability in mechanical properties [99]. In contrast, powder metallurgy processing [99–103] and SPS [102] have been more successful. P/M processing has been used to produce uniform microstructures with evenly distributed TiB reinforcements.

The early work was done by Sastry [89], Satry et al. [90], and Soboyejo, Lederich, and Sastry [12], who used powder atomization techniques to produce α/β in situ TMCs with uniform

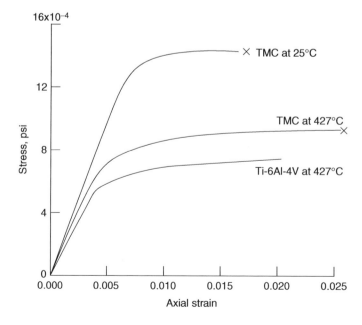

FIGURE 11.21 Stress–strain response of titanium matrix at room temperature and 427°C as compared with Ti–6Al–4V at 427°C. Strain rate, 0.001 (Thesken et al. 2002).

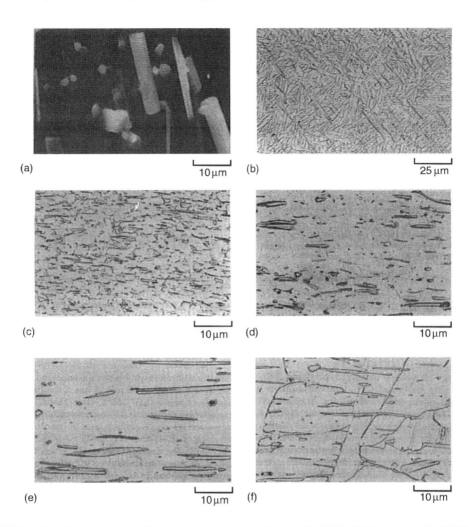

FIGURE 11.22 Microstructure of in situ titanium composites. (a) Ti–8.5Al–1B (as-extruded); (b) Ti–6Al–4V–0.5B (1200°C/1 h/AC); (c) Ti–7.5Al–4V–0.5B (704°C/24 h/AC); (d) Ti–6Al–0.5B(704°C/24 h/AC); (e) Ti–6Al–0.5B (1200°C/1 h/FC + 600C/24 h/AC); and (f) Ti–7.5Al–4V–0.5B (1200°C/1 h/AC). FC = furnaced coole and AC = Air cool.

distributions of TiB reinforcements. The in situ reinforcements were subsequently aligned during extrusion to produce composite microstructures such as those shown in Figure 11.22a–f. Note that hybrid composites, consisting of Ti_5Si_3 particulates and TiB whiskers, have also been made simply by alloying the powders with Si and B. A typical micrograph of such a hybrid composite is shown in Figure 11.22a.

P/M in situ TMCs have been produced from a wide range of in situ TMCs with different microstructures (Figure 11.22a–f). The matrix microstructures are controlled simply by heat treatment in the β and α/β phase fields, and cooling at the appropriate rates to obtain the appropriate equiaxed or transformed α/β microstructures. As with normal α/β titanium alloys, α/β in situ TMCs exhibit attractive combinations of strength (Figure 11.23), ductility, creep resistance (Figure 11.24); fracture toughness (40–60 MPa m$^{1/2}$) (Figure 11.25); and fatigue resistance (Figure 11.26) [13].

The improved strength of in situ TMCs has been attributed largely to the deformation restraint provided by the TiB whiskers. This has been modeled successfully using simple rule-of-mixture

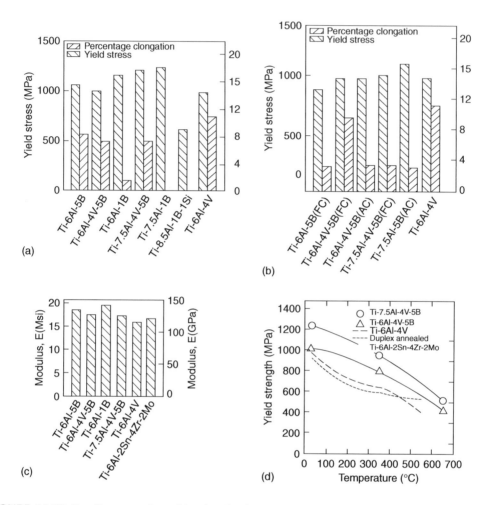

FIGURE 11.23 Tensile properties of in situ titanium composites. (a) After stabilization anneal at 704°C/24 h/AC and testing at 25°C; (b) After two-stage anneal at 1200°C/1 h/FC + 600°C/24 h/AC and testing at 25°C; (c) Effect of temperature after stablization anneal at 704°C/24 h/AC; and (d) Improvements in modulus due to TiB reinforcement Mill anneal Ti–6Al–4V data included for comparison. (From Soboyejo, W. O. Lederich, R. J., and Sastry, S. M. L., *Acta Metall. Mater.*, 42(8), 2579, 1994.)

and shear lag theory [12,100]. Similarly, the incremental improvement in fatigue crack growth resistance has been explained largely by the role of crack bridging provided by the aligned TiB whiskers (Figure 11.27). These give rise to the reduction of crack-tip stresses and an improvement in fracture toughness and fatigue crack growth resistance. This has been modeled successfully in prior work by Soboyejo et al. [13].

The improved creep behavior of the in situ TMCs has been explained by considering the deformation restraint (to flow) provided by the relatively stiff TiB whiskers. However, these may be relaxed by diffusional processes [13]. Such diffusional processes may reduce the overall extent of creep strengthening, if the whisker aspect ratio (length to diameter ratio) is not sufficiently high. The creep performance of in situ TMCs is, therefore, favored by increased whisker aspect ratios.

Similarly, property improvements have been engineered at Toyota [93], where significant progress has been made in the fabrication of automotive components such as valves, gears,

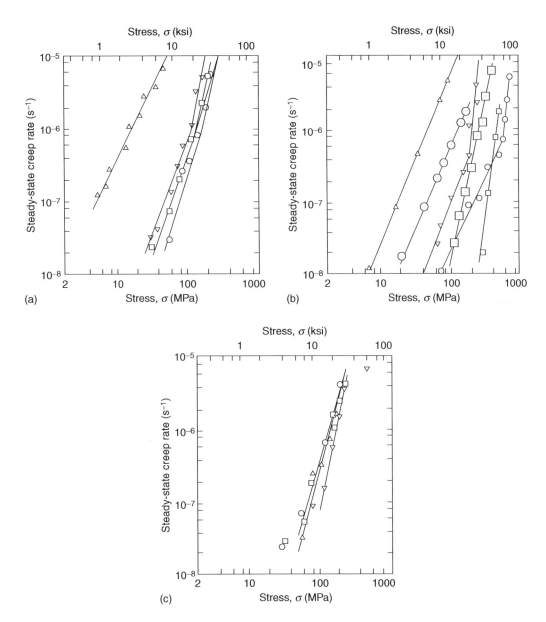

FIGURE 11.24 Stress dependence of steady state creep of Ti–6Al–4V–0.5B, Ti–7.5Al–4V–0.5B, mill annealed Ti–6Al–4V, and duplex annealed Ti–6Al–2Sn–4Zr–2Mo. (a) ($\triangle$) Milled annealed Ti–6Al–4V, (∇) duplex annealed Ti–6Al–2Sn–4Zr–2Mo, ($\bigcirc$) Stabilization annealed Ti–7.5Al–4V–0.5B, and ($\square$) Stabilization annealed Ti–7.5Al–4V–0.5B tested at 540°C; (b) ($\triangle$) Mill annealed Ti–6Al–4V, ($\blacksquare$) as-extruded PM Ti–7.5Al–4V, and (∇) duplex annealed Ti–6Al–2Sn–4Zr–2Mo tested 600°C; and ($\square$) Ti–6Al–4V–0.5B ($\bigcirc$) Ti–7.5Al–4V–0.5B tested at 650°C after stabilization annealing (704°C/24 h/AC); and (c) Effect of heat treatment on creep properties of *in situ composites tested at* 650°C ($\triangle$)Ti–6Al–4V–0.5B annealed at 1200°C/1 h/FC+600C/24 h/AC; ($\square$) Ti–6Al–4V–0.5B annealed at 704°C/24 h/AC; ($\triangle$) Ti–7.5Al–4V–0.5B annealed at 1200°C/1 h/FC+600C/24 h/AC; and ($\bigcirc$)Ti–7.5Al–4V annealed at 704°C/24 h/AC. AC = air cool and FC = furnace cool. (From Soboyejo, W. O. Lederich, R. J., and Sastry, S. M. L., *Acta Metall. Mater.*, 42(8), 2579, 1994.)

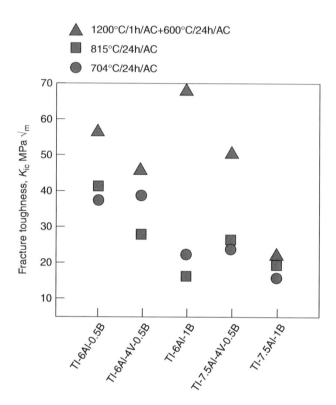

FIGURE 11.25 (a) Fracture toughness of Ti–B composites annealed at 704°C/24 h/AC; (b) 815°C/24 h/AC; and (c) 1200°C/1 h/FC + 600C/24 h/AC. FC = furnace cool and AC = air cool. (From Soboyejo, W. O. Lederich, R. J., and Sastry, S. M. L., *Acta Metall. Mater.*, 42(8), 2579–259, 1994.)

crankshafts, and connecting rods (Figure 11.28). These have been made using a solid-state in situ synthesis technique. Inlet and exhaust values produced form in situ TMCs have been made by Toyota Motor Corporation in Japan [93]. These have been used in recent models of Toyota, such as the Toyota Alzetta [93]. The applications of such valves have resulted in improved performance due to the reduction in the weight of the drive train (in situ TMCs have half the density of alloy steels used in such applications and higher temperature capability). In situ TMCs also result in modest reductions in fuel consumption ($\sim$ 1–3 miles per gallon improvements).

11.3.8.3 Fiber-Reinforced TMCs

Although significant efforts have been made to develop fiber-reinforced TMCs, real applications have been slow to emerge [96,97]. This has been due largely to the thermal expansion mismatch between the carbon-coated silicon carbon fibers (SCS-6 and SCS-9) and the α/β titanium matrices. These give rise to residual stresses and thermal stresses that can cause premature failure during thermal/thermomechanical cycling due to elevated temperatures [87]. Furthermore, fiber-reinforced TMCs are relatively expensive, due largely to the cost of the fibers and the composite processing methods.

In most cases, the composite are fabricated by the foil/fiber/foil method. The stacked foils and fibers are then hot-isostatically-pressed (HIPed) to produce composites with a wide range of architectures (stacking sequences). In situ chemical reactions occur between the titanium matrices and carbo-coated SiC fibers, giving rise to the formation of multilayered interfaces that include

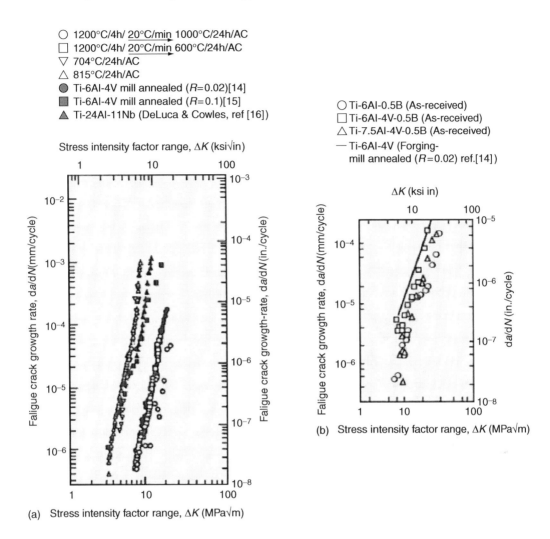

○ 1200°C/4h/ 20°C/min 1000°C/24h/AC
□ 1200°C/4h/ 20°C/min 600°C/24h/AC
▽ 704°C/24h/AC
△ 815°C/24h/AC
● Ti-6Al-4V mill annealed (R=0.02)[14]
◼ Ti-6Al-4V mill annealed (R=0.1)[15]
▲ Ti-24Al-11Nb (DeLuca & Cowles, ref [16])

○ Ti-6Al-0.5B (As-received)
□ Ti-6Al-4V-0.5B (As-received)
△ Ti-7.5Al-4V-0.5B (As-received)
— Ti-6Al-4V (Forging-
mill annealed (R=0.02) ref.[14])

(a) Stress intensity factor range, ΔK (MPa√m)

(b) Stress intensity factor range, ΔK (MPa√m)

FIGURE 11.26 Effects of heat treatment and alloying on fatigue crack growth rates. (a) Ti–8.5Al–1B–1Si and (b) Lower Al in situ Ti composites. (From Soboyejo, W. O. Lederich, R. J., and Sastry, S. M. L., *Acta Metall. Mater.*, 42(8), 2579–259, 1994.)

titanium carbide, titanium silicide, and multicomponent layers. These affect the interfacial strength and degrade the resistance to cyclic loading [42].

In the worst cases, the cracks initiate from the interfaces and propagate across the matrix. This is followed by fiber pullout and fiber failure during the final stages of damage under monotonic or cyclic mechanical and/or thermal loading [42,87]. Such complex damage modes make the potential applications of fiber-reinforced TMCs very difficult in the short/medium term. However, it is possible that future efforts to control interfacial reactions and interfacial sliding characteristics could give rise to improved composites with the potential for long-term applications in aerospace vehicles such as the X-33.

11.3.9 BIOMEDICAL APPLICATIONS

Titanium alloys, such as Ti–6Al–4V, are used extensively in biomedical applications due to their attractive combinations of biocompatibility [104,105], corrosion resistance [105,106], and

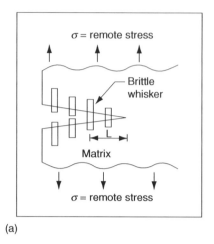

(a)

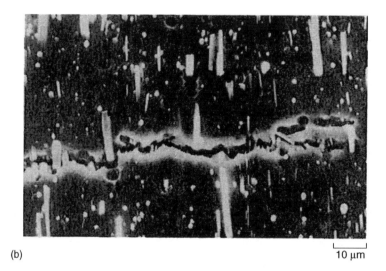

(b)   10 μm

FIGURE 11.27 Elastic crack bridging in Ti–8.5Al–1B–1Si. (a) Schematic illustration and (b) Actual crack-tip region. (From Soboyejo, W. O. Lederich, R. J., and Sastry, S. M. L., *Acta Metall. Mater.*, 42(8), 2579–259, 1994.)

structural properties [104–106]. Since titanium is one of the two metals shown not to induce any cyto-toxic reactions in the body (zirconia is the other), it is used extensively in orthopedic and dental applications such as those shown in Figure 11.29 through Figure 11.31.

In the case of the orthopedics, the applications include the stem segments of total hip replacements (Figure 11.29), screws, and plates used in several parts of the body. Titanium alloys are also used extensively in dental screws (Figure 11.30 and Figure 11.31) that attach crowns to the jawbone, as shown in Figure 11.31. These are subjected to relatively low loads/stresses compared with those in the stem sections of the hip implant, which are known to support between 5 and 10 times the body weight [107–111].

The surfaces of the titanium hip implants are often textured (Figure A) by roughening to improve their integration with bone [107–111]. Porous hydroxyapatite (HA) [112–114] and trabecular metal [115] have also been used successfully to improve the integration of the titanium implants with bone. However, further work is needed [104,108–111] to develop micro-groove textures that facilitate osseointegration and reduce scar tissue formation.

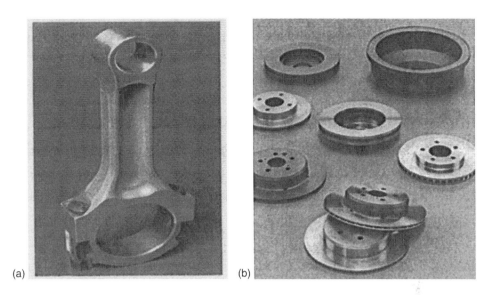

FIGURE 11.28 Examples of MMC application in the automotive industry. (a) Prototype aluminum MMC connecting rod and (b) Aluminum MMC brake rotors (Hunt and Miracle 2001).

There are currently three major concerns with the existing biomedical titanium alloys. These include (i) the cylotoxicity of the Al and V alloying elements [104], (ii) the stress-shielding of bone due to the modulus mismatch between titanium and bone [105], and (iii) the detachment that often occurs between bone and titanium implants after ~15–30 years of presence in the body [111].

11.3.10 CYLOTOXICITY IN BIOMEDICAL TI ALLOYS

In the case of the cylotoxicity, this is typically managed by the body's physiological processes [104,105]. Nevertheless, there have been efforts to develop novel Ti alloys that reduce the amounts of Al and V, or substitute them with less toxic α and β stabilizers. In the case of β stabilizer substitute, Nb has been used increasingly in biomedical titanium alloys. Zirconium,

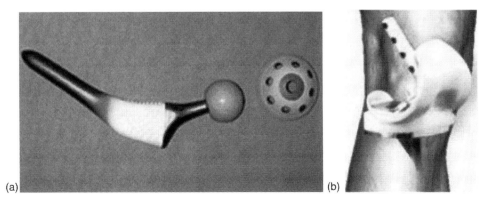

FIGURE 11.29 (a) Prototype total knee replacement. Consists of titanium upper and lower structural component and zirconia wear surface, which articulates against a UHMWPE insert on the lower section and (b) Total knee replacement prothesis including screws, pins, and plates used for replacing bones. (From Brunette, D. M., Tengvall, P., Textor, M., and Thomsen, P., Eds., *Titanium in Medicine: Material Science, Surface Science, Engineering, Biological Responses and Medical Applications*, 1st ed., Springer, Germany, 2001.)

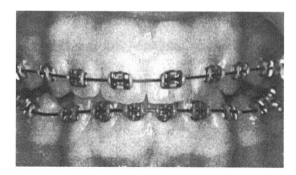

FIGURE 11.30 Titanium brackets and archwires in patient undergoing treatment. (From Brunette, D. M., Tengvall, P. Textor, M, and Thomsen, P., Eds., *Titanium in Medicine: Material Science, Surface Science, Engineering, Biological Responses and Medical Applications*, 1st ed., Springer, 2001.)

the other non-cylotoxic metal has also been used increasingly as an alloying element in biomedical alloys. This has given rise to increasing applications of Ti–Zr and Ti–Zr–Nb alloys with an attractive balance of mechanical properties. A summary of the mechanical properties of some of the biomedical titanium alloys is presented below.

It is important to note here that the biomedical applications of Ti benefited tremendously from the alloy development work done in the aerospace industry from the 1950s and beyond. Similarly, the Ti–Zr and Zr–Ti-based alloys have been in the nuclear industry.

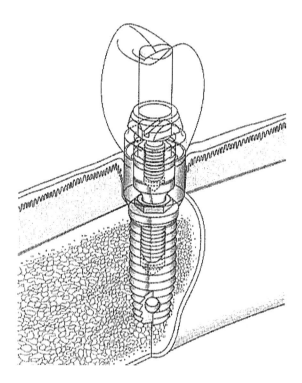

FIGURE 11.31 A schematic of titanium Branemark implant with titanium and connecting screw. (From Brunette, D. M., Tengvall, P., Textor, M., and Thomsen, P., Eds., *Titanium in Medicine: Material Science, Surface Science, Engineering, Biological Responses and Medical Applications*, 1st ed., Springer, 2001.)

11.3.11 STRESS SHIELDING IN BIOMEDICAL TITANIUM ALLOYS

Conventional α/β titanium alloys have a Young's modulus of ~ 110 GPa. This is much greater than the modulus of ~ 15–30 GPa that has been reported for bone [116,117]. This causes stress-shielding in the bone, since most of the applied loads are supported by the stiffer titanium alloys that are typically used to fabricate titanium implants. Since bone generally adapts its structure to increase its density when the supported loads are greater, stress shielding can lead to reduced bone density and increased susceptibility to fractures. However, the failures of most titanium implants are not typically due to stress-shielding [104].

11.3.12 BONE DETACHMENT FROM TI IMPLANTS BY OSTEOLYSIS

Osteolysis [105,106,117] is the third and probably the most important failure mode for titanium implants. It is caused by wear debris such as ultra-high-molecular-weight-polyethylene (UHMWPE) that formed due to contact between the femoral head and the UHMWPE cup in total hip implants (Figure 11.29). These give rise to micron-scale wear debris transported to the interface between the titanium stem and the bone in the femur. These particles induce osteolysis (bone loss) that leads ultimately to the detachment of the stem of the implant from the bone. This reduced the service lives of earlier implants to 10–15 years.

However, in recent years, implants with longer lives have been engineered by using ceramic femoral heads that liberate much lower levels of wear debris. These include alumina and zirconia femoral heads with ultra-small grain sizes [118–120]. Such femoral heads have much lower wear rates in contact with UHMWPE and are expected to give rise to much greater clinical lives. However, there are still some durability concerns in the case of the ceramic femoral heads due to their susceptibility to brittle fracture [119].

The microstructures of the yttria-stabilized zirconia femoral heads, which consist of tetragonal and monochromic phases, have also been shown to be unstable in the presence of body fluids. Such microstructural instability has resulted in recent failures of zirconia femoral heads attached to titanium alloy stems.

For these reasons, fine-grained alumina femoral heads, attached to titanium stems, have become increasingly popular in recent years. However, since these generally support between 5 and 10 times the body weight, they can be susceptible to failure in the case of larger and heavier patients.

In any case, the stems of hip implants and screws in dental and orthopedic implants are increasingly being fabricated from α/β titanium alloys. Surface texturing and porosity (HA or traebecular metal) are also being used to improve the integration between the titanium stem and bone. Similarly, laser textures and rough surface textures (produced by blasting with Al_2O_3 pellets) have also been used to improve the integration between dental screws and the jawbone [109].

11.4 CONCLUSIONS

The study of fatigue behavior of titanium and its alloys spans more than five decades. Each year, millions of dollars are provided for titanium research and its fatigue behavior in advanced structural components. Despite these efforts, our ability to use titanium in structural components is impeded by the absence of mechanistic-based life prediction models. Secondly, our current understanding of the influence of microstructural variability, (including the role of microstructural processing defects such as hard α phase), micro-textural effects, and the role of titanium anisotropy on dwell fatigue life is still limited.

Furthermore, mechanisms associated with the early stages of dwell fatigue crack growth, especially the mechanism of facet formation, are not well understood. The prediction of surface/subsurface crack growth is also limited by the lack of data on crack shape evolution in the different regimes of crack growth. It is evident from the foregoing discussion that the quest to develop advanced titanium-based structural materials has been met with mixed success. In the case of advanced composites, incremental improvements have been achieved with in situ whisker-reinforced titanium matrix composites. These are now being used in limited applications (valves) in automobile engines. However, efforts to develop titanium matrix composites for airframes or aero-engines are still ongoing.

In contrast, titanium alloys have been applied successfully in medicine and dentistry. The success has been due largely to the excellent combination of biocompatibility, corrosive resistance, and mechanical properties. However, further work is required to engineer improved interface for the integration of Ti surface into the biological environment. This represents an interesting area for future interdisciplinary research.

REFERENCES

1. Williams, J. C., In *Phase Transformation in Titanium Alloys: A Review*, Jaffee, R. I. and Burte, H. M., Eds. *Titanium Science and Technology*, Vol. 3, Plenum Press, New York, p. 1433, 1973.
2. Williams, J. C. and Blackburn, M. J., A comparison of phase transformation in three commercial titanium alloys, *ASM Quart. Trans.*, 60, 373, 1967.
3. Eylon, D. and Froes, F. H., *Titanium Technology: Present Status and Future Trends*, Titanium Development Association, Dayton, OH, 1985.
4. Collings, E. W., *Physical Metallurgy of Titanium*, ASM, Materials Park, OH, 1994.
5. Irving, P. E. and Beevers, C. J., Microstructural influences on fatigue crack growth in Ti–6Al–4V, *Mater. Sci. Eng.*, 14, 229–238, 1974.
6. Yoder, G. R. and Eylon, D., *Metall. Trans.*, 10A, 1808, 1979.
7. Yoder, G. R., Cooley, L. A., and Crooker, T. W., *Metall. Trans.*, 8A, 1737, 1977 (a).
8. Yoder, G. R., Cooley, L. A., and Crooker, T. W., *J. Eng. Mater. Technol.*, 99, 313–318, 1977 (b).
9. Yoder, G. R., Cooley, L. A., and Crooker, T. W., *Eng. Fract. Mech.*, 11, 805–816, 1979.
10. Yoder, G. R., Cooley, L. A., and Crooker, T. W, In *Fatigue 84*, Beevers, C., Ed., Engineering Materials Advisory Services Ltd., West Midland, UK, pp. 351–360.
11. Boyce, B. L. and Ritchie, R. O., Effect of load ratio and maximum stress intensity on the fatigue threshold in Ti–6Al–4V, *Eng. Fract. Mech.*, 68, 129–147, 2001.
12. Soboyejo, W. O., Lederich, R. J., and Sastry, S. M. L., Mechanical behavior of damage tolerant TiB whisker-reinforced in situ titanium matrix composites, *Acta Metall. Mater.*, 8, 2579, 1994.
13. Soboyejo, A. B. O., Foster, M., Shademan, S., Katsube, N., and Soboyejo, W. O., *Fatigue Eng. Mater. Struct.*, 24, 225–241, 2001.
14. Shen, W., Soboyejo, A. B. O., and Soboyejo, W. O., Microstructural effects on fatigue and dwell-fatigue crack growth in α/β Ti–6Al–2Sn–4Zr–2Mo–0.1Si, *Met. Trans. A*, 35A, 163, 2004 (a).
15. Shen, W., Soboyejo, W. O., and Soboyejo, A. B. O., An investigation of fatigue and dwell fatigue crack growth in Ti–6Al–2Sn–4Zr–2Mo–0.1Si, *Mech. Mater.*, 36, 117–140, 2004 (b).
16. Soboyejo, W. O., Rabeeh, B. M., Zhang, J., and Katsube, N., Tensile damage in a symmetric $[0/90]_{2s}$ silicon–carbide fiber-reinforced titanium-matrix, *Compos. Sci. Technol.*, 58(6), 915–931, 1998.
17. Shademan, S. and Soboyejo, W. O., An investigation of short and long crack growth in Ti–6Al–4V with colony microstructures, *Mater. Sci. Eng.*, A335, 116–127, 2002.
18. Shademan, S. et al., A physically-based model for the prediction of long fatigue crack growth in Ti–6Al–4V, *Mater. Sci. Eng.*, A315, 1–120, 2001.
19. Sinha, V. and Soboyejo, W. O., *Mater. Sci. Eng.*, A287, 30–42, 2000.
20. Shankar, P. and Ravichandran, K. S., Fatigue of beta annealed Ti–10V–2Fe–3Al alloy: Effects of aging and mechanisms of crack nucleation, *Proceedings Fatigue*, 2002, EMAS, Warley, Blom, A. F., Ed., pp. 1789–1796, 2002.
21. Beevers, C. J. and Halliday, M. D., Non-closure of cracks and fatigue crack growth in β heat-treated Ti–6Al–4V, *Int. J. Fract.*, 15, R27–R30, 1979.
22. Suresh, S. and Ritchie, R. O., *Int. Met. Rev.*, 29, 445, 1984.
23. Sinha, V., Ph.D. Thesis, The Ohio State University, 1999.
24. Taylor, D. and Knott, J. F., Fatigue crack propagation behavior of short cracks: The effects of microstructure, *Fatigue Eng. Mater. Struct.*, 4, 147–155, 1981.

25. Soboyejo, A. B. O., Foster, M., Shademan, S., Katsube, N., and Soboyejo, W. O., *Fatigue Eng. Mater. Struct.*, 24, 225–241, 2001.
26. Eylon, D. and Hall, J. A., *Met. Trans. A*, 8A, 981, 1977.
27. Evans, W. J. and Gostelow, C. R., *Met. Trans. A*, 10A, 1837, 1979.
28. Hack, J. E. and Leverant, G. R., *Met. Trans. A*, 13A, 1729, 1982.
29. Hack, J. E. and Leverant, G. R., Leverant, *Scripta Met.*, 14, 437–441, 1980.
30. Wojcik, C. C., Chan, K. S., *and* Koss, D. A., *Acta Metall.*, 36(5) 1261–1270, 1988.
31. Bache, M. R. and Evans, W. J. et al., *J. Mater. Sci.*, 32(13), 3435–3442, 1997.
32. Bache, M. R., Davies, H. M., and Evans, W. J., *Titanium '95: Science and Technology*, The Institute of Materials, London, UK, p. 1347, 1996.
33. Spence, S. H. Evans, W. H. and Cope, M., Advances in fracture research, *Proceedings of 9th International Conference on Fracture*, Pergamon, p. 1571, 1997.
34. Nikbin, K. and Radon, J., Prediction of fatigue interaction from static creep and high frequency fatigue crack growth data advances in fracture research, *Proceedings of the 9th International Conference on Fracture*, 1–5 April 1997, Pergamon, Sydney, Australia, p. 423, 1997.
35. Woodfield, A. P., *Dwell Fatigue Cracking in Titanium Alloys*, Presentation to AIAA Rotor Integrity Subcommittee, October 28th, 1998.
36. Woodfield, A. P. et al., Effect of microstructure on the dwell fatigue behavior of Ti-6242, *Titanium '95: Science and Technology*, pp. 1116–1123, 1995.
37. Goswami, T., Low cycle fatigue—dwell effects and damage mechanisms, *Int. J. Fatigue*, 21, 55–76, 1999.
38. Saxena, V. K. and Radhakrishnan, V. M., Effect of phase morphology on fatigue crack growth behavior of α–β titanium alloys—a crack closure rationale, *Metall. Mater. Trans. A*, 29A (January), 245–261, 1998.
39. Koss, D. A., Eylon, D., and Boyer, R. R., Eds. *Beta Titanium Alloys in the 1990s*, Times of Acadiana Pr Inc., December 1993.
40. Lampman, S., Wrought titanium and titanium alloys, *ASM Handbook*, pp. 595–629, 1990.
41. Fujishiro, S. et al., *Metallurgy and Technology of Practical Titanium Alloys: Proceedings*, Times of Acadiana Pr Inc., March 1995.
42. Rabeeh, B. M., Ramasudaram, P., and Soboyejo, W. O., The tensile behavior of silicon carbide fiber-reinforced titanium matrix composite, *Appl. Compos. Mater.*, 3, 215–247, 1996.
43. Kim, Y.-W., Ordered intermetallic alloys, part III: gamma titanium aluminides, *J. Met.*, 30–39, 1994.
44. MIL-HDBK-5H 1, Dec 1998.
45. Donachie, M. J., Ed, *Titanium: A Technical Guide* 2nd ed., ASM International, Metals Park, OH, 2000.
46. Suri, S., Hou, D.-H., Neeraj, T., Daehn, G. S., Scott, J. M., Hayes, R. W., and Mills, M. J., Creep of titanium alloys at lower temperatures, *Mater. Sci. Eng., A*, A234–236, 996, 1997.
47. Suri, S., Neeraj, T., Daehn, G. S., Hou, D.-H., Scott, J. M., Hayes, R. W., and Mills, M. J., Mechanisms of primary creep in titanium alloys at lower temperatures, In *Creep and Fracture of Engineering Materials and Structures*, Earthman, J. C. and Mohamed, F. A., Eds., TMS Publications, Warrendale, PA, p. 119, 1997.
48. Suri, S., Viswanathan, G. B., Neeraj, T., and Mills, M. J., Room temperature deformation and microstructural characterization of two phase α/β titanium alloys, *Acta Mater.*, 47, 1019–1034, 1999.
49. Dubey, S., Soboyejo, A. B. O., and Soboyejo, W. O., An investigation of the effects stress ratio and crack closure on the micromechanisms of fatigue crack growth in Ti–6Al–4V, *Acta Metall. Mater.*, 45, 2777–2787, 1997.
50. Chestnutt, J., Thompson, A. W., and Williams, J. C., *Titanium 80: Science and Technology*, AIME, Warrendale, PA, p. 1875, 1980.
51. Chestnutt, J., Thompson, A. W., and Williams, J. C., In *Fatigue 84*, Beevers, C., Ed., Engineering Materials Advisory Services Ltd., West Midland, UK, Vol., pp. 314–350, 1984.
52. Hicks, M. J. and Pickard, A. C., *Int. J. Fract.*, 20, 91–101, 1982.
53. Gray, G. T., Williams, J. C., and Thompson, A. W., *Metall. Trans. A*, 14A, 421–433, 1983.
54. Ravichandran, K. S., *Acta Metall.*, 39, 401, 1991.

55. Ravichandran, K. S., Jha, S. K., and Shankar, P. S., Fatigue of beta titanium alloys: crack initiation, growth and fatigue life, In *Proceedings of Fatigue*, Blom, A. F., Ed., Vol. 3, EMAS, Warley, Cradley Heath, UK, pp. 1751–1762, 2002.

56. McEvily, A. J., Current aspects of fatigue, *Met. Sci.*, 11, 260–274, 1977.

57. McEvily, A. J. and Boettner, R. C., On the fatigue crack propagation in FCC metals, *Acta Metall.*, 11, 725–744, 1963.

58. McBagonluri, F., Akpan, E., Mercer, C., Shen, W., and Soboyejo, W. O., An investigation of the effects of microstructure on fatigue crack growth in Ti-6242, *J. Eng. Mater. Technol.*, 127(1), 46–57, 2005.

59. McBagonluri, Ph.D. Thesis, University of Dayton, OH, 2005.

60. McBagonluri, F., Akpan, E., Mercer, C., Shen, W., and Soboyejo, W. O, An investigation of fatigue and dwell fatigue in Ti-6242, *Mater. Sci. Eng., A*, 405(1–2), 111–134, 2005.

61. Evans, W. J. and Bache, M. R., *Int. J. Fatigue*, 16, 443–452, 1994.

62. Evans, W. J., Jones, J. P., and Bache, M. R., High temperature fatigue/creep/environment interactions in compressor alloys, *J. Eng. Gas Turbines Power*, 125, 246–251, 2003.

63. Ritchie, R. O. et al., High fatigue and time-dependent in metallic alloys for propulsion system, AFOSR F49620-96-1-0478, 2001.

64. Soboyejo, W. O., *Mechanical Properties of Engineered Materials*, Marcel Dekker, New York, 2004.

65. Katcher, M. and Kaplan, M, Effects of *R*-ratio and crack clistes on the fatigue crack growth in aluminum and titanium alloys, fracture toughness and slow-stable cracking, ASTM STP 559, American Society of Testing and Materials, pp. 264–282, 1974.

66. Doker, H. and Bachmann V, Determination of crack opening load by use threshold behavior, Mechanics of Fatigue Crack Closure, ASTM STP 982, Newman, J. C. and Elber, W., Eds. American Society of Testing and Materials, pp. 247–259, 1988.

67. Suresh, S., Fatigue crack deflection and facture surface contact: micromechanical models, *Metall. Trans.*, 16A, 249–260, 1985.

68. Ritchie, R. O., Gilbert, C. J., and McNaney, J. M., Mechanics and mechanisms of fatigue damage and crack growth in advanced materials, *Int. J. Solids Struct.*, 2000, 311–329, 2000.

69. Sansoz, F. and Ghonem, H., Fatigue crack growth mechanisms in Ti6242 lamellar microstructure: influences of loading frequency *and* temperature, *Met. Trans. A*, 35A, 2565, 2003.

70. Bache, M. R. and Evans, W. J. et al., Characterization of mechanical anisotropy in titanium alloys, *Mater. Sci. Eng., A: Struct. Mater. Prop. Microstruct. Process.*, A257(1), 139–144, 1998.

71. Allison, J. E., The measurement of crack closure during fatigue crack growth, In *Fracture Mechanics, 18th Symposium*, ASTM STP 945, Read, D. T. and Reed, R. P., Eds. American Society of Testing and Materials, Philadelphia, PA, pp. 913–933, 1998.

72. Sehitoglu, H. and Garcia, A. M., Contact of crack surfaces during fatigue, *Metal. Trans.*, 28A, 2263–2289, 1997.

73. Newman, J. C., Finite element analysis of fatigue crack closure, In *Mechanics of Fatigue Crack Growth*, ASTM STP 590, American Society of Testing and Materials, Philadelphia, PA, pp. 281–301, 1976.

74. Budiansky, B. and Hutchinson, J. W., Analysis of closure in fatigue crack growth, *J. Appl. Mech.*, 45, 267–276, 1978.

75. Fleck, N. A., Finite element analysis of plasticity-induced crack closure under plane strain conditions, *Eng. Fract. Mech.*, 25, 411–449, 1986.

76. Thompson, A. W., William, J. C., Frandsen, J. D., and Chesnutt, J. C. The effect of microstructure on fatigue crack propagation rate in Ti–6Al–4V, In *Titanium and Titanium Alloys*, Vol. 1, pp. 691–704, 1982.

77. Thompson, A. W., Williams, J. C., Frandsen, J. D., and Chesnutt, J. C., In *Titanium and Titanium Alloys: Scientific and Technological Aspects*, Williams, J. C. and Belov, A. F., Eds., Plenum Press, New York, p. 691, 1982.

78. Hall, I. W. and Hammond, C., Fracture toughness and crack propagation in titanium alloys, *Mater. Sci. Eng.*, 32, 241–253, 1978.

79. Boyce, B. L. and Ritchie, R. O., Effect of load ratio and maximum stress intensity on the fatigue threshold in Ti–6Al–4V, *Eng. Fract. Mech.*, 68, 129–147, 2001.

80. Chu, H. P., MacDonald, B. A, and Arora, O. P., *Titanium Science and Technology*, Munich, FRG 4, pp. 2395, 1984.

81. Bache, M. R., Cope, M., Davies, H. M., Evans, W. J., and Harrison, G., *Int. J. Fatigue*, 19, S1, 1997, see also, page S83.

82. Kassner, M. E., Kosaka, Y., and Hall, J. A., *Met. Trans. A*, 30A, 2383, 1999.

83. Davidson, D. L. and Eylon, D., *Met. Trans. A.*, 11A, 837, 1980.

84. Trantina, G. G., deLorenzi, H. G., and Wilkening, W. W., Three-dimensional elastic-plastic finite element analysis of small surface cracks, *Eng. Fract. Mech.*, 18(5), 925–938, 1983.

85. Shang, J.-K. and Ritchie, R. O., Monotonic and cyclic crack growth in a TiC-particulate-reinforced Ti–6AI–4V metal-matrix composite, *Scripta Metall. Mater.*, 24(9), 1691–1694, 1990.

86. Shang, J. K. and Liu, G., Role of composite interface in fatigue crack growth. *Proceedings. of Conference. Control of Interfaces, Metal and Ceramics Composites*, San Francisco, CA, pp. 187–196, 1994.

87. Majumdar, B. S., Nawaz, G. M., and Ellis, J. R., *Met. Trans.*, 24A, 1597–1610, 1993.

88. Lerch, B. A. and Saltsman, J., Tensile deformation damage in SiC reinforced Ti–15V–3Cr–3Al–3Sn, *NASA Technical Memorandum 103620*, NASA-Lewis Research Center, Cleveland, OH, 1991.

89. Sastry, S. M. L., *In* Rapid Solidification Processing, Principles and Technologies, Mehrabian, R, Parrish, P. A., Eds., Claitor's, Baton Rouge, LA, p. 165, 1988.

90. Sastry, S. M. L., O'Neal, J. E., and Peng, T. C., In *International Metallographic Society*, Blum, M. E., French, P. M., Middleton, R. M., and Van der Voor, G. F., Eds., Elsevier, Amsterdam, p. 275, 1986.

91. Soboyejo, W. O., Investigation of the effects of matrix microstructure and interfacial properties on the fatigue and fracture behavior of a Ti–15V–3Cr–3Al–3Sn/SCS9, *Mater. Sci. Eng., A*, 183(1–2), 49–58, 1994.

92. Saito, T. and Furuta, T., Toyota Motor Corporation, Private Communication, Japan, 1998.

93. Saito, T., Furuto, T., and Yamaguchi, T., *Developments of Low-Cost Titanium Matrix Composites*, The Minerals, Metals and Materials Society, Warrendale, PA, 1995.

94. Saito, T., Takamiya, T. H., and Furuta, T., Thermomechanical properties of P/M β titanium metal matrix composite, *Mater. Sci. Eng., A*, 243(1–2, 15), 273–278, 1998.

95. The Materials Institute, The Selection and Use of Titanium, A Design Guide, Materials Information Service, Carlton House Terrace, London, p. 119.

96. Harrigan, W. C., In *Handbook of Metallic Composites*, Ochai, S., Ed., Marcel Dekker Inc., New York, pp. 759–773, 1994.

97. Clyne, T. W., Withers, P. J., Clarke, D. R., Suresh, S., and Ward, I. M., Eds., *An Introduction to Metal Matrix Composites (Cambridge Solid State Science Series)*, Cambridge University Press, London, UK, 1993.

98. Grosse, S. and Miracle, D. B., Mechanical properties of Ti–6Al–4V/TiB composites with randomly oriented and aligned TiB reinforcements, *Acta. Mater.*, 51(9), 2427–2442, 2003.

99. Lederich, R. J., Soboyejo, W. O., and Srivatsan, T. S., Preparing damage-tolerant titanium-matrix composites, *JOM*, 46(11), 68–71, 1994, see also, page 83

100. Dubey, S., Srivatsan, T. S., and Soboyejo, W. O., Fatigue crack growth in an in-situ titanium matrix composites, *Int. J. Fatigue*, 22, 161–174, 2000.

101. Yang, J. M. and Jeng, S. M., *J. Met.*, 44, 53–57, 1992.

102. Jeng, S. M., Shih, C. J., Kai, W., and Yang, J. M., *Mater. Sci. Eng., A*, 144, 189–196, 1989.

103. Rangarajan, S., Aswath, P. B., and Soboyejo, W. O., Microstructure development and fracture of in-situ reinforced Ti–8.5Al–1B–1Si, *Scripta Mater.*, 35(2), 239–245, 1996.

104. Brunette, D. M., Tengvall, P., Textor, M., and Thomsen, P., Eds., *Titanium in Medicine: Material Science, Surface Science, Engineering, Biological Responses and Medical Applications (Engineering Materials) (Hardcover)* 1st ed., Springer, Heidelberg, Germany, 2001.

105. Ratner, B. D., Hoffman, A. S., Schoen, F. J., and Lemons, J. E., Eds., *Biomaterials Science: An Introduction to Materials in Medicine*, Academic Press, New York, 1996.

106. Ratner, B. D., New ideas in biomaterials science—a path to engineered biomaterials, *J. Biomed. Mater. Res.*, 27837–27850, 1993.

107. Thomas, K. A., Kay, J. F., Cook, S. D., and Jarcho, M., The effect of surface texture on the mechanical strengths and histologic profiles of titanium implant materials, *J. Biomed. Mater. Res.*, 21, 1395–1414, 1987.

108. Ricci, J. L. and Alexander, H., Laser microtexturing of implant surfaces for enhanced tissue key engineering materials, *Mater. Sci. Eng.*, 198–199, pp. 179–202, 2001.
109. Ricci, J. L., Spivak, J. M., Blumenthal, N. C., and Alexander, H., Modulation of bone ingrowth by surface chemistry and roughness, In *The Bone-Biomaterial Interface*, Davies, J. E., Ed., University of Toronto Press, Toronto, Canada, pp. 334–349, 1991.
110. Ricci, J. L., Charvet, J., Frenkel, S. R., Chang, R., Nadkarni, P., Turner, J., and Alexander, H., Bone response to laser microtextured surfaces, In *Bone Engineering*, Davies, J. E., Ed., Em Inc., Toronto, Canada, pp. 1–11, 2000, Chapter 25.
111. Soboyejo, W. O., Nemetski, B., Allameh, S., Mercer, C., Marcantonio, N., and Ricci, J., On the interaction between MC-3t3 cells and textured Ti–6Af–4Vr surfaces, *J. Biomed. Mater. Res.*, 62, 56–72, 2002.
112. Ripamonti, U., Osteoinduction in porous hydroxyapatite implanted in hererotopic sites of different animal models, *Biomaterials*, 17, 31–35, 1996.
113. Hing, K. A., Best, S. M., and Bonfield, W., Characterisation of porous hydroxyapatite, *J. Mater. Sci.: Mater. Med.*, 10, 135, 1999.
114. Litsky, A. S., Sandhage, K. H., Saw, E., Briggs, S., and Gallagher, P. K., Near net-shape fabrication of hydroxyapatite and ha-alloy composites, Key Engineering Materials, *Mater. Sci. Eng.*, 198–199, 101–114, 2001.
115. Hayakawa, T., Yoshinari, M., Kiba, H., Yamamoto, H., Nemoto, K., and Jansen, J. A., Trabecular bone response to surface roughened and calcium phosphate (Ca–P) coated titanium implants, *Biomaterials*, 23(4), 1025–1031, 2002.
116. Frankel, V. H. and Nordin, M., Eds., *Basic Biomechanics of the Musculoskeletal System* 3rd ed., Lippincott Williams and Wilkins, Philadelphia, PA, 2001.
117. Thorngren, K.-G. and Poitout, D., Eds., *Biomechanics and Biomaterials in Orthopedics* 1st ed., Springer, Heidelberg, Germany, 2004.
118. Fruh, H.-J. and Willmann, G., Tribological investigations of the wear couple alumina-CFRP for total hip replacement, *Biomaterials*, 19(13), 1145–1150, 1998.
119. Masonis, J. L., Bourne, R. B., Ries, M. D., McCalden, R. W., Salehi, A., and Kelman, D. C., Zirconia femoral head fractures: A clinical and retrieval analysis, *J. Arthroplasty*, 19(7), 898–905, 2004.
120. Saikko, V., Wear of polyethylene acetabular cups against zirconia femoral heads studied with a hip joint simulator, Wear, 176(2), 207–212, 1994.

12 Niobium Alloys and Composites

Kwai S. Chan
Southwest Research Institute, Mechanical and Materials Engineering Division

CONTENTS

12.1 Introduction 401
12.2 Alloy Types, Compositions, and Microstructures 402
    12.2.1 Nb–Si–X Alloys 403
    12.2.2 Nb–Cr–X Alloys...404
    12.2.3 Nb–Al–X 406
    12.2.4 Nb–Si–Cr–Al–X 407
12.3 Physical and Tensile Properties 408
12.4 Fracture Resistance...412
    12.4.1 Solid Solution Alloys 412
    12.4.2 In Situ Composites 416
12.5 Fatigue Crack Growth Resistance...417
    12.5.1 Solid Solution Alloys 417
    12.5.2 In Situ Composites 419
12.6 Creep Behavior...422
12.7 Oxidation Properties...427
12.8 Applications..431
12.9 Summary...432
Acknowledgments 432
References 432

12.1 INTRODUCTION

Considerable efforts have been made to develop new high-temperature materials for gas turbine engine applications. These efforts have been motivated by the desire to increase the thrust to weigh ratio by increasing the operating temperature of the engine. Since Ni-based superalloys are currently used at about 0.8–0.9 of the melting point, any increase in the engine operating temperature requires new structural materials that can withstand the higher temperature requirements [1].

Several Nb-based multiphase alloys have been developed as possible high-temperature materials for the future. The compositions of these Nb-based materials generally contain Si [2–17], Cr [18–31], Al [18,31–35], Ti [5–9,12–17,19–35], Hf [12,17,33], Ge [17], Fe [17] and among other alloying elements, Sn [17]. These Nb-based alloys are often referred to as refractory metal intermetallic composites or in situ composites because they can be processed to exhibit a multiphase microstructure containing a significant amount of one or more intermetallic phases embedded a metallic matrix or an intermetallic matrix with ductile metallic particles. Such a multicomponent, multiphase microstructure is formed through composition control but resembles those of the ductile-phase toughened composites obtained by phase mixing [36–38], and hence, the name in situ composites.

Recent Nb-based in situ composites include those based on silicides [2–17,28], Laves phase, [18–22] or both [5–7,12,28]. Substantial efforts have been made to develop Nb_5Si_3/Nb [2–5]; $NbCr_2$/Nb [18–21]; and Nb_3Al/Nb systems [18]; ternary variants [9,19,20,29–34]; as well as multi-component variants [5–17,26–28,35] of these materials systems. Since these alloys are intended for high-temperature structural applications, alloy development efforts have focused on developing high strength materials with high creep resistance, improved oxidation resistance, and enhanced fracture and fatigue properties.

Extensive research has demonstrated that alloy addition can alter material properties in the Nb solid solution [19,20,26,39–43]; silicides [5–9,12–17,38]; and Laves phases [5–7,17–28,40]. Depending on the alloy composition, as many as four silicides (Nb_5Si_3, Nb_3Si, Ti_5Si_3, Ti_3Si) and two Laves phases (C14 and C15 $NbCr_2$) in alloyed forms can exist in the microstructure of Nb-based in situ composites [5–9,12–17,28]. The large number of potential alloying elements and constituent phase makes the discovery of beneficial alloy additions a daunting task if undertaken via empirical means. Thus, there have been considerable interests in developing computational tools [26,44–53] for designing Nb-based materials with desired composition, microstructure, and performance. Computational models for predicting fracture toughness [26,27,53,54]; creep [55] and oxidation resistance [28,56,57] were used to predict alloy compositions based on three considerations: (a) the desired Nb solid solution alloy, and (b) the volume fraction of intermetallics, and (c) the properties of candidate Nb-based in situ composites. These models have also been used to develop a fundamental understanding of the microstructure/property relationships in Nb-based in situ composites.

This chapter summarizes recent advances in the development of high-temperature Nb-based alloys and composites and current understandings of the structure/property relationships in this new class of structural alloys. Section 12.1 introduces the alloy system, while Section 12.2 presents a summary of the alloy types, nominal compositions, and microstructures. Most of these materials are experimental alloys that are still under development. Section 12.3 discusses the physical and tensile properties. Section 12.4 through Section 12.6 highlight the fracture toughness, fatigue crack growth resistance, and creep behavior, respectively. Section 12.7 summarizes the oxidation behavior of current Nb-based alloys. The applications of this class of emerging structural alloys are discussed in Section 12.8. Section 12.9 closes the chapter by providing a summary of the state-of-the-art of the alloy system.

12.2 ALLOY TYPES, COMPOSITIONS, AND MICROSTRUCTURES

Recent Nb-based multiphase alloys are derivative of the Nb–Si, Nb–Cr, and Nb–Al binary systems. These alloys can be divided into four main categories according to the strengthening phases, which are (a) Nb–Si–X [2–17], which are strengthened by silicides; (b) Nb–Cr–X [18–31], which are strengthened by Laves phases; (c) Nb–Al–X [18–35,51,52], which are strengthened by aluminides; and (d) Nb–Cr–Si–Al–X [5–7,12,17,28], which are strengthened by both silicides and Laves phases. The parameter, X, represents a list of alloying elements that include Ti [5–17,19–28]; Hf [14,28,51,52,58]; Ge, Fe, Sn, Ta, W, Mo, and among others, Pd [51,52]. Among the elements in this list, Ti is the top alloying element as Ti addition is common to all four types of multiphase Nb alloys. These alloys have been fabricated via a number of processing methods, including conventional arc-melting [18–20,28]; powder-metallurgy technique [2]; casting and extrusion [12]; mechanical alloying [25]; rapid solidification [24]; directional solidification [15]; as well as investment casting [17]. A summary of the chemical compositions of selected Nb-based alloys is shown in Table 12.1. Figure 12.1a–d show the ternary phase diagrams for Nb–Ti–Si [57], Nb–Ti–Cr [57], Nb–Cr–Si [57], and Nb–Ti–Al [33,60], respectively, to illustrate the various Nb-based alloys. The salient characteristics of each of these four classes of multiphase Nb alloys are discussed in the next four subsections.

TABLE 12.1
Composition and Microstructures of Selected Nb-Based Alloys and In Situ Composites

Alloy	Microstructure	References
Nb–10Si	$\beta + Nb_5Si_3$	[3,5]
Nb–16Si	$\beta + Nb_5Si_3$	
Nb–10Ti–10Si	$\beta + (Nb,Ti)_5Si_3 + (Nb,Ti)_3Si$	[8]
Nb–15Ti–10Si	$\beta + (Nb, Ti)_3Si$	
Nb–45Ti–15Si	$\beta + (Ti,Nb)_3Si$	
Nb–40Ti–5Al–15Si	$\beta + (Nb,Ti)_5(Si,Al)_3 + (Ti,Nb)_5(Si,Al)_3$	
Nb–30Cr	$\beta + Cr_2Nb$	[19,21]
Nb–50Cr	$\beta + Cr_2Nb$	
Nb–37Ti–13Cr	β	
Nb–40Ti–18Cr	β	
Nb–12Ti–8Cr	β	
Nb–29Ti–29Cr	$\beta + Cr_2Nb$	
Nb–27Ti–36Cr	$\beta + Cr_2Nb$	
Nb–18Al	$\beta + Nb_3Al$ (A15)	[18]
Nb–40Ti–15Al	B2	[5]
Nb–25Ti–15Al	$B2 + Nb_3Al$	
Nb–10Ti–15Al	$B2 + Nb_3Al$	
Nb–40Ti–10Cr–10Al	$\beta/B2(+Orthorhombic)$	[5]
Nb–45Ti–10Al	B2	
Nb–47Ti–14Al–7Cr	B2	
Nb–27Ti–4Hf–2.5Cr–1Si–3Ge (Nbx)	β + small amounts of Ge-rich intermetallics	[28]
Nb–27Ti–4Hf–12Cr–10Si–5Ge (M1)	$\beta + D8_1 (Nb,Ti)_5Si_3 + D8_8(Ti,Nb)_5Si_3 + C14$ Laves	
Nb–22Ti–4Hf–16Cr–17Si–5Ge (M2)	$D8_1 (Nb,Ti)_5Si_3 + C14$ Laves	
Nb–22Ti–4Hf–12Cr–15Si–5G3 (AX)	$\beta + D8_1 (Nb,Ti)_5Si_3 + D8_8 (Ti,Nb)_5Si_3 + C14$ Laves	
Nb–22Ti–2Hf–7Cr–9Si–5Ge–4Fe–2Al–1Sn (CNG-1B)	$\beta + D8_1 (Nb,Ti)_5Si_3 + D8_8 (Ti,Nb)_5Si_3 + C14$ Laves	

12.2.1 Nb–Si–X Alloys

Binary Ni–Si alloys typically have a Si content ranging from 10 to 25 at.% Si [5–17]. Unless stated otherwise, all compositions in this chapter are in atomic percent. The microstructure comprises a metallic solid solution phase embedded in an intermetallic matrix of niobium silicides, which include Nb_5Si_3, Nb_3Si, or both [5–9,12–17]. The volume fraction of silicides, which increase with Si content, typically ranges from 40 to 65% [12–17]. Metallic phase is body-centered cubic (A2), and is desig-nated as β, while Nb_5Si_3 has the Cr_5B_3 (D8$_1$, tI32) structure and Nb_3Si manifests the PTi$_3$ (tP32) structure. In Nb–Ti–Si alloys [8,9], Ti forms an Nb(Ti) solid solution and $(Nb,Ti)_5Si_3$ and $(Nb,Ti)_3Si$ silicides. The main beneficial effect of Ti addition is to enhance the fracture resistance of the Nb solid solution [22,26], Nb$_{ss}$, and the silicides [28]. The effects of Hf addition appear to be similar to those of Ti [28]. Cr and Al additions promote cleavage fracture and reduce fracture toughness [26,61]. Cr addition, however, is beneficial for oxidation and creep resistance [17,56,57]. Additions of Ge, Sn, and Fe are intended to improve oxidation resistance [17].

The microstructure of niobium silicide-based in situ composites depends on the alloy compo-sition. In binary Nb/Nb_5Si_3 materials such as Nb–10Si and Nb–16Si, the microstructure comprises large primary Nb particles embedded within a matrix of continuous niobium solid solution and Nb_5Si_3 eutectic [4–7,12] (Figure 12.2a). In Nb–Ti–Si alloys heat-treated at 1500°C, the equilibrium phases are beta Nb$_{ss}$, with 0.5–1.5% Si in solution, and alloyed $(Ti,Nb)_3Si$. The $(Ti,Nb)_3Si$ phase

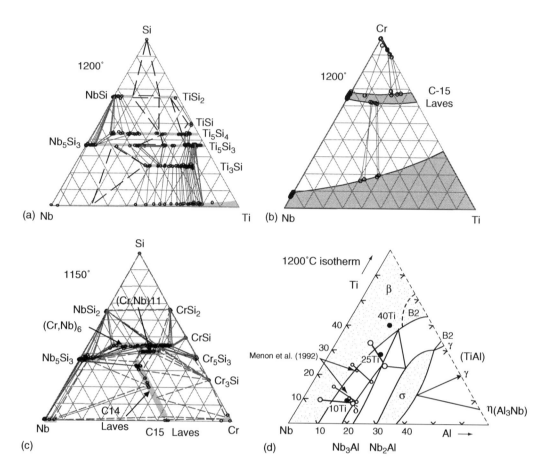

FIGURE 12.1 Ternary phase diagrams depicting the compositions of various types of Nb–Ti–Cr–Si and Nb–Ti–Al based multiphase alloys. (a) Nb–Ti–Si [59]; (b) Nb–Ti–Cr [59]; (c) Nb–Cr–Si [59]; and (d) Nb–Ti–Al[33,60]. Figure 12.1a and c. (From Zhao, J.-C., Bewlay, B. P., Jackson, M. R., and Peluso, L. A., *Structural Intermetallics 2001*, Hemker, K. J., Dimiduk, D. M., Clemens, H., Darolia, R., Inui, H., Larsen, J. M., Sikka, V. K., Thomas, M., and Whittenberger, J. D., Eds., TMS, Warrendale, PA, pp. 483–491, 2001. With permission.). Figure 12.1d. (From Ye, F., Mercer, C., and Soboyejo, W. O., *Metall. Mater. Trans. A*, 29A, 2361–2374, 1998. With permission; Das, S., Jewett, T. J., and Perepezko, J. H., *Structural Intermetallics*, Darolia, R., Lewandowski, J. J., C.T., Martin, P. L., Miracle, D. B., and Nathal, M. U., Eds., TMS, Warrendale, PA, pp. 35–52, 1993.)

(PTi$_3$, tP32) is formed instead of Ti$_5$Si$_3$ (D8$_8$, hP16 structure) or (Ti,Nb)$_5$Si$_3$ because the high-temperature PTi$_3$ (tP32) structure is stabilized by Nb addition. In Nb–Ti–Si–Al alloys containing 2–20 at.% Al and 15–30 at.% Si, the microstructures heat-treated at 1200–1500°C generally comprise (Nb,Ti)$_5$(Si,Al)$_3$, (Ti,Nb)$_5$(Si,Al)$_3$, and beta Nb(Ti,Al) with 0.6% to 1.65% Si in solution. In both Nb–Ti–Si and Nb–Ti–Si–Al alloys, the beta Nb$_{ss}$ and the (Nb,Ti)$_5$(Si,Al)$_3$ silicide appear as co-continuous phases (Figure 12.2b).

12.2.2 Nb–Cr–X Alloys

The Laves phase Cr$_2$Nb (or NbCr$_2$) exhibits a very low creep rate and an inherently high oxidation resistance [62,63]. The Cr$_2$Nb compound has a melting point of 1720°C, a C14 crystal structure

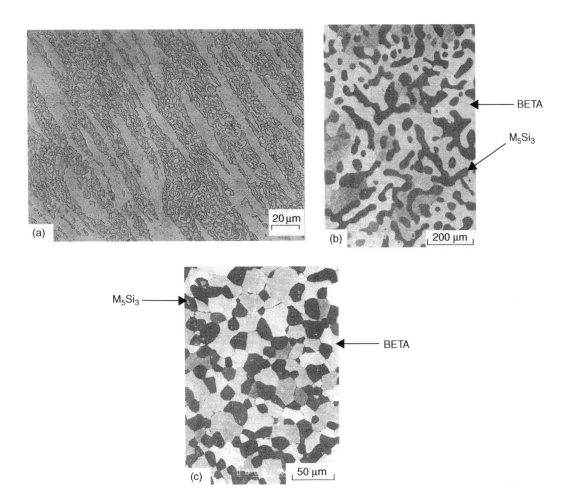

FIGURE 12.2 Microstructures of a binary Nb–Si alloy and an Nb–Ti–Si–Al multiphase alloy. (a) Nb–10Si[20]; (b) Nb–3Ti–8Cr–10Al–13Si[12]; and (c) Nb–27Ti–8Cr–9Al–15Si[12]. Figure 12.2a. (From Chan, K. S. and Davidson, D. L., *Metall. Mater. Trans. A*, 30A, 925–939, 1999[12]. With permission). Figure 12.2b and c. (From Subramanian, P. R., Mendiratta, M. G., Dimiduk, D. M., and M. A. Stucke, *Mater. Sci. Eng.*, A239–A240, 1–13, 1997. With permission.)

above 1650°C but a C15 crystal structure below 1650°C and a density of 7.7 g/cm^3 at ambient temperature. This intermetallic compound is extremely brittle at ambient temperature, exhibiting a fracture toughness of 1.5 MPa m$^{1/2}$ only. Anton and Shah showed by microhardness indentation that the concept of ductile phase toughening for the Cr_2Nb/Nb system was possible [18]. Depending on the composition, Nb-based in situ composites containing Cr_2Nb can be obtained in the form of (a) Nb solid solution particles embedded within a Cr_2Nb matrix, and (b) Nb(Cr) or a Cr(Nb) solid solution matrix containing Cr_2Nb particles. The fracture resistance of $Cr_2Nb/Nb(Cr)$ system is relatively low because Cr embrittles the Nb solid solution phase. Like Cr additions, Al addition also embrittles the Nb solid solution phase. Davidson et al. showed that Ti can be used as an alloying element for imparting fracture toughness in Nb–Cr–Ti alloys [19,20,30,31]. In these alloys, Ti forms solid solution with Nb and substitute for Nb to form alloyed $Cr_2(Nb,Ti)$ Laves phase [19,20,30,31]. Figures 12.3a and b show the microstructures of Nb–Ti–Cr solid solution alloy and in situ composite, respectively.

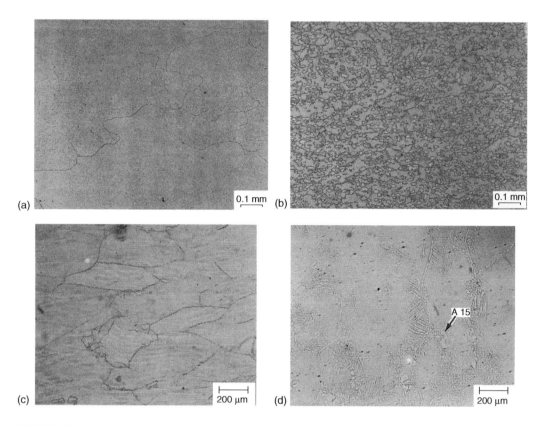

FIGURE 12.3 Microstructures of Nb–Ti–Cr and Nb–Ti–Al alloys. (a) Solid solution alloy, Nb–37Ti–13Cr[20]; (b) In situ composite, Nb–37Ti–36Cr[20]; (c) Solid solution alloy, Nb–40Ti–15Al[33]; and (d) In situ composite, Nb–25Ti–15Al[33]. Figure 12.3a and b. (From Chan, K. S. and Davidson, D. L., *Metall. Mater. Trans. A*, 30A, 925–939, 1999[20]. With permission). Figure 12.3c and d. (From Ye, F., Mercer, C., and Soboyejo, W. O., *Metall. Mater. Trans. A*, 29A, 2361–2374, 1998. With permission.)

12.2.3 Nb–Al–X

Binary Nb–Al alloys (e.g., 18 at.% Al) typically show a microstructure of Nb solid solution and Nb$_3$Al [18,29]. Experimental evidence indicates that B2 ordering occurs in the Nb(Al) solid solution in binary Nb–Al alloys and in some Nb–Ti–Al alloys [32–34,42,60]. The B2 phase has an ordered bcc structure, while Nb$_3$Al has the ordered cubic A15 structure. Because the Nb$_3$Al compound is extremely brittle, Nb alloys containing the A15 compound are also very brittle. However, Ti addition improves the tensile ductility and fracture toughness of the Nb–Ti–Al solid solution, and hence, the corresponding in situ composites containing an Nb(Ti,Al) matrix. Nb–Ti–Al alloys typically contain 15 at.% Al and 10–40 at.% Ti. The microstructure contains either 100% B2 (Nb–40Ti–15Al) or a combination of B2 and Nb$_3$Al (e.g., Nb–25Ti–15Al and Nb–10Ti–15Al). The microstructure of Nb–Ti–Al solid solution is shown in Figure 12.3c white that of an in situ composite is shown in Figure 12.3d.

Some of the Nb–Ti–Al–Hf and Nb–Ti–Al–Cr alloys have a microstructure of Nb solid solution (β) and alloyed sigma phases of (Nb, Ti, Hf)$_2$Al and (Nb, Ti Cr)$_2$Al [5]. An area of current research is the development of Nb–Pd–Hf–Al alloys that contain a microstructure of Nb$_{ss}$ (β) and an ordered bcc aluminide phase, which can be B2 or L2$_1$ (Heusler) phase [51]. This class of alloys is referred to as "Noburnium" and is designed via computational material sciences methods, rather than by trial-and-error, to improve the oxidation resistance of Nb-based alloys by five orders of magnitude [52].

12.2.4 Nb–Si–Cr–Al–X

This class of multicomponent alloys has been studied extensively as it shows the most promising properties for high-temperature structural applications [5–7,12,17,28]. These alloys generally contain a substantial amount of Ti addition, usually in the range of 20–26% Ti for improving fracture resistance. In addition, Hf addition in the range of 2–4% is often used to increase the effective Ti content since Hf and Ti additions produce very similar effects. The Si content, which ranges from 12 to 18%, determines the amount of silicides in the alloy. A small amount of Ge, usually less than 5%, is added to increase the effective Si content. The Cr content is typically in the range of 12–16%, which is sufficient large to cause the formation of a Laves phase. A high Cr content generally improves the oxidation resistance [17,28]. A small amount of Al is used to improve oxidation resistance, but the Al content is usually kept below 2% since a higher Al content can promote cleavage fracture in the Nb solid solution phase [23,61]. Minor alloying elements of Fe and Sn are used to improve low-temperature oxidation resistance [17].

The microstructures of Nb–Ti–Cr–Al–Si in situ composites either with or without Hf are entirely different from those in the binary Nb–Si system [12]. The typical microstructure in Nb–Ti–Hf–Cr–Al–Si alloys consists of a continuous or near continuous bcc (β) Nb-matrix with dispersed refractory-metal silicides, Laves phases or both, depending on alloy composition [12]. This is illustrated in Figure 12.4 [28], which shows the microstructure of an Nb–Ti–Hf–Cr–Si–Ge in situ composite heat-treated at 1350°C for 100 h. The microstructure of this alloy is comprises an Nb solid solution matrix (light phase) with dispersed silicides (gray phase) and Laves phases (dark phase). The volume percents of the Nb solid solution, silicides, and Laves phases are 54, 34, and 12%, respectively. The morphology of the silicide and Laves phase ranges from equiaxed to rodlike with an aspect ratio as high as 5. The alloyed silicides can be M_3Si or M_5Si_3, where M represents Nb, Ti, or Hf, and the Laves phases can be of the C14 or C15 structure, depending on composition. The silicide phases include $(Nb,Ti)_5Si_3$, $(Nb,Ti)_3Si$, $(Ti,Nb)_5Si_3$, and $(Ti,Nb)_3Si$ [28]. It is noted that $(Nb,Ti)_5Si_3$ and $(Nb,Ti)_3Si$ are Nb_5Si_3 and Nb_3Si alloyed with Ti additions which exhibit $D8_l$ (tI32) and PTi_3 (tP32) structures, respectively. In comparison, $(Ti,Nb)_5Si$ and $(Ti,Nb)_3Si$ are Ti_5Si_3 and Ti_3Si alloyed with Nb which have $D8_8$ and structures, respectively. In Hf-containing alloys, Hf can substitute for Ti in the silicides. The Cr content is generally sufficiently large to cause the formation of a C14 Laves phase [15,17,28,59]. The C14 is generally a high-temperature phase but it can be stabilized by Al and Si additions to form at room temperature instead of the usual C15 Laves phase [28,59]. The solubility of Ge in Nb is about 3–4%. In some

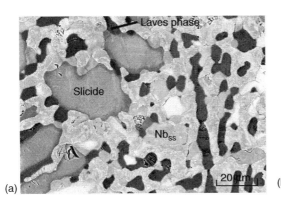

FIGURE 12.4 Microstructure of Nb–Ti–Hf–Cr–Al–Si–Ge multiphase alloys[28]. (a) Nb–22Ti–4Hf–12Cr–15Cr–5Ge and (b) Nb–22Ti–2Hf–7Cr–9Si–5Ge–4Fe–2Al–1Sn (CNG-1B). (From Chan, K. S. and Davidson, D. L., *Metall. Mater. Trans. A*, 34A, 1833–1849, 2003. With permission.)

alloys, Ge tends to segregate to the grain boundaries during solidification and can form one or more low-melting metastable intermetallic phases along grain boundaries [28]. The low-melting phases can be removed by heat-treating at an appropriate temperature to allow dissolution of the intermetallic phases and the diffusion of Ge away from grain boundaries.

12.3 PHYSICAL AND TENSILE PROPERTIES

Elastic moduli of Nb-based alloys, silicides, and Laves phases for selected temperatures are summarized in Table 12.2 [15]. The modulus of Nb-based in situ composites is 165 GPa at room temperature (RT) and is 140 GPa at 1200°C. The modulus of polycrystalline Nb_5Si_3 ranges from 170 to 210 GPa and Cr_2Nb has an elastic modulus of about 220 GPa at RT [19,21].

Table 12.2 also presents the coefficient of thermal expansion (CTE) for the temperature range of RT to 539°C for various constituent phases in Nb-based in situ composites [15]. In general, the CTEs of Nb_{ss}, silicides, Nb_5SiB_2 (T2 phase) and Laves phases are similar. Thus, thermal racheting in constituent phases is expected to be minimal during thermal cycling of these composites [15].

The yield strength of conventional Nb alloys is on the order of 150–300 MPa at ambient temperature and it decreases with increasing temperatures to about 50 MPa at 1200 K, Figure 12.5 [5]. The yield strength of the multicomponent Nb-based alloys and in situ composites are significantly higher than conventional Nb alloys (particularly at elevated temperatures) because of solid solution and particle strengthening mechanisms. Figure 12.5 shows a comparison of conventional Nb alloys, multicomponent Nb-based solid solution alloys, and in situ composites processed by several different means against those of an Ni-based single crystal superalloy (PWA 1480). Because of solid solution strengthening, the yield strength of beta Nb–Ti–Cr–Al alloys is as high as 1000 MPa at 298 K, but it decreases with increasing temperatures to less than 100 MPa above 1273 K. Cast and heat-treated Nb–Ti–Cr–Al–Si in situ composites exhibit higher yield strength than the beta solid solution alloys at elevated temperatures. Extruded and heat-treated materials show a yield strength maximum as high as 1200 MPa at 1073 K, but it decreases drastically at higher temperatures. Directionally solidified multicomponent Nb–Ti–Hf–Cr–Al–Si and Nb/Nb_5Si_3 in situ composites exhibit higher yield strength than other Nb-based materials and PWA 1480 at temperature above 1273 K. At temperatures below 1273 K, the Nb-based alloy and in situ composites are not as strong as PWA 1480.

TABLE 12.2
Elastic Properties at Room Temperature (RT) and Thermal Expansion for a Range of Intermetallic Phases up to 1200°C

Phase	Nb (a/o)	Ti	Hf	Si	Cr	Al	B	RT-1200°C Expansion (%)	E(MPGa)
Laves	21.0	11.0	5.5	8.5	53.0	1.0	—	1.07	220
M_5Si_3 Silicide	38.5	16.0	6.0	37.0	1.0	1.0	0.5	0.78	170–220
T2	41.5	13.0	3.0	12.5	4.0	0.5	25.5	0.94	—
Nb	55.3	28.2	2.0	1.0	10.0	3.5	—	1.06	—
M_3Si Silicide	49.0	18.2	7.8	25.0	—	—	—	1.05	165

Source: From Bewlay, B. P., Jackson, M. R., and Gigliotti, M. F. X, *Intermetallic Compounds Principles and Practice*, 26, Westbrook, J. H. and Fleischer, R. L., Eds., John Wiley, pp. 541–560, 2002.

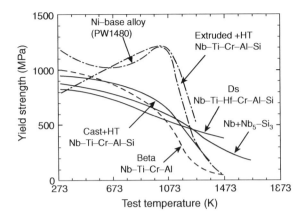

FIGURE 12.5 Yield strength of Nb alloys and in situ composites compared against that of PWA 1480. (From Subramanian, P. R., Mendiratta, M. G., and Dimiduk, D. M., *JOM*, 48, 33–38, 1996. With permission.)

The temperature dependence of the ultimate tensile strength, notch strength, and impact energy of Nb alloys is summarized by Lewandowski and coworkers [64,65]. The review by Lewandowski et al. [64] also indicates that the yield and the ultimate tensile strengths of Nb alloys are relatively insensitive to the grain size in the range of 5–500 μm. On the other hand, the brittle-to-ductile fracture transition temperature decreases with reducing grain size. There is also a tendency for Nb alloys to cleave in coarse-grained materials at low temperatures.

The presence of an Nb solid solution, Nb_{ss}, phase in an Nb-based in situ composite does not automatically guarantee ambient temperature ductility because this property depends critically on the composition of the solid solution phase, microstructure, defect structure, and interstitial contents [19,20]. Extensive work by Begley [39] on binary Nb solid solution alloys has established that alloying additions such as Cr, Al, W, Mo, V, Zr, and Re cause embrittlement in Nb and raise the brittle-to-ductile transition temperature (BDTT). From Begley's study [39], Ti and Hf are the only elements known not to increase the BDTT of Nb. More recent results and theoretical modeling [26] have indicated that Ti enhances both the tensile ductility and fracture toughness of Nb solid solution alloys [19,20,22,26,41–43], while Cr, Al, and W exert the opposite effects [23,26,41–43,53,58,61]. The dependence of tensile ductility of Nb–Ti–Cr–Al solid solution on Ti alloy addition is shown together with theoretical calculation in Figure 12.6 [26].

The prevailing view on brittle-to-ductile fracture transition is that it is a competition between cleavage fracture and emission of dislocation from the crack tip [66]. The propensity to cleavage fracture is often measured in terms of the surface energy. In contrast, dislocation emission from the crack tip can be controlled by the nucleation of dislocations from the crack tip [66] or the mobility of dislocations moving away from the dislocation nucleation sites either at the crack tip or other non-tip sources [67,68] as illustrated schematically in Figure 12.7 [69]. The propensity of crack-tip dislocation nucleation (Figure 12.7a) can be described in terms of the unstable stacking energy (γ_{us}) [66] (Figure 12.7b), while the mobility of dislocations moving away from the crack tip, Figure 12.7c, is represented in terms of the Peierls–Nabarro (P–N) barrier energy, $U_{P–N}$ [70,71] and a generalized stacking fault energy, γ_F (Figure 12.7d).

As shown in a recent analysis [26] the P–N barrier energy is related to the unstable stacking energy and their relationship can be derived from the misfit energy of a dislocation. The misfit energy, $E_D(\alpha)$, of a dislocation in an anisotropic elastic solid is given by [26]

$$E_D(\alpha) = 4\pi^2 \gamma_{us} \left[\frac{\sinh \psi \cosh \psi}{\cosh^2 \psi - \sin^2 \alpha \pi} \right]$$

(12.1)

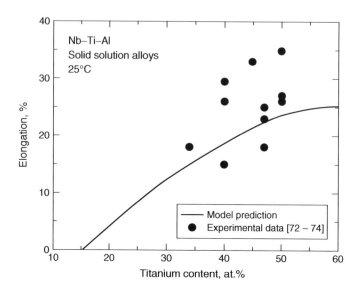

FIGURE 12.6 Dependence of tensile ductility of Nb–Ti–Cr–Al solid solution alloys on alloy addition. (From Chan, K. S., *Metall. Mater. Trans. A*, 32A, 2475–2487, 2001. With permission.)

with

$$\gamma_{us} = \frac{\kappa b^2}{2\pi^2 d} \tag{12.2}$$

where ψ is the lattice phase angle and α represents the fraction of the Burgers vector by which the dislocation is displaced. The P–N barrier energy, U_{P-N}, is the energy encountered by the dislocation when it moves through the periodic lattice from the position of $\alpha = 0$ to $\alpha = 1/2$. Applying this condition to Equation 12.1 leads one to [26]

$$U_{P-N} = \frac{8\pi^2 \gamma_{us}}{\sinh 2\psi} \tag{12.3}$$

and the lattice phase angle,

$$\psi = \frac{\pi \kappa d}{cb}, \tag{12.4}$$

is a function of the slip plane spacing (d), the magnitude of Burgers vector (b), the anisotropic shear modulus (c), and the anisotropic dislocation line energy factor (κ).

The effects of alloying on tensile ductility and BDTT of Nb-base solid solution alloys can be understood in terms of the ratios of γ_s/γ_{us} and γ_s/U_{P-N}. Both γ_s and γ_{us} are insensitive to composition, while the P–N energy is very sensitive to alloy additions because c and κ depend on alloy compositions. The appropriate value of γ_s/U_{P-N} for use as the brittle-to-ductile fracture transition was determined from experimental data of tensile ductility. Specifically, experimental data of tensile ductility of Nb, Nb–Ti–Al, Nb–Ti–Cr, and Nb–Ti–Cr–Al alloys [33,41–43] were correlated with calculated values of the ratio of γ_s/U_{P-N} in Figure 12.8 [26]. There is considerable scatter in the correlation; nonetheless, the overall trend is that tensile ductility and fracture toughness both increase with increasing values of the γ_s/U_{P-N}. Figure 12.8 shows that the γ_s/U_{P-N} must exceed

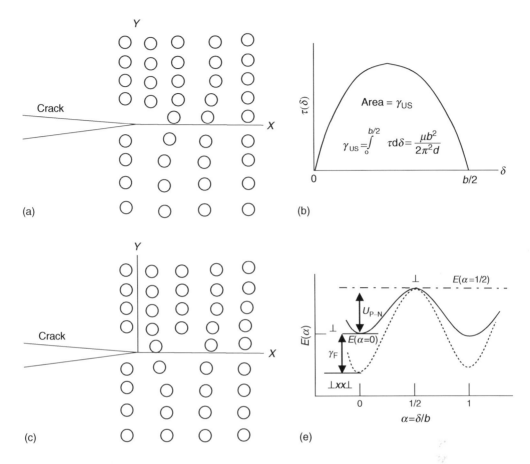

FIGURE 12.7 Dislocation processes controlling brittle-to-ductile fracture transition. (a) Crack-tip dislocation nucleation; (b) Unstable stacking energy, γ_{us}, required to nucleate a dislocation at the crack tip; (c) Dislocation mobility away from the crack tip; and (d) Peierls–Nabarro barrier energy, U_{P-N}, and the generalized stacking fault energy, γ_F, that a dislocation with an unspecified fault must overcome before it can move away from the crack tip. (From Chan, K. S., *Metall. Mater. Trans. A*, 34A, 2315–2328, 2003. With permission.)

14 to order to obtain a 4% elongation. No appreciable elongation can be attained in an alloy with a γ_s/U_{P-N} value less than 10 (Figure 12.8).

Increase in tensile ductility is achieved by a reduction of U_{P-N} by alloy addition. A systematic analysis [26] of the various components that contribute to the P–N barrier energy indicates that the $\sinh 2\psi$ term in Equation 12.3 is very sensitive to changes due to alloy addition and to the number of $d+s$ electrons in the alloy. The P–N barrier energy is high when the lattice phase angle, ψ, is small, and vice versa. Figure 12.9a shows a plot of the dislocation misfit energy, $E_D(\alpha)$, as a function of α, which is atom displacement normalized by the Burgers displacement, for various numbers of $d+s$ electrons and Ti additions [26,72]. A reduction in the charge density ($d+s$ electrons) by Ti addition is seen to reduce the P–N energy barrier by increasing $E(\alpha=0)$ and reducing $E_D(\alpha=1/2)$, mainly through a reduction in the anisotropic shear modulus. At $d+s=4.2$, $E_D(\alpha=1/2)=E_D(\alpha=0)$ so that $U_{P-N}\approx 0$ and dislocation motion is unimpeded by the crystal lattice. The opposite effect is seen in Figure 12.9b, which shows the dislocation misfit energy, $E_D(\alpha)$, as a function of α for various numbers of $d+s$ electrons and Cr additions [26,72]. An increase in the charge density by Cr addition is seen to increase the P–N barrier energy by reducing $E_D(\alpha=0)$ and increasing $E_D(\alpha=1/2)$ due to increasing shear moduli in the slip direction.

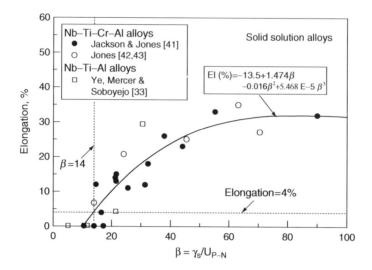

FIGURE 12.8 Dependence of tensile ductility on the ratio of surface energy, γ_s, to the Peierls–Nabarro barrier energy, U_{P-N}. (From Chan, K. S., *Metall. Mater. Trans. A*, 32A, 2475–2487, 2001. With permission.)

12.4 FRACTURE RESISTANCE

The salient characteristic of Nb-based in situ composites is a microstructure of hard intermetallic particles embedded in an Nb solid solution, Nb_{ss}, matrix or vice versa—a microstructure of ductile particles dispersed in a continuous intermetallic matrix. Hard intermetallics, which include $NbCr_2$ [18–20]; Nb_3Al [32–35]; and Nb_5Si_3 [2–17] are intended to impart high temperature strength, oxidation and creep resistance [5–8,12–17,28], while the Nb solid solution phase is intended to impart ambient temperature fracture toughness and tensile ductility through a number of ductile phase toughening mechanisms such as crack blunting, branching, deflection, and bridging [73,74]. The presence of a ductile Nb solid solution phase, however, does not always produce a high fracture resistance in the in situ composites. The reasons are (a) the Nb solid solution might not be ductile because of reduced dislocation mobility resulting from solid solution strengthening, and (b) the brittle intermetallics are non-deformable and induce high plastic constraints in the solid solution phase. Current understanding of the effects of alloy addition on the fracture resistance in Nb-base solid solution alloys and in situ composites are presented in the next two subsections.

12.4.1 SOLID SOLUTION ALLOYS

Ti addition has been found to improve the fracture resistance of Nb-based solid solution alloys and in situ composites [19–21], while Al and Cr additions tend to promote cleavage [23,41–43,61] and lead to a reduction in fracture toughness or tensile ductility. Enhancement in the fracture toughness of Nb-based in situ composites by Ti addition has been reported by Davidson et al. [19], Subramanian et al. [8], and Bewlay et al. [9]. Figure 12.10 shows that the fracture toughness of Nb–Cr–Ti solid solution alloys increase with Ti additions. Crack-tip micromechanics experiments performed the Nb–Ti–Cr [22] and Nb–Ti–Cr–Al [23,61] solid solution alloys inside a scanning electron microscope revealed that a high Ti content promotes the emission of (110) slip from the crack tip, as illustrated in Figure 12.10c [22]. This slip emission process is instrumental in suppressing the emission of cleavage cracks from the crack tip and the corresponding plastic dissipation leads to a high fracture toughness in Nb–Cr–Ti alloy. In contrast, a high Cr and Al content promotes the formation of cleavage cracks ahead of the crack tip, as shown in Figure 12.10a [23]. These cleavage micro-cracks tend to joint quickly, thereby leading to a premature onset of rapid crack growth and a

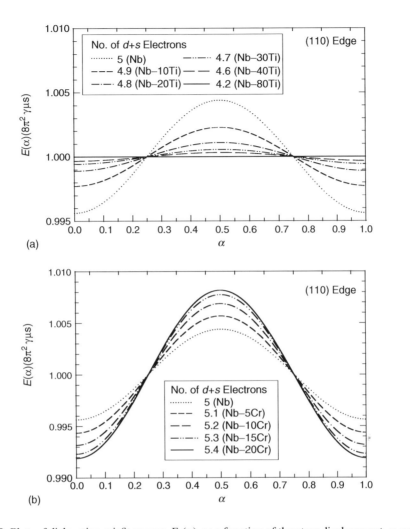

FIGURE 12.9 Plots of dislocation misfit energy, $E_d(\alpha)$, as a function of the atom displacement, α, normalized by the magnitude, b, of the Burgers vector showing the dependence of the Peierls–Nabarro barrier energy, where $U_{P-N} = E_d(\alpha = 1/2) - E_d(\alpha = 0)$, on the number of $d+s$ electrons/atoms. (a) Decrease in U_{P-N} and the number of $d+s$ electrons/atoms resulting from increasing Ti additions and (b) increase in U_{P-N} and the number of $d+s$ electrons resulting from increasing Cr additions in Nb. (From Chan, K. S., *Metall. Mater. Trans. A*, 32A, 2475–2487, 2001. With permission.)

low fracture resistance in a Nb–Ti–Cr–Al alloy. A comparison of the fracture-resistance curves of these alloys is shown in Figure 12.10b.

The correlation of theoretical and experimental results indicates that the fracture toughness in the tough Nb–Cr–Ti alloy originates from extensive dislocation emission that suppresses cleavage crack propagation from the crack tip. The effects of alloying on fracture toughness have been identified by performing theoretical calculations of the unstable stacking energy (USE) and the Peierls–Nabarro (P–N) energy as a function of Ti, Cr, and Al contents in the Nb–Ti–Cr–Al alloys [22]. These computations indicated Ti addition lowers the P–N energy and stress, but has little effect on the unstable stacking energy [22]. Thus, the dislocation emission process in the Nb–Cr–Ti alloys appears to be controlled by the P–N energy, which influences dislocation mobility, rather than by the USE, which influences dislocation nucleation. Figure 12.11 shows the fracture toughness of Nb–Ti–Cr–Al solid solution alloys increases with increasing values of the γ_s/U_{P-N}.

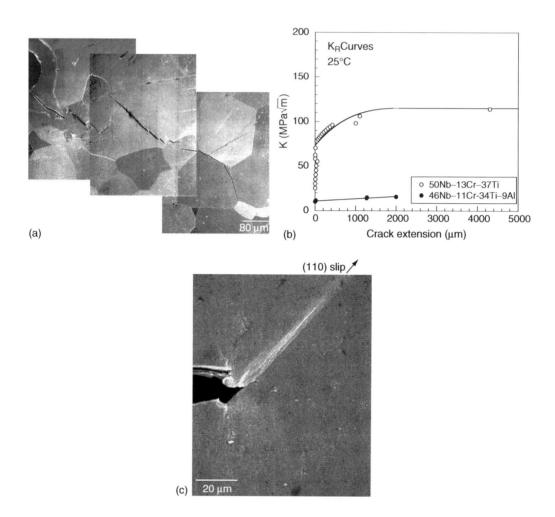

FIGURE 12.10 Effects of alloy addition on the near-tip fracture process and the fracture resistance in Nb–Ti–Cr alloys. (a) Decrease in fracture resistance by cleavage crack formation ahead of the crack tip resulting from Cr and Al additions; (b) K-resistance curves of two Nb–Ti–Cr–Al alloys; and (c) increase in fracture resistance by emission of (110) slip at the crack tip resulting from Ti addition. (From Davidson, D. L. and Chan, K. S., *Metall. Mater. Trans. A*, 30A, 2007–2018, 1999.) and (From Chan, K. S. and Davidson, D. L., *Metall. Mater. Trans. A*, 30A, 925–939, 1999. With permission.)

Ti addition increases the fracture toughness of Nb-Cr-Ti alloys by decreasing the P–N energy, U_{P-N}, which leads to increases in dislocation mobility and dislocation emission from the crack tip [22,26]. Conversely, the computational results showed that alloy additions such as Cr and Al additions lead to an increase in the P–N barrier energy, a corresponding decrease in dislocation mobility, and an increased propensity to emission of cleavage cracks from the crack tip [22,23,26,61]. Therefore, Al and Cr are undesirable and should be avoided in Nb solid solution alloys because reduced dislocation mobility inevitably leads to reductions of tensile ductility and fracture toughness [23,26,61].

Experimental results also shed new light on the importance of electronic effects on fracture resistance in Nb-based alloys. As indicated earlier, Davidson et al. [9] discovered that Ti addition enhances the fracture toughness of Nb-based solid solution alloys, Nb$_{ss}$, and in situ composites containing Nb$_{ss}$ and (Nb,Ti)Cr$_2$ Laves Phases. They observed that the fracture toughness of these

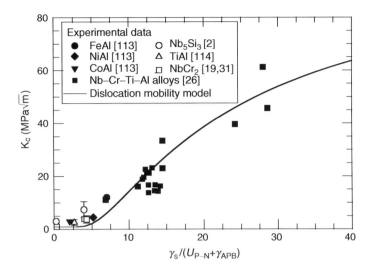

FIGURE 12.11 Dependence of fracture toughness, K_C, of Nb–Ti–Cr–Al solid solution alloys on the ratio of surface energy, γ_s, to the Peierls–Nabarro barrier energy, U_{P-N}. (From Chan, K. S., *Metall. Mater. Trans. A*, 34A, 2315–2328, 2003. With permission.)

Nb-based materials increase with decreasing number of $d+s$ electrons per atom, as shown in Figure 12.12. Subsequent work [22] demonstrated that Ti addition enhances fracture resistance by reducing the Peierls–Nabarro barrier energy [70,71] and improving the dislocation mobility. A theoretical analysis [26,72] revealed that a reduction of the number of $d+s$ electrons

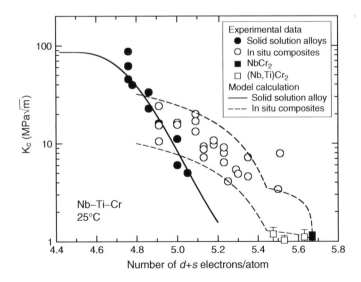

FIGURE 12.12 Comparison of measured [19,31] and predicted fracture toughness values of Nb-Ti-Cr–Al solid solution alloys and in situ composites as a function of the number of $d+s$ electrons/atoms in the alloy system. The inverse relationship observed in Nb–Ti–Cr–Al solid solution alloys is the consequence of increasing fracture toughness, K_C, resulting from reductions of anisotropic shear modulus and Peierls–Nabarro energy, U_{P-N}, with decreasing numbers of $d+s$ electrons when Ti addition in increased fracture toughness of (Nb, Ti)Cr_2 Laves phase is independent of $d+s$ electrons/atoms because the anisotropic shear modulus and U_{P-N} are not altered by Ti addition. (From Chan, K. S., *Phil. Mag. A*, 85, 239–259, 2005. With permission.)

per atom enhance fracture toughness and dislocation mobility by reducing the anisotropic shear modulus and the P–N barrier energy. Thus, electronic bonding and alloy addition affect dislocation mobility and fracture resistance by altering the anisotropic shear modulus which, in turn, influences the P–N barrier energy.

The relationship between fracture resistance and dislocation mobility in metals and intermetallics was recently formulated on the basis of thermally activated slip in the crack-tip region by emitting a perfect dislocation or a pair of partial dislocations separated by a generalized stacking fault [69]. For a crack tip under small-scale yielding, the J-integral, J_D, which represents the plastic work per unit crack extension dissipated by dislocations moving away from the crack-tip, is given by [69]

$$J_D = J_0 \exp\left[-\frac{b^2}{kT}(U_{P-N} + \gamma_F)\right] \qquad (12.5)$$

where k is Boltzmann's constant, T is absolute temperature, and γ_F is the generalized stacking fault energy. The pre-exponent term, which is a function of the density, ρ, of mobile dislocations and the plastic zone size, h, is given by [69]

$$J_0 = \left(\frac{kT}{b^2}\right)\rho\pi h^2 \qquad (12.6)$$

and J_0 can be interpreted as the maximum plastic dissipation attained per unit area when dislocations are activated at all possible nucleation sites and the P–N barriers are overcome. Equation 12.5 can be related to the stress intensity factor, K_C, at fracture, leading [75,76]

$$K_C = \sqrt{\frac{E(2\gamma_s + J_D)}{(1 - \nu^2)}} \qquad (12.7)$$

for plane-strain fracture of isotropic materials with Young's modulus, E, Poisson's ratio, ν, and surface energy, γ_S.

Equation 12.5 and Equation 12.7 enable one to compute the fracture resistance of Nb–Ti–Cr alloys and (Nb,Ti)Cr$_2$ Laves phases on the basis of the P–N barrier energy values. The results for Nb–Ti–Cr alloys are presented in Figure 12.12 [72], which shows the critical stress intensity fracture, K_C, increases with decreasing numbers of $d+s$ electrons/atom. The observed inverse relationship between K_C and the number of $d+s$ electrons is the result of a decrease of the P–N barrier energy, the anisotropic shear modulus, and the number of $d+s$ electrons with increasing Ti contents in the Nb–Ti–Cr solid solution alloys. In contrast, the fracture toughness for the (Nb,Ti)Cr$_2$ Laves phase is 0.97 MPa m [31] and is insensitive to the number of $d+s$ electrons, Figure 12.12, which is consistent with the observation that the P–N barrier energy of (Nb,Ti)Cr$_2$ is high and is not substantially reduced by Ti addition [72].

12.4.2 IN SITU COMPOSITES

A ductile Nb solid solution phase can enhance the initiation toughness, which is the stress intensity at the onset of crack extension and an intrinsic material property. In addition, a ductile Nb phase can impart fracture resistance in Nb-based in situ composites through extrinsic toughening mechanisms such as crack branching, deflection, and bridging mechanisms [73,74,77,78]. On the other hand, hard intermetallic particles in the multiphase composites influence, adversely, the fracture resistance by creating a high plastic constraint in the microstructure [78]. The plastic constraint originates from triaxial hydrostatic stresses induced by the hard phases in the composites

[79,80]. The level of matrix constraint increases with increasing volume fractions of the hard phase. The fracture toughness, K_C, of a two-phase material containing a hard phase of toughness, K_b, and a ductile phase of toughness, K_d, is given by [27]

$$K_C = K_b \left[1 + \sqrt{1 - V_f} \left(\left[\frac{K_d}{K_b} \right]^2 P_C(V_f) - 1 \right) \right]^{1/2} \tag{12.8}$$

where V_f is the volume fraction of the hard phase. The plastic constraint term, P_C, was evaluated from finite-element analyses of the buildup of triaxial stresses in a composite unit cell containing various volume fractions of hard particles in a ductile phase. An analytical expression derived from the FEM results gives [27]

$$P_C(V_f) = \exp \left\{ -\frac{8q'}{3} \left[\frac{V_f}{1 - V_f} \right] \right\} \tag{12.9}$$

where q' is an indicator of the plastic constraint condition at the particle/matrix interface. The q' parameter has a value of zero when the interface is totally un-bonded, but has a value of unity when the interface is fully bonded and constrained plastically. Equation 12.9 is applicable for two-phase materials of which volume fraction of hard particles, V_f, is less than the critical volume fraction, V_{crit}, for the occurrence of a continuous brittle layer consisting of contiguous hard particles ($V_f < V_{crit}$). When a continuous brittle phase forms at $V_f \geq V_{crit}$, the fracture toughness of the in situ composites is then given by Equation 12.8 and Equation 12.9 with $V_f = V_{crit}$ [27].

Experimental data [19] of the fracture toughness (K_C) of Nb–Ti–Cr in situ composites are plotted as a function of the volume percent of Laves phase $NbCr_2$ in Figure 12.13. The K_C data show significant variations because of different Ti contents. Equation 12.8 and Equation 12.9 were used to compute the fracture toughness of Nb-based in situ composites as a function of volume fraction of $NbCr_2$. These calculations were performed using $K_b = 1$ MPa m$^{1/2}$ for $NbCr_2$, $K_b = 10$ and 32 MPa m$^{1/2}$ for the Nb solid solution phase, and $q' = 1$ for fully constrained particle/matrix interface. The dashed curve in Figure 12.13 gives the lower bound fracture toughness for the in situ composites with lower Ti contents, while the solid curve gives the upper bound for in situ composites with higher Ti content. In both cases, the fracture toughness decreases with increasing volume fractions of $NbCr_2$. The breaks in the computed curves at high volume percents of $NbCr_2$ were the consequence of fracture along a continuous brittle phase.

12.5 FATIGUE CRACK GROWTH RESISTANCE

12.5.1 SOLID SOLUTION ALLOYS

The fatigue crack growth behaviors of Nb-based solid solution alloys have been studied by Davidson and co-workers [23,61,81] and by Lewandowski et al. [10,64,65]. Figure 12.14 shows a comparison of the da/dN data for various Nb solid solution alloys to illustrate the effects of alloy addition on the fatigue crack growth response. In general, Ti addition increases the fracture toughness of Nb solid solution alloys. This increase in fracture resistance produces two effects on the da/dN response: (a) a reduction in the fatigue crack growth exponent in the Paris regime, and (b) an increase in the critical stress intensity factor at which unstable fatigue crack growth commences. Both reduce da/dN and enhance the fatigue resistance. In contrast, Al addition promotes cleavage fracture in the Nb phase. The reduction in the fracture toughness results in an increase in the fatigue crack growth exponent and a premature onset of rapid crack growth and unstable fracture, as shown in Figure 12.14.

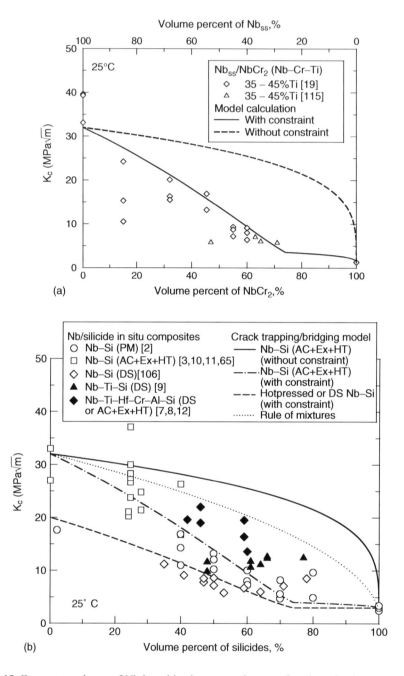

FIGURE 12.13 Fracture toughness of Nb-based in situ composites as a function of volume percent of inter-metallic particles. (a) Nb/NbCr$_2$ and (b) Nb/Nb$_5$Si$_3$. (From Chan, K. S. and Davidson, D. L., *Metall. Mater. Trans. A*, 32A, 2717–2727, 2001. With permission.)

Direct observations of the fatigue crack growth process in a scanning electron microscope revealed that fatigue crack growth occurred in an intermittent and sporadic manner in an Nb-alloy containing high levels of Al and Cr contents [61]. The sporadic behavior originated from cleavage fracture in selected Nb grains located ahead of the tip of the fatigue crack. Cleavage-like facets (Figure 12.15a) were observed on the fracture surfaces and the fracture path was

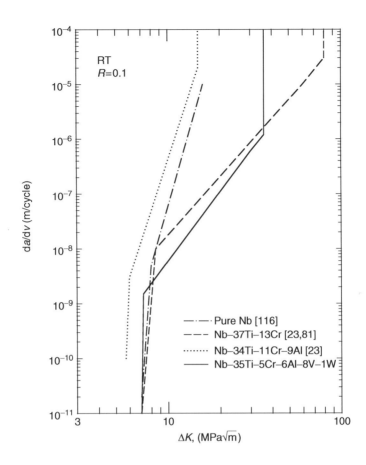

FIGURE 12. 14 Fatigue crack growth rate data of Nb-based slid solution alloys. (From Davidson, D. L., Chan, K. S., Loloee, L., and Crimp, M. A., *Metall. Mater. Trans. A*, 31A, 1075–1084, 2000. With permission.)

tortuous, crystallographic, and appeared to favor {100} and {112} planes (Figure 12.15b) [61]. The intermittent and erratic fatigue crack growth behavior was observed in Nb34Ti–11Cr–9Al, whose fracture toughness was only 10–15 MPa m$^{1/2}$, as well as in Nb–35Ti–6Al–5Cr–8V–1W–0.3Hf–0.5Mo, whose fracture toughness is 35 MPa m$^{1/2}$ [61]. The erratic, rapid cleavage crack growth during fatigue is a concern because, despite a K_{IC} value of 35 MPa m$^{1/2}$, it leads to uncertainties of the crack growth response in the Paris regime. Neither the amount of crack extension accompanied cleavage nor the arrest mechanism can be predicted. In particular, there is no certainty that during subcritical fatigue crack growth, a cleavage crack, once nucleated, would not grow to the critical length for fracture. The cleavage-like fracture behavior appears to be promoted by a high Al content in the Nb solid solution phase. Thus, the Al content in the Nb-based solid solution alloys and in situ composites should be kept low, e.g., less than 2%.

12.5.2 IN SITU COMPOSITES

Fatigue crack growth in Nb-based in situ composites generally occurs at higher growth rates than for unalloyed Nb or Nb-based solid solution alloys at equivalent ΔK. The ΔKth threshold, the da/dN in the Paris regime, and the K_C at fracture are reduced for the in situ composites. The increase in the crack growth rate increases with the volume fraction of intermetallics in the composite, but differences in composite microstructure make this comparison difficult. In general, the presence of an intermetallic greatly alters the deformation characteristics in the near-tip region.

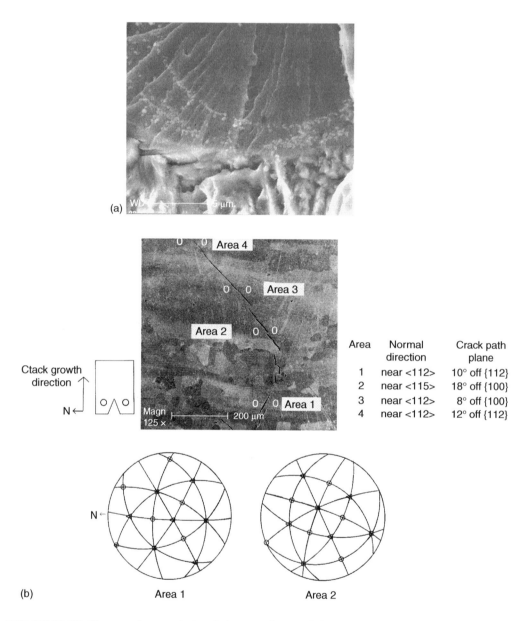

FIGURE 12.15 Cleavage fracture during fatigue crack growth in an Nb–Ti–Al–Cr solid solution alloy. (a) Flat cleavage facets show river lines and fine particles on the surface, and (b) {100} and {112} cleavage planes identified by EBSP. (From Davidson, D. L. Chan, K. S., Loloee, L., and Crimp, M. A., *Metall. Mater. Trans. A*, 31A, 1075–1084, 2000. With permission.)

In composites containing relatively large intermetallic particles, the large particles are usually fractured by an advancing fatigue crack tip, leading to increased crack growth rates. Figure 12.16 illustrates the fatigue crack growth process in an Nb–Ti–Cr in situ composite containing $NbCr_2$ in an Nb solid solution matrix [20]. Fracture of the $NbCr_2$ particles results in the formation of un-cracked ligaments between the main crack and the particle micro-cracks. The un-cracked ligaments do not remain intact but become broken during subsequent fatigue loading. The overall effect is that particle fracture generally leads to a lower ΔK_{th}, higher da/dN, and a lower

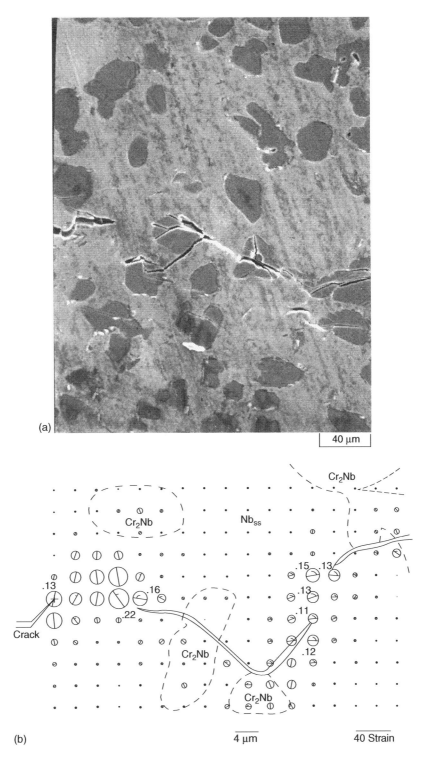

FIGURE 12.16 Fracture of intermetallic (Laves phase) particles and crack bridging by uncracked ligaments during fatigue crack growth in an Nb-based in situ composite. (a) Near-tip fracture processes, and (b) near-tip strain distribution determined by the DISMAP technique. (From Chan, K. S., *Metall. Mater. Trans. A*, 27A, 2518–2531, 1996. With permission.)

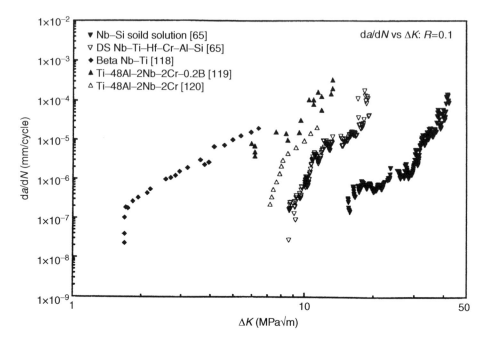

FIGURE 12.17 Comparison of the fatigue crack growth rate, da/dN, of several Nb-based alloys [65, 118] and in situ composites [65] against these of TiA/alloys [119,120]. (From Subramanian, P. R., Mendiratta, M. G., Dimiduk, D. M., and M. A. Stucke, *Mater. Sci. Eng.*, A239–240, 1–13, 1997. With permission of Elsevier, Oxford, U.K.; Zinsser, W. A. and Lewandowski, J. J., *Metall. Mater. Trans A*, 29A, 1749–1757, 1998; Bewlay, B. P., Lewandowski, J. J., and Jackson, M. R., *J. Metals*, 49, 44–45, 1997; Venkateswara Rao, K. T., Odette, G. R., and Ritchie, R. O., *Acta Metall. Mater.*, 42:893–911, 1994; Chan, K. S. and Shih, D. S., *Metall. Mater. Trans.*, 28A, 79–90, 1997; Larsen, J. M., Worth, B. D., Balsone, S. J., and Jones, J. W., In *Gamma Titanium Aluminides*, Kim, Y. W., et al., Eds., TMS, Warrendale, PA, pp. 821–834, 1994).

K_C in the in situ composites. Thus, a refinement of the intermetallic particles is required to improve the fatigue crack growth resistance in Nb-based in situ composites.

The silicide phases in Nb-based in situ composites generally are smaller in size and are less likely to fracture compared with the Laves phases. Figure 12.17 shows a comparison of the da/dN curves of a number of Nb-based in situ composites including those containing Laves phase and silicides [65]. Extruded and directionally solidified Nb silicide composites often contain elongated primary Nb grains that act as bridging ligaments [10,20,65] during fatigue loading (Figure 12.18) [20]. Like un-cracked ligaments in Nb/NbCr$_2$, these bridging ligaments of primary Nb grains degrade by fatigue and do not provide a significant improvement in fatigue crack growth resistance, even though the bridgi ng ligaments improve the critical stress intensity factor at fracture. The da/dN curves of Nb-based in situ are sensitive to the stress ratio, R, because a high mean stress promotes cleavage fracture in the Nb solid solution phase and brittle fracture of intermetallic particles. As pointed out earlier, these two brittle fracture mechanisms increase the crack growth rate exponent in the Paris regime and reduce the critical stress intensity at unstable fracture.

12.6 CREEP BEHAVIOR

The creep resistance of niobium silicide in situ composites are very sensitive to compositions and microstructure [55,82–89]. In binary Nb/Nb$_5$Si$_3$ materials such as Nb–10Si and Nb–16Si, the microstructure comprises large primary Nb particles embedded within a matrix of continuous

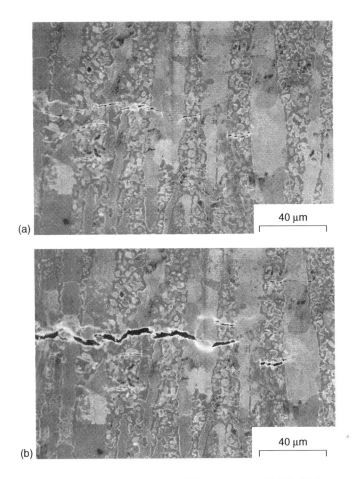

(a)

40 μm

(b)

40 μm

FIGURE 12.18 Crack bridging by elongated primary Nb grains in an Nb/Nb$_5$Si$_3$ in situ composite. (From Chan, K. S., *Metall. Mater. Trans. A*, 27A, 2518–2531, 1996. With permission.)

niobium solid solution and Nb$_5$Si$_3$ eutectic [12]. Creep resistance of niobium silicide in situ composites with this microstructure was characterized by Subramania et al. [82]. Subsequently, Henshall and coworkers [84–87] investigated the primary and steady state creep response of ductile-phase toughened Nb$_5$Si$_3$/Nb in situ composites through a combination of analytical modeling and numerical simulation, using experimental data generated by Subramanian et al. [82]. These studies, which treated the Nb$_5$Si$_3$ as the continuous phase, demonstrated that the creep behavior of the in situ composites is dominated by the silicide phase, which is stronger and bears a higher load than the weaker Nb solid solution. At a given stress, the steady state creep rate of the niobium silicide in situ composites [83] is significantly lower than that of the Nb$_{ss}$ phase, even though it is higher than that exhibited by monolithic silicides [82]. The creep exponent of the in situ composite is about 2 [83], compared with $\approx$ 1 for Nb$_5$Si$_3$ [82] and 5.8 for Nb [84,85] or Nb(Si) solid solution [87]. Despite several attempts, Henshall et al. [84–87], were not successful in simulating the correct creep exponent observed in the in situ composites using continuum methods. These authors attributed the discrepancy to a possible change in the creep mechanisms in the in situ composites and in monolithic Nb$_5$Si$_3$.

The microstructures of Nb–Ti–Cr–Al–Si in situ composites either with or without Hf are entirely different from those in the binary Nb–Si system [12]. The typical microstructure in Nb–Ti–Hf–Cr–Al–Si alloys consists of a continuous or near continuous bcc (β) Nb-matrix with dispersed refractory-metal silicides, Laves phases or both, depending on alloy composition [12]. The creep response of Nb-base alloys with this type of microstructure differs significantly from that with a continuous intermetallic phase. Bewlay [83] reported the steady state creep rates of a number of niobium-silicide based in situ composites that contained Ti, Hf, and Mo alloy additions and a continuous Nb solution phase matrix. Figure 12.19 compares the results of Bewlay et al. [83] for Nb–Ti–Hf–Si materials against those for Nb–Si [83–85], Nb [84], and monolithic Nb_5Si_3 [82]. The creep behavior of the Nb–Ti–Hf–Si material exhibited a wider variation compared with those of the binary Nb–Si materials. For example, the creep exponent of the Nb–Ti–Hf–Si materials varies from 1.08 to 11. In contrast, the creep exponent is on the order of 2–3 for Nb–Si and about 1 for Nb_5Si_3. The various creep behaviors observed in these niobium silicide in situ composites have been attributed to both compositional and microstructural effects.

The effects of alloy additions such as Hf, Ti, Cr, Al, Mo, and Si on the compression creep behavior of Nb-based in situ composites were investigated by Bewlay et al. [15]; Bewley et al. [83]; Bewley et al. [88]; Briant et al. [89]. Both Ti and Hf additions have been found to increase the creep rate. In general, the creep rate increases with increasing Hf additions and in particular, Ti addition

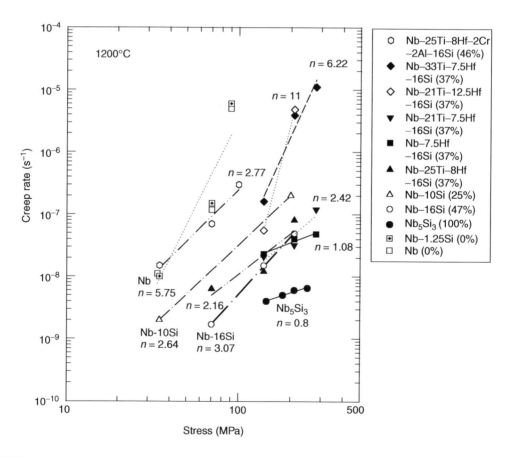

FIGURE 12.19 Comparison of the steady state creep rates of Nb-based alloys, in situ composites, and monolithic silicides. (From Chan, K. S., *Mater. Sci. Eng.*, A337, 59–66, 2002. With permission; Subramanian, P. R., Parthasarathy, T. A., Mendiratta, M. G., and Dimiduk, D. M., *Scripta Metall. et. Mater.*, 32(8), 1227–1232, 1995. See also, Mendiratta, M. G. et al., Materials Development Research, WL-TR-96-4113, WPAFB, Dayton, OH, 1996; Belway, B. P., Whiting, P. W., Davis, A. W., and Briant, C. L., *Symposium Proceedings*, MRS, 552, KK6.11.1–KK6.11.5, 1999.)

has a detrimental effect on creep performance at stress levels above 210 MPa. Because of the increased creep rates, Bewley et al. [15] suggested that the Ti:Hf ratio in Nb-based alloys should be less than 3, and the concentration of Ti should be kept below 21%. The enhanced creep rates in Nb-based in situ composites appear to originate from the stabilization of Ti_5Si_3 type silicides in preference of Nb_5Si_3 type silicides when Ti and Hf concentrations are high, since Ti_5Si_3 type silicides exhibit poor creep resistance and have a detrimental effect on the creep performance of the composites [15,89].

Si addition affects the creep resistance of in situ composites by influencing the amounts of silicides in the microstructure. A systematic investigation of Si concentration on the creep response of quaternary composites indicated that the volume fractions of the metallic phase changes from 0.3 to 0.7 when the Si content increases from 12 to 22%. As shown in Figure 12.20 [15,89], the creep rate of the composite decreases with increasing Si content and volume fraction of silicides until it reaches a minimum value at 18% Si, which corresponds to a volume fraction of silicides of 0.6. Beyond this minimum, the creep rate increases again with increasing Si content and volume fraction of silicides. At Si contents less than 12%, the composite creep resistance is dominated by the creep behavior of the solid solution phase and is, therefore, relatively poor. At Si contents greater than 20%, the composite creep properties are controlled by creep damage in the silicides.

The creep response of monolithic silicides and Laves phases in Nb-based in situ composites vary with compositions and crystal structure. Belway et al. [89] identified the compositions and crystal structure of alloyed silicides and Laves phases in selected Nb-based in situ composites and performed creep test to determine the creep characteristics on these compounds. Table 12.3 shows the study creep rates and creep exponent for selected monolithic phases investigated by Bewlay et al. [89].

In Table 12.3, the monolithic phases generated from alloys are given the postscript 3. For example, C14 Nb(Cr,Si) is referred to as Laves-3. In comparison, a postscript C signifies monolithic phases that were generated from quaternary and higher-order alloys, e.g., silicide-C represents a multicomponent Nb_5Si_2-based silicide containing more than 4 alloying elements. Similarly, T_2–C refers to a multicomponent Nb_5SiB_2-based silicide with the $D8_1$ crystal structure and (Nb)-3

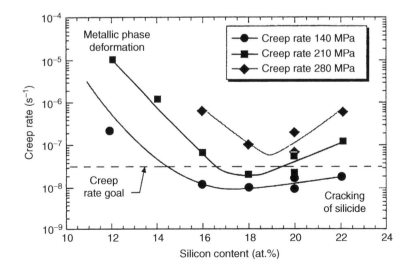

FIGURE 12.20 Effects of Si content on the steady creep rates of Nb-based in situ composites at 1200°C for stresses of 140–280 MPa. A minimum creep rate is observed in materials containing 18% Si. (From Bewlay, B. P., Briant, C. L., Davis, A. W., and Jackson, M. R., *Proceedings on High Temperature Ordered Intermetallic Alloys IX*, MRS, 646, N2.7.1–N2.7.6, 2000. With permission.)

TABLE 12.3
Power Law Constants (B and n in $\dot{\varepsilon}_{ss} = B\sigma^n$) for Steady-State Creep Rate of Monolithic Phases in Nb-Based In Situ Composites

Phase	Stress Range (MPa)	Temperature (°C)	B (s^{-1})	Exponent, n
Nb$_5$Si$_3$	100–280	1200	6.16×10^{-11}	1.0
Nb–10Si	70–140	1200	6.57×10^{-12}	1.9
Silicide–C	70–140	1200	3.84×10^{-11}	1.5
Nb$_3$Si–C	70–140	1200	1.07×10^{-14}	3.6
HP16–C	140–210	1200	—	>3.6
Laves-3	70–140	1200	1.11×10^{-9}	1
T2–C	70–140	1200	2.15×10^{-9}	0.9
(Nb)-3	3–80	1200	1.9×10^{-10}	3.3
(Nb)	3–80	1200	4.7×10^{-14}	2.9

Source: From Bewlay, B. P., Briant, C. L., Sylven, E. T., Jackson, M. R., and Xiao, G., *Proceedings on High Temperature Ordered Intermetallic Alloys IX*, MRS, 646, N2.6.1–N2.6.6, 2000.

is a ternary Nb solid solution phase. Table 12.3 shows that the creep exponent (n) varies among the intermetallic phases. A creep exponent of ≈ 1 is observed in Nb$_5$Si$_3$, Laves-3, and T$_2$–C. A creep exponent of 3.6 occurs in Nb$_3$Si–C, which is comparable to that of the Nb solid solution phase. The highest creep exponent is observed in hP16–C, a multicomponent Ti$_5$Si$_3$-based silicide, whose n value is greater than 3.6 [89].

The creep response of Nb-based in situ composites has been analyzed [55] using a creep model for metal-matrix composites [90]. Model calculations [55] revealed that the creep exponent of the in situ composites is significantly influenced by the creep behavior of the stronger reinforcement (silicide or Laves) phase (Figure 12.19). The wide range of creep exponents (1–11) observed in Nb–Ti–Hf–Si in situ composites can be explained on the basis of the rigid or creeping behavior of the silicide (or Laves) phase during creep of the in situ composites. The creep model [55] predicts that a rigid phase in an Nb-based in situ composite would result in a creep curve that is a translation of creep curve of the metal phase to a higher stress. For composites with creeping matrix and reinforcements, the major microstructural parameters are the creep exponent and the volume fraction of the reinforcement phase. In particular, creeping silicides are beneficial for creep resistance of the in situ composites, as long as the creep exponent of the creep silicides is low (e.g., $n = 1$). Rigid particles or creeping particles with a high creep exponent ($n > 3$) are both undesirable for creep resistance. For optimum creep resistance in composites, the desirable characteristics are reinforcement phases that are rigid at low stresses but creep with a low stress exponent ($n = 1$) at higher stresses. These model calculations are shown in Figure 12.21. Thus, diffusional creep [91–93] and Harper–Dorn Creep [94] in the silicide or Laves phases with $n = 1$ (e.g., Nb$_3$Si$_3$, Laves-3, and T2–C (in Table 12.3)) are preferred to power-law creep with $n > 1$ (e.g., Nb$_3$Si–C) in Table 12.3. The wide range of creep exponents observed in Nb–Ti–Hf–Si in situ composites may be the consequence of the presence of rigid and creeping silicides (intermetallics and Laves phases) in these composites. Alloying addition is known to alter the Peierls–Nabarro stress [70,71] of Nb solid solutions [22,26] and possibly silicide and Laves phases [28]. Recent theoretical work [95] has established a link between the Peierls–Nabarro (P–N) stress and the creep exponent of $n = 1$ in Harper–Dorn creep. A high P–N stress can make a silicide or Laves phase act as a rigid particle, and leads to unfavorable creep characteristics. With proper alloying addition to reduce the P–N stress, it may be possible to induce Harper–Dorn [92]

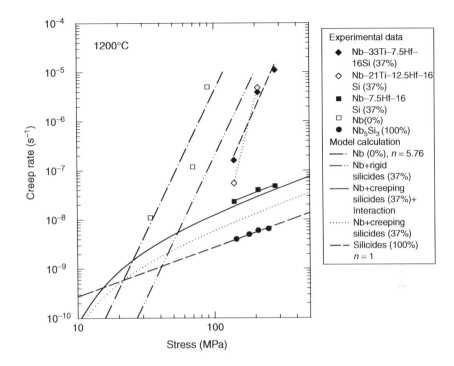

FIGURE 12.21 Comparison of measured and computed creep curves showing optimum creep resistance is obtained in materials with intermetallic phases that are rigid at low stresses but creep with a low stress exponent ($n = 1$) at higher stresses. (From Chan, K. S., *Mater. Sci. Eng.*, A337, 59–66, 2002. With permission.)

or diffusional [91–93] creep in the silicide or Laves phase in order to attain a low creep exponent, leading to more favorable creep characteristics.

12.7 OXIDATION PROPERTIES

Extensive work has been undertaken to develop niobium-based structural alloys for high temperature applications. The early studies focused mostly on the use of alloying addition to reduce the oxidation kinetics of Nb alloys by modifying the oxidation products. The early efforts, which were reviewed extensively [39,96–99], were largely unsuccessful because of the formation of Nb_2O_5, a non-protective oxide, which was too dominant to be altered by alloying addition. Subsequent studies [100–105] were concentrated on an aluminide-based system and the use of Al addition to improve oxidation resistance of Nb-based alloys through the formation of a protective alumina layer. While the formation of a protective alumina was feasible, the resulting alloys had low melting points and were too brittle to be used as structural materials [102].

Recent efforts have focused mostly on materials that contain substantial amounts of niobium silicides and Laves phases. In these Nb-based alloys, Ti and Hf additions are intended for improvement in fracture [19,22,106] and oxidation resistance [12,16]. Cr, Al, and Ge additions [12,16] are also intended for enhancing oxidation resistance, while Si is intended to provide oxidation and creep resistance [16] through the formation of various refractory-metal silicides (e.g., alloyed Nb_5Si_3, Nb_3Si, Ti_5Si_3, and Ti_3Si) and Laves phases (e.g., alloyed $NbCr_2$). The silicides and Laves phases are intended to provide high temperature strength and oxidation resistance. The oxidation resistance of these multiphase Nb-based alloys, which is substantially better than that of conventional Nb alloys [5,6,12,107,108], is comparable to that of Ni-based superalloys at

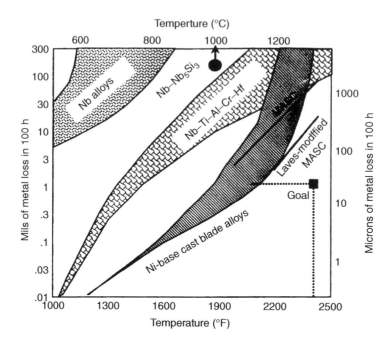

FIGURE 12.22 A summary of the oxidation resistance of Nb-based alloys and in situ composites compared with those of Ni-based superalloys. (From Bewlay, B. P. and Jackson, M. R., In *Comprehensive Composite Materials,* 3, *Metal Matrix Composites,* Kelly, A. and Zweben, C., Eds., 579–613, 22, ed., T. W. Clyne, 2000. With permission.)

1000°C, as shown in Figure 12.22. On the other hand, the oxidation resistance of the multiphase Nb-based alloys is still inadequate for structural materials in the 1200–1400°C range, as the oxides that formed on these materials are nonprotective [5,6,12,107,108].

Menon et al. [109] reported the static oxidation behavior of a number of Nb-silicides containing multicomponent alloys for the temperature range of 600–1300°C. The oxidation products observed in these alloys include $TiNb_2O_7$, $CrNbO_4$, cristoballite SiO_2, $3Nb_2O_5 \cdot TiO_2$, and Nb_2O_5. Their results also indicated that the Nb solid solution phase oxidized selectively, while the silicide phase in the microstructure did not oxidize. At temperatures below 900°C, internal oxidation caused numerous micro-cracks that laid parallel to the surfaces just below the oxide layer. These micro-cracks resulted in spallation of the oxide scales and breakaway oxidation at low temperature.

Cyclic oxidation tests performed on multiphase Nb-based alloys containing silicide, Laves, and Nb solid solution phases. The results indicated that Nb-based alloys formed a mixture of $CrNbO_4$, Nb_2O_5, and $Nb_2O_5 \cdot TiO_2$, with possibly small amounts of SiO_2 or GeO_2 [56,57]. The oxidation resistance was improved when $CrNbO_4$ formed instead of Nb_2O_5 and $Nb_2O_5 \cdot TiO_2$. For illustration, Figure 12.23 compares the cyclic oxidation curves of several Nb-based alloys subjected to 22-hour thermal cycles at a peak temperature of 1100°C [56]. Those alloys are designated as Nbx, M1, M2, and AX, whose compositions are listed in Table 12.1. Alloys AX and UES-AX have similar compositions. Among all the alloys studied, M2 contained the highest Cr and Si contents and this alloy exhibited the best oxidation resistance at 1100°C, Figure 12.23b. At this temperature, the oxidation products were mostly $CrNbO_4$ with small amounts of Nb_2O_5 and $Nb_2O_5 \cdot TiO_2$. At other temperatures as well as in other alloys, the oxidation resistance decreased with increasing amounts of Nb_2O_5 and $Nb_2O_5 \cdot TiO_2$ in the oxide mixtures. For these alloys, the worst oxidation resistance occurred at 1400°C and the oxide product mixture contained mostly Nb_2O_5 and $Nb_2O_5 \cdot TiO_2$ with little $CrNbO_4$. The intensity of the XRD pattern for $CrNbO_4$ occurs at $2\theta = 27.3°$,

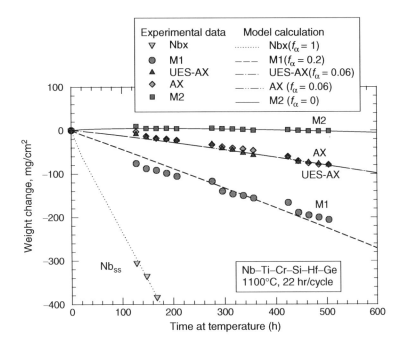

FIGURE 12.23 Cyclic oxidation curves of Nb-based in situ composites containing various volume fractions of Nb solid solution and intermetallic phases. (From Chan, K. S., *Oxidation of Metals*, 61 (3/4), 65–194, 2004. With permission.)

while those for Nb_2O_5 and $Nb_2O_5 \cdot TiO_2$ occur at $2\theta = 23.9°$. The ratio of the intensity, I_{CrNbO_4}, of the $CrNbO_4$ peak at $2\theta = 27.3°$ to the intensity, $I_{Nb_2O_5 \cdot TiO_2}$, at $2\theta = 23.9°$ can be used as a measure of the relative amounts of $CrNbO_4$, Nb_2O_5 and $Nb_2O_5 \cdot TiO_2$ in the oxidation product. The values of this relative intensity ratio were obtained for several Nb alloys tested at various temperatures. The results are shown as a function of volume percents of Nb solid solution phase in the alloy in Figure 12.24 [56,57]. The relative intensity ratio is seen to decrease with increasing volume percent of Nb solid phase in the alloy. A high $I_{CrNbO_4}/I_{Nb_2O_5 \cdot TiO_2}$ ratio was observed only in M2 at 1100 and 1315°C.

The relationship between the microstructure and oxidation resistance of Nb-based in situ composites is illustrated in Figure 12.24, which shows the relative intensity ratio and metal recession as a function of volume percent Nb solid phase after 500 hours of cyclic oxidation at a peak temperature of 1100°C. Three general trends can be deduced from Figure 12.24: (a) high material recession is associated with the formation and spallation of Nb_2O_5 and $Nb_2O_5 \cdot TiO_2$, (b) low material recession is associated with the formation and spallation of $CrNbO_4$, and (c) the formation and spallation of Nb_2O_5 and $Nb_2O_5 \cdot TiO_2$ are favored in alloys containing high volume percents of Nb solid solution phase, while $CrNbO_4$ formation and spallation are favored in alloys with high volume percents of silicide and Laves phase. The lowest oxidation resistance was observed in Nbx, which contained the lowest Cr content, while M2, which contained the highest Cr content, exhibited the highest oxidation resistance. Unfortunately, $CrNbO_4$ does not offer long-term protection and the oxidation resistance of M2, which varies widely with temperatures, is limited to 500 h at 1100°C and is less at other temperatures.

Since Nb-based multiphase alloys are intended for high-temperature structural applications, it is of interest to compare the oxidation resistance of M2 against those of Ni-based single crystal superalloys. Such a comparison is presented in Figure 12.25 [56], which shows that the oxidation resistance of M2 at 1200°C is as good as, if not better, than those of Rene N5 and N6 [110] for

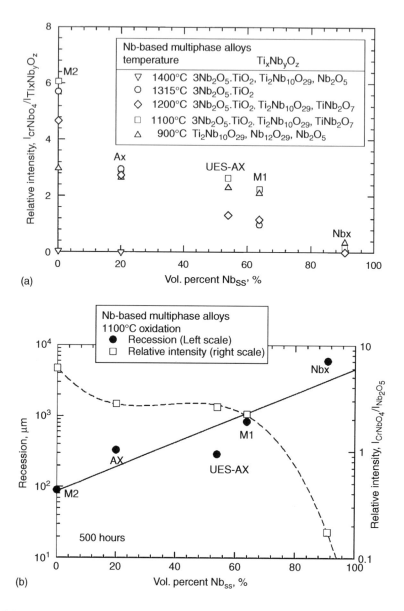

FIGURE 12.24 Metal recession and intensity of $CrNbO_4$ XRD peak normalized by that of $Nb_2O_5 \cdot TiO_2$ peak as a function of volume percent of Nb solid solution, Nb_{ss}, in the Nb-based composites. The results show increased metal recession and $Nb_2O_5 \cdot TiO$ formation with increasing volume fraction of Nbss. Formation of $CrNbO_4$ reduces metal recession and improves oxidation resistance of Nb-based in situ composites. (From Chan, K.S., 37*Oxidation of Metals*, 61 (3/4), 65–194, 2004. With permission.)

oxidation times up to 150 hours. Furthermore, M2 exhibits better oxidation resistance than those of PWA 1484 and CMSX-10Ri at 1177°C [110]. The weight change data of the Ni-based single crystal superalloys have been computed by dividing the weight loss results reported in Walston et al. [110], by the surface area (7.6 cm^2) [111]. This observation is consistent with recent reviews that show state-of-the-art Nb-based multiphase alloys exhibit higher oxidation resistance than cast Ni-based alloys at temperatures above ≈ 1100°C [12,16]. Current Nb-based alloys and composites tend to have poor oxidation resistance at temperatures at or below 900°C [56,57,109].

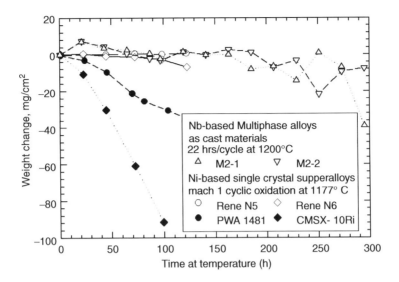

FIGURE 12.25 A comparison of the cyclic oxidation curves of an Nb-based in situ composite against those of several Ni-based single-crystal superalloys $\approx 1200°C$. (From Chan, K. S., *Oxidation of Metals*, 61, 3/4, 65–194, 2004. With permission.)

As indicated earlier, an area of current research is the development of Nb–Pd–Hf–Al alloys that contain a microstructure of Nb_{ss} (β) and an ordered bcc aluminide phase, which can be B2 or $L2_1$ (Heusler) phase [51]. This class of alloys is referred to as "Noburnium" and is designed via computational material sciences methods, rather than by trial-and-error, to improve the oxidation resistance of Nb-based alloys by five orders of magnitude [52]. The approach is to identify alloy additions through computational material science methods for enhancing selective oxidation of aluminum to form a continuous protective Al_2O_3 scale.

12.8 APPLICATIONS

In a recent overview article, Wojcik summarized current high-temperature applications of conventional Nb alloys [112]. The most common application for niobium alloys is in sodium lamp, which contains a small tubing of Nb–1Zr (in wt.%) made by deep drawing. The sodium light bulbs are used throughout the world for highway lighting because of their high electrical efficiency and long life, which is typically in excess of 25,000 h.

Commercial Nb alloy C-103 (Nb–10Hf–1Ti in wt.%) is used in aerospace applications as thrust cones, and high temperature valves. Silicide coatings are used extensively to protect C-103. Most of these applications in propulsion systems operate for relatively short times at temperatures between 1200 and 1400°C and in less oxidizing environments than normal atmosphere. Coated C-103 is also used as thrust augmenter flaps in one of the PWA turbine engines. These flaps typically reach 1200–1300°C and last for about 100 h of afterburner time. Nb alloys have also been considered for heat-pipe applications in space nuclear reactors and spacecrafts. Nb alloy heat-pipes were successfully manufactured into hypersonic leading ledges and nose cones, as well as tested under various high heat flux conditions.

Nb-based in situ composites have been developed as possible high-temperature blade materials for applications in the 1200–1400°C. Current Nb-based in situ composites do not have adequate oxidation resistance for airfoil applications in this temperature range yet. However, some progress has been made toward making niobium silicide airfoils. Under Air Force funding, the team of

General Electric, and Rolls-Royce Advanced Development Company (formerly Allison Advanced Development Company) performed a design trade study to provide alloy developers with detailed material properties for specific airfoil applications [16,17]. A preliminary set of material properties for a generic niobium silicide alloy was used as input and preliminary design tools were used to size the airfoil and establish temperature profiles [17]. The results helped set the creep, oxidation, and coating requirements for Nb airfoil design and application. Recently, researchers at General Electric reported results of investment casting thin sheets of selected niobium silicide composites using a hybrid arc-melting and drop casting technique [17]. Investment-cast plates with thickness ranging from 3 to 8 mm have been manufactured for a number of Nb-based in situ composites, including ternary Nb–Ti–Si and multicomponent Nb–Ti–Hf–Cr–Al–Si alloys. A prototype airfoil has also been fabricated from a niobium silicide composite using the investment casting technique [17]. These researchers also reported results on the development of both bond coats and thermal barrier coatings for Nb-silicide airfoils. They have developed a series of oxidation-resistant coatings for Nb-silicide coatings at temperatures of up to 1370°C for more than 100 h [17].

12.9   SUMMARY

This chapter summarizes recent advances in the development of Nb-based solid solution alloys and in situ composites. This class of emerging high-temperature structural alloys can be demarcated into four types on the basis of the strengthening mechanisms, including (a) Nb-based silicides, (b) Nb-based Laves phases, (c) Nb-based aluminides, and (d) Nb-based materials containing both silicide and Lave phases. The compositions and microstructures of selected alloys are summarized and their properties highlighted. Depending on compositions, Nb-based in situ composites exhibit a multiphase microstructure that comprises an Nb solid solution and one or more silicides, Laves phases, or aluminides. Extensive alloying additions and microstructural controls are required in order to enhance and optimize material performance. Current understanding of alloying effects and microstructure/property relations are elucidated for various properties of Nb-based in situ composites, including tensile strength, tensile ductility, fracture toughness, fatigue crack growth resistance, creep, and oxidation resistance. Potential applications of Nb-based in situ composites are discussed and contrasted against those of conventional Nb alloys.

ACKNOWLEDGMENTS

Preparation of this manuscript was supported by Southwest Research Institute, SwRI. The author's work on Nb-based in situ composites was supported by the Air Force Office of Scientific Research through Contract No. F3962001-C-0016, Dr. Craig S. Hartley, Program Manager. Clerical assistance by Ms. L. Salas, Ms. A. Mathews, and Ms. P. Soriano, of SwRI, in the preparation of the manuscript is appreciated.

REFERENCES

1. Dimiduk, D. M. and Miracle, D. B., *Mater. Res. Soc. Symp. Proc.*, Vol. 133, MRS, Pittsburgh, PA, pp. 349–359, 1988.
2. Nekkanti, R. M. and Dimiduk, D. M., *Mater. Res. Soc. Symp. Proc.*, Vol. 194, MRS, Pittsburgh, PA, pp. 175–182, 1990.
3. Mendiratta, M. G., Lewandowski, J. J., and Dimiduk, D. M., *Metall. Mater. Trans. A*, 22A, 1573–1583, 1991.
4. Miracle, D. B. and Mendiratta, M. G., In *Intermetallic Compounds Principles and Practice*, Fleischer, R. L. and Westbrook, J. H., Eds., Vol. 2, John Wiley, Chichester, UK, Chap. 13, pp. 287–300, 1995.

5. Subramanian, P. R., Mendiratta, M. G., and Dimiduk, D. M., *JOM*, 48(January), 33–38, 1996.
6. Jackson, M. R., Bewlay, B. P., Rowe, R. G., Skelly, D. W., and Lipsitt, H. A., *JOM*, 48(January), 39–44, 1996.
7. Bewlay, B. P., Jackson, M. R., and Lipsitt, H. A., *Metall. Mater. Trans. A*, 27A, 3801–3808, 1996.
8. Subramanian, P. R., Mendiratta, M. G., and Dimiduk, D. M., *Mater, Res. Soc. Symp. Proc.*, Vol. 322, MRS, Pittsburgh, PA, pp. 491–502, 1994.
9. Bewlay, B. P., Jackson, M. R., Reeder, W. J., and Lipsitt, H. A., *Mater. Res. Soc. Symp. Proc.* Vol. 364, MRS, Pittsburugh, PA, pp. 943–948, 1995.
10. Rigney, J. D., Singh, P. M., and Lewandoski, J. J., *Acta Metall. Mater.*, 43, 1955–1967, 1995.
11. Kajuch, J., Short, J., and Lewandowski, J. J., *JOM*, 44(8), 36–41, 1992.
12. Subramanian, P. R., Mendiratta, M. G., Dimiduk, D. M., and Stucke, M. A., *Mater. Sci. Eng.*, A239–A240, 1–13, 1997.
13. Lipsitt, H. A., Blackburn, M., and Dimiduk, D. M., In *Intermetallic Compounds Principles and Practice*, Westbrook, J. H. and Fleischer, R. L., Eds., Vol. 3, John Wiley, New York, Chap. 2 pp. 471–499, 2002.
14. Bewlay, B. P., Jackson, M. R., In *Comprehensive Composite Materials*, Kelly, A. and Zweben, C., Eds., *Volume 3: Metal Matrix Composites*, Clyne T. W., Ed., Vol. 3, Elsevier, Chap. 22, pp. 579–613, 2000.
15. Bewlay, B. P., Jackson, M. R., and Gigliotti, M. F. X., In *International Compounds Principles and Practice*, Westbrook, J. H. and Fleischer, R. L., Eds., Vol. 3, John Wiley, New York, Chap. 26, pp. 541–560, 2002.
16. Balsone, S. J., Belway, B. P., Jackson, M. R., Subramanian, P. R., Zhao, J.-C., Chatterjee, A., and Hefferman, T. M., In *Structural Intermetallics*, Hemker, K. J., Dimiduk, D. M., Clemens, H., Darolia, R., Inui, H., Larsen, J. M., Sikka, V. K., Thomas, M., and Whittenberger, J. D., Eds., TMS, Warrendale, PA, pp. 99–108, 2001.
17. Bewlay, B. P., Jackson, M. R., Zhao, J.-C., and Subramanian, P. R., *Metall. Mater. Trans. A*, 34A, 2043–2052, 2003.
18. Anton, D. L. and Shah, D. M., *Mater. Res. Soc. Symp. Proc.*, Vol. 194, MRS, Pittsburgh, PA, pp. 45–52, 1990.
19. Davidson, D. L., Chan, K. S., and Anton, D. L., *Metall. Mater. Trans. A*, 27A, 3007–3018, 1996.
20. Chan, K. S., *Metall. Mater. Trans. A*, 27A, 2518–2531, 1996.
21. Chan, K. S. and Davidson, D. L., *JOM*, 48(9), 62–68, 1996.
22. Chan, K. S. and Davidson, D. L., *Metall. Mater. Trans. A*, 30A, 925–939, 1999.
23. Davidson, D. L. and Chan, K. S., *Metall. Mater. Trans. A*, 30A, 2007–2018, 1999.
24. Chan, K. S., Davidson, D. L., and Anton, D. L., *Metall. Mater. Trans. A*, 28A, 1797–1808, 1997.
25. Davidson, D. L. and Chan, K. S., *Metall. Mater. Trans. A*, 401–416, 2002.
26. Chan, K. S., *Metall. Mater. Trans. A*, 32A, 2475–2487, 2001.
27. Chan, K. S. and Davidson, D. L., *Metall. Mater. Trans. A*, 32A, 2717–2727, 2001.
28. Chan, K. S. and Davidson, D. L., *Metall. Mater. Trans. A*, 34A, 1833–1849, 2003.
29. Davidson, D. L. and Anton, D. L., *Mater. Res. Soc. Symp. Proc.*, Vol. 288, Pittsburugh, PA, pp. 807–813, 1992.
30. Thoma, D. J., Ph.D. Dissertation, University of Wisconsin, Univ. Microfilms, Ann Arbor, MI, 1992.
31. Thoma, D. J., Nibur, K. A., Chen, K. C., Cooley, J. C., Dauelsberg, L. B., Hults, W. L., and Kotula, P. G., *Mater. Sci. Eng.*, A329–A331, 408–415, 2002.
32. Hou, D,-H., Shyue, J., Yang, S. S., and Fraser, H. L., In *Alloy Modeling and Design*, Stocks, G. M. and Turchi, P. E. A., Eds., TMS, Warrendale, PA, pp. 291–302, 1994.
33. Ye, F., Mercer, C., and Soboyejo, W. O., *Metall. Mater. Trans. A*, 29A, 2361–2374, 1998.
34. Guan, D. L., Brooks, C. R., and Liaw, P. K., *Intermetallics*, 10, 441–458, 2002.
35. Soboyejo, W. O., Di Pasquale, J., Ye, F., Mercer, C., Srivatsan, T. S., and Konitzer, D. G., *Metall. Mater. Trans. A*, 30A, 1025–1038, 1999.
36. Elliott, C. K., Odette, G. R., Lucas, G.E., and Scheckherd, J.W., *Mater. Res. Soc. Symp. Proc.*, Vol. 120, MRS, Pittsburgh, PA, pp. 95–101, 1988.
37. Odette, G. R., Chao, B. L., Sheckherd, J. W., and Lucas, G. E., *Acta Metall. Mater.*, 40, 2381–2389, 1992.
38. Venkateswara Rao, K. T., Odette, G. R., and Ritchie, R. O., *Acta Metall. Mater.*, 42, 893–911, 1992.

39. Begley, R. T., Columbium alloy development at Westinghouse, In *Evolution of Refractory Metals and Alloys*, Dalder, E. N., Grobstein, T., and Olsen, C. S., Eds., TMS, Warrendale, PA, pp. 29–48, 1993.
40. Chan, K. S., *Mater. Sci. Eng.*, A329–A331, 513–522, 2002.
41. Jackson, M. R. and Jones, K. D., In *Refractory Metals: Extractions, Processing and Applications*, Nona, K., Kiddell, C., Sadoway, D. R., and Bautista, R. G., Eds., TMS, Warrendale, PA, pp. 311–319, 1990.
42. Jones, K. D., Master's Thesis, Rensselaer Polytechnic Institute, Troy, NY, 1990.
43. Jones, K. D., Jackson, M. R., Larsen, M., Hall, E. L., and Woodford, D. A., In *Refractory Metals: Extraction, Processing and Applications*, Liddell, K. C., Sadoway, D. R., and Bautista, R. G., Eds., TMS, Warrendale, PA, pp. 321–334, 1990.
44. Broughton J., Bristowe, P. J., and Newsam, J., Eds., *Materials Theory and Modeling*, Broughton, Materials Research Society, Vol 291, Pittsburgh, PA, 1993.
45. *MRS Bulletin*, Materials Research Society, Pittsburgh, PA, Vol. 21(2), 1996.
46. *MRS Bulletin*, Materials Research Society, Pittsburgh, PA, Vol. 25(5), 2000.
47. Sutton, A. P., Godwin, P. D., and Horsfield, A. P., *MRS Bull.*, 21(2), 42–48, 1996.
48. Waghmare, U. V., Kaxiras, E., Bulatov, V. V., and Duesberry, M. S., *Model. Simul. Mater. Sci. Eng.*, 6, 493–506, 1998.
49. Farkas, D., *Mater. Sci. Eng.*, A249, 249–258, 1998.
50. Chu, F., Šob, M., Siegl, R., Mitchell, T. E., Pope, D. P., and Chen, S. P., *Phil. Mag. B*, 70, 881–892, 1994.
51. Misra, A., Bishop, R., Ghosh, G., and Olson, G. B., *Metall. Mater. Trans. A*, 34A, 1771–1782, 2003.
52. Olson, G. B., Presented at TMS Annual Meeting, Charlotte, NC, 2004; Misra, A., Ghosh, G., and Olson, G. B., *J. Phase Equlibra and Diffusion*, 25, 507–514, 2004.
53. Chan, K. S., In *Mechanisms and Mechanics of Fracture: The John F. Knott Symposium*, Soboyejo, W. O., Lewandowski, J. J., and Ritchie, R. O., Eds., TMS, Warrendale, PA, pp. 143–148, 2002.
54. Chan, K. S., *Mater. Sci. Forum*, 426–432, 2059–2064, 2003.
55. Chan, K. S., *Mater. Sci. Eng.*, A337, 59–66, 2002.
56. Chan, K. S., *Oxid. Met.*, 61(3/4), 165–194, 2004.
57. Chan, K. S., *Metall. Mater. Trans. A*, 34A, 589–598, 2003.
58. Zhao, J,-C., Belway, B. P., and Jackson, M. R., *Intermetallics*, 9, 681–689, 2001.
59. Zhao, J,-C., Bewlay, B. P., Jackson, M. R., and Peluso, L. A., In *Structural Intermetallics*, Hemker, K. J., Dimiduk, D. M., Clemens, H., Darolia, R., Inui, H., Larsen, J. M., Sikka, V. K., Thomas, M., and Whittenberger, J. D., Eds., TMS, Warrendale, PA, pp. 483–491, 2001.
60. Das, S., Jeweth, T. J., and Perepezko, J. H., *Structural Intermetallics*, Darolia, R., Lewandowski, J. J., Liu, C. T., Martin, P . L., Miracle, D. B., and Nathal, M. V., Eds., TMS, Warrendale, PA, pp. 35–52, 1993.
61. Davidson, D. L., Chan, K. S., Loloee, L., and Crimp, M. A., *Metall. Mater. Trans. A*, 31A, 1075–1084, 2000.
62. Anton, D. L. and Shah, D. M., *Mater. Res. Soc. Symp. Proc. Vol.* 213, MRS, Pittsburg, PA, PP. 733–738, 1991.
63. Shah, D. M. and Anton, D. L., *Mater. Sci. Eng.*, A153, 402–409, 1992 See also. Anton, D. L. and Shah, D. M., *Mater. Sci. Eng.*, A153, 410–415, 1992.
64. Lewandowski, J. J., Padhi, D., and Solvyev, S., In *Structural Intermetallics*, Hemker, K. J., Dimiduk, D. M., Clemens, H., Darolia, R., Inui, H., Larsen, J. M., Sikka, V. K., Thomas, M., and Whittenberger, J. D., Eds., TMS, Warrendale, PA, pp. 371–380, 2001.
65. Zinsser, W. A. and Lewandowski, J. J., *Metall. Mater. Trans. A*, 29A, 1749–1757, 1998.
66. Rice, J. R., *J. Mech. Phys. Solids*, 40, 239–271, 1992.
67. Hirsch, P. B. and Roberts, S. G., *Phil. Mag. A*, 64, 55–80, 1991.
68. Roberts, S. G. and Booth, A. S., *Acta Mater.*, 45, 1045–1053, 1997.
69. Chan, K. S., *Metall. Mater. Trans. A*, 34A, 2315–2328, 2003.
70. Peierls, R. E., *Proc. Phys. Soc.*, 52, 34–37, 1940.
71. Nabarro, F. R. N., *Proc. Phys. Soc.*, 59, 236–394, 1947.
72. Chan, K. S., *Phil. Mag. A*, 85, 239–259, 2005.
73. Chan, K. S., *Mater. Res. Soc. Symp. Proc.*, *Vol.* 364, MRS, Pittsburgh, PA, pp. 469–480, 1995.

74. Bowen, A. F. and Ortiz, M., *J. Mech. Phys. Solids*, 39, 815–858, 1991.
75. Irwin, G. R., *Fracture Dynamics*, ASM, Cleveland, OH, pp. 147–166, 1948.
76. Rice, J. R., *J. Appl. Mech.*, 379–386, 1968.
77. He, M. Y., Heredia, F. E., Wissuchek, D. J., Shaw, M. C., and Evans, A. G., *Acta Metall. Mater.*, 41, 1223–1228, 1993.
78. Ashby, M. F., Blunt, F. J., and Bannister, M., *Acta Metall.*, 37, 1847–1857, 1989.
79. Hutchinson, J. W. and McMeeking, R. M., In *Fundamentals of Metal-Matrix Composites*, Suresh, S., Mortensen, A., and Needleman, A., Eds., Butterworth–Heinemann, Boston, MA, Chap. 9, pp. 158–173, 1993.
80. Lin, G. and Chan, K. S., *Metall. Mater. Trans. A*, 30A, 3239–3251, 1999.
81. Davidson, D. L., *Metall. Mater. Trans. A*, 28A, 1297–1314, 1997.
82. Subramanian, P. R., Parthasarathy, T. A., Mendiratta, M. G., and Dimiduk, D. M., *Scripta Metall. et. Mater.*, 32(8), 1227–1232, 1995. See also, Mendiratta, M. G., et al., Materials Development Research, WL-TR-96-4113, WPAFB, Dayton, OH, 1996.
83. Belway, B. P., Whiting, P. W., Davis, A. W., and Briant, C. L., *Mater. Res. Soc. Symp. Proceedings*, Vol. 552, MRS, Pittsburgh, PA, pp. KK6.11.1–KK6.11.5, 1999.
84. Henshall, G. A. and Strum, M. J., *Symposium Proceedings*, Vol. 364, MRS, Pittsburgh, PA, pp. 937–942, 1995.
85. Henshall, G. A., Strum, M. J., Subramanian, P. R., and Mediratta, M. G., *Scripta Metall. et. Mater.*, 30(7), 845–850, 1994.
86. Henshall, G. A. and Strum, M. J., *Acta Mater.*, 44(8), 3249–3257, 1996.
87. Henshall, G. A., Subramanian, P. R., Strum, M. J., and Mendiratta, M. G., *Acta Mater.*, 45(8), 3135–3142, 1997.
88. Bewlay, B. P., Briant, C. L., Davis, A. W., and Jackson, M. R., *Proceedings on High Temperature Ordered Intermetallic Alloys IX*, Vol. 646, MRS, Pittsburgh, PA, pp. N2.7.1– N2.7.6, 2000.
89. Bewlay, B. P., Briant, C. L., Sylven, E. T., Jackson, M. R., and Xiao, G., *Proceedings on High Temperature Ordered Intermetallic Alloys IX*, Vol. 646, MRS, Pittsburgh, PA, pp. N2.6.1–N2.6.6, 2000.
90. Kelly, A., Street, K. N., *Proc. R. Soc. Lond. A*, 328, 283–293, 1972.
91. Nabarro, F. R. N., Report of a Conference on the Strength of Solids, The Physical Society, London, p. 75, 1948.
92. Herring, C. J., *J. Appl. Phys.*, 21, 437–445, 1950.
93. Coble, R. L., *J. Appl. Phys.*, 34, 1679–1682, 1963.
94. Harper, J. and Dorn, J. E., *Acta Metall.*, 5, 654–665, 1957.
95. Wang, J. N., *Acta Mater.*, 44(3), 855–862, 1996.
96. Stringer, J. F., High Temperature Corrosion of Aerospace Alloys, AGARD-AG-200, Advisory Group for Aerospace Research and Development, NATO, August 1975.
97. Inoye, H., In *Proceedings of the International Symposium on Niobium*, Stuart, H., Ed., TMS, Warrendale, PA, pp. 615–636, 1984.
98. Begley, R. T., *Evolution of Refractory Metals and Alloys*, Dalder, E. N. C., Grobstein, T., and Olsen, C. S., Eds., TMS, Warrendale, PA, pp. 29–48, 1994.
99. Perkins, R. A. and Meier, G. H., *JOM*, 42, 17–21, 1990.
100. Svedberg, R. C., In *Proceedings of the Symposium on Properties of High Temperature Alloys*, Foroulis, Z. A. and Pettit, F. S., Eds., The Electrochemical Society, Princeton, NJ, pp. 331–362, 1976.
101. Perkins, R. A., Chiang, K. T., and Meier, G. H., *Scripta Met.*, 22, 419–424, 1988.
102. Perkins, R. A., Chiang, K. T., Meier, G. H., and Miller, R., In *Oxidation of High Temperature Intermetallics*, Grobstein, T. and Doychak, J., Eds., TMS, Warrendale, PA, pp. 157–169, 1988.
103. Lee, J. S., Stephens, J. J., and Nieh, T. G., In *High Temperature Niobium Alloys*, Stephens, J. J. and Amad, I., Eds., TMS, Warrendale, PA, pp. 143–155, 1991.
104. Perkins, R. A. and Meier, G. H., *Microscopy of Oxidation*, Institute of Metals, London, UK, 183–192, 1991.
105. Meier, G. H., *Mater. Corros.*, 47, 595–618, 1996.

106. Bewlay, B. P., Lipsitt, H. A., Reeder, W. J., Jackson, M. R., and Sutliff, J. A., In *Processing and Fabrication of Advanced Materials IV*, Ravi, V. A., Srivatsan, T. S., and Moore, J. J., Eds., TMS, Warrendale, PA, pp. 547–565, 1994.

107. Jackson, M. R., Jones, K. D., Huang, S. C., and Peluso, L. A., In *Refractory Metals: Extrusion, Processing and Applications*, Liddell, K. C., Sadoway, D. R., and Butista, R. G., Eds., TMS, Warrendale, PA, pp. 335–346, 1990.

108. Jackson, M. R., Rowe, R. G., and Skelly, D. W., *Mater. Res. Soc. Sym. Proc.*, Vol. 364, Pittsburugh, PA, pp. 1339–1344, 1995.

109. Menon, E. S. K., Mendiratta, M. G., and Dimiduk, D. M., In *Structural Intermetallics*, Hemker, J. J., Dimiduk, D. M., Clemens, H., Darolia, R., Inui, H., Larsen, J. M., Sikka, V. K., Thoma, M., and Whittenberger, J. D., Eds., TMS, Warrendale, PA, pp. 591–600, 2001.

110. Waltson, W. S., O'Hara, K. S,, Ross, E. W., Pollock, T. M., and Murphy, W. H., In *Superalloy*, Kissinger, R. D., Deye, D. J., Anton, D. L., Cetel, A. D., Nathal, M. V., Pollock, T. M., and Woodford, D. A., Eds., TMS, Warrendale, PA, pp. 27–34, 1996.

111. Walston, W. S, General Electric Aircraft Engines, Cincinnati, OH, Private communication, July, 25 2003.

112. Wojcik, C. C., In *High Temperature Niobium Alloys*, Stephens, J. and Ahmad, I., Eds., TMS, Warrendale, PA, pp. 1–12, 1991.

113. Chang, K., Darolia, R., and Lipsitt, H., *Acta Metall. Mater.*, 1, 2727–2737, 1992.

114. Chan, K. S., Onstott, J., and Kumar, K. S., *Metall. Mater. Trans. A*, 31A, 71–80, 2000.

115. Fleischer, R. L. and Zabala, R. J., *Metall. Mater. Trans. A*, 21A, 2149–2154, 1990.

116. Fariabi, E., Collins, A. L. W., and Salama, K., *Metall. Mater. Trans. A*, 14A, 701–707, 1983.

117. Bewlay, B. P., Lewandowski, J. J., and Jackson, M. R., *J. Metals*, 49, 44–45, 1997.

118. Venkateswara Rao, K. T., Odette, G. R., and Ritchie, R. O., *Acta Metall. Mater.*, 42, 893–911, 1994.

119. Chan, K. S. and Shih, D. S., *Metall. Mater. Trans.*, 28A, 79–90, 1997.

120. Larsen, J. M., Worth, B. D., Balsone, S. J., and Jones, J. W., In *Gamma Titanium Aluminides*, Kim, Y. W., Wagner, R., and Yamaguchi, M., Eds., TMS, Warrendale, PA, pp. 821–834, 1994.

13 Mo–Si–B Alloys for Ultrahigh Temperature Applications

J. H. Perepezko and R. Sakidja
Department of Materials Science and Engineering, University of Wisconsin-Madison

K. S. Kumar
Department of Materials Science, Brown University

CONTENTS

13.1 Introduction 437
13.2 Materials Selection 439
13.3 Phase Equilibria 441
    13.3.1 Phase Relations in the Metal-Rich Binary RM-B and RM-Si
         and Ternary RM-Si-B Systems 441
    13.3.2 Phase Stability in the Mo-Si-B System
         and Defect Structures in the T_2 Phase 441
13.4 Alloying Strategy in Refractory Metal Silicides and Borosilicides............442
    13.4.1 Structural Stability of the T_2 Phase in RM-Si-B 443
    13.4.2 Silicide Phase Structures 446
    13.4.3 Alloying Behavior in Quaternary Mo-RM-Si-B Systems 448
13.5 Materials Processing Strategies 450
    13.5.1 Solidification Processing 450
    13.5.2 Powder Processing 453
13.6 Thermodynamic Evaluation of the Phase Equilibria 453
13.7 Diffusion Behavior 454
13.8 Oxidation Behavior of Refractory Metals and Refractory Metal Silicides 455
13.9 Mechanical Properties of Mo-Si-B Alloys 459
    13.9.1 Single-Phase Alloys..460
    13.9.2 Two-Phase Alloys 461
    13.9.3 Three-Phase Alloys 465
13.10 Concluding Remarks 468
Acknowledgments 468
References..469

13.1 INTRODUCTION

In an elevated temperature environment, the essential requirements for a structural material, aside from the obvious one of high melting point, include high temperature strength, stiffness, and oxidation resistance at low density, but adequate room temperature, mechanical properties, and cost effectiveness are also critical factors. Intermetallic alloys, especially the titanium aluminide alloys, have attracted a great deal of interest in these applications. However, the main applications for titanium aluminides are limited to temperatures below 1000°C. The most significant payoffs in terms of advanced aerospace applications and engine uses require temperatures approaching the range of 1300–1500°C [1,2].

The increase in operating temperature represents a huge challenge that can be put into perspective by examining the development trend data on the use of Ni-alloy technologies in gas-turbine engines. As shown in Figure 13.1, one measure of both material and engine performance is the amount of horsepower produced as a function of the turbine rotor inlet temperature, the $T4_1$ temperature of an engine. This relationship is known for an ideal gas-turbine engine and is shown on the figure as an analytical expression and by the green line. Blue bullets in Figure 13.1 represent several Pratt and Whitney engines developed from the 1940s through the early 1990s. In each case, the actual engine performance falls short of the ideal line because of inefficiencies in the engine. As a result of continuous effort, structural and aerothermal design advances minimized leakage in the pressurized systems, optimized aerothermal flows of the engine, and took advantage of highly refined cooling and coating schemes to prevent materials from melting in the high-pressure turbine (note the fine cooling hole configuration on the airfoil inset in Figure 13.1). The blue shading on the lower left region of Figure 13.1 depicts the incremental gain in service temperatures from such design refinements. The orange and purple shaded regions of Figure 13.1 depict the advantages to the engine from efficient design and the use of advanced materials in the turbine.

Current engines expose superalloys to temperatures approaching 1150°C, while the turbines are able to operate at considerably higher temperatures by distributing cooling air through the advanced single-crystal airfoils and by using thermal barrier coatings. However, closer inspection of the data shown for the PW2037 through PW4084 engines reveals a new and disturbing trend. While the rotor inlet temperatures rose with each of these engines in turn, the deviation from ideal performance increased, principally as the result of necessarily increased cooling airflow. With

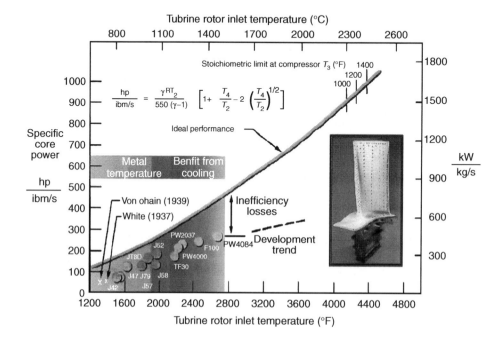

FIGURE 13.1 Core horsepower versus turbine inlet temperature for selected gas turbine engines. Data for specific engines spans about 70 years and is compared with the ideal or theoretical limit [3].

modern superalloys, metal temperatures have essentially stagnated and cannot increase significantly due to the limit set by melting; thus, significant advancement requires new materials for the high-pressure turbine.

In the search for new materials for ultrahigh temperature service, the required thermal stability of the constituent phases can be considered from several points of view. First and foremost is survivability, which clearly requires an inherently high melting temperature. In addition, a limited interaction with the service environment and compatibility with oxidation resistant coatings or with reinforcement phases in composites are often basic requirements for survivability of a component at high temperature. Moreover, application induced microstructural modification due to thermal cycling and the effect of creep under varying stress states are essential to consider for a useful component. The underlying basis for alloy synthesis, processing, and the assessment of thermal stability is established by the relevant phase equilibria in all systems. Kinetic data are also required to understand and control the possible interphase/interface reactions and the evolution of microstructure during phase transformations. With this approach a systematic and knowledge-based assessment of advanced materials may be pursued which is essential for successful development [4].

As an important background to the focus in this chapter, it is valuable to consider several critical issues in materials selection and design such as phase stability, multiphase options, oxidation behavior, and high temperature mechanical properties that result in the identification of Mo–Si–B alloys as a key material for ultrahigh temperature applications. From these background highlights, a perspective can be developed to identify the attributes of Mo–Si–B alloys as well as the areas where further effort is necessary to enhance the performance.

13.2 MATERIALS SELECTION

Requirements of elevated temperature use limit the available materials choices to a few specific classes. There are a number of ceramic materials, intermetallic compounds and several refractory metals with melting points (T_m) above 1500°C. As single components, the ceramic and intermetallic phases are known to suffer from severe embrittlement problems at low temperatures. Similarly, refractory metals such as molybdenum and niobium have sufficient ductility, but are sensitive to oxidation problems [5]. Nevertheless, the melting point criterion is a very important one because high melting temperature tends to signify high elastic modulus as indicated by correlation developed by Fleischer [6]. In addition, the limiting creep rate that defines a maximum operating temperature tends to increase with increases in melting temperatures and is often estimated to be between $T_m/2$ and $2/3T_m$. Similarly, the expansion coefficients tend to vary inversely with melting temperature and low values are often useful in thermal cycling considerations.

The successful application of high temperature structural materials is reflected in a variety of mechanical and thermochemical properties: strength, ductility, toughness, fatigue resistance, creep resistance, and oxidation resistance. Since a single phase component is unlikely to satisfy all of these demanding and diverse requirements for structural integrity at high temperatures, multiphase designs are essential. Moreover, history has strongly indicated that multiphase designs in a monolithic material are the most effective and reliable components for structural applications. Indeed, the experience with nickel base superalloys and titanium aluminides based on TiAl [7] and Ti_3Al [8] indicates that multiphase designs are again the most promising strategy [9]. With this in mind, a multiphase design based on a refractory metal and an intermetallic offers an effective strategy for the required balance of properties and performance.

With the establishment of refractory metal plus intermetallic as a starting point, the melting temperature criteria can be satisfied. In terms of multiphase designs, it is useful to consider briefly some of the characteristics that can be used for the separate choice of a refractory metal. A listing of some distinguishing properties is provided in Table 13.1 for Nb, Ta, Mo, and W [10,11]. Although

TABLE 13.1
Physical Properties of the Refractory Metals [10,11]

Property	Nb	Ta	Mo	W
Melting point °C	2470	3000	2610	3410
Density (g/cm^3)	8.57	16.6	10.2	19.3
Elastic modulus, GPa (at RT)	110	186	289	358
Linear thermal expansion, 10^{-6}/°C	7.1	5.9	5.4	4.5
Thermal conductivity, w/m-K	52.3	54.3	142.1	166
Interstitial solubility, ppm by weight[a]				
Oxygen	1000	300	1.0	1.0
Nitrogen	300	1000	1.0	<0.1
Carbon	100	70	<1.0	<0.1
Hydrogen	9000	4000	0.1	N.D.

[a] Based on estimates of the equilibrium solubility at the temperature where $D = 10^{-11}$ cm^2/s.

V and Cr could be included also in the listing, they have been omitted because 1500°C is about 80% of T_m, so rapid diffusion would produce excessive creep rates and microstructural instabilities. For structural applications, both Ta and W have high densities, and Ta has a relatively low elastic modulus and high solubility for an interstitial solute. While Nb has the lowest density of the refractory metals in Table 13.1, it also has the lowest elastic modulus, E, so that E/ρ for Nb is only 1.8 but for Mo it is 4.1. In addition, Mo has a relatively low expansion coefficient and almost three times the thermal conductivity of Nb. There is a major difference in interstitial solubility between Nb and Mo, which will impact solution strengthening and also ductility. As a result, Mo base alloys offer a number of attractive property characteristics for structural applications at high temperature.

The next level of selection focuses on environmental stability and in particular oxidation resistance. In this regard, refractory metals cannot perform in an uncoated form at a temperature above 1000°C [12,13,14]. However, several intermetallics may be used; in particular, $MoSi_2$ is a proven oxidation resistant material that finds wide application commercially [15]. Moreover, the oxidation protection is in the form of a self-healing, adhesive silica glass layer on the surface, which is also non-spalling and allows for long life spans. However, single phase $MoSi_2$ has a ductile–brittle transition temperature of about 900–1000°C and a clear method has not been identified for an effective toughening strategy [2,16,17,18].

In terms of options for combinations between refractory metals based on Mo and Nb and intermetallic phases, the potential selections can be limited effectively. For example, carbide and nitride intermetallics can be excluded because of their poor oxidation resistance compared with $MoSi_2$ and because any possible multiphase mixture with a refractory metal would tend to embrittle the refractory metal since it must be saturated with carbon or nitrogen to be in equilibrium with the intermetallic [5]. Moreover, ternary phases (e.g., carbonitrides) can develop which do not confer crucial oxidation resistance [19]. Of course, oxidation and thermal barrier coatings can be used for components, but without an inherent oxidation resistance and high thermal stability, failure of the coating will lead to catastrophic failure of the component. There are several situations where refractory metals are in equilibrium with binary silicide and boride phases. There are also a few selected cases where a refractory metal is in equilibrium with a ternary intermetallic. Most notably Me_5SiB_2 (T_2) + Me two-phase regions coexist as reported in the literature where Me is niobium or molybdenum [20,21]. In terms of materials selection, the combination of Mo or Nb with a silicide or borosilicide phase appears to be most favorable.

13.3　PHASE EQUILIBRIA

13.3.1　PHASE RELATIONS IN THE METAL-RICH BINARY RM-B AND RM-Si AND TERNARY RM-Si-B SYSTEMS

There is a reasonably complete database available for the phase equilibria in binary refractory metal silicides and borides, but the extension of this base to ternary systems is incomplete and the reported information is of uncertain reliability. Most of the information that is available was developed as part of a systematic study of ternary structures in Me–Si–B systems. Since the crystal structure was the main focus, the associated phase equilibria were given only cursory attention. With this caveat, it is still of value to review briefly the available information to provide a basis to identify important features of ternary alloy behavior and unresolved or outstanding issues. An inspection of the summary of phase equilibria and crystal structure information in the binary systems of the RM–B, RM–Si and ternary RM–Si–B reveals several points that concern ternary alloying behavior.

The high melting BCC phase in the binary systems is in equilibrium with the high-melting, but metal-rich boride or silicide phases. For example, the Mo–B system forms two-phase field of Mo + Mo_2B. In most cases, there is a stable two-phase field between a BCC phase with the metal rich silicide phases such as Mo_3Si. However, there is typically a low oxidation resistance for these silicide phases. Relative to the 3-1 and 2-1 phases, there is a notably lighter high melting compound of the 5-3 phase (i.e., RM_5Si_3) that is in equilibrium with the BCC phase in some of the binary RM–Si systems. Unfortunately, these phase mixtures do not exhibit a high enough oxidation resistance even though they do exhibit high thermal stability. The addition of B into certain 5-3 phases, most notably Mo_5Si_3, has been shown to improve the oxidation resistance [22]. However, in the case of Mo_5Si_3, there is no Mo + Mo_5Si_3 two-phase field existing in the Mo–Si binary system.

In the several ternary RM–Si–B systems, there is a stable two-phase field of the BCC phase with the ternary-bases refractory-metal borosilicides of RM_5SiB_2 (T_2) "5-1-2" phase, most notably, the Mo–Si–B, Nb–Si–B, V–Si–B, and W–Si–B systems. The higher content of Si and B in the T_2 phase also significantly enhances the oxidation resistance. In addition, the melting point does not appear to be reduced significantly by the addition of two metalloids in the RM–Si–B systems. Mo_5SiB_2 for instance has a melting point in excess of 2100°C [23]. Thus, based on the overall balance of initial materials properties, the two-phase mixture of BCC and T_2 should serve as a basis for the multi-phase alloy design.

13.3.2　PHASE STABILITY IN THE MO-Si-B SYSTEM AND DEFECT STRUCTURES IN THE T_2 PHASE

Based upon the EPMA examination [24] of phase compositions in long-term annealed samples and x-ray diffraction determination of phase identity, the Mo-rich portion of the phase diagram isotherm for the Mo–Si–B system at 1600°C has been established as shown in Figure 13.2. The main emphasis is on the phase equilibrium bordered by the Mo–MoB–Mo_5Si_3 region. Within this region, the central role of the T_2 phase as a key constituent in the phase stability is evident. Recently, the Mo-rich phase diagram isotherm for the Mo–Si–B system at 1600°C has been extended to the Mo–MoB–$MoSi_2$ region. EPMA examination on three-phase equilibrium (MoB, Mo_5Si_3, and $MoSi_2$) samples provided the phase boundary composition information. The homogeneity region of the boride phases (Mo_2B and MoB) at 1,600°C generally indicates a very small Si solubility (less than 0.25 at.%). Mo(ss) phase has negligible B solubility, but a Si solubility up to about 3 at.% Si. Mo_3Si phase has a limited homogeneity region that ranges from about 24 to 25 at.% Si and a negligible B solubility [25]. The compositional boundary of Mo_5Si_3 (T_1 phase) on the Mo-rich side extends to ∼37 at.% Si and on the Si-rich side to ∼39 at.% Si. The B solubility in the T_1 phase has been found to be much more limited than that reported by Nowotny [20] and is in agreement with a recent assessment [26].

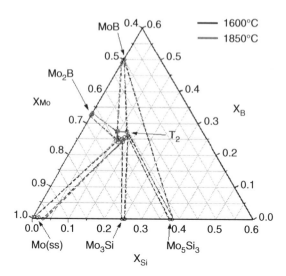

FIGURE 13.2 Isothermal sections of Mo–Si–B ternary system at 1600 and 1850°C [27].

The apparent operation of anti-site B–Si substitution over a portion of the T_2 stability range is consistent with the observations from other ternary systems that exhibit the ternary-based T_2 phase. In fact, in the Nb–Si–B system for instance it is possible to accommodate a large solid solution mixing of the metalloid constituents in the T_2 phase as manifested by the wide span of B and Si homogeneity range. It is worth noting here that while a large degree of mixing in the metal sublattice is common for the metal-rich silicides phases, only the T_2 phase appears to also display a large degree of metalloid mixing.

In addition to the anti-site substitution defect mechanism, there appears to be a different defect structure that influences the temperature dependence of Mo solubility in the T_2 phase. Formation of the Mo(ss) precipitates in the T_2 primary phase [27,28] suggests that there is an additional role of a vacancy-based defect structure that both promotes the precipitation process and determines the magnitude of the Mo solubility limit variation with temperature.

13.4 ALLOYING STRATEGY IN REFRACTORY METAL SILICIDES AND BOROSILICIDES

A common feature of the refractory metals is the high mutual solid solution solubility. This trend exists not only in the BCC solutions, but also extends to many intermediate phases (e.g., silicides, borides, and aluminides). At the same time, the sensitivity of the phase stability of silicides in general and T_2 phase in particular to atomic size and off-stoichiometric site substitution suggests that refractory metal substitution for Mo can be an effective approach to controlling phase reactions and a basis to formulate an effective alloying strategy. Previous studies on the alloying behavior of the silicide phases suggested that the incorporation of selected transition metals (W, Nb, V, and Cr) results in the formation of a continuous T_2 phase and the two-phase field of BCC + T_2 in the respective quaternary systems [27,29]. The observed alloying trends also appear to highlight a number of fundamental geometrical factors that influence the relative stability of the T_2 phase and provide a basis to develop modified multiphase designs [4,27]. There is, however, limited evaluation on the possible alloying extension of potentially critical additional transition metal (TM) elements for molybdenum alloys such as Group IVB (such as Ti, Zr, and Hf) and Group VIIB metals (such as Re). Titanium and Zirconium are the major additives to commercial Mo-based alloys such as "TZM" which is composed of a BCC phase and dispersoids of transition metal carbides to enhance the high temperature strength. However, there is no known ternary-based T_2 phase reported in TM–Si–B systems (TM = Ti, Zr, and Hf) [30]. The addition of rhenium (Re) into a Mo BCC solid solution has been well known to markedly lower the Ductile Brittle Transition Temperature (DBTT) of the BCC phase resulting in a higher ductility and toughness [31].

In general, the size factor metrics such as the atomic radius of the metal components or the atomic radius ratio between metal to metalloids have been shown to play an essential role in alloying extension as exemplified by the Hume–Rothery 15% rule. Similarly, the effect of chemical bonding within the structure as expressed by parameters such as valence electron concentration per atom (e/a) has been shown to be essential in combination with the size factors to define the phase stability. In particular, it has been shown previously that the e/a criteria can be successfully applied to transition metals and transition metal-based compounds [32,33,34,35,127]. Moreover, the e/a criteria can also be correlated directly to some of the characteristics in the electronic structure of the compounds that favor a high cohesive energy [36]. For example, for a stable and high melting transition metal-based BCC phase, the favorable e/a values range between 5.0 (corresponding to that of Group VB metals such as Nb) and 6.0 (corresponding to that of Group VIB metals such as Mo) with the highest cohesive energy (at the ground state) corresponding to an e/a of 5.5 [13,37,38]. The optimum e/a value and consequently the maximum cohesive energy are characterized by the electronic structure of the BCC phase showing that all of the bonding states have been filled and none of the anti-bonding states which are at higher energy level are occupied. Furthermore, the extent of solid solution among the transition metals can also be estimated from a similar analysis [13,37,38].

13.4.1 Structural Stability of the T_2 Phase in RM-Si-B

Structural stability of the T_2 phase is central to the alloying design especially as related to the refractory metal substitution. Therefore, it is useful to examine the crystal structure at the stoichiometric composition. The Mo_5SiB_2 phase has the $D8_1$ structure which has a body-centered tetragonal unit cell (space group I4/mcm) as shown in Figure 13.3. The unit cell contains 32 atoms, which means 20 Mo, 4 Si, and 8 B atoms are situated in layered arrangements. Three types of layers can be identified: layer A with only Mo atoms, layer B with only Si atoms, and layer C with a mixture of Mo and B atoms. The structural arrangement of these layers in T_2 has been viewed as the means to achieve an efficient atomic packing between metal atoms such as Mo and

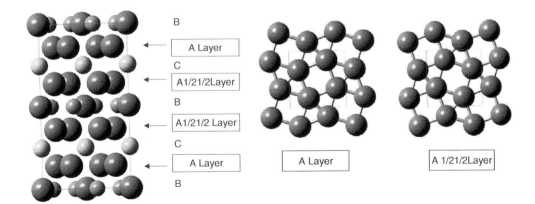

FIGURE 13.3 T_2 Crystal structure comprising stacking of Mo only atomic layer (A layer), Mo + B atomic layer (B) and Si only atomic layer (C). Atomic arrangement of the A layer and $A_{1/2\ 1/2}$ layer are illustrated in b and c.

metalloid constituents (Si and B in this case) [39,40]. Based on the radius ratio of the metal atom and the metalloid atoms, variations in the successive stacking can be constructed. The relatively large difference in atomic radius of the two types of metalloids necessitates stacking arrangements of the A layers that would yield two distinct sites. Layer arrangements of A–A$_{1/2\ 1/2}$–A$_{1/2\ 1/2}$–A–A– in the [001] direction are therefore developed in the T$_2$ structure. The A$_{1/2/1/2}$ layer refers to the A layer that has been translated by half the base diagonal relative to neighboring layers. With the A–A $_{1/2\ 1/2}$ or A$_{1/2\ 1/2}$ –A arrangements, a cubic anti-prismatic site is created and filled by Si atoms forming the layer B (see Figure 13.3). The B atoms on the other hand are situated in the trigonal prismatic hole generated by sandwiching two symmetrically oriented A layers (the A–A or A$_{1/2\ 1/2}$– A$_{1/2/1/2}$ layer arrangements). In this site, the B atoms are capped by two triangular arrangements of Mo atoms along the c axis and one B and two more Mo atoms forming an intermediate layer (layer C). The two Mo atoms fill the remaining available hole created by the A–A layer arrangements that accordingly constitutes the largest hole available (the cubic prismatic hole). Thus, the limited ability to stabilize the T$_2$ phase in the Si-rich region in the Mo–Si–B system may be interpreted as the difficulty in situating Si atoms in the B sublattice in the trigonal prismatic hole. On the other hand, there is a ready accommodation of B atoms in the Si lattice position, which is also indicated by the reduction of the cell volume for B-rich T$_2$ compositions. The limitation in the enriching the T$_2$ phase with Si appears to be related to the limited available volume of the B sublattice.

A further important consequence of the defect structure has been identified as the development of a precipitation reaction involving the formation of Mo precipitates in the T$_2$ phase. Usually, it is observed that intermetallic phases precipitate within disordered solid solution phases. However, due to the non-stoichiometric nature of the T$_2$ phase field, Mo-rich compositions are accommodated by the formation of metalloid (i.e., Si and/or B) constitutional vacancies. The extent of non-stoichiometry increases with increasing temperature. Due to sluggish solid-state diffusion, even modest cooling rates allow for retention of a T$_2$ phase that has a Mo supersaturation. Upon annealing the Mo phase precipitates with a well-defined orientation relationship and can offer a means of toughening [28].

The Mo$_5$Si$_3$ (T$_1$ phase) on the other hand has a radius ratio of metal to metalloid below that of T$_2$ phase. The normal A–A$_{1/2\ 1/2}$ atomic stacking to fill in Si atoms is accommodated by replacing part of the Mo "A" layers with vacancies and Si atoms to form a modified A layer. Hence, in the T$_1$

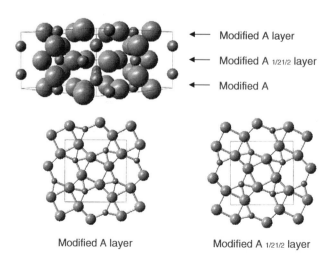

FIGURE 13.4 Crystal structure of Mo$_5$Si$_3$ (T$_1$) phase showing the alternating sequence of modified A and A $_{1/2\ 1/2}$ layers. The atomic arrangements of the modified A layer and A$_{1/2\ 1/2}$ layer are illustrated.

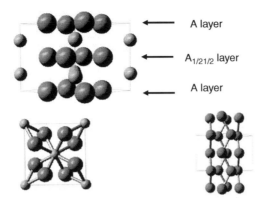

The A- A $_{1/2/1/2}$ layered arrangement yield the Mo-B clusters in Mo2B

Mo-Mo chains in Mo2B phase with an interatomic distance of 0.267 nm

FIGURE 13.5 Crystal structure of Mo_2B phase showing the alternating sequence of the modified A and $A_{1/2\ 1/2}$ layers bordering the B only layers. The crystal structure is also characterized by the presence of Mo–Mo vertical chain similar to that of the T_1 phase (Figure 13.4).

phase, there are two types of the Si atoms, one sandwiched between the modified A and $A_{1/21/2}$ layers and the one within the modified A layers (see Figure 13.4 and Figure 13.5). In both Mo_2B and T_1 phases, there is no stabilization of the A–A layered arrangement.

In this context, it is worthwhile to note an other crystal variant of the T_2 crystal structure, namely the $D8_8$ phase (Figure 13.6). Similar to the Mo_2B and T_1 phases, the $D8_8$ phase also has alternating atomic layer stacking of "modified" A layer and $A_{1/2\ 1/2}$ layer. The difference is that the modified A layer forms a hexagonal base symmetry instead of a tetragonal one (like in Mo_2B, T_1, and T_2 phases). In addition, only half the interstitial sites available from this type of configuration are filled by the metal constituent, i.e., Ti in Ti_5Si_3. Unlike the T_2 phase, the $D8_8$ phase is most stable when the base metal is the Group IVB such as Ti or Hf.

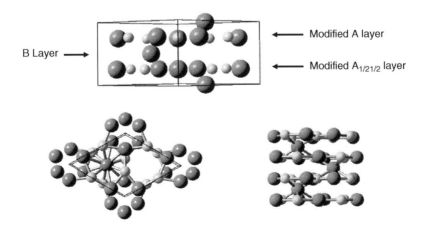

B Layer →

Modified A layer

Modified $A_{1/21/2}$ layer

FIGURE 13.6 Crystal structure of Ti_5Si_3 ($D8_8$) phase showing the alternating sequence of the modified A and $A_{1/2\ 1/2}$ layers bordering the half-filled Mo-only B' layer.

13.4.2 SILICIDE PHASE STRUCTURES

The geometric rule relating to structural stability, however, does not address the fact that there is an unusually large solubility of the refractory metals in the Mo sites even with elements such as Hf or Ti for which the respective RM–Si–B ternary systems exhibit a stable $D8_8$ phase. In fact, the alloying behavior is apparently quite comparable to the alloying behavior of RM substitutions in the Mo BCC phase. The similarity in the alloying between the metal sites in the T_2 and BCC (A2) phases may be traced to the fact that similar to the BCC crystal structure, the T_2 structure also maintains a relatively high coordination number (CN) of metal–metal atomic contacts to retain a relatively close-packed structure [39,41]. Furthermore, the T_2 crystal structure (I4/mcm) retains a body-centered symmetry for the refractory metals as exemplified by the similar atomic surroundings of Mo atoms at the (0,0,0) and (1/2, 1/2, 1/2) positions in the lattice. The BCC-like environment is quite evident by examining the surroundings of the Mo atomic sites in Figure 13.7, which shows that the CN of Mo-center and Mo-edge contacts in the Mo clusters in the T_2 crystal structure is actually eight, and is the same as that of the BCC lattice. In fact, the T_2 structure can be viewed as constructed from vertical chains of the BCC-like corner-sharing Mo clusters connected by the anti-prismatic hole filled by Si. Examination of the inter-atomic distances further supports the resemblance [42]. The Mo–Mo inter-atomic distances associated with the nearest neighbors are close to those in the BCC lattice as well. The shortest Mo–Mo interatomic distance yields a value of 0.2737 nm (at room temperature) which is quite similar to that of the Mo–Mo interatomic distance in the BCC lattice (0.272 nm). Indeed, the consequence of the BCC-like environment in the T_2 phase can be directly linked to the solubility behavior (in comparison with that of Mo_2B and Mo_5Si_3 phases) and the electronic factors as represented by the e/a ratio.

There have been limited studies on the relationship between alloying behavior and the electronic structure of metal rich silicides and borosilicides [43,44]. The most recent work has been concentrated mainly on the effect of alloying on the physical properties such as thermal expansion and elastic moduli. In contrast, there have been several phase stability studies that focused on the rare earth (RE) and alkaline-earth (AE) based T_2 phases such as La_5Si_3 and Ca_5Ge_3 where the phase stability criteria may not be similar to that of the transition-metal based T_2 phase. The RE and

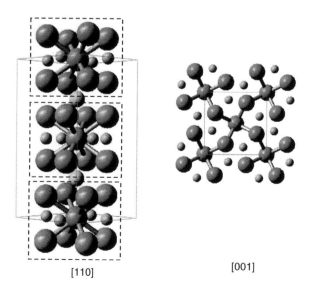

[110]   [001]

FIGURE 13.7 The Mo–Mo BCC-like clusters embedded in the T_2 tetragonal crystal surrounded by the Mo–Si and Mo–B polyhedra as shown from [110] to [001] directions.

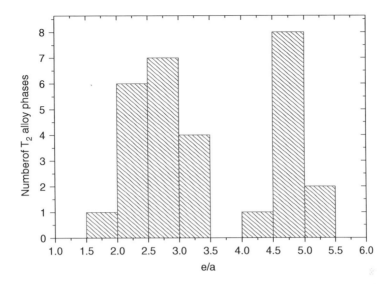

FIGURE 13.8 Plot of the occurrences of T_2 phases as a function of e/a showing two distinct groups.

AE-based T_2 phases can be distinguished from the transition-metal based T_2 phase by the value of the valence electron per atom (e/a). A plot of occurrences of the known T_2 compounds versus the e/a value is shown in Figure 13.8. Clearly, there is a distinct grouping for the T_2 phase. The low e/a range (e/a of 2–3) T_2 phase occurs with the metal constituent as the rare earth or alkaline earth metals and the high e/a range T_2 phase (e/a of 4–5) occurs with a transition metal as the metal atom constituent. It is noteworthy that Mo_5SiB_2 (e/a value 5.0) is positioned at the high end of the e/a range.

Combined criteria that include geometry and electronic factors to define the stability of the T_2 phase are developed in Figure 13.9a and b which present plots of the valence electron concentration per atom (e/a) versus the a parameter lattice unit as well as the atomic radius ratio between the metal (r_M) to metalloid/simple metal (r_X) constituent of the T_2 crystal structure.

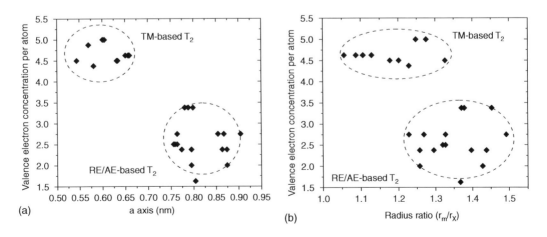

FIGURE 13.9 (a) Plot of e/a versus lattice unit and (b) e/a versus atomic radius ratio. The two groups of T_2 phases also have different geometric characteristics; the low e/a, RE-based T_2 phase occurs with a relatively larger lattice unit cell and a higher radius ratio than that of the high e/a, TM-based T_2 phase.

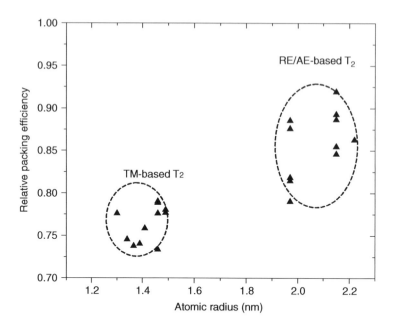

FIGURE 13.10 Plot of packing volume fraction of alloys with the T_2 phase structure showing the distinct grouping of the TM-based and RE-based T_2 phases.

The two domains of T_2 phase can clearly be discerned. More importantly, there are characteristics defining the two groups of T_2 phase. Namely, the low e/a T_2 phase generally has larger lattice dimensions and also a larger atomic radius ratio in comparison to that for the high e/a T_2 phases. The larger lattice unit is directly correlated to the fact that the metal constituent of the low e/a T_2 phase is composed of either rare earth or alkaline earth metals, which have a larger atomic radius than that for the transition metals. This further confirms a definite correlation between the volumes of the T_2 unit cell with the size of the metal atoms. The relatively high value of the atomic radius ratio of low e/a T_2 phases has another important geometric characteristic. An evaluation of the packing efficiency of the T_2 phase shows that the low e/a T_2 phase also has a much higher volume density per unit cell than that of low e/a T_2 phases (see Figure 13.10). Thus, there is a general tendency for the low e/a value T_2 phases to achieve a relatively high packing efficiency and consequently high atomic radius ratio.

13.4.3 ALLOYING BEHAVIOR IN QUATERNARY MO–RM–Si–B SYSTEMS

A common feature of the refractory metals is their high mutual solid solution solubility. This trend exists not only in the BCC solutions, but also extends to many intermediate phases (e.g., silicides, borides, and aluminides). At the same time, the sensitivity of the T_2 phase stability to atomic size and off-stoichiometric site substitution suggests that refractory metal substitution for Mo can be an effective approach to controlling phase reactions.

This behavior has been examined by determining the effect of refractory metal substitution on the extent of the Mo(ss) $+ T_2$ two-phase field and the influence of alloying on the solubility of metalloid constituents in T_2. There are three distinct types of behavior of metalloid solubility that can be discerned with refractory metal substitution [29]. Alloying with Nb, for instance, tends to shift the metalloid solubility towards Si-rich values by lowering the B to Si ratio with increasing Nb levels. On the contrary, Cr substitution stabilizes the T_2 phase by altering the metalloid content toward the B-rich side resulting in a continuous increase of B to Si ratio in the T_2 phase with an increase in Cr substitution. Substitution with W or V on the other hand does not appear to cause a major shift in the B to Si ratio.

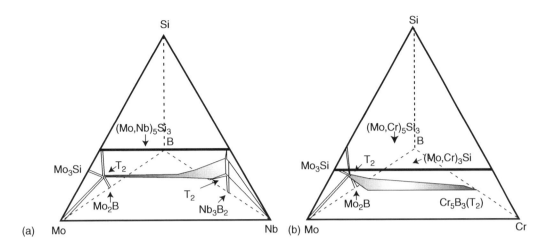

FIGURE 13.11 Illustrative diagrams showing the extension of the T_2 phase field in the (a) Mo–Nb–Si–B and (b) Mo–Cr–Si–B quaternary systems. To aid the visualization only relevant phases are indicated.

It has been shown previously [40] that most of the binary alloy T_2 (Me_5X_3) phases maintain an atomic radius ratio (r_{Me}/r_X) in a range of 1.2–1.5. This behavior leads to a size factor concept that is based upon a geometrical constraint to allow for optimum packing of the transition metal constituents with the metalloids. In ternary T_2 phases ($Me_5Xa_1Xb_2$) where the radius ratio is $r_{Me}/(1/3\ r_{Xa} + 2/3\ r_{Xb})$, an increase in average metallic radius in the metal sites necessitates an expansion of the metalloid sites. This may be accomplished by enrichment of Si filling in the B lattice site at the trigonal prismatic hole, as exemplified in the case of Nb substitution. In fact, the relatively limited two-phase field of Mo(ss) + T_2 in the Mo–Si–B ternary system can also be enlarged by Nb substitution due to the continuous solution of both the (Mo,Nb) phase and the T_2 phase in the Mo–Nb–Si–B quaternary system as illustrated in Figure 13.11a [27].

A continuous metallic solid solution that results in a reduction in the metallic atomic radius in the T_2 phase alters the solubility behavior of B and Si to accommodate the volume reduction in the metalloid sites. For example, it is expected that there is a continuous increase of B solubility with a corresponding increase in Cr addition in the metal sites of the T_2 phase as schematically illustrated in Figure 13.11b. In fact, Si atoms can be totally replaced forming the $(Cr,Mo)_5B_3$ phase. Similar to the Nb–Si–B system [45], a large two-phase field is also present in the Cr–Mo–B system [45]. It is worth noting also that there is a linear correlation between the corrected atomic size of the metal constituents and the cube root of the unit cell volume over a wide range of binary-based T_2 phases after including a necessary atomic size correction based on the Miedema electronegativity [40]. This observation apparently can also be extended to the ternary-based Me_5SiB_2 phases where Me is a transition metal as shown in Figure 13.12a. Furthermore, it has been shown in binary-based T_2 phases that there is also a preferential dilation of the a parameter relative to the c parameter (Figure 13.12b) with the expansion of the crystal due to an increase in radius of the metal atoms. Hence, the solubility behavior that can be judged in terms of the size factor concept provides useful guidance in designing effective alloying strategies for the T_2 phase. Furthermore, the ability to predict the structural stability of the T_2 phase with refractory metal substitution for Mo can be utilized to optimize the overall materials properties. In this context, the most critical issue in terms of taking full advantage of the geometrical concept is to what extent the size factor can be implemented optimally for the phase stability analysis. This is exemplified in the effect of refractory metal substitution on the solidification pathways and segregation in the Mo-rich Mo–Si–B alloys.

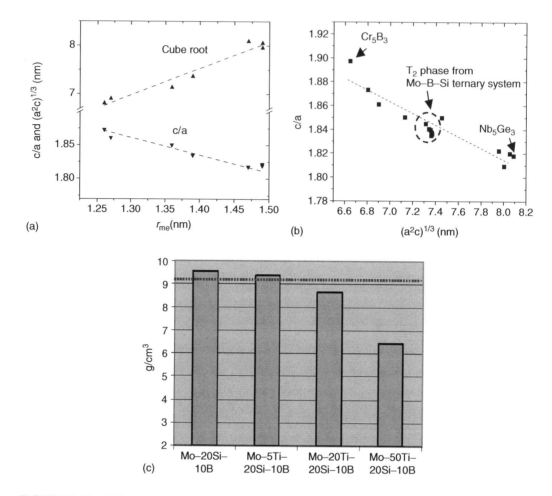

FIGURE 13.12 (a) Plot indicating the correlation between r_{Me} the metal atomic radius in Me_5SiB_2 (T_2) phases and c/a and $(a^2c)^{1/3}$, (b) Plot showing the correlation between the c/a and $(a^2c)^{1/3}$, and (c) the benefit in weight density reduction of Mo–Si–B alloys by Ti substitution for Mo. Dotted horizontal line represents the typical density for advanced Ni-based superalloys.

Benefits gained from the use of the refractory metal substitution for Mo can also be utilized to reduce the weight density of the Mo-based alloys, which is typically about 10% above that of the Ni-based alloys. The big advantage of the extended alloying behavior has been the capability to alloy Mo with a number of lighter transition metals such as Ti, V, and Nb. The addition of transition metals such as Ti for instance reduces the weight density of the Mo–Si–B alloys even below that of Ni-based alloys as depicted in Figure 13.12c. Further work needs to be conducted in alloying design that will reduce the weight density without compromising the performance in oxidation resistance and mechanical properties [46].

13.5 MATERIALS PROCESSING STRATEGIES

13.5.1 SOLIDIFICATION PROCESSING

For the solidification processing of the Mo–Si–B alloys, the main challenge has been the large extent of solidification segregation in solidifying the Mo-rich Mo–B–Si alloys as manifested by the presence of non-equilibrium primary boride (Mo_2B and MoB) products in alloys with compositions

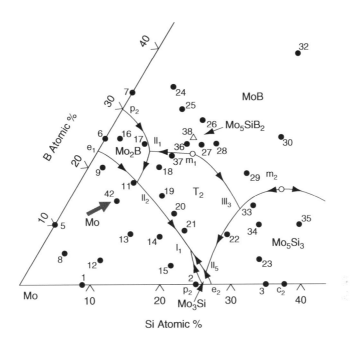

FIGURE 13.13 Liquidus projection in the Mo-rich region of Mo–Si–B System. The arrow indicated the nominal composition of Mo–14Si–7B.

in the $Mo(ss) + T_2$ two-phase field. The Mo_2B primary solidification precludes the attainment of eutectic $Mo(ss) + T_2$ alloy microstructures as shown in Figure 13.13, which presents the liquidus projection [47] and as illustrated by the as-cast structure of Mo–7Si–14B in Figure 13.14. To avoid the Mo_2B primary phase, the liquidus projection suggests that at least the formation of the four-phase equilibria of a Class II reaction $(Mo_2B + L \Rightarrow Mo(ss) + T_2)$ must be avoided. Since the equilibrium point for this reaction is outside the two-phase field, the compositions used will always be within the three-phase field of $Mo(ss) + T_2 + Mo_3Si$. This may present some difficulties particularly since the preceding Class I reaction of $(L \Rightarrow Mo(ss) + T_2 + Mo_3Si)$ may be bypassed by a slight undercooling due to the relatively shallow liquidus surfaces for Mo_3Si and T_2, and as a consequence, the monovariant eutectic structure of $Mo_3Si + T_2$ can form instead. The presence of a relatively brittle $MoSi_3$ constituent may not be desirable. It is therefore necessary to alter the primary Mo_2B solidification event.

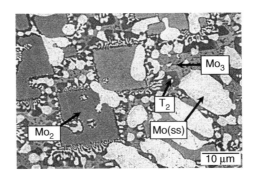

FIGURE 13.14 BSE SEM image of as-cast Mo–14Si–7B showing the extensive solidification segregation.

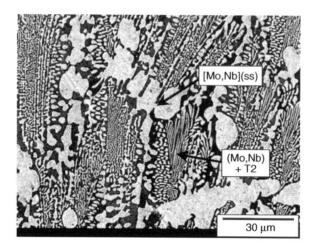

FIGURE 13.15 BSE SEM image of as-cast Mo–20Nb–14Si–7B showing the two-phase microstructures and the reduced solidification segregation.

One strategy to produce the Mo(ss)+T$_2$ two-phase microstructure directly from the melt is to suppress the primary boride phases [48]. Rapid solidification processing (RSP) can yield a fine two-phase microstructure by bypassing the formation of the boride phase. The solidification path of the undercooled melt is simplified into either Mo(ss) or T$_2$ primary solidification followed by the Mo(ss)+T$_2$ eutectic formation. Due to the very short diffusion distances in this fine microstructure, subsequent annealing at a relatively low temperature such as 1200°C for 150 h is sufficient to yield a uniform submicron two-phase Mo(ss)+T$_2$ microstructure [48].

Another effective design strategy for altering the solidification pathway is to apply selected quaternary additions. In fact, the systematic substitution of Mo by Nb has been shown to reduce the extent of Mo$_2$B primary formation [27]. The reduction in the Mo$_2$B liquidus extension may be due to the limited amount of Nb substitution for Mo that is possible in the Mo$_2$B phase[49]. On the other hand, a large degree of Nb substitution for Mo in both the bcc Mo(ss) phase as well in the T$_2$ phase has been established (Figure 13.11). As a result, two-phase [Mo,Nb] (ss)+T$_2$ microstructures can be produced directly from the melt as shown in Figure 13.15.

In alloys where the nominal composition is no longer in the Mo(ss)+T$_2$ two-phase field, but rather in the three-phase field of Mo(ss)+T$_2$+Mo$_3$Si, refractory metal (RM) substitution may also affect the solidification pathways significantly. The substitution can modify the final solidification eutectic reaction that yields the Mo$_3$Si phase. As for the case of the Mo$_2$B phase, the liquidus surface extension for the Mo$_3$Si phase into the quaternary system with refractory metal substitutions such as Nb has been found to be quite limited. This is due to the limited solubility of Nb in the Mo$_3$Si phase [50]. In contrast, the T$_2$ as well as the T$_1$ phase form a continuous solid solution with Nb and W substitution and hence, a three-phase field of (Mo,RM)+T$_2$+T$_1$ can be stabilized in the quaternary system of Mo–Nb–Si–B. Accordingly, the modifications in the phase equilibria with increasing refractory metal substitution are reflected by the formation of a three-phase eutectic of (Mo,RM)+T$_2$+T$_1$ as exemplified in Figure 13.16. With increasing substitution of Mo by Nb, the Mo–Si–B based three-phase eutectic of (Mo,RM)$_3$Si+T$_2$+(Mo,RM) terminates and the solidification pathway proceeds with the formation of a five-phase reaction of L+ (Mo,RM)$_3$Si$\Rightarrow$(Mo,RM)+T$_2$+T$_1$. A significant consequence of this reaction and the new phase equilibrium is the synthesis of a microstructure with a uniform dispersion of (Mo,RM) ductile phase together with compatible high melting T$_2$ and T$_1$ phases [29]. The new multiphase configuration stabilized by these refractory metals may be optimized to take advantage of the benefits of high oxidation and creep resistance of T$_1$ and T$_2$ phases and the toughness and ductility of the bcc phase.

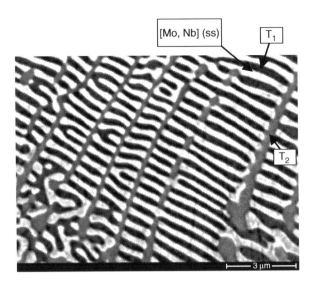

FIGURE 13.16 BSE SEM image of the three-phase eutectic composed of BCC $+ T_2 + T_1$ in the Mo–Nb–Si–B alloys.

13.5.2 Powder Processing

There have been similarly extensive efforts in producing the alloys through solid state processing (i.e., powder metallurgy and subsequent thermomechanical treatments) that also show promising processing routes [51,52,53]. The challenge has been to produce a uniform distribution of inter-metallic phases within the α-Mo matrix. Annealing procedures at temperatures in excess of 1600°C are typically needed to obtain a continuous Mo phase in the Mo–Si–B microstructures so that adequate room-temperature fracture toughness can be achieved [54]. However, there will be a trade-off between the room-temperature fracture toughness and the high-temperature oxidation resistance. Further work is needed to optimize powder processing for both nominal composition homogeneity and microstructural uniformity.

13.6 THERMODYNAMIC EVALUATION OF THE PHASE EQUILIBRIA

In order to provide a basis for the reliable analysis of the measured phase stability and to provide guidance in the further analysis and interpretation of the character of the defect structure in the T_2 phase, an evaluation of the thermodynamic properties is necessary. Currently, reliable thermodynamic data for the T_2 phase are not available. However, the limiting tangent method [27] that uses the data on the phases coexisting in equilibrium with the T_2 phase can be applied to establish reasonable bounds on the free energy of formation, ΔG_f of the T_2 phase. As a first level of analysis, all phases are treated as line phases without any mutual solubility. Next, the available thermodynamic information on the binary phases such as the free energy of formation, ΔG_f, is considered together with the established ternary phase equilibrium relationships. In this regard, the free energy of formation as a function of temperature is evaluated as $\Delta G_f \cong -100.34 - 0.505T$ kJ/mole. At the present stage, there are clear limitations on the calculation method since it treats phases as line compounds and employs an approximate value for ΔG_f for the T_2 phase. However, the approach illustrates the potential for a full analysis once measurements of the solubilities and analysis of the defect structures of the intermediate phases are used to develop solution models.

It is necessary to examine the T_2 structure carefully in order to develop a reliable thermodynamic model for the T_2 phase that can serve not only to provide a consistent analysis of the phase stability in the Mo–Si–B system, but also to serve as a guide for the influence of alloying on the equilibria. Figure 13.3 shows the T_2 crystal structure, which consists of four types of layers: layer A with only Mo atoms, layer $A_{1/21/2}$ with only Mo atoms, layer B with only Si atoms, and layer C with a mixture of Mo and B atoms. The $A_{1/21/2}$ layer refers to the A layer that has been translated by half the base diagonal relative to neighboring layers. Each element has different neighboring atoms and coordination numbers. According to Aronsson [58ARO], a Mo_I atom (Mo atom in layer A or $A_{1/21/2}$) is surrounded by two B atoms (interatomic distance $= 0.234$ nm), two Si atoms (0.256 nm), two Mo_{II} atoms (Mo atom in layer C, 0.272 nm) and nine Mo_I atoms (0.282–0.318 nm). A Mo_{II} atom is surrounded by four B atoms (0.238 nm), two Si atoms (0.277 nm), and eight Mo_I atoms (0.272 nm). A Si atom is surrounded by two Mo_{II} atoms (0.277 nm) and eight Mo_I atoms (0.256 nm). A boron atom is surrounded by two Mo_{II} atoms (0.238 nm), 6 Mo_I atoms (0.234 nm), and one B atom (0.213 nm). This indicates that the structure of Mo_5SiB_2 in Figure 13.3 consists of actually four different sublattices of Mo_I, Mo_{II}, Si, and B. The defect structure of T_2 indicates that the major defects are the constitutional vacancy in the Mo-rich side and the anti-site Si and B atoms in the Mo-lean side [78]. Based upon the detailed consideration of the T_2 structure and the established defect structure, it is evident that a simplified two sublattice thermodynamic model is inadequate to account for the known behavior of the T_2 phase over the entire homogeneity range. While there has been recent work on the Mo–Si–B thermodynamic model based on two sublattices [55], only the four-sublattice model provides the full range of analysis that is essential for a reliable thermodynamic model.

13.7 DIFFUSION BEHAVIOR

In light of the central role of the T_2 phase as a key constituent in high-temperature applications, the formation of the T_2 phase was studied through diffusion couples experiments between Mo_2B and Mo_5Si_3 at temperature ranges of 1600–1800°C. The back-scattered electron (BSE) image of the cross section of the heat-treated Mo_2B/Mo_5Si_3 diffusion couple taken by the secondary electron microscopy (SEM) is shown in Figure 13.17. The diffusion reaction produces two product phases— T_2 and Mo_3Si—and the phase sequence is $Mo_2B/T_2/Mo_3Si/Mo_5Si_3$. Composition measurements with an electron probe microanalysis (EPMA) system elucidates that silicon atoms move from Mo_5Si_3 to Mo_2B and boron atoms from Mo_2B to Mo_3Si. Due to the negligible boron solubilities of Mo_3Si and Mo_5Si_3, most boron atoms react with Mo_3Si at the T_2/Mo_3Si interface to form T_2. The T_2 phase initiates and grows in a columnar structure parallel to diffusion direction toward the Mo_2B phase side, with a tendency for growth to be approximately normal to [001].

Due to limited homogeneity ranges of composition, the interdiffusion coefficients for Si and B have constant values. It should be noted that the interdiffusion coefficients with approximately constant values over homogeneity ranges of composition can be expressed with an effective

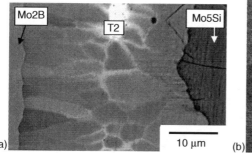
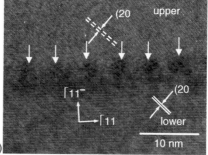

FIGURE 13.17 (a) BSE image of cross section of the Mo2B/Mo5Si3 diffusion couple annealed at 1600°C for 400 h, (b) HREM image of the columns of low-angle growth boundary in T_2 phase from the diffusion couple in (a) with a periodic of misfit dislocations.

integrated interdiffusion coefficient, which is defined as the total interdiffusion flux over the diffusion zone to the corresponding concentration difference [59,60]. The integrated interdiffusion coefficients for Si and B, $\tilde{D}_{Si}^{int}$ and $\tilde{D}_{H_B}^{int}$, obtained from concentration profiles of the Mo_2B/ Mo_5Si_3 diffusion couples annealed between 1600 and 1800°C are in the order of $10^{-15} - 10^{-16}$ m^2/s. An Arrhenius plot of the integrated interdiffusion coefficients indicates that the activation energies for Si and B diffusion are 355 and 338 kJ/mol, respectively. Therefore, the fact that the self-diffusion of Mo in pure Mo (bcc structure) is 1.5×10^{-17} m^2/s at 1600°C [61] demonstrates the sluggish diffusional behavior of the T_2 phase. A more detailed analysis of the diffusion behavior with respect to the defect structure in the T_2 phase is necessary.

13.8 OXIDATION BEHAVIOR OF REFRACTORY METALS AND REFRACTORY METAL SILICIDES

A key performance capability for elevated temperature materials is environmental resistance. A prime requirement for structural materials is the achievement of some degree of non-catastrophic environmental resistance so that the integrity of a component is not entirely dependent on an oxidation resistant coating. Under oxidizing conditions, the development of an adherent oxide scale is the foundation of the environmental resistance. Typically, for high-temperature oxidation resistance this may be expressed in terms of a metal recession rate where an acceptable level is of the order of 2.5 μm/h or less including all forms of loss due to evaporation, internal oxidation, and the potential depth of interstitial enbrittlement [62].

For Ni-base superalloys, the protective oxide scale is based upon an Al_2O_3 surface layer, which also exhibits a small parabolic rate constant, K_s. However, K_s is temperature dependent and at temperatures above about 1300°C the K_s for an SiO_2 layer is smaller than that for Al_2O_3. Indeed, this transition in surface oxide growth behavior is part of the design advantage provided by the use of SiO_2 layers for oxidation protection. Moreover, the relatively poor environmental resistance of the refractory metals is one of the strongest underlying factors for pursuing intermetallic and in particular silicides/borides as coexisting phases.

Thermodynamic stabilities and vapor pressures of the oxides of the refractory metals vary considerably. As a result, the ability of the various silicides to form silica films by selective oxidation differs greatly. Oxides fall into two classes: those that are much less stable than SiO_2, e.g., MoO_2 and WO_2 and those which are only slightly less stable than SiO_2, e.g., Ta_2O_5, NbO, and Cr_2O_3. Silica should be stable on silicides of Mo and W, even at quite low silicon activities, while this is not the case for silicides of Ta, Nb, and Cr [63].

There are also substantial differences between the volatilities of the oxides of various metals. There are high partial pressures of $(MoO_3)_3$ and MoO_3 in equilibrium with $MoO_{2(g)}$ and $MoO_{3(l)}$ (the W-O system is similar), while the partial pressures in equilibrium with Ta_2O_5 are negligible (the Nb-O system is similar) [63]. It can therefore be concluded that transient oxides likely to form on Mo and W silicides are relatively unstable and are likely to evaporate at high temperatures. On the other hand, the transient oxides on the silicides of Ta and Nb are very stable, and once formed, will remain on the surface. Also for the Si–O system, the oxide vapor pressures are low at high oxygen pressures [63].

Various oxidation modes are possible for silicides depending upon the particular silicide as well as the environment to which it is exposed [64]. Total oxidation occurs producing a mixture of metal oxides and SiO_2 when the inward diffusion rate of oxygen through the passive oxide layer exceeds the outward silicon diffusion rate through the silicide sublayer containing less silicon. Conversely, selective oxidation of silicon will occur if the transport of silicon through the silicide is higher than the transport of oxygen through SiO_2. At low oxygen partial pressure, active oxidation due to the formation of gaseous SiO will occur because the oxygen transport to the substrate surface is too low. In some cases, passive oxidation with the formation of SiO_2 is expected based upon thermodynamic and kinetic data.

In terms of environmental resistance, a number of refractory metal silicides and especially disilicides offer a good level of oxidation resistance. However, only $MoSi_2$ is known to form pure SiO_2 upon oxidation while most other refractory disilicides form mixed oxides, which can lead to excessive recession rates. At elevated temperature from 1000 to 1700°C $MoSi_2$ exhibits excellent oxidation resistance due to the formation of a thin continuous protective SiO_2 layer. This layer exhibits no spalling or cracking over extended periods. Numerous studies have shown that the rate of oxidation of $MoSi_2$ at high temperature follows parabolic kinetics with the rate-limiting step in high temperature oxidation being represented by the diffusion of oxygen through the SiO_2 coating, which is sluggish.

To evaluate the suitability of alloys for high-temperature exposure in oxidizing environments, two quantitative metrics of oxidation damage are used: weight change per unit area and time; and recession rate, or rate of thickness loss from a flat specimen per unit time. The $(Mo(Si,B) + Mo_3Si + (T_2)Mo_5SiB_2$ alloys exhibit an interesting oxidation response as determined by these two measures. Beyond an initial transient stage during oxidation in air, which is attributed mostly to the initial mass loss due to MoO_3 and B_2O_3 evaporation, they exhibit parabolic rate kinetics for selected temperatures (in the weight-change experiment). The kinetic rate attained at temperatures above $\sim 1000°C$ is sufficiently slow that the materials are attractive for engineering design [56,57]. However, unlike most high-temperature materials, this alloy class exhibits challenges with oxidation control at lower temperatures where the oxidation rates can be linear. Below temperatures $\sim 500°C$ the oxidation rates for these alloys are so slow that they are of little consequence to engineering structures. However, between ~ 500 and 650°C, a $MoO_3(Si,B)$ scale forms that is non-protective. Between 650 and 750°C there is a transient wherein $MoO_3(Si,B)$ begins to evaporate, and $SiO_2(B)$ does not yet form (Figure 13.18). This is the regime of rapid oxidation that is most critical for engineering use of these alloys.

The steps in the oxidation of Mo–Si–B alloys lead initially to MoO_3 formation, but this layer offers no protection against continued oxidation. The high pressure of MoO_3 (g) indicates that MoO_3 is the most volatile species in the range of between 10^{-7} and 1.01×10^5 (Pa) oxygen pressure. The initially formed MoO_3 layer can be transient because it sublimates readily at

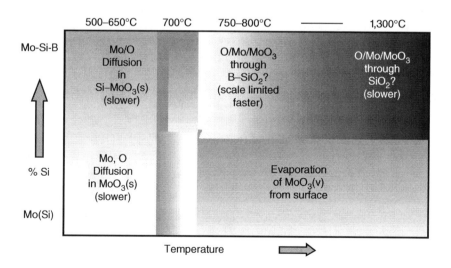

FIGURE 13.18 Oxidation mechanism map in temperature-composition space for selected Mo and Mo–Si–B alloys. For Mo–Si–B alloys, at low temperatures solid MoO_3 scales will form. At slightly higher temperatures, MoO_3 is a gaseous phase and evaporates, but temperatures are too low for $SiO_2(B)$ scale formation, thus oxidation rates are high. Above $\sim 750°C$ a borosilicate glass scale will form that offers varying levels of oxidation protection as boron increasingly evaporates from the scale at progressively higher temperatures [58].

temperatures above 900°C to leave a surface that is enriched in Si and B. The enriched surface then develops an protective SiO_2 layer that also contains B_2O_3. However, the $SiO_2 + B_2O_3$ surface does restrict oxygen transport and provides a reduced oxygen activity so that MoO_2 forms at the base alloy surface [65]. Further, the high level of SiO_2 in the amorphous layer (i.e., about 80 mole%) indicates that B_2O_3 is more volatile at this temperature. In fact, it has been documented that at operating temperature between 1100 and 2300 K boron loss can occur and due to evaporation [66].

Following oxidation in air, the SiO_2 (+B_2O_3) layer produced is naturally protective, but it does not completely block O_2 transport so that with continued oxidation exposure, the thickness of the oxide increases along with a recession of the alloy base. In order to minimize the alloy recession several strategies that can modify the $SiO_2 + B_2O_3$ scale are being examined to reduce O_2 mobility. One simple but effective way to do this is shown in Figure 13.19, where an additional thickness of SiO_2 is spray deposited on the alloy surface followed by oxidation exposure at 1000°C. Following oxidation exposure, this treatment considerably reduced the underlying MoO_2 layer thickness.

Several oxidation studies have been done using thermogravimetric analysis to examine the behavior of Mo–Si–B alloys over longer periods of time. One study focused on the T_2 composition

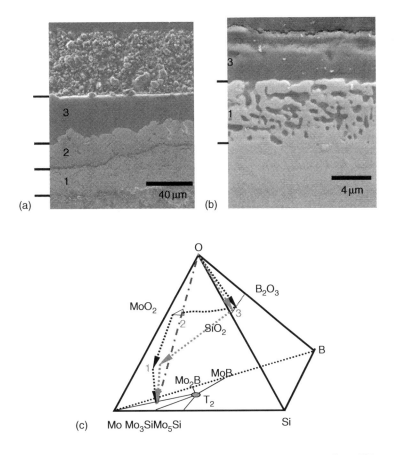

FIGURE 13.19 Cross section BSE image of (a) crystal SiO_2 powder and (b) amorphous SiO_2 powder sprayed Mo–14.2Si–9.6B (at.%) alloy following oxidation at 1200°C for 100 h. (The numbers in each figure indicate (1) Mo(ss) phase with internal oxide precipitates, (2) MoO_2, and (3) borosilicate layer. The region below (1) is the alloy substrate.) The diffusion pathways of oxygen for the two cases are shown in (c).

(Mo–12.5Si–25B at.%, B/Si = 2) and involved testing the alloy at several different temperatures [67]. The testing range was between 700 and 1400°C for up to 24 h. At low temperatures, the alloy initially gained mass, but then started to lose mass as MoO_3 began to evaporate at a significant rate. At 1000°C, the alloy lost a slight amount of mass initially, but then did not change mass for about 15 h, at which time it again lost a slight amount of mass, and appeared not to change mass for the remainder of the test. At all higher temperatures, the alloy lost some mass continuously during the test, indicating that molybdenum is being oxidized and is evaporating away concurrently with the oxidation of silicon and boron. Mass loss was quite pronounced at 1400°C and was about three times faster than the rate of mass loss at 1300°C. This indicates that while the scale that formed on the sample was somewhat protective, it did not reduce the partial pressure of oxygen to the point where molybdenum would not oxidize. This means that the oxygen transport through the scale was appreciable and that it is too rich in boron oxide.

Mendiratta et al. [57] tested six different Mo–Si–B alloy compositions in a cyclic oxidation environment that cycled from 1200°C to room temperature to determine behavior as a function of B/Si. The Mo–11Si–11B (at.%) alloy performed best in this type of test, showing a very small mass change over 100 h of testing. The Mo–11Si–9B alloy performance was superior and showed very little mass loss. Alloys that did not perform well tended to have high levels of boron and lower levels of silicon. The worst performing alloy was Mo–4Si–10B, with Mo–7Si–11B improving slightly, then Mo–7Si–13B further improving, and Mo–7Si–14B showing the best performance of the silicon deficient alloys. This indicates that in the absence of silicon, the addition of boron may improve oxidation resistance slightly, but not significantly enough to be considered protective. It was also found that none of the alloys were passivated at lower temperatures, and all suffered from catastrophic oxidation between 700 and 800°C. Bubbles were observed in the Mo–11Si–11B alloy scale at 800°C, indicating that oxidation of molybdenum was occurring and that the MoO_3 bubbles were moving through the scale. This behavior illustrates the ease with which material may be transported across the protective scale if the alloy chemistry is not optimized.

In addition, work performed by Yoshimi [67] and Mendiratta [57] examined pure Mo_5Si_3 in comparison with previous works and other alloys. As expected, it was found that without boron additions, Mo_5Si_3 will pest and oxidize catastrophically above 700°C. There was some protective behavior observed at time less than 50 h at 1000°C, but beyond this time, mass loss was observed in all tests. There have been a few reports on oxidation of B doped Mo_5Si_3 [22]. During high temperature exposure (above 800°C) a protective SiO_2 layer develops, but at low temperature, a non-protective scale forms consisting of MoO_3 together with SiO_2. Recently, it has been documented that a three-phase alloy composed of the Mo_5Si_3 phase as the major constituent together with $MoSi_2$ and MoB yielded oxidation products of MoO_2 and SiO_2 at 800°C, and Mo and SiO_2 at higher temperatures.

From the current observations, the common feature in the oxidation behavior for Mo-rich compositions in the Mo–Si–B system that include alloys with $(Mo + T_2)$, $(Mo + Mo_3Si + T_2)$ and $(Mo_3Si + T_1 + T_2)$ microstructures is the initial transient stage which is attributed to the mass loss from the evaporation of MoO_3 as well as the boron oxide. The oxidation process then proceeds with the eventual formation of a steady state, which is signified by the development of a borosilicate layer at the outermost portion of the oxide structure. The initial onset temperature for the steady state oxidation is highly dependent on the composition of the Mo–Si–B substrate alloys. Furthermore, the B/Si ratio in the substrate that develops into the oxide scale controls the nature of the transient stage and the overall behavior. Clearly, this is an area where some progress has been made, but more work is necessary. Additionally, the role of minor elements on the viscosity of the borosilicate glass has not been explored in detail. While there have been a few experimental results indicating the benefit of minor additions of Fe, Hf, and/or Ti on the high-temperature oxidation resistance of Mo–Si–B alloys [56,68], further work is needed to elucidate the mechanism. In this respect, it is worth noting the recent progress that has been made on coatings strategies for Mo–Si–B alloys for the oxidation protection not only under isothermal condition but also under thermal cycles [69–73]. The key components of the coating strategy are the identification of oxygen diffusion pathways into the substrate and the application of high-temperature coatings made of silica, silicides, borosilicides, aluminides, or combinations thereof to alter the path to extend the lifetime of the coatings [71–74].

13.9 MECHANICAL PROPERTIES OF MO–Si–B ALLOYS

Molybdenum and its alloys (for example, TZM, MHC, and Mo–Re) are routinely used for a variety of applications including die and punch components, furnace elements, and furnace components, and occasionally in the aerospace industry for rocket propulsion applications (see for example, [53]). As a natural outcome of these applications, mechanical property data are available at elevated temperatures (1000–1500°C) for these alloys [75–83]. These properties have also been evaluated as a function of microstructure (e.g., recrystallized versus unrecrystasllized micro-structures), and processing routes and alloy compositions have been modified to suppress recrystallization when desired. In the late 1970s and early 1980s, Mo-based alloys were examined in Europe for their potential as turbine blade and hot duct materials for high-capacity helium turbines for electricity generation [84,85]. In this extensive study, forged TZM (nominal composition: Mo–0.5Ti–0.1Zr–0.012C in weight percent) was compared in creep, fatigue (HCF and LCF), corrosion resistance in helium, and in terms of fabricability to several Ni-base superalloys such as Nimonic-80A, 713LC, Inconel-625 and 617 and Hastelloy-S. In the same program, developmental work was reported for precision forging rotor blades of vacuum arc-melted and powder metallurgy processed TZM that weighed ∼6 kg and were ∼20 cm long. Microstructure and mechanical properties including creep and bending fatigue behavior as a function of temperature were evaluated using specimens from such blades. Further, TZM in sheet metal form intended for hot gas ducts was examined for weldability by considering electron beam welding, friction welding, and tungsten inert gas welding.

While technological developments of Mo alloys for a variety of high-temperature applications (dies, punches, furnace components, and heat sinks in electronic devices) have continued, in the early 1990s, a new thrust emerged in evaluating the compound $MoSi_2$ as a potential high-temperature material for aerospace and aircraft engine components (static and rotating), both as a single phase material, and as a second, but majority phase in synthetically composed two-phase alloys (intermetallic–matrix composites). The excellent oxidation resistance of $MoSi_2$ and the high melting temperature of the compound were the driving forces behind the research efforts, and the metallic second phase was perceived to provide the much needed damage tolerance to this otherwise brittle compound. A large number of studies on the mechanical properties (isothermal and cyclic oxidation, ductile–brittle transition, creep, and fatigue response) of these materials were conducted, and sections of several symposia were dedicated to the subject [81,86,87]. Numerous original and review articles have been published on the subject as well, and the interested reader is referred to them [88–90]. The efforts, however, slowly decreased in magnitude as it became increasingly clear that $MoSi_2$ and its composites were not capable of providing the desired level of damage tolerance while retaining high temperature properties and oxidation resistance.

In the mid 1990s, the realization that the phase equilibria in the ternary Mo–Si–B system permits the T_2 phase (Mo_5SiB_2) to be in equilibrium with the Mo solid solution spurred new alloy design concepts and interest in Mo-based alloys grew once again. The first part of this chapter has dealt with the phase equilibria, diffusion paths, alloying effects, and microstructures, and oxidation resistance of these alloys. Here, we primarily review the mechanical properties of alloys in this system. In doing this, we first examine properties of the constituent phases where available (i.e., T_2 phase, Mo_3Si phase and the Mo_5Si_3 phase; the Mo solid solution phase), then consider two- and three-phase alloys.

13.9.1 SINGLE-PHASE ALLOYS

Ito et al. [91,92] grew single crystals of the T_2 phase using the floating zone process and characterized its physical and mechanical properties as well as the operating slip systems that enable high temperature plastic deformation. Deformation of the single crystals was carried out at 1500°C in five different orientations ([001], [010], [110], [443], and [021]). Plastic deformation was observed for [021] and [443]. Dislocation analysis on specimens oriented in the [021] orientation confirmed that slip occurred on the [001]{010} system but not on the <100>(001) system as previously reported [22]. The steady state creep rate was reported as being low and of the order of 3×10^{-8} s^{-1} at 1500°C for a stress level of 432 MPa. The activation energy and stress exponent were reported as ~740 kJ/mol and 6.8, respectively, (Figure 13.20a and b).

The creep behavior of single crystal Mo_5Si_3 has been examined in the [314] orientation, (an activation energy of 510 kJ/mol and a stress exponent of 6 were reported [93]), and this compound has been shown to be less creep resistant than T_2 [92,94]. Yoshimi et al. [94] have examined the yielding and flow characteristics of single crystals of Mo_5Si_3 in compression in the temperature range of 1200–1450°C using a nominal strain rate of 1.7×10^{-4} s^{-1}. Brittle failure was noted for the near-[101] oriented sample at 1200°C whereas at 1250°C and above, a large yield drop was

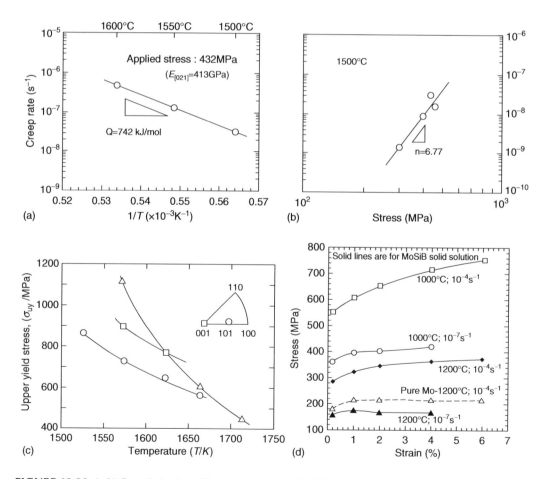

FIGURE 13.20 (a, b) Creep behavior of T_2 single crystals in the [021] orientation [91,92]—(a) minimum creep rate variation with temperature at 432 MPa and (b) stress dependence at 1500°C. In (c), the temperature dependence of the upper yield point for three orientations of single crystal Mo_5Si_3 [94] is shown while in (d), the high temperature compression stress–strain response of Mo solid solution is compared with that of pure Mo [95].

observed, producing an upper and a lower yield point. The lower yield stress and flow stress were reported to be strongly strain rate sensitive. Similar behavior was also noted for specimens with [101] orientation but not for the [001] orientation. The variation of upper yield stress for the three orientations with test temperature was determined. A stress exponent of 6 and an activation enthalpy of 433 kJ/mole were reported for the {101} orientation and these high values were attributed to the high Peierls stress expected of the complex crystal structure (Figure 13.20c).

There are at least two papers that have reported on the physical and mechanical properties of the Mo_3Si phase [25,96]. In the first of these two papers [83], Rosales and Schneibel arc-melted binary Mo–Si alloys with Si levels ranging from 22 to 28 at.% and drop cast the molten metal into water-cooled copper molds. Specimens were annealed at 1600°C for 24 h and slow cooled to room temperature. Whereas the specimen containing 22 at.% Si contained α-Mo precipitates at grain boundaries of the Mo_3Si matrix, the specimen with 25 at.% Si contained small particles of Mo_5Si_3 within the matrix. The alloy with the composition Mo-24 at.% Si was single phase. Notched three-point bend bar specimens (without a fatigue precrack) yielded a toughness value of ∼3 MPa m. Compression tests at 1400°C over the strain rate regime 10^{-3} s^{-1} to 10^{-5} s^{-1} yielded 0.2% offset stress values ranging from 600 to 100 MPa, respectively. Subsequently, single crystal specimens of the Mo_3Si phase were grown using an optical float zone furnace [106] with a growth direction near <102>. Nanoindentation studies yielded elastic constants of $C_{11} = 505 \pm 35$ GPa, $C_{12} = 80 \pm 60$ GPa and $C_{44} = 130 \pm 15$ GPa. Non-contact AFM images obtained from the regions around the indents revealed pileup of material that appeared consistent with {100}(010) slip.

Little is known about the mechanical properties of the α-Mo (containing Si and B) solid solution phase. A survey of the early literature indicates that Si is an extremely potent solid solution strengthener of Mo while simultaneously drastically reducing the ductility [97,98]. Furthermore, it was shown that the solid solution strengthening effect was substantially retained up to about 800°C. More recently, Alur et al. [51] used microhardness measurements to examine the extent of solid solution hardening of the Mo solid solution by Si and B in a ternary Mo–Si–B alloy and compared the response to pure Mo. They noted that the hardness at room temperature for pure Mo was 212 VHN, which is in good agreement with that previously reported [97] whereas in comparison, the matrix hardness in the two-phase Mo–Si–B alloy in a well-annealed condition was 375 VHN. This number is somewhat higher than that reported for the binary Mo–Si solid solution in the literature (325 VHN) [97] and may be due to B in solid solution, residual dislocations arising from thermal mismatch between the matrix and the T_2 phase that then act as obstacles to gliding dislocations and/or due to constraint effects imposed by the second phase T_2 particles.

Jain et al. [95] have reported on the high-temperature compression behavior of a ternary Mo–Si–B alloy whose composition was chosen so that it contained a low volume fraction (∼3%) of the T_2 phase. They argued that this would enable the Mo to be saturated in Si and B and therefore be representative of the matrix solid solution in the two-phase and three-phase alloys while the low volume fraction and fairly coarse particle size of the T_2 phase would ensure that the contribution to yield strength and early-stage work hardening from the presence of the T_2 particles was negligible. Their study demonstrated that small levels of Si and B in solid solution in Mo provide substantial strengthening at temperatures as high as 1000°C and 1200°C and over a range of strain rates (Figure 13.20d) and can account for a major part of the high strength and work hardening response of the Mo solid solution $+ T_2$ two-phase alloys examined by Alur and co-workers [51].

13.9.2 Two-Phase Alloys

Two-phase alloys composed of the Mo solid solution and the T_2 phase have been examined for their mechanical properties [51,99,100,101,102]. The specific alloy composition examined in [69,101] by Ito et al. was Mo–9Si–18B (at.%) and contained approximately 28 volume percent of the Mo solid solution phase, the rest being the T_2 phase. The alloy is part of a eutectic system and was obtained by directional solidification at a growth rate of less than 5 mm/h. The solid solution phase exhibits two types of microstructures: coarse particles and short rod-type morphology. T_2 is the continuous phase. Compression tests conducted on the Mo–9Si–18B (at.%) alloy at a nominal strain rate of 1×10^{-4} s^{-1} confirmed brittle failure at stress levels of 1.5–2.0 GPa between RT and 1000°C. Above 1000°C, the material displayed compressive plasticity with a peak flow stress of ~ 1 GPa in the 1000–1200°C regime followed by flow softening. At 1500°C, the peak stress dropped to about 700 MPa. Even after the 1500°C deformation, the matrix T_2 phase was reported to show significant microcracking. Three-point bend tests at room temperature using notched specimens provided K_Q values of ~ 11 MPa m, which is significantly higher than the value of 1–2 MPa m reported for single phase T_2 by Ihara et al. [103].

Yoshimi et al. [126] have published the results of a systematic investigation of the effect of the volume fraction of the Mo solid solution phase in Mo $+ T_2$ two-phase alloys on thermal expansion, compressive yield strength and oxidation resistance. The alloys were produced by arc melting and the resulting buttons were not further hot worked. The volume fractions of Mo solid solution phase in the four alloy compositions they examined were 0.72, 0.52, 0.34, and 0.2. They found that at a given test temperature, compressive yield strength progressively increased with decreasing volume fraction of the Mo solid solution phase; however, the alloy containing 0.34 (and 0.2) volume fraction of the solid solution phase did not show compressive ductility up to temperatures as high as 1500°C.

Recently, a comprehensive study by Alur et al. [51] of the compression response of a two-phase (Mo solid solution $+ T_2$) Mo-6.1Si-7.9B (at.%) alloy as a function of strain rate in the 1000–1400°C range confirmed that deformation in the temperature-strain rate space evaluated was matrix-controlled, yielding an activation energy of ~ 415–445 kJ/mol. Furthermore, a limited number of tests conducted on a three-phase alloy with a composition of Mo-8.6Si-8.7B (at.%) that lies in the three-phase field, Mo solid solution $+ T_2 + Mo_3Si$, illustrated that the response of the three-phase material overlaps that of the two-phase material (Figure 13.21). A comparison with an

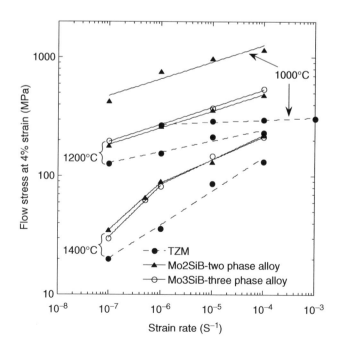

FIGURE 13.21 A comparison of the flow stress (at 4% strain) of TZM (MT104), a two-phase MoSiB alloy (Mo–6Si–8B in at.%), and a three-phase MoSiB alloy (Mo–8.6Si–8.7B in at.%) in the temperature interval 1000–1400°C [51].

off-the-shelf, powder processed TZM alloy illustrated the superior strength of these Mo–Si–B alloys over the temperatures and strain rate regimes examined. Examination of the deformed microstructure illustrates that recovery, and in a few instances, partial recrystallization occurred in the Mo solid solution matrix whereas the T_2 phase either cracked or deformed plastically depending on the temperature and strain rate. Finite element analysis assuming an elastic–plastic matrix and an elastic second-phase illustrated that strain localized in the matrix, the extent being more severe when the work-hardening rate in the matrix was low (as would be at high temperatures and slow strain rates), while the T_2 particles were highly stressed. However, if plastic deformation was permitted in the T_2 particles, the strain distribution was homogenized substantially, and the level of stress build up in the T_2 particles diminished by an order of magnitude.

Alur and co-workers [100] have also examined (a) the variation of fracture toughness with temperature in the 20–1400°C regime, (b) the S–N behavior at room temperature and at 1200°C of the and (c) the cyclic crack growth response ($R = 0.1$) over a range of temperatures in vacuum (573–1673 K) and in air (300–873 K) for a two-phase Mo-6.1Si-7.9B (at.%) alloy. Some of these results are presented in Figure 13.22 through Figure 13.25. These data confirm that the two-phase alloy has a fracture toughness of 8–11 MPa m in the 300–873 K regime which is a little lower than that reported by Ito et al. [101] for the directionally solidified two-phase material containing only 28% of the Mo solid solution phase. This may be due to the fact the specimens did not include a sharp pre-crack in [101]. At higher temperatures (873–1673 K), the toughness increases from 11 to 25 MPa $\sqrt{m}$ with increasing temperature.

Detailed examination of the crack interaction with surface microstructure as well as fracture surfaces showed that the matrix solid solution failed by brittle cleavage at room temperature, and the cracks predominantly ran along the matrix/T_2 interface, although on several occasions, secondary cracking of the T_2 particles was readily recognized on the fracture surface. In spite of this fairly brittle behavior, a fatigue response was observable at room temperature (Figure 13.23 and Figure 13.24a). Crack growth experiments confirmed a threshold ΔK at room temperature that was of the order of 5 MPa $\sqrt{m}$ and a Paris slope that was very high (~ 20), features reminiscent of the

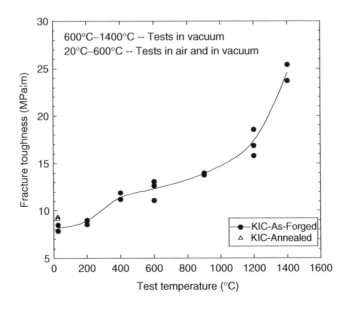

FIGURE 13.22 Fracture toughness variation with test temperature from 20–1400°C for a two-phase Mo–6Si–8B (in at.%) alloy, obtained using notched and fatigue pre-cracked three-point bend specimens [100].

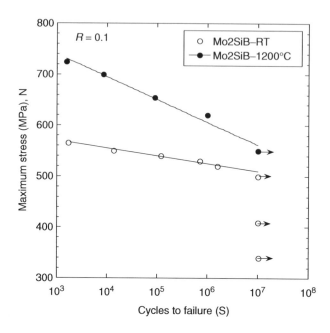

FIGURE 13.23 S–N (Fatigue life) response of a two-phase Mo–Si–B alloy (Mo–6Si–8B in at.%) obtained using an R ratio of 0.1 in tension–tension mode at 20 and 1200°C [100].

behavior of ceramics. At 1473 K in vacuum, the threshold ΔK did not change substantially, but the Paris slope dropped to ~6; the corresponding values at 1673 K were 4.8 MPa $\sqrt{m}$ and 3 (Figure 13.24b and Figure 13.25).

Fatigue-creep interactions are an important aspect of elevated temperature deformation behavior. Alur et al. [99] performed a series of cyclic crack growth tests in vacuum at $R = 0.1$ using a trapezoidal waveform at 1200 and 1400°C. The trapezoidal waveform enabled different

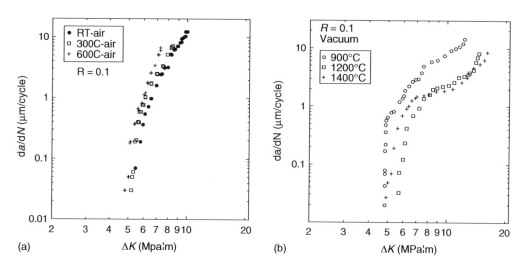

FIGURE 13.24 Cyclic crack growth response of a two-phase Mo–Si–B alloy (Mo–6Si–8B in at.%) using $R = 0.1$. (a) $\mathrm{d}a/\mathrm{d}N$-ΔK in air at 20, 300, and 600°C, and (b) $\mathrm{d}a/\mathrm{d}N$-ΔK in vacuum in the temperature range 900–1400°C [100].

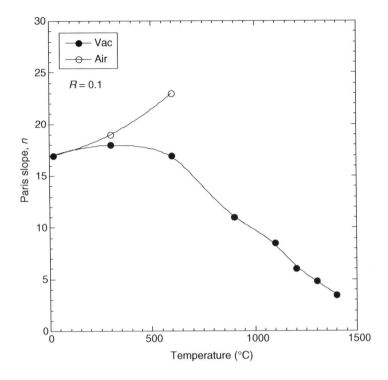

FIGURE 13.25 Variation of Paris slope with test temperature in air (20–600°C) and in vacuum (20–1400°C) for a two-phase Mo–Si–B alloy (Mo–6Si–8B in at.%) [100].

dwell times to be incorporated in the loading cycle, thereby permitting the examination of time-dependent effects such as creep; the trapezoidal waveform tests were, however, restricted to the Paris regime (due to the extremely long duration of these tests) as determined by the sinusoidal waveform tests [100]. The fatigue crack growth data in the form of da/dN versus ΔK curves, for the Mo–Si–B alloy, using a sinusoidal waveform as well as trapezoidal waveforms with various dwell times at the maximum stress, are shown in Figure 13.26a for tests conducted at 1200°C and in Figure 13.26b for tests conducted at 1400°C. Whereas at 1200°C, a dwell time of 5 s using a trapezoidal wave form does not influence the crack growth response (i.e., crack growth rate and Paris slope) relative to the sinusoidal waveform, an increase in the hold time to 10s causes the crack growth rate for a given value of ΔK in the Paris regime to increase although the Paris slope decreases. When the dwell time in each cycle at the maximum stress is further increased to 15 s, there is an additional enhancement in the crack growth rate and an increase in the Paris slope (Figure 13.26a). At 1400°C (Figure 13.26b), even a dwell time of 5 s per cycle leads to a visible increase in the crack growth rate (at a given DK in the Paris regime) as well as an increase in the Paris slope. When the dwell time is increased the 10 s per cycle, fatigue crack growth behavior deteriorates substantially. Deterioration in crack growth response was attributed to the combined effect of the development of "pockets" of recrystallized zones in the vicinity of the crack tip and ahead of it and formation of creep cavities at recrystallized grain boundary triple junctions as well as at T_2 particles/matrix interfaces.

13.9.3 THREE-PHASE ALLOYS

Investigations have also been conducted on the monotonic and cyclic properties of three-phase alloys in the Mo–Si–B system with and without further alloying. These studies can be subdivided

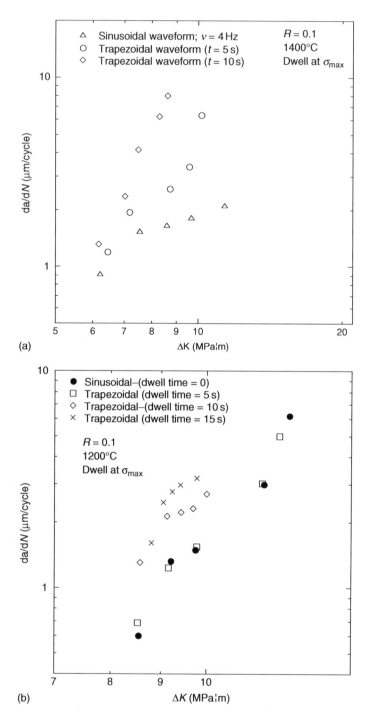

FIGURE 13.26 Effect of dwell time at maximum stress on cyclic response of a two-phase Mo–2Si–1B (Mo–6Si–8B in at.%) alloy in the Paris crack growth regime illustrated using plots of da/dN against ΔK. (a) At 1200°C and (b) At 1400°C [99].

into two groups: those that pertain to the Mo solid solution $+T_2+Mo_3Si$ phase field and others that correspond to the adjacent three-phase field composed of $Mo_3Si+T_2+T_1$ (Mo_5Si_3). In the interest of brevity, we focus our attention only on the first of these two groups—that relating to the alloys lying in the Mo solid solution $+T_2+Mo_3Si$ phase field.

Schneibel and co-workers have published the results of their efforts assessing the oxidation resistance, fracture toughness, and creep resistance of several three-phase alloys with varying levels of the Mo solid solution phase [52,54,98,104,105]. Their alloys were produced using the casting route as well as by powder processing. The results demonstrate that the fracture toughness at room temperature increases from about 3 MPa m for a material containing negligible volume fraction of the Mo solid solution phase to about 15 MPa m for a material containing 38% of the Mo solid solution phase when a powder processing route is adopted. They have also studied the effect of alloying and claim that Nb is a better alloying substitution for Mo than W in terms of improving creep resistance. Ito et al. [46,69,101], however, have demonstrated that directionally solidified three-phase alloys with the eutectic morphology derived from quaternary compositions of the type Mo–30.4Nb–19.5Si–4.5B and Mo–38.8Nb–19.5Si–3B have lower fracture toughness at room temperature (5–6 MPa $\sqrt{m}$) than a ternary two-phase alloy like Mo–9Si–18B (11 MPa $\sqrt{m}$).

The only high temperature evaluation of these alloys using tensile specimens that appears to be reported in the literature is that of Nieh et al. [106]. The composition examined was Mo–9.4Si–13.8B (at.%) and the alloy was produced by powder processing (mixing of elemental powders and hot pressing). Tensile specimens were electrodischarge machined and tested at strain rates between $10^{-4}\,s^{-1}$ and $5\times10^{-3}\,s^{-1}$ in the temperature interval 1350–1550°C. A stress exponent of 2.8 and an activation energy value of 740 kJ/mol were reported. At a strain rate of $10^{-4}\,s^{-1}$ and a temperature of 1400°C, a tensile elongation of 150% was realized. The tensile ductility, however, was reported to decrease rapidly with increasing strain rate and decreasing temperature; thus at 1350°C for the same strain rate, tensile elongation to failure was reported as 25% and it dropped to 20% when the strain rate was increased by an order of magnitude.

To the best of our knowledge, there has been only one group as yet that has reported on the cyclic loading response of three-phase Mo–Si–B alloys. Choe et al. [108] examined the crack growth behavior under cyclic loading of a three-phase Mo–12Si–8.5B alloy (38% of the Mo solid solution phase) using disk-shaped compact tension specimens that contained large cracks (>3 mm); they cycled them at $R=0.1$ using a sinusoidal wave form and conducted such tests at 25, 800, 1200, and 1300°C. At room temperature, the ΔKth was ~5 MPa m, and the Paris slope was characteristic of a brittle material (m ~60). At 1200°C, the threshold increased to ~7 MPa m, and the Paris slope was still reported to be high at ~55. A variety of mechanisms was invoked to explain the fatigue crack growth process. It is worth noting that these values are substantially different from those reported by Alur et al. [100] for the two-phase alloy, particularly at the high-temperature end, although the alloy in that study [100] contained a significantly higher proportion of the matrix solid solution phase. Subsequently, Choe et al. [109] characterized the fracture toughness and fatigue responses of four Mo–Si–B three-phase alloys produced by ingot metallurgy and powder metallurgy routes and demonstrated that both properties were superior for ingot metallurgy materials; they ascribed this difference to the coarser distribution of the Mo solid solution phase in the ingot metallurgy processed material.

As reported in an earlier section of this chapter, Perepezko and co-workers have examined the role of quaternary alloying additions on the microstructure of these Mo–Si–B alloys. Noteworthy are some recent high-temperature compression test data [3] for a Nb-containing quaternary alloy with a three-phase eutectic microstructure (Figure 13.16). The compressive yield strengths of this alloy at two different strain rates at temperatures in the range 1573–1873 K are compared with the ternary two-phase and three-phase alloys Mo–6.1Si–7.9B and Mo–8.6Si–8.7B in Figure 13.27, and the superior high-temperature strength of the quaternary alloy is evident.

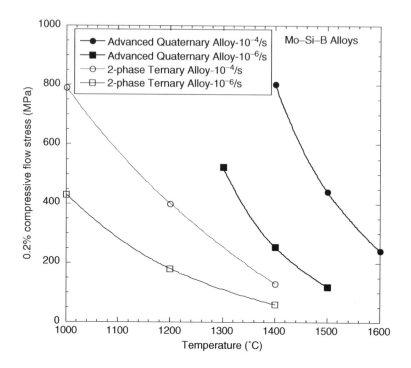

FIGURE 13.27 A three-phase lamellar eutectic microstructure (from Figure 13.16) in a quaternary Mo–Nb–Si–B alloy showing superior high-temperature strength as compared with the ternary two-phase and three-phase MoSiB alloys at strain rates of $10^{-4}\,\mathrm{s}^{-1}$ and $10^{-6}\,\mathrm{s}^{-1}$ [107].

13.10 CONCLUDING REMARKS

There are multiple challenges involved in the development of ultrahigh temperature materials designs for structural applications in a gas turbine engine environment. Besides the obvious requirements for high melting temperature, there is the basic requirement of a satisfactory level of inherent oxidation resistance. Once these essential conditions are addressed satisfactorily, there are the requirements of elevated temperature strength, creep resistance, and fatigue resistance. It is evident that alloy compositions in the Mo–Si–B system have been identified that can be treated to develop microstructures that either satisfy or approach the goals of the multiple requirements. While further enhancement and evaluation of the key performance requirements are necessary, the available performance metrics are attractive. At the same time, there is at present no other materials system that approaches the established performance of Mo–Si–B alloys for high temperature structural applications.

ACKNOWLEDGMENTS

At the University of Wisconsin-Madison, the phase stability studies have been supported by the AFOSR (F49620-03-1-0033) and the examination of oxidation resistant coatings is supported by the ONR (N00014-02-1-0004). At Brown University, the research on the analysis of high temperature mechanical properties is supported by the ONR (N00014-00-1-0373). The authors are most grateful to Dr. S. Fishman and Dr. D. Shifler for their continuous support and encouragement. The discussion in this chapter covers developments up to the time of submission in fall 2005.

REFERENCES

1. Heppenheimer, T. A., Launching the aerospace plane, *High Technology*, 6(7), 46–51, 1986.
2. Viars, P. R., Report #AIAA-892137, American Institute of Aeronautics and Astronautics, Systems and Operation Conf. Seattle, WA. 31 July – 2 August, 1988.
3. Dimiduk, D. M. and Perepezko, J. H., Mo-Si-B alloys: Developing a revolutionary turbine-engine material, *MRS Bulletin*, 28(9), 639–645, 2003.
4. Dimiduk, D. M., Miracle, D. B., and Ward, C. H., Development of intermetallic materials for aerospace systems, *Materials Science and Technology*, 8(4) , 367–375, 1992.
5. Banerjee, S., Physical metallurgy of metals belonging to groups VA and VIA of the periodic table, *High Temperature Materials and Processes*, 11(1–4), 1–33, 1993.
6. Fleischer, R. L., High-temperature, high-strength materials—an overview, *Journal of Metals*, 37(12), 16–20, 1985.
7. Kim, Y.-W. and Dimiduk, D. M., Progress in the understanding of gamma titanium aluminides, *Journal of Metals*, 43(8), 40–47, 1991.
8. Ward, C. H., Microstructure evolution and its effect on tensile and fracture behaviour of Ti-Al-Nb a2 intermetallics, *International Materials Reviews*, 38(2), 79–101, 1993.
9. Destefani, J. D., Advances in inter-metallics, *Advanced Materials and Processes*, 135(2), 37–41, 1989.
10. Buckman, R. W., Alloying of refractory metals, In *Alloying*, Chapter 12, Walter, J. L., Jackson, M. R., Sims, C. T., Eds., ASM. Metals Park, Ohio, pp. 419–445, 1988.
11. Calderon, H. A., Kostorz, G. and Ullrich, G., Microstructure and plasticity of two molybdenum-base alloys (TZM), *Materials Science & Engineering A*, A160(2), 189–199,1993.
12. Meschter, P. J., Low-temperature oxidation of molybdenum disilicide, *Metallurgical Transactions A*. 23A(6), 1763–1772, 1992.
13. Pettifor, D. G., Theory of energy bands and related properties of 4d transition metals. I. Band parameters and their volume dependence, *Journal of Physics F*, 7(4), 613–633, 1977.
14. Wittenauer, J., Refractory Metals 1990: Old Challenges, New Opportunities, *Journal of Metals*, 42, 7, 1990.
15. Kumar, K. S., Silicides: Science, Technology and Applications, In *Intermetallic Compounds: v.2, Practice*, Chapter 10, Westbrook, J. H. and Fleischer, R. L., Eds., John Wiley & Sons Ltd., New York, pp. 211–235, 1994.
16. Boettinger, W. J., Perepezko, J. H. and Frankwicz, P. S., Application of ternary phase diagrams to the development of $MoSi_2$-based materials, *Materials Science & Engineering A*, A155(1-2), 33–44, 1992.
17. Schlichting, J., Molybdenum disilicide as a modern component high temperature bonded material, *High Temperatures -High Pressures*, 10(3), 241–269, 1978.
18. Xiao, L., Kim, Y. S., Abbaschian, R. and Hecht, R. J., Processing and mechanical properties of niobium-reinforced $MoSi_2$ composites, *Materials Science & Engineering A*, A144, 277–285, 1991.
19. Goldschmidt, H. J., *Interstitial Alloys*, Plenum Press, New York, 1967.
20. Nowotny, H., Kieffer, R. and Benesovsky, F., Combination of silicides and borides (silicoborides) of transition metals vanadium, niobium, tantalum, molybdenum and tungsten, *Planseeberichte Fuer Pulvermetallurgie*, 5, 86–93, 1957.
21. Nowotny, H., Benesovsky, F., Rudy, E. and Wittmann, A., Aufbau und Zunderverhalten von Niob-Bor-Silicium Legierungen, *Monatshefte Fuer Chemie.*, 91, 975–990, 1960.
22. Meyer, M., Kramer, M. J. and Akinc, M., Boron-doped molybdenum silicides, *Advanced Materials*, 8(1), 85–88, 1996.
23. Katrych, S., Grytsiv, A., Bondar, A., Rogl, P., Velikanova, T. and Bohn, M., Structural materials: metal-silicon-boron. On the melting behavior of Mo-Si-B alloys, *Journal of Alloys and Compounds*, 347(1-2), 94–100, 2002.
24. Fournelle, J. H., Donovan, J. J., Kim, S. and Perepezko, J. H., Analysis of Boron by epma: Correction for dual Mo and Si interferences for phases in the Mo-B-Si system, *Int. Phys. Conf. Ser. No. 165: Symp. 14*, In *Proceedings of the 2nd Conference of the International Union of Microbeam Analysis Societies*, IOP Publishing Ltd., Kailua-Kona, Hawaii, Jul 9–13, pp. 425–426, 2000.

25. Rosales, I. and Schneibel, J. H., Stoichiometry and mechanical properties of Mo_3Si, *Intermetallics*, 8(8), 885–889, 2000.

26. Huebsch, J. J., Kramer M. J., Zhao H. L. and Akinc, M., Solubility of boron in $Mo_{5+y}Si_{3-y}$, *Intermetallics*, 8(2),143–50, 2000.

27. Perepezko, J. H., Sakidja, R., Kim, S., Dong, Z. and Park, J. S Multiphase microstructures and stability in high temperature Mo-Si-B alloys, In *Structural Intermetallics 2001, Proceedings of the Third International Symposium on Structural Intermetallics (ISSI-3)*, Kevin J. Hemker and Dennis M. Dimiduk, Eds., TMS, Warrendale, PA, pp. 505–514, 2001.

28. Sakidja, R., Sieber, H. and Perepezko, J. H., Formation of Mo precipitates in a supersaturated Mo_5SiB_2 intermetallic phase, *Philosophical Magazine Letters*, 79(6), 351, 1999.

29. Sakidja, R., Myers, J., Kim, S. and Perepezko, J. H., The effect of refractory metal substitution on the stability of Mo(ss)+T2 two-phase field in the Mo-Si-B system, *International Journal of Refractory Metals & Hard Materials*, 18(4-5), 193–204, 2000.

30. Maex, K., Ghosh, G., Delaey, L., Probst, V., Lippens, P., Van den Hove, L. and De Keersmaecker, R. F., The effect of refractory metal substitution on the stability of Mo(ss)+T_2 two-phase field in the Mo-Si-B system, *Journal of Materials Research*, 4 (5), 1209–1217, 1989.

31. Klopp, W. D., A review of chromium, molybdenum, and tungsten alloys, *J. Less-Common Met.* 42(3), 261–278, 1975.

32. Fu, C. L., Freeman, A. J. and Oguchi, T., Prediction of strongly enhanced two-dimensional ferromagnetic moments of metallic overlayers, interfaces, and superlattices, *Physical Review Letters*, 54(25), 2700–2703, 1985.

33. Ohnishi, S., Freeman, A. J. and Weinert, M., Surface magnetism of Fe (001), *Physical Review B* 28(12), 6741–6748, 1983.

34. Xu, J.-H., Oguchi, T. and Freeman, A. J., Crystal structure, phase stability, and magnetism in Ni_3V, *Physical Review B*, 35(13), 6940–6943, 1987.

35. Xu, J.-H. and Freeman, A. J., Band filling and structural stability of cubic trialuminides: YAl_3, $ZrAl_3$, and $NbAl_3$, *Physical Review B*, 40(11), 11927–11930, 1989.

36. De Boer, F. R., Boom, R., Mattens, W. C. M., Miedema, A. R. and Niessen, A. K., *Cohesion in Metals: Transition Metal Alloys*, Amsterdam, North-Holland, 1989.

37. Harrison, W. A., *Electronic Structure and the Properties of Solids*, New York, Dover, 1989.

38. Williams, A. R., Gelatt, C. D., Connolly, J. W. D. and Moruzzi, V. L., Cohesion, compound formation and phase diagrams from first principles, In *Alloy Phase Diagram*, Bennett, L. H., Massalski, T. B., and Giessen B. C., Eds., North-Holland, New York, pp. 17–28, 1983.

39. Aronsson, B. The crystal structure of Mo_5SiB_2, *Acta Chemica Scandinavica*, 12, 31–37, 1958.

40. Franceschi, E. A. and Ricaldone, F., Intermetallic binary phases of 5:3 composition, *Revue de Chimie Minerale*, 21, 202–220, 1984.

41. Aronsson, B. and Lundgren, G., X-ray investigations on Me-Si-B systems (Me =Mn, Fe, Co). I. Some features of the Co-Si-B system at $100\overline{0}$ C. Intermediate phases in the Co-Si-B and Fe-Si-B systems, *Acta Chemica Scandinavica*, 13(3), 433–441, 1959.

42. Rawn, C. J., Schneibel, J. H., Hoffmann, C. M. and Hubbard, C. R., The crystal structure and thermal expansion of Mo_5SiB_2, *Intermetallics*, 9, 209–216, 2001.

43. Fu, C. L. and Wang, X., Thermal expansion coefficients of Mo-Si compounds by first-principles calculations, *Philosophical Magazine Letters*, 80(10), 683–690, 2000.

44. Fu, C. L., Wang, X., Ye, Y. Y. and Ho, K. M., Phase stability, bonding mechanism, and elastic constants of Mo_5Si_3 by first-principles calculation, *Intermetallics*, 7, 179–184, 1999.

45. Vilars, P., Prince, A. and Okamoto, H., *Handbook of Ternary Alloy Phase Diagrams*, ASM International: Materials Park, Ohio, 1995.

46. Sakidja, R. and Perepezko, J. H., Microstructure development of high-temperature Mo-Si-B Alloys, In *Materials for Space Applications, MRS. Symp. Proc. No. 851*, Chipara, M., Edwards, D. L., Benson, P. and Phillips, S., Eds., MRS, Pittsburgh, PA, p. NN11.11, 2005.

47. Nunes, C. A., Sakidja, R., Dong, Z. and Perepezko, J. H., Liquidus projection for the Mo-rich portion of the Mo-Si-B ternary system, *Intermetallics*, 8, 327–337, 2000.

48. Sakidja, R., Wilde, G., Sieber, H. and Perepezko, J. H., Microstructural development of Mo(ss) + T_2 two-phase Alloys, *Mater. Res. Symp. Proc.*, 545 KK6.3.1-6.3.6, 1999.

49. Kuz'ma, Y. B., X-ray diffraction investigations of the Nb-Ti-B and Nb-Mo-B systems, *Poroshkovaya Metallurgiya*, 4, 54–57, 1971.

50. Savitskii E. M., Baron, V. V., Bychkova, M. I., Bakuta, S. A., and Galdysheskiy, E. I. *Russian Metallurgy*, 2, 91, 1965.

51. Alur, A. P., Chollacoop, N. and Kumar, K. S., High-temperature compression behavior of Mo-Si-B alloys, *Acta Materialia*, 52(19), 5571–5587, 2004.

52. Schneibel, J. H., Kramer, M. J., Unal, O. and Wright, R. N., Processing and mechanical properties of a molybdenum silicide with the composition Mo-12Si-8.5B (at.%), *Intermetallics*, 9(1), 25–31, 2001.

53. www.plansee.com

54. Schneibel, J. H., Easton, D. S., Choe, E. and Ritchie, R. O., Fracture toughness, creep strength and oxidation resistance of Mo-Mo$_3$Si-Mo$_5$SiB$_2$ molybdenum silicides, In *Structural Intermetallics 2001, Proceedings of the Third International Symposium on Structural Intermetallics (ISSI-3)*, Kevin J. Hemker and Dennis M. Dimiduk, Eds., TMS, Warrendale, PA, pp. 801–809, 2001.

55. Yang, Y. and Chang, Y. A. Thermodynamic modeling of the Mo-Si-B system, *Intermetallics*, 13(2), 121–128, 2005.

56. Berczik, D. M., United Technologies Corporation, Method for enhancing the oxidation resistance of a molybdenum alloy, and a method of making a molybdenum alloy, *United States Patent No. 5,595,616*, Jan. 21, 1997; Berczik, D. M., United Technologies Corporation, Oxidation Resistant Molybdenum Alloys, *United States Patent No. 5,693,156*, Dec. 2, 1997.

57. Mendiratta, M. G., Parthasarathy, T. A. and Dimiduk, D. M., Oxidation behavior of aMo-Mo$_3$Si-Mo$_5$SiB$_2$ (T2) three phase system, *Intermetallics*, 10(3), 225–232, 2002.

58. Xu, J.-H. and Freeman, A. J., Phase stability and electronic structure of ScAl$_3$ and ZrAl$_3$ and of Sc-stabilized cubic ZrAl$_3$ precipitates, *Physical Review B*, 41(18), 12553–12561, 1990.

59. Dayananda, M. A., Average effective interdiffusion coefficients and the Matano plane composition, *Metallurgical and Materials Transactions A*, 27A(9), 2504–2509, 1996.

60. Tortorici, P. C. and Dayananda, M. A., Growth of silicides and interdiffusion in the Mo-Si system, *Metallurgical and Materials Transactions A*, 30A(3), 545–550, 1999.

61. Maier, K., Mehrer, H. and Rein, G., Self-diffusion in Molybdenum, *Zeitschrift fuer Metallkunde*, 70(4), 271–276, 1979.

62. Shah, D. M., Berczik, D., Anton, D. L. and Hecht, R., Appraisal of other silicides as structural materials, *Materials Science & Engineering A*, 155(1-2), 45–57, 1992.

63. Berztiss, D. A., Cerchiara, R. R., Gulbransen, E. A., Pettit, F. S. and Meier, G. H., Oxidation of MoSi$_2$ and comparison with other silicide materials, *Materials Science and Engineering A*, 155(1-2), 165–181, 1992.

64. Grunling, H. W. and Bauer, R. The role of silicon in corrosion-resistant high temperature coatings, *Thin Solid Films*, 95(1), 3–20, 1982.

65. *Mechanisms of Oxidation and Hot Corrosion of Metals and Alloys at Temperatures of 1150 and 1450K under Flow*. Materials Research Symposium, National Bureau of Standards.

66. Singh, M. and Wiedemeier, H., Chemical interactions in diboride-reinforced oxide-matrix composites, *Journal of the American Ceramic Society*, 74(4), 724–727, 1991.

67. Yoshimi, K., Nakatani, S., Suda, T., Hanada, S. and Habazaki, H., Oxidation behavior of Mo$_5$SiB$_2$-based alloy at elevated temperatures, *Intermetallics*, 10(5), 407–414, 2002.

68. Woodard, S. R., Raban, R., Myers, J. F. and Berczik, D. M., United Technologies Corporation, Oxidation resistant molybdenum, *United States Patent No. 6,652,674.*, Nov. 25, 2003.

69. Ito, K., Murakami, T., Adachi, K. and Yamaguchi, M. Oxidation behavior of Mo-9Si-18B alloy pack-cemented in a Si-base pack mixture, *Intermetallics*, 11(8), 763–772, 2003.

70. Ito, K., Yokobayashi, M., Murakami, T. and Numakura, H., Oxidation protective silicide coating on Mo-Si-B alloys, *Metallurgical and Materials Transactions A*, 36(3A), 627–636, 2005.

71. Park, J. S., Sakidja, R. and Perepezko, J. H., Coating designs for oxidation control of Mo-Si-B alloys, *Scripta Materialia*, 46(11), 765–770, 2002.

72. Perepezko, J. H., Park, J. S., Sakidja, R. and Kim, S., Elevated Temperature Coating Layer Designs for Mo-Si-B Alloys, In *High Temperature Corrosion and Materials Chemistry IV, Proc. of the 203rd Meeting of the Electrochemical Society*, Paris, 27 April 2-May 2003, PV 2003-16, Opila, E., Hou, P., Maruyama, T., Pieraggi, B., McNallan, M., Shifler, D. and Wuchina, E., Eds., The Electrochemical Society, Pennington, NJ, 310, 2003.

73. Sakidja, R., Park, J. S., Hamann, J. and Perepezko, J. H., Synthesis of oxidation resistant silicide coatings on Mo-Si-B alloys, *Scripta Materialia*, 52(6), 723–728, 2005.

74. Sakidja, R., Rioult, F., Werner, J., and Perepezko, J. H., Aluminum pack cementation of Mo-Si-B alloys, *Scripta Materialia*, 55(10), 903–906, 2006.

75. Boehling, H. and Burman, R., Moly and Moly Alloys — Properties and Applications, In *Progress in Powder Metallurgy*, Nayar, H. S., Kaufman, S. M. and Meiners K. E., Eds., MPIF, Princeton, NJ., Vol. 39, 491–505, 1984.

76. Briggs, J. Z. and Barr, R. Q., Arc-cast molybdenum-base TZM alloy: Properties and applications, *High Temperatures-High Pressures*, 3(4), 363–409, 1971.

77. Ferro, A., Mazzetti, P. and Montalenti, G., On the effect of the crystalline structure on fatigue: comparison between body-centred metals (Ta, Nb, Mo and W) and face-centred and hexagonal metals, *Philosophical Magazine*, 12, 867–875, 1965.

78. Patrician, T. J., Sylvester, V. P. and Daga, R. L., High-temperature PM-molybdenum alloys, *International Journal of Refractory & Hard Metals*, 4(3), 133–137, 1985.

79. Shi, H.-J., Korn, C. and Pluvinage, G., High temperature isothermal and thermomechanical fatigue on a molybdenum-based alloy, *Materials Science and Engineering A*, A247(1-2), 180–186, 1998.

80. Terent'yev, V. F., Kogan, I. S. and Orlov, L. G., Fatigue Failure Mechanism of the Molybdenum Alloy TsM-10, *Physics of Metals and Metallography (English translation of Fizika Metallov i Metallovedenie)*, 42(6), 128–134, 1976.

81. Terent'yev, V. F., Savitskaya, E. E., Kogan, I. S. and Orlov, L. G., Stress cycling of the molybdenum alloys Mo-Ti-Zr-C and Mo-B-Al-C at room temperature, *Izvestiya Akademii Nauk SSSR, Metally*, 6, 129–136, 1976.

82. Tuominen, S. M. and Biss, V., In *Modern Developments in Powder Metallurgy*, Properties of PM Molybdenum-Base Alloys Strengthened by Carbides, *Proceedings of the 1980 International Powder Metallurgy Conference, Vol. 14 - Special Materials*, Hausner, H. H., Antes, H. W. and Smith, G. D., Eds., MPIF, Princeton, NJ, 215–230, 1981.

83. Vitorskiy, Ya. M., Ivashchenko, R. K., Kaverina, S. N., Lotsko, D. V., Mil'man, Yu. V. and Trefilov, V. I., Influence of Deformation Temperature on the Structure and Mechanical Properties of Low-Alloyed Molybdenum, *Physics of Metals and Metallography (English translation of Fizika Metallov i Metallovedenie)*, 31(5), 181–189, 1971.

84. Jakobeit, W., PM Mo-TZM Turbine Blades - Demands on Mechanical Properties, *International Journal of Refractory and Hard Metals*, 2(3), 133–136, 1983.

85. Jakobeit, W., Pfeifer, J.-P and Ullrich, G., Evaluation of high-temperature alloys for helium gas turbines, *Nuclear Technology*, 66(1), 195–206, 1984.

86. Briant, C. L., Petrovic, J. J., Bewlay, B. P., Vasudevan, A. K. and Lipsitt, H. A., Eds., *High Temperature Silicide and Refractory Alloys*, MRS Symp. Proc. Vol. 322, Materials Science Society, Pittsburgh, PA, 1994.

87. Miracle, D. B., Anton, D. L. and Graves, J. A, Eds., *Intermetallics Matrix Composites II*, MRS Symp. Proc. Vol. 273, Materials Science Society, Pittsburgh, PA, 1992.

88. Vasudevan, A. K. and Petrovic, J. J., Comparative overview of molybdenum disilicide composites, Materials Science & Engineering A., 155 (1-2), 1-17. In *Proceedings of the First High Temperature Structural Silicides Workshop, Gaithersburg, MD, USA, 4-6 Nov. 1991*, Vasudevan, A. K. and Petrovic, J. J., Eds., 1992.

89. Sadananda, K., Feng, C. R., Jones, H. and Petrovic, J., Creep of molybdenum disilicide composites, *Materials Science and Engineering A*, 155, 227–239, 1992.

90. Sadananda, K., Feng, C. R., Mitra, R. and Deevi, S. C., Creep and fatigue properties of high temperature silicides and their composites, *Materials Science and Engineering A*, 261(1-2), 223–238, 1999.

91. Ito, K., Ihara, K., Fujikura, M. and Yamaguchi, M. High temperature deformation of T_2 single crystals in Mo-Si-B system, In *Structural Intermetallics 2001, Proceedings of the Third International Symposium on Structural Intermetallics (ISSI-3)*, Kevin J. Hemker and Dennis M. Dimiduk, Eds., TMS, Warrendale, PA, 441–448, 2001.

92. Ito, K., Ihara, K., Tanaka, K., Fujikura, M. and Yamaguchi, M., Physical and mechanical properties of single crystals of the T_2 phase in the Mo-Si-B system, *Intermetallics*, 9(7), 591–602, 2001.

93. Mason, D. P. and Van Aken, D. C., On the creep of directionally solidified $MoSi_2$-Mo_5Si_3 eutectics, *Acta Metallurgica et Materialia*, 43(3), 1201–1210,1995.

94. Yoshimi, K., Yoo, M. H., Wereszczak, A. A., Borowicz, S. M., George, E. P. and Zee, R. H., Yielding and flow behavior of Mo_5Si_3 single crystals, *Scripta Materialia*, 45(11), 1321–1326, 2001.

95. Jain, P., Alur, A. P. and Kumar, K. S., High temperature compressive flow behavior of a Mo-Si-B solid solution alloy, *Scripta Materialia.*, 54, 13, 2006.

96. Swadener, J. G., Rosales, I. and Schneibel, J. H., Elastic and Plastic Mechanical Properties of Mo3Si Determined by Nanoindentation, In *High-Temperature Ordered Intermetallic Alloys IX*, MRS proceedings, Vol. 646, Schneibel, J. H., Hanada, S., Hemker. K. J., Noebe, R. D. and Sauthoff, G. eds., Materials Research Society, Warrendale, PA, pp. N4.2.1–N4.2.6, 2001.

97. Northcott L, *Metallurgy of the Rarer Metals -5*, Butterworth Scientific Publications, London, 1956.

98. Schneibel, J. H., Kramer, M. J. and Easton, D. S., A Mo-Si-B intermetallic alloy with a continuous a-Mo matrix, *Scripta Materialia*, 46(3), 217–221, 2002.

99. Alur, A. P., *PhD dissertation*, Brown University, October 2005.

100. Alur, A. P. and Kumar, K. S., Monotonic and cyclic crack growth response of a Mo-Si-B alloy, *Acta Materialia*, 54(2), 385–400, 2006.

101. Ito, K., Kumagai, M., Hayashi, T. and Yamaguchi, M., Room temperature fracture toughness and high temperature strength of T_2/Mo_{ss} and $(Mo,Nb)_{ss}$/T_1/T_2 eutectic alloys in the Mo-Si-B system, *Scripta Materiala*, 49(4), 285–290, 2003.

102. Yoshimi, K., Nakatani, S., Nomura, N. and Hanada, S., Thermal expansion, strength and oxidation resistance of Mo/Mo_5SiB_2 in-situ composites at elevated temperatures, *Intermetallics*, 11(8), 787–794, 2003.

103. Ihara, K., Ito, K., Tanaka, K. and Yamaguchi, M., Mechanical properties of Mo_5SiB_2 single crystals, *Materials Science and Engineering A*, 329–331, 222–227, 2002.

104. Schneibel, J. H., Liu, C. T., Heatherly, L. and Kramer, M. J., Assessment of processing routes and strength of a 3-phase molybdenum boron silicide (Mo_5Si_3-Mo_5SiB_2-Mo_3Si), *Scripta Materialia*, 38(7), 1169–1176, 1998.

105. Schneibel, J. H., High temperature strength of Mo-Mo_3Si-Mo_5SiB_2 molybdenum silicides, *Intermetallics*, 11(7), 625–632, 2003.

106. Nieh, T. G., Wang, J. G. and Liu, C. T., Deformation of a multiphase Mo-9.4Si-13.8B alloy at elevated temperatures, *Intermetallics*, 9(1), 73–79, 2001.

107. Alur, A. P., Sakidja, R., Kumar, K. S. and Perepezko, J. H., Unpublished results, 2005.

108. Choe, H., Chen, D., Schneibel, J. H., and Ritchie, R. O., Ambient to high temperature fracture toughness and fatigue-crack propagation behavior in a Mo - 12Si - 8.5B (at.%) intermetallic, *Intermetallics*, 9, 319–329, 2001.

109. Choe, H., Schneibel, J. H. and Ritchie, R. O., On the fracture and fatigue properties of Mo-Mo_3Si-Mo_5SiB_2 refractory intermetallic alloys at ambient to elevated temperatures (25°C to 1300°C), *Metallurgical and Materials Transactions*, 34A, 225–239, 2003.

14 Nickel-Base Alloys

W. O. Soboyejo

Department of Mechanical and Aerospace Engineering, Princeton Institute for the Science and Technology of Materials (PRISM), Princeton University

CONTENTS

14.1   Introduction 475
14.2   Introduction to Nickel and Nickel Alloys 476
14.3   Ni–Cu Alloys..477
14.4   Ni–Cr Alloys 479
14.5   Ni–Fe–Cr Alloys 480
14.6   Nickel-Base Superalloys 481
14.7   Fatigue and Fracture of Nickel-Base Superalloys 484
14.8   The Future Beyond Nickel-Base Superalloys..490
14.9   Summary and Concluding Remarks 491
References 491

14.1   INTRODUCTION

Nickel-base alloys are currently the materials of choice for structural applications at intermediate/high temperatures between $\sim 500°C$ and $1100°C$. This is due largely to their attractive combinations of strength, ductility, fracture toughness, hot corrosion resistance, creep resistance, and high temperature oxidation resistance. Due to their excellent combinations of properties, they are used especially in high temperature turbines and combustors for aeroengines and land-based engines, heat exchangers, and a wide range of thermostructural applications.

The historical evolution of the applications of nickel-base superalloys provides one of the best examples of how advances in alloying and microstructure development have influenced the temperature range that can be achieved in thermostructural applications. The earliest applications of nickel-base alloys involved the use of polycrystalline structures that had limited creep resistance. This limited the service temperature ranges of the systems that were designed and built in the 1940s and 1950s. However, with the progress in our understanding of microstructural effects on creep, it became clear that significant improvements in creep resistance could be achieved via increased grain size. This evolved in the 1980s into the development of directionally solidified (DS) nickel-base superalloys, and finally in the 1990s in the introduction of internally cooled nickel alloy single crystals used in the hottest sections of today's aeroengines, where the surface temperatures can be as high as ~ 1100–$1150°C$ after internal cooling.

Since improvements in the maximum turbine temperature translate into improvements in the engine thrust and efficiency, there are still significant efforts to increase the service temperatures of nickel-base superalloys further. However, the intrinsic limits in the high temperature creep and oxidation resistance limit the potential to achieve higher temperature via alloying. Hence, the interest has shifted in recent years to the development of high temperature thermal barrier coatings (TBCs) that increase the overall surface temperatures that can be achieved. Such TBCs are currently made from zirconia coatings. They have the potential to improve the overall high temperature capability of nickel alloys by ~100–150°C. However, they are also prone to cracking. Consequently, engine manufacturers and aircraft operators have been reluctant to use them above ~1150°C.

This chapter presents an introduction to the structure, properties, and applications of nickel and nickel-base alloys and is divided into 9 sections. In Section 14.2 through Section 14.6, the structure and basic properties of nickel and nickel alloys are reviewed, followed by Section 14.7 in which the fatigue and fracture behavior of nickel and its alloys are presented. Salient conclusions arising from this work are summarized in Section 14.9, after presenting possible materials systems beyond Ni in Section 14.8. In each section, the relationships between structure, properties, and applications are highlighted.

14.2 INTRODUCTION TO NICKEL AND NICKEL ALLOYS

Pure nickel has a face centered cubic (f.c.c.) crystal structure with a lattice parameter, $a = 3.5232$ Å. This structure has a melting point of 1455°C, which suggests potential applications at elevated temperatures. However, the elastic modulus is only moderate. The anisotropic elastic moduli are $\langle 100 \rangle$ 138 GPa, $\langle 110 \rangle$ 215 GPa, and $\langle 111 \rangle$ 262 GPa. Typical strengths of pure Ni (99.9% pure nickel) range from ~350 to 665 MPa at room temperature, depending on impurity content and the degree of cold work. Annealed pure nickel can also have a ductility of ~451, at room temperature compared with a typical value of ~4% for pure Ni in the cold worked condition. Ni also has high stacking fault energy (SFE), which is consistent with high levels of cross slip.

The ductility of Ni means that it is formable by hot-working processes. It also has good oxidation and corrosion resistance. However, it is relatively expensive. Also, pure nickel is used only in cases that require corrosion resistance. Furthermore, since Ni in its refined form contains ~0.5% Co, most commercial applications consider Ni+Co as Ni. Two forms of commercially pure Ni exist

$$Ni200: Ni-0.13Cu-0.2Fe-0.08C-0.2Mn-0.05Si$$

and

$$Ni201: Ni-0.13Cu-0.2Fe-0.01C-0.2Mn-0.005Si$$

where the only real difference is the level of carbon. In general, the lower carbon content results in increased ductility and reduced risk of stress corrosion cracking, which is typically associated with the precipitation of graphite at grain boundaries.

Within the electronics industry, there is a strong interest in Ni205 (Ni–0.1Cu–0.1Fe–0.02C–0.2Mn–0.04Mg). This is typically used as internal supports and lead wires in vacuum. Also, Ni207 (Ni–0.01C) is an exceptionally pure form of Ni that does not contain any inclusions. This is generally suitable for heavy cold working operations without intermediate annealing.

In general, refining Ni to reduce the carbon content is relatively expensive. Hence, it is more common to improve the properties of Ni by alloying. Common alloying elements include Cu, Fe, Cr, Mo, Nb, Ta, Ti, and Al. ThO_2 is also added to dispersion strengthened Ni alloys. Typical mechanical properties of Ni and its alloys are compared with the properties of Co superalloys in Table 14.1. The data show clearly that Ni alloys are strengthened significantly by alloying and aging. However, ductility is also degraded by aging. In most cases, carbide precipitation has beneficial effects on microstructure that can translate into improved strength and ductility. However, Co superalloys are somewhat stronger than Ni-based superalloys.

TABLE 14.1
Compositions, Properties, and Applications of Selected Nickel and Cobalt Alloys

Material	Tensile Strength (psi)	Yield Strength (psi)	% Elongation	Strengthening Mechanism	Applications
Pure Ni (99.9% Ni)	50,000	16,000	45	Annealed	Corrosion
	95,000	90,000	4	Cold worked	Resistance
Ni–Cu alloys					
Monel 400 (Ni-31.5%, Cu)	78,000	39,000	37	Annealed	Valves, pumps, heat exchangers
Monel K-500 (Ni-29.5%, Cu-2.7%, Al-0.6%, Ti)	150,000	110,000	30	Aged	Shafts, springs, impellers
Ni superalloys					
Inconel 600 (Ni-15.5%, Cr-8%, Fe)	90,000	29,000	49	Carbides	Heat treatment equipment
Hastelloy B-2 (Ni-28%, Mo)	130,000	60,000	61	Carbides	Corrosion resistance
Hastelloy G (Ni-20%, Cr-20%, Fe-7%, Mo+Nb, Ta)	100,000	47,000	50	Aged	Chemical processing
MAR-M 246 (Ni-10%, Co-9%, Cr-10%, W+Ti, Al, Ta)	140,000	125,000	5	Aged	Jet engines
DS–Ni (Ni-2%, ThO_2)	71,000	48,000	14	Dispersion	Gas turbines
Fe–Ni superalloys					
Incoloy 800 (Ni-46%, Fe-21%, Cr)	89,000	41,000	37	Carbides	Heat exchangers
Co superalloys					
Haynes 25 (50% Co-20%, Cr-15%, W-10%, Ni)	135,000	65,000	60	Carbides	Jet engines
Stellite 6B (60%, Co-30%, Cr-4.5%, W)	177,000	103,000	4	Carbides	Abrasive wear resistance

Source: Adapted from *Metals Handbook*, Vol. 3, 9th ed., ASM International, Materials Park, OH, 1980.

The overall temperature-dependence of the tensile strengths of Ni and selected Ni alloys is presented in Figure 14.1. Note that the tensile strength of pure Ni falls at temperatures above $\sim 375°C$. However, the elevated temperature strengths are retained to higher strength levels, for Monel 400, Inconel 600, and Monel K-500, respectively. Furthermore, the strength/hardness of nickel increases with increasing cold work (Figure 14.2).

14.3 Ni–Cu ALLOYS

The phase diagram for the Ni–Cu system is presented in Figure 14.2. This shows clearly a binary isomorphous system with a magnetic transformation line between points C and D. Ni–Cu alloys have good strength and ductility, and good resistance to corrosion in the atmosphere and in seawater. Originally, the alloy Ni-30 at.%Cu was produced and referred to as "Monel." More recently, it has been designated as Monel 400. In any case, Monel 400 has a composition Ni–30Cu–2.5Fe(max)–1Mn–0.15C. In the wrought condition (hot rolled and annealed), it has a yield strength of 200 MPa, a tensile strength of 550 MPa, and an elongation to failure of 48% at room temperature. In the cost condition, Si is generally added to increase the fluidity during casting.

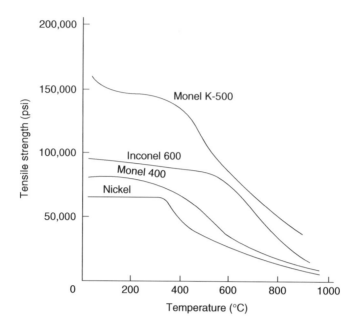

FIGURE 14.1 Effect of temperature on the tensile strength of nickel and selected nickel alloys.

Under such conditions, the yield stress varies between 150 and 240 MPa, the tensile strength ranges from 430 to 620 MPa, and the elongation to failure is between 10% and 16% (Figure 14.3).

Monel 400 cannot be hardened by heat treatment. However, it can be work hardened. In the cold drawn and stress-relieved condition, it has a yield stress of 380–700 MPa, a tensile strength of 580–830 MPa, and an elongation to failure between 22% and 35%. The strengths are retained up to 400°C, but drop significantly above this value (Figure 14.1).

Monel 400 is resistant to seawater, brine, superheated steam, non-oxidizing acids, neutral and alkali salts, alkalis, and ammonium sulfate. However, it is not resistant to oxidizing acids (HNO_3, HNO_2, and H_2SO_4) and hypochlorites.

Furthermore, age hardening can be affected by alloying with small amounts of Al and Ti. For example, the alloy designated as Monel 500 contains about 3 at.%Al and 0.5 at.%Ti. Hardening is achieved in this alloy by aging at 580°C for 16 hours after hot working. This results in a yield stress

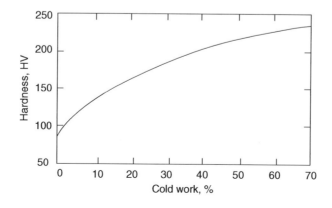

FIGURE 14.2 Effect of cold work on the Vickers hardness of nickel.

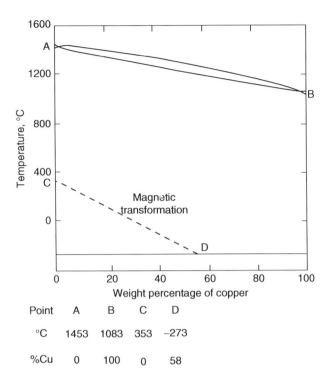

Point	A	B	C	D
°C	1453	1083	353	−273
%Cu	0	100	0	58

FIGURE 14.3 Ni–Cu phase diagram.

of 760 MPa, a tensile strength of 1100 MPa, and an elongation to failure of $\sim 20\%$. Aging of the cold worked alloy results in a yield strength of ~ 1200 MPa.

The combination of high strength, corrosion resistance, and non-magnetic properties enables applications in non-sparking tools and marine propeller shafts. Ni–52Cu (Constantan) is also used in thermocouple wires due to its very low temperature coefficient of $\sim 0.00002°C^{-1}$ and its resistivity of 49×10^{-8} Ωm at 20°C.

14.4 Ni–Cr ALLOYS

The binary Ni–10Cr alloy is often used as a heating element or a thermocouple. One element of the thermocouple is Ni–10Cr (chromel) while the other element is Ni–5Al (alumel). These can be used at temperatures up to 1100°C. However, above this, the oxidation of Cr and Al (during service) causes the calibration to change.

In general, Ni–Cr alloys are prone to sulfur attack at elevated temperature. This results, initially, in a protective (Cr_2S_3 layer) until the Cr is exhausted. Once this occurs, a low melting point Ni–Ni-sulfide system forms. The resulting liquid phase results in rapid attack and characteristic wart-like growth of corrosion products. In general, however, the resistance to this kind of corrosion increases with increasing Cr content.

Table 14.2 summarizes the effects of Cr content on tensile properties. Since these alloys are two-phase, and the Cr phase is brittle, there is a loss of ductility with increasing Cr content. The addition of 1.5 Nb can improve the overall levels of strength and ductility. Also, small additions of Ti e.g., Ni–48Cr–0.35Ti (IN671) can soak up the oxygen to form oxides and promote improved ductility.

TABLE 14.2
Mechanical Properties of Ni–Cr Alloys

Cr (%)	Tensile strength (MPa)	Elongation (%)
35	480	62
50	540–680	7–24
60	800–1000	1–2

14.5 Ni–Fe–Cr ALLOYS

The Ni–Fe–Cr phase diagram is shown in Figure 14.4. There is clearly a wide range of alloys that are based on the f.c.c. γ phase. These industrially important alloys are characterized by their good combinations of corrosion resistance and mechanical strength at elevated temperature. In general, there are two groups of alloys: Inconels based on the composition Ni–16Cr–7Fe (IN600) and incoloys (IN800 family) with the basic composition Ni–20Cr–48Fe. The latter are typically used in scenarios in which hot corrosion resistance is required. In general, the high Fe content increases the solubility of interstitials, thereby reducing the possibility of forming embrittling carbides.

In the case of the first group of alloys, the strengths are retained up to ∼600°C. Increased strengths can also be achieved by increased Cr and alloying with Mo and Nb in solid solution. This results in increased elevated-temperature strength and improved ductility. A typical alloy of this type if IN625 (Ni–22Cr–9Mo–4Nb–5Fe–0.05C) in which Nb increases the strength and C forms carbides that reduce the extent of grain boundary sliding. In the second group of alloys, small

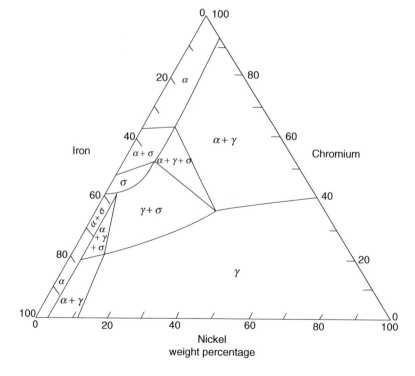

FIGURE 14.4 Ni–Fe–Cr phase diagram.

additions of Al and Ti can promote age hardening. These can result ultimately in tensile strengths of 600–720 MPa and elongations to failure of ~ 40–50%.

14.6 NICKEL-BASE SUPERALLOYS

One of the most significant achievements in alloy development has been the emergence of nickel-base superalloys. These are remarkable alloys that have excellent combinations of strength and creep resistance. In most cases, the strengths are based on the ability to age harden due to the presence of the γ' Ni_3Al phase. This is an ordered f.c.c. phase with the $L1_2$ crystal structure. The remarkable properties of these alloys have really been the key to the development of modern gas turbine engines for applications in aeroengines and land-based engines for power generation.

The historical evolution of the elements used in the Ni-based "superalloy stew" is shown in Figure 14.5. Table 14.3 also shows the evolution of compositions and the applications of nickel-base superalloys. Note that the alloys have evolved from wrought polycrystalline alloys in 1935 to cast single crystals in the hottest sections of today's modern engines. The compositions of the wrought alloys have also evolved toward IN718, which is currently a popular material used in the disk sections of aeroengines.

A typical micrograph of cast single crystal Ni superalloys is shown in Figure 14.6. Note that the cuboidal structure may evolve into raft structures under creep loading (Figure 14.7). In any case, experimental single crystals have also been produced from IN718 in which Ni_3Nb γ'' precipitates provide the strengthening (Figure 14.8). The properties of this alloy will be discussed later in Section 14.7. In any case, a wide range of phase is observed in today's superalloys. A panorama of the different phases that can be encountered in nickel alloys is presented in Figure 14.9. This includes both useful and deleterious phases. It also shows the historical progression of alloy development. These phases give rise to remarkable elevated temperature-strength retention and

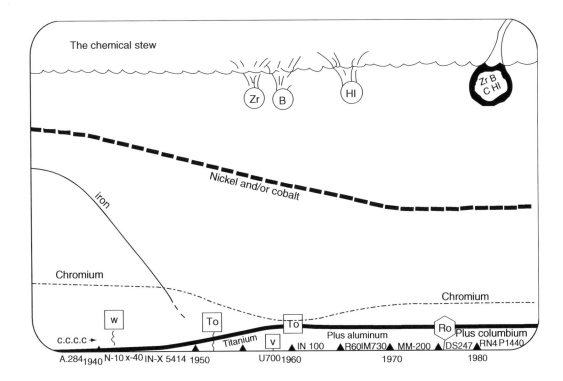

FIGURE 14.5 Ni chemical stew.

TABLE 14.3
The Evolution and Applications of Ni Superalloys

			Fe	Ni	Co	Cr	Al	Ti	Ta	Mo	W	Hf	Zr	C	Other
1935	A-286	Wrought Fe-base Bucket alloy	65.5	26	—	15	0.2	2.0	—	1.3	—	—	—	0.05	0.015B
	K 42 B	Wrought Ni–Co–Fe–Base blade alloy	13.0	43	22.0	18	0.2	2.1	—	—	—	—	—	0.05	
1985	FSX-414	Cast Co-base Vane alloy	10	10	52.5	29	—	—	—	—	7.5	—	—	0.25	
	CM SX-2	Cast Ni-base Single crystal alloy	—	66.5	4.6	8.0	5.6	0.9	5.8	0.6	7.9	<.1	<.01	0.005	<10ppm S.N.O
	MA 6000E	Wrought Ni-base ODS Blade alloy	—	70	—	15	4.5	2.5	2.0	2.0	4.0	—	0.15	0.05	1.1Y_2O_3: 0.01B
	IN-718	Forged NI-Fe-base Wheel alloy	18.5	52.5	—	19	0.5	0.9	—	3.0	—	—	—	0.04	SCb: 0.005B

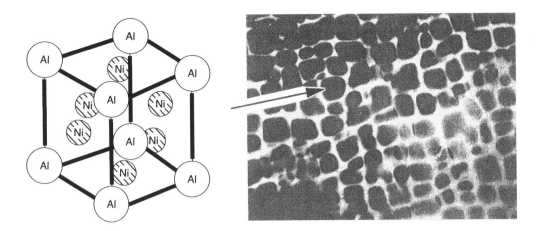

FIGURE 14.6 Microstructure of cast Ni single crystal alloy with cuboidal Ni$_3$Al γ' precipitates.

creep resistance in nickel-base superalloys (Figure 14.10). In the case of the wrought alloys, these generally limit the applications to the disk sections of gas turbines. The most commonly used disk alloys are (by wt.%)

IN718: Ni–0.4Al–0.9Ti–18.5Fe–5.0Nb–3.1Mo–18.6Cr–0.2Mn–0.3Si–0.04C (802°C capability)

Udimet700: Ni–4.3Al–3.5Ti–5.2Mo–18.5Co–15.0Cr–0.03B–0.08C (960°C capability)

In the case of the cast directionally solidified and single crystal structures, the improved creep resistance enables applications up to 1100°C (Figure 14.10). Such remarkably high temperature capability is also enabled by their phase stabilities and tendency to form protective Cr$_2$O$_3$ scales. However, above 1100°C, high temperature oxidation becomes serious. Within this regime, thermal barrier coatings (Figure 14.11) are often used to increase resistance to oxidation. These zirconia-based coatings are generally deposited using plasma spray or vapor deposition techniques [19]. They can reduce the surface temperatures in the Ni by $\sim$150–200°C. The temperatures within the Ni blades are also reduced by internal air cooling passages (Figure 14.12). These increase the overall thrust and efficiencies that can be achieved in modern aeroengines.

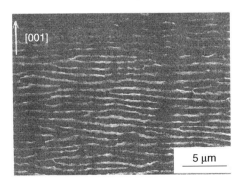

FIGURE 14.7 Rafting under creep loading.

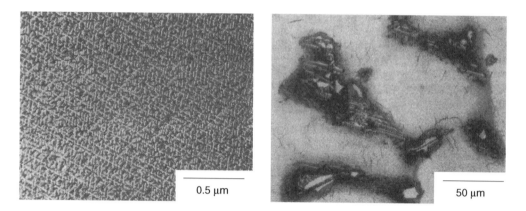

FIGURE 14.8 γ/γ'' Structure of IN718 single crystal.

14.7 FATIGUE AND FRACTURE OF NICKEL-BASE SUPERALLOYS

Nickel-base superalloys have fracture toughnesses that are relatively high. In the case of IN718, the fracture toughness is ~ 100 MPa m$^{1/2}$, while IN706 has a fracture toughness of ~ 130 MPa m$^{1/2}$. However, failure typically occurs by subcritical crack growth processes such as fatigue and creep crack growth. In spite of the widespread applications of Ni-base alloys, most of the prior work on fatigue has focused on the mechanisms of fatigue crack initiation [1,2]. However, there have also been studies of fatigue crack growth in nickel-base alloys by a number of researchers [3–7]. Selected information from these studies will be presented in this section.

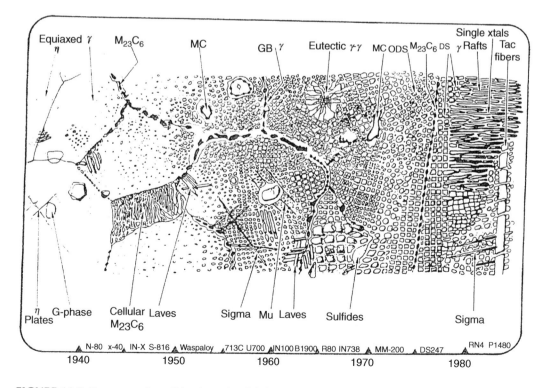

FIGURE 14.9 Panorama of possible phases in nickel-base superalloys.

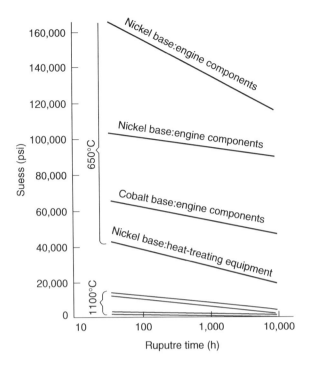

FIGURE 14.10 Stress-rupture behavior of selected superalloys at 650°C and 1100°C.

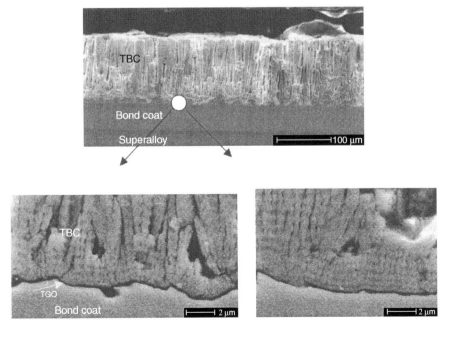

FIGURE 14.11 Thermal barrier coating of zirconia on Ni superalloy.

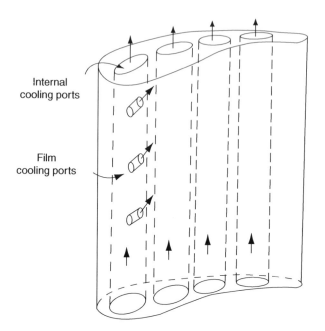

FIGURE 14.12 Schematic of air-cooled blades.

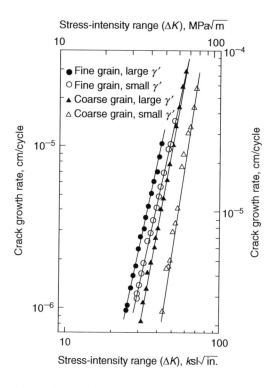

FIGURE 14.13 Fatigue crack growth rate data for Waspaloy tested at 25°C.

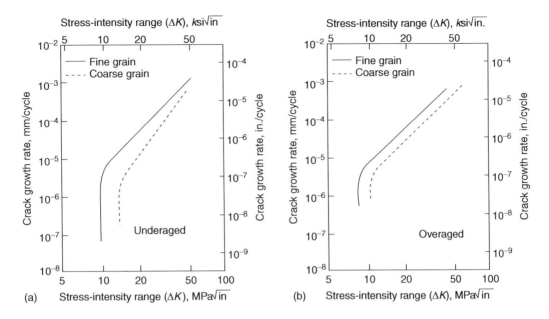

FIGURE 14.14 Fatigue crack growth rate data for polycrystalline IN718 tested at 425°C. (a) Under-aged condition and (b) Over-aged condition.

Much of the work on the fatigue of nickel-base superalloys has been done on polycrystalline IN718 and Waspalloy. Typical fatigue crack growth rate data obtained for these materials are presented in Figure 14.13 and Figure 14.14. These show strong effects of microstructure on fatigue crack growth rates. In general, slower crack growth rates are associated with increasing γ grain size and decreasing γ′ size.

The underlying mechanisms of fatigue crack growth have also been studied by Mercer et al. [3,4]. Typical fatigue fracture modes are shown in Figure 14.15a–d for forged polycrystalline IN718. These show a near-threshold crystallographic fatigue fracture mode in the near-threshold regime (Figure 14.15a). A crystallographic fracture mode is also observed in the lower Paris regime (Figure 14.15b), prior to a transition to a fatigue striation mode in upper Paris regime (Figure 14.15c). A combination of fatigue striations and dimples is observed in the high ΔK regime (Figure 14.15d) before final overload fracture by a ductile dimpled fracture mode (Figure 14.15e). The ranges in the fatigue crack growth rates corresponding to these different fracture modes are shown in Figure 14.16. Similar trends have also been reported for single crystal and polycrystalline IN718 [3,4]. However, the incidence of cleavage-like crystallographic fracture modes is greater in the single crystal IN718 alloy [3,4].

The fatigue crack-tip deformation modes have also been studied in IN718 using crack-tip transmission electron microscopy (TEM) techniques. The results are shown in Figure 14.17a–d. Slip lines and fine dislocation networks are observed in the γ matrix (Figure 14.17a and b), along with stacking faults (Figure 14.17c). Stacking faults are also observed in the Ni_3Nb precipitates, as shown in Figure 14.17d. The results suggest fatigue irreversibility associated with dislocation interactions and dislocation reactions [4].

A typical fatigue mechanism map obtained for IN718 polycrystals is shown in Figure 14.18. This shows the fatigue and fracture modes associated with different values of ΔK and K_{max} that were applied in fatigue tests conducted at stress ratios between 0.1 and 0.8. Note that, for a given stress ratio, the trends are similar to those described earlier for the stress ratio of 0.1 (Figure 14.15 and Figure 14.16). However, the incidence of static fracture modes increases with increasing stress ratio. Similar results have been obtained in other studies on single crystal IN718 [4] and Ti–6Al–4V [8].

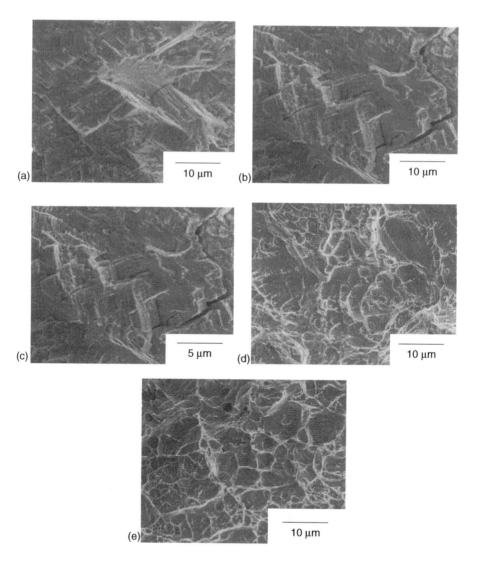

FIGURE 14.15 Typical fatigue crack growth modes in a polycrystalline IN718 alloy. (a) Near-threshold regime; (b) lower Paris regime; (c) upper Paris regime; (d) high ΔK regime, and (e) tensile overhead region, for a stress ratio of 0.1.

The concept of a fatigue mechanism map, therefore, appears to be one that is generally applicable to structural metallic materials [9].

In the case of the single crystal PW1480 Ni single crystal alloy, DeLuca and Watkins [7] have studied the mechanisms of fatigue crack growth. Typical fatigue fracture modes are presented in Figure 14.19, along with the fatigue crack growth rate data obtained for this material. Fatigue crack growth in PW1480 is influenced significantly by interactions with cubidal γ' precipitates. Two types of crack growth are observed: crack growth between the cuboids and crack growth across the cuboids. At low ΔK levels, crack growth occurs between the cuboids, which results in a flat crystallographic fracture mode. In the regime where the plateau occurs in the fatigue crack growth rate data, a deflected crack path is observed on octahedral planes. A more tortuous crack path is then observed at higher ΔK levels, as the cracks are deflected between the cuboidal γ' precipitates. In the high ΔK regime, a noncrystallographic fatigue fracture mode is observed. Evidence of crack branching is also observed at higher ΔK levels where the crack growth rates are fastest (Figure 14.19).

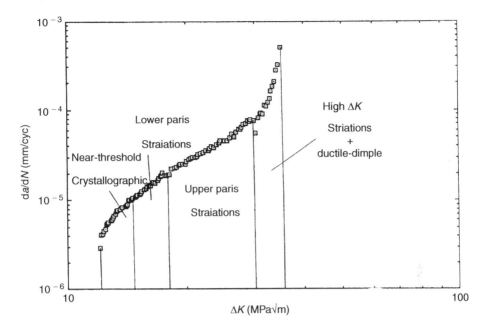

FIGURE 14.16 Fatigue crack growth curves for IN718 showing positions of fatigue regimes corresponding to fatigue fracture modes in Figure 14.15.

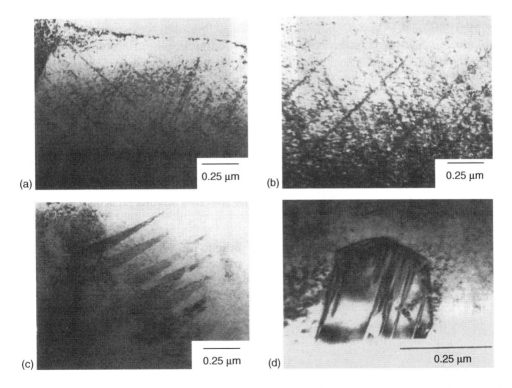

FIGURE 14.17 Transmission electron microscopy (TEM) images showing (a,b) slip lines and fine dislocation networks in γ matrix; stacking faults in the γ matrix, and (d) stacking faults in Ni_3Nb precipitates in polycrystalline IN718.

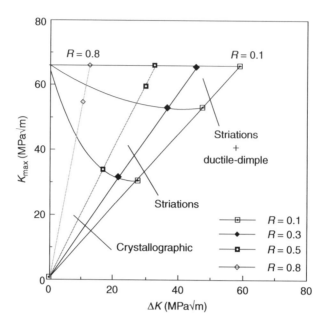

FIGURE 14.18 Fatigue mechanism map for forged IN718.

14.8   THE FUTURE BEYOND NICKEL-BASE SUPERALLOYS

Before closing, it is of interest to present a perspective of high temperature materials beyond nickel-base superalloys. Quite clearly, there are a number of efforts to develop lower density, high temperature materials for high temperature structural applications. However, in the short-to-medium term, the potential for actual structural applications appears to be limited. This is because the materials that can support the loads at room temperature cannot resist oxidation and creep effects at high temperature, and vice versa.

In the temperature range of $\sim 700°C$, gamma-based titanium aluminides have been proposed for potential applications in aero- and land-based engines, and even in the airframes of supersonic vehicles [10]. This is due to their attractive combinations of elevated-temperature oxidation

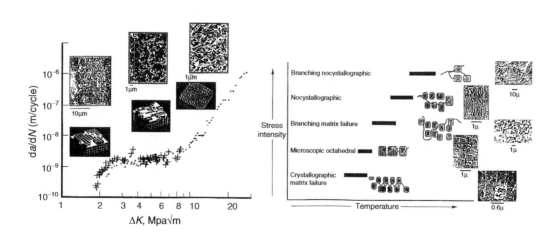

FIGURE 14.19 Fatigue crack growth rates and fatigue fracture modes in PW1480 single crystal.

resistance, moderate creep resistance, and moderate density (~ 4.5 g/cm^3). However, their limited room-temperature ductility ($\sim 1-3\%$), fracture toughness ($\sim 10-35$ MPa m$^{1/2}$) and fatigue crack growth resistance remain as major obstacles to their applications [11–18].

Niobium-based alloys [19–21] and niobium-based intermetallics [22,23] have also been considered for intermediate-temperature applications. Due to their limited oxidation resistance, niobium-based alloys and intermetallics are typically limited to applications below $\sim 700°C$ [19–23]. However, unlike the titanium aluminide, the niobium alloys and niobium aluminide intermetallics have excellent combinations of fracture toughness (40–100 MPa m$^{1/2}$) and fatigue crack growth resistance [22].

For applications at higher temperatures ($\sim 1100-1350°C$), significant efforts have been made to toughen molybdenum disilicides [24,25], niobium disilicides [26], and nickel aluminides [27–30]. Nevertheless, these systems remain brittle, in spite of the significant efforts to toughen them over the past two decades. Similarly, ceramic composites remain brittle [31,32]. However, thermal barrier coatings (TBCs) based on ceramic coatings, have been commercialized successfully for applications in high temperature gas turbines [33,34]. Further work is clearly needed to develop coated multifunctional structures for high temperature structural applications.

14.9 SUMMARY AND CONCLUDING REMARKS

This chapter presented an introduction to nickel and its alloys with a brief review of the applications, basic physical and mechanical properties, and a description of fatigue and fracture behavior. The underlying mechanisms of failure were also discussed, along with their implications for thermostructural applications. Some of the materials that might be used beyond Ni/Ni alloys were also presented.

REFERENCES

1. King, J. E., *Fatigue and Fracture of Engineering Materials and Structures*, 5, 177, 1982.
2. Lerch, B. A. and Antolovich, S. D., *Metallurgical Transactions*, 21A, 2169, 1990.
3. Mercer, C. and Soboyejo, W. O., *Acta Metallurgica*, 45, 961–971, 1997.
4. Mercer, C., Soboyejo, A. B. O., and Soboyejo, W. O., *Materials Science and Engineering*, A270, 308–322, 1999.
5. Suresh, S., *Fatigue of Materials*, 2nd ed., Cambridge University Press, Cambridge, 1998.
6. Soboyejo, A. B. O., Shademan, S., Foster, M., Katsube, N., and Soboyejo, W. O., *Fatigue and Fracture of Engineering Materials and Structures*, 24, 225–241, 2001.
7. DeLuca, D. P. and Watkins, T., Selected properties of anisotropic PWA1472, In *Proceedings of The Symposium on Superalloys 718, 625, 706 and Various Derivatives*, Loria, E. A., Ed., TMS, Warrendale, PA, 1997.
8. Shademan, S., Soboyejo, A. B. O., Knott, J. F., and Soboyejo, W. O., *Materials Science and Engineering A*, A315, 1–10, 2001.
9. Soboyejo, W. O., *Mechanical Properties of Engineered Materials*, Marcel Dekker, New York, NY, 2003.
10. Kim, Y,-W. and Dimiduk, D. M., *Journal of Metals*, 43, 40, 1991.
11. Soboyejo, W. O., Venkateswara-Rao, K. T., Sastry, S. M. L., and Ritchie, R. O., *Metallurgical and Materials Transactions*, 24A, 585–600, 1993.
12. Soboyejo, W. O., Mercer, C., Lou, K., and Heath, S., *Metallurgical and Materials Transactions*, 26A, 2275–2291, 1995.
13. Soboyejo, W. O. and Mercer, C., *Scripta Metallurgica*, 30, 1515, 1994.
14. Srivatsan, T. S., Soboyejo, W. O., and Aswath, P. B., *Materials Science and Engineering*, A138, 95, 1991.
15. Davidson, D. L. and Campbell, J. B., *Metallurgical and Materials Transactions*, 24A, 1555, 1993.

16. Lou, J., Mercer, C., and Soboyejo, W. O., *Materials Science and Engineering*, A319–321, 618–624, 2001.
17. Lou, J. and Soboyejo, W. O., *Metallurgical and Materials Transactions*, 32A, 325–337, 2001.
18. Mercer, C., Lou, J., Allameh, S. M., and Soboyejo, W. O., *Metallurgical and Materials Transactions*, 32A, 2781–2794, 2001.
19. Soboyejo, W. O., Brooks, D., and Chen, L,-C., *Journal of The American Ceramic Society*, 78, 1481–1488, 1995.
20. Kim, Y-W. and Carneiro, T., *Niobium: High Temperature Applications*, TMS, Warrendale, PA, 2004.
21. Bordignano, P. J. P. et al., *Niobium Science and Technology*, TMS, Warrendale, PA, 2001.
22. Ye, F., Mercer, C., and Soboyejo, W. O., *Metallurgical and Materials Transactions*, 29A, 2361–2374, 1998.
23. Li, Y., Soboyejo, W., and Rapp, R., *Metallurgical and Materials Transactions*, 30B, 495–504, 1999.
24. Lou, J., Ye, F., Li, M., and Soboyejo, W. O., *Journal of Materials Science*, 37, 3023–3034, 2002.
25. Shaw, L. and Abbaschian, R., *Acta Metallurgica Materials*, 42, 213–223, 1994.
26. Kajuch, J., Short, J., and Lewandowski, J., *Acta Metallurgica Materials*, 43, 1955–1967, 1995.
27. Soboyejo, W. O., Ye, F., Dipasquale, J., and Pine, P., *Journal of Materials Science*, 34, 3567–3575, 1999.
28. Ramasundaram, P., Ye, F., Bowman, R. R., and Soboyejo, W. O., *Metallurgical and Materials Transactions A*, 29A, 493–505, 1998.
29. Li, M., Wang, R., Katsube, N., and Soboyejo, W. O., *Journal of Engineering Materials and Technology*, 121, 453–459, 1999.
30. Noebe, R. D., Misra, R., and Gibala, R., *Iron Steel Institute of Japan International*, 31, 1172–1185, 1991.
31. Evans, A. G., *Journal of The American Ceramic Society*, 73, 187–206, 1990.
32. Evans, A. G. and Marshall, D. B., *Acta Metallurgical Materials*, 32, 2567–2583, 1989.
33. Mumm, D. R., Walter, M., Popoola, O., and Soboyejo, W. O., *Durable Surfaces*, Trans Tech Publications, Zurich, Switerzerland, 2001.
34. Walter, M. E., Onipede, B., Soboyejo, W., and Mercer, C., *Journal of Engineering Materials and Technology*, 122, 333–338, 2000.

Index

A

ACMC. *See* Advanced Composite Materials Corporation (ACMC)
Actuation techniques, 81–83
Actuators, 150, 152–154
Additions
  aluminum and aluminum alloys, 232–234
  NiTi alloys, 157
  ternary alloys, 157–158
Additive cold-welding technique, 26
Adhesion. *See* Small scale contact and adhesion
Advanced Composite Materials Corporation (ACMC), 290, 293
Advanced materials
  aeroengines, 2–4
  applications, 2–11
  automotive applications, 8–11
  balanced properties engineering, 11–12
  fundamentals, 1–2, 12–13
  human prosthetic devices, 8
  national aerospace plane, 4–8
  sporting goods, 8
Aeroengines applications, 2–4
AFM. *See* Atomic force microscopy (AFM)
Agere Systems, 39
Air bags, 38, 41–42
Air Force applications, 341
Al-based alloys. *See* Glass formation abilities (GFA), Al-based alloys
ALCOA (Aluminum Company of America), 291–292
Al-Gd-Fe ternary systems, 134
Allameh studies, 61, 65–95
Allen's postulation, 217–218
Allison Advanced Development Company, 432
Alloying additions effects, 157–158
Alloying behavior, 448–450
Alloying elements, 363
Alloys. *See specific type*
Al-Ni-Gd systems, 135
Al-RE systems, 131
Aluminum alloy, 1100, 322
Aluminum alloy, X2080, 306
Aluminum alloy, 1XXX, 244
Aluminum alloy, 2XXX
  ductility, 316
  elasticity modulus, 309

  elevated temperature properties, 262
  fracture toughness, 322–323
  historical developments, 228–229
  powder metallurgy, 291
  stress and strain amplitude, 335–336
  tensile behavior, 254
Aluminum alloy, 3XXX, 244
Aluminum alloy, 5XXX, 244
Aluminum alloy, 6XXX
  elasticity modulus, 309
  fracture toughness, 322–323, 329
  powder metallurgy, 291
  stress and strain amplitude, 332–333
Aluminum alloy, 7XXX
  developments, 264
  ductility, 314, 316
  elasticity modulus, 309
  elevated temperature properties, 261
  fracture toughness, 322–323, 325, 329
  historical developments, 229
  stress and strain amplitude, 336–337
Aluminum alloy, 8XXX, 254, 322–323
Aluminum and aluminum alloys
  additions, 232–234
  alloys, 231
  aluminum-lithium alloys, 263–266
  aluminum macro-laminates, 266
  availability, 227
  building and construction applications, 268–269
  castings, 266–267
  classification, 268–269
  conductor alloys, 269
  corrosion, 261, 263
  cracks, 257–260
  creep characteristics, 261
  damage tolerance, 229–230
  deformation, 234–242
  durability, 229–230
  early alloys, 228–230
  electrical applications, 269
  elevated temperatures, 261–262
  engineered products, 266–268
  extrusions, 268
  fatigue, 256–260
  forgings, 268
  fracture resistance, 230
  fracture toughness, 254–255
  fundamentals, 226–227, 270

heat-treatable, systems, 250–251
historical developments, 228–230
household appliance applications, 269
impurities role, 232–234
kinetics, 242
low temperatures, 263
metallurgy, 231–234
microstructure role, 251–263
physical metallurgy, 231–234
powder metallurgy and alloys, 245–248, 268
precipitation, 236–242
production, 227–228
properties, 242–243
solid solution strengthening, 235–236
temper designation system, 249–251
tensile behavior, 252–254
trace elements, 242
wrought aluminum alloys, 243–244, 263
wrought non-heat-treatable aluminum alloys,
    244–245
Aluminum Association, 231, 249
Aluminum Company of America (ALCOA),
    291–292
Aluminum-lithium alloys, 263–266
Aluminum macro-laminates, 266
Alur studies, 461–464, 467
Ambient pressure chemical vapor deposition, 74
Ambient pressure chemical vapor deposition
    (APCVD), 73
American Society for Testing and Materials
    (ASTM), 221, 318
Amorphous glass alloy, 8
Analog Devices, 38
Anderson studies, 384
Andrews, Gibson, and Ashby studies, 122
Anisotropic etching, 71, 76–77
Annealed (O) foams, 108, 114–116
Anton and Shah studies, 405
Applications
    actuation techniques, 81–83
    actuators, 152–154
    advanced materials, 2–11
    aeroengines, 2–4
    automotive, 8–11, 342–345
    automotive applications, 90
    biochemical sensors, 88–90
    biomedical applications, 221–222
    building applications, 268–269
    cobalt alloys and composites, 219–222
    commercial products, 346
    conductor alloys, 269
    conductrimetric gas sensors, 89
    constrained recovery, 152
    construction applications, 268–269
    damping, 155

electrical, 269
electronic packaging, 345–346
flow sensors, 88
force sensors, 85–86
free recovery, 151
high-temperatures, 219, 221
household appliances, 269
human prosthetic devices, 8
inertial sensors, 86–88
life science applications, 91–92
magnetic devices, 221
mechanical sensors, 84–88
metal matrix composites, 340–346
national aerospace plane, 4–8
niobium alloys, 431–432
NiTi-based high-temperature shape-memory
    alloys, 151–155
photonics applications, 90–91
potentiometric devices, 90
pressure sensors, 84–85
RF applications, 92
sensing techniques, 79–81
sensors and actuators, 79–84
silicon-based microelectromechanical systems,
    79–92
smart sensors, 84
space, 341–342
sporting goods, 8
superelastic behavior, 154
thin films materials, 38–39
titanium alloys, 362, 391–395
torque sensors, 85–86
wear resistance, 155
work production, 152–154
APVCD. *See* Ambient pressure chemical vapor
    deposition (APCVD)
ARALL laminates, 263, 266
Arnold, Steve, 180
Aronsson studies, 454
Arora, Chu, MacDonald and, studies, 379
Arsenault, Taya and, studies, 312
Arsenault studies, 312
As-fabricated (F) foams, 108, 114–116
Ashby, Andrews, Gibson and, studies, 122
Ashby, Gibson and, studies, 106
Ashby studies, 52
ASTM. *See* American Society for Testing and
    Materials (ASTM)
Atomic force microscopy (AFM), 29, 50
Atomic mobility, 136
Atomic size difference, 129
Atomization, spray
    co-deposition processing, 296–298
    deposition processing, 295–296
    premixed MMCs, 298–299

Automotive applications
  advanced materials, 8–11
  aluminum alloy products, 268–269
  cost, 342–343
  dispersion strengthened metal matrix
      composites, 280
  fatigue, 332
  MEMS applications, 38
  metal matrix composites, 342–345
  silicon-based microelectromechanical systems, 90
Availability aluminum and aluminum alloys, 227
Avitzur studies, 306

B

Bache, Davies and, Evans studies, 374
Bache, Evans and, studies, 374, 381–382
Bache and Cope studies, 376, 380
Bache and Evans studies, 372, 379
Balanced properties engineering, 11–12
Basquin relationship, 337–338
BDT. *See* Brittle-to-ductile-transition (BDT)
      temperature
Begley and Hutchinson studies, 55
Begley studies, 409
Berkovich tip and indenter, 17–18, 50
Bewlay studies, 412, 424–425
Biles studies, 145–180
Binary systems, 441
Biochemical sensors, 88–90
Biological adhesion, 21–26. *See also* Small scale
      contact and adhesion
Biomedical applications
  bone detachment, 395
  cobalt alloys and composites, 221–222
  cyclotoxicity, 393–395
  fundamentals, 391–393
  implants, 395
  stress shielding, 395
  titanium alloys, 391–395
Bio-systems. *See* Small scale contact and adhesion
Blazer studies, 239
Bodner-Partom unified model, 217
Boeing, 777, 230, 278
Boltzmann's constant, 416
Bone detachment, 395
Boriskina and Kenina studies, 161
Boron, 12
Boyce studies, 57
Brenner studies, 281
Briant studies, 424
Brittle-to-ductile-transition (BDT) temperature, 94
Brotzen studies, 43
Bruggeman studies, 307
Buehler studies, 156

Building and construction applications, 268–269
Bulge test, 44–45
Bulk micromachining, 37–38
Burgers vector
  nanoindentation methods, 18
  niobium alloys, 410
  plasticity size effects, 53
  precipitation hardening, 237, 241
  strain gradient plasticity, 52

C

Cai studies, 164
Calculation of phase diagrams. *See* CALPHAD
      models
Caliper Life Sciences, 92
CALPHAD models
  glass formation abilities, 131–132
  heat of mixing, 136, 138
  quasi-kinetics, primary crystallization, 134
Cao studies, 15–30
Capacitive pressure sensors, 85
Capacitive sensors, 86
Capacitive techniques, 80
Carpenter, Bernie, 180
Castings, 10, 266–267
Cell mechanical properties, 28–30
Cemented carbides, 222–223
Chan and Koss, Wojcik, studies, 370, 372, 378, 382
Chan studies, 401–432
Chemical short range ordering (CSRO), 136
Chemical vapor deposition (CVD)
  fiber reinforcement, 284–285
  fundamentals, 73
  polysilicon deposition, 73–74
  silicon-based microelectromechanical systems,
      73–74
  silicon nitride deposition, 74
  silicon oxide deposition, 74
Chen studies, 209
Choe studies, 467
Cho studies, 57–58
Christenson studies, 56
Chu, MacDonald and, Arora studies, 379
Clark and Howe studies, 87
Classifications, 268–269, 363
Clustering, reinforcement, 325–327
Cobalt alloys and composites
  applications, 219–222
  balanced properties engineering, 12
  biomedical applications, 221–222
  cemented carbides, 222–223
  corrosion resistance, 205–207
  end uses, 191
  extraction, 191

fatigue behavior, 208–219
fundamentals, 188
high-cycle fatigue behavior, 208–209
high-temperature applications, 219, 221
high-temperature resistance, 198–204
low-cycle fatigue behavior, 209–211
magnetic device applications, 221
mechanical properties, 192–218
occurrence, 191
phase equilibria, 188–191
physical properties, 192–218
processing, 191
structure, 188–191
temperature evolution, fatigue, 212–219
wear resistance, 194, 196–198
Co-Cr-Mo alloy, 222
Co-Cr-W-Ni alloy, 222
Co-deposition, 296–298. *See also* Deposition
Coefficients of thermal expansion (CTE)
actuation techniques, 82
electronic packaging, 346
silicon, 70
Coffin-Mason equation and relationship, 209,
    337–338
Cohen, Gregg and, studies, 235–236
Cold forming/forging, 303
Cold-welding techniques, 26
Commercial accelerometers, 87
Commercial gyroscopes, 88
Commercial products, 346
Compositions, 402–408
Conductimetric gas sensors, 89
Conductivity, discontinuous reinforcements,
    307–308
Conductor alloys, 269
Conductrimetric gas sensors, 89
Co-Ni-Cr-Mo alloy, 222
Constrained recovery, 149–150, 152
Contact mechanics, 19–21
Cooley and Crooker, Yoder, studies, 368
Cope, Bache and, studies, 376, 380
Cope, Spence, Evans and, studies, 374, 376
Corrosion
behavior, aluminum and aluminum alloys, 263
fatigue, microstructure, 261
resistance, cobalt alloys and, composites,
    205–207
Cost
automotive applications, 8–9
composites, 342–343
MEMS failure, 41–42
nickel-based alloys, 477–478
powder metallurgy alloys, 248
titanium matrix composites, 10
Cottrell interaction, 236

Coupled heat-transfer formula, 218
Cracks and crack growth behavior. *See also* Fatigue
closure, 373
deflection, 373
fatigue behavior, 373
growth, 258–260, 338–340
initiation, 257
mechanisms, 381–384
metal matrix composites, 338–340
niobium alloys, 417–422
short growth, 373–374
three-phase alloys, 467
titanium alloys, 373–374
whisker reinforcement, 388
Creep
cobalt, 193
fatigue synergism, 379–380
low-cycle fatigue, 210
microstructure role, 261
niobium alloys, 422–426
porous metallic materials, 120–123
resistance, 366–368
silicon MEMS, 94
single-phase alloys, 460
in situ composites, 425–426
time dependent strain accumulation, 376
titanium alloys, 366–368, 379–380
titanium matrix composites, 10
Criterion, glass formation abilities, 127
Critical resolved shear stress (CRSS), 236, 239
Crooker, Yoder, Cooley and, studies, 368
Crowe studies, 328
CRSS. *See* Critical resolved shear stress (CRSS)
Crystallization
glass formation abilities, 127
quasi-kinetics, glass formation abilities, 134–135
titanium alloys, 362
Crystallography, 372–373, 378
CSRO. *See* Chemical short range ordering (CSRO)
CTE. *See* Coefficients of thermal expansion (CTE)
CVD. *See* Chemical vapor deposition (CVD)
Cyclic strain accumulation, 116–118
Cyclotoxicity, 393–395

D

Damage tolerance, 229–230
Damping, 155
Davidson and Eylon studies, 379
Davidson studies, 405, 412, 414, 417
Davies and Evans, Bache, studies, 374
dc-glow discharge, 72
DC plasma etching, 77
Deep reactive ion etching (DRIE), 71, 78–79
Defect structures, 441–442

Deformation
    aluminum and aluminum alloys, 234–242
    porous metallic materials, 110–116
    precipitation hardening, 236–241
    precipitation kinetics, 242
    solid solution strengthening, 235–236
Deleterious intermetallic compounds, 201
Delft University, 266
Deloro Stellite, Inc., 198
DeLuca and Watkins studies, 488
Density
    aeroengines, 4
    balanced properties engineering, 11
    discontinuous reinforcements, 307–308
Deposition, 295–296, 300. *See also* Co-deposition
Depth sensing indentation (DSI), 16
Derjaguin-Muller-Toporov (DMT) model, 19–21
DIC. *See* Digital image correlation (DIC) analysis
Diffusion behavior, 454–455
Digital image correlation (DIC) analysis, 110
Discontinuous reinforcements
    atomization, spray, 295–298
    co-deposition processing, 296–298
    conductivity, 307–308
    crack growth behavior, 338–340
    density, 307–308
    deposition processing, 295–296, 298–300
    ductility, 312–318
    elasticity modulus, 309
    fatigue behavior, 332–340
    forging, 302–305
    fracture toughness, 318–338
    heat capacity, 307–308
    heat treatment effects, 323–325
    high-velocity oxy-fuel spraying, 301–302
    hydrostatic extrusion, 305–306
    low pressure plasma deposition, 300
    matrix microstructure effects, 323–325
    mechanical properties, 309–340
    mixed mode toughness, 327
    modified gas welding technique, 300–301
    modulus of elasticity, 309
    notch root radium change effects, 320
    particle hardened materials, 328–329
    physical properties, 307–308
    powder metallurgy, 290–294
    premixed, 298–299
    primary processing, 289–302
    properties, 307–340
    reinforcements, 325–327
    secondary processing, 302–306
    solid phase processes, 290–294
    spray processing, 294–299
    strain amplitude, 332–338
    strength, 310–312
        stress amplitude, 332–338
        toughened MMCs, 329–332
        volume fraction change effects, 320–322
Dispersion strengthened metal matrix composites,
        279–280. *See also* Metal matrix composites
        (MMCs)
Divecha studies, 281
DMT. *See* Derjaguin-Muller-Toporov (DMT) model
DNA analysis, 92
Doerner and Nix studies, 16
Donkersloot and Van Vucht studies, 165–166
Double-sided lithography, 76
Dowling studies, 384
DRIE. *See* Deep reactive ion etching (DRIE)
Dry etching, 77–79
DSI. *See* Depth sensing indentation (DSI)
Dubey studies, 373
Ductility
    discontinuous reinforcements, 312–318
    LIGA technology, 41
    nickel-base alloys, 476, 479
    (Ni,Pt)Ti alloys, 168
Dugdale model. *See* Maugis-Dugdale (MD) model
Duocell aluminum foams
    fabrication, 105
    fundamentals, 104
    stress-strain behavior, 108
    tetrakaidecahedron unit cell model, 107
Dupre equation, 22–23
Durability, 229–230
Duralumin, 228–229
Dwell, titanium alloys
    behavior, 374–381
    fundamentals, 361
    hold time, 381
    mechanisms, 381–384
    modeling, 384–385
    short crack growth, 374
    time dependent strain accumulation, 376

E

Early alloys, 228–230
EBSD. *See* Electron back scattered diffraction
        (EBSD)
Eckelmeyer studies, 165, 174
Ecklemeyer studies, 174
Elasticity modulus. *See also* Young's moduli
    discontinuous reinforcements, 309
    nanoindentation methods, 17
    temperature, 218
Electrical applications, 269
Electrical properties, 68–69
Electric power-assisted steering systems (EPAS), 86
Electron back scattered diffraction (EBSD), 379

Electronic packaging, 345–346
Elestrostatic actuation, 81–82
Elevated temperature properties, 261–262
Embedded particle tracking method, 30
Empirical rules, 127
End uses, cobalt alloys and composites, 191
Engineered products, 266–268
Engineering applications, titanium alloys, 362
Enthalpy, glass formation abilities, 129–131
Entran, 43
Entropy estimation, 131
Environmental effects
    dwell behavior, 378
    fatigue behavior, 370
    oxidation behavior, 456
    silicon MEMS, 94
Environmental Protection Agency (EPA), 8, 10
EPAS. *See* Electric power-assisted steering systems
    (EPAS)
Epitaxy, topical films growth, 72
Eshelby's elastic inclusion model, 130
Etching, 71, 76–79
Eurocopter, 341
Evans, Bache, Davies and, studies, 374
Evans, Bache and, studies, 372, 379
Evans, Stölken and, studies, 47, 54
Evans and Bache studies, 374, 381–382
Evans and Cope, Spence, studies, 374, 376
Evans and Gostelow studies, 374, 376, 382
Evaporation, topical films growth, 73
Extraction, cobalt alloys and composites, 191
Extrusions, 268. *See also* Hydrostatic extrusion
Eylon, Davidson and, studies, 379
Eylon and Hall studies, 373–374, 381

F

Fabrication methods, 37–38
Face-centered-cubic form. *See* FCC (face-centered-
    cubic) form
Failure mechanism assessment, 95
F-16 aircraft, 341
Faraday's Principle, 88
Fatigue. *See also* Cracks and crack growth behavior
    cobalt alloys and composites, 208–219
    corrosion, 261
    cracks, 257–260, 338–340
    creep synergism, 379–380
    cyclic strain accumulation, 116–118
    growth of cracks, 258–260
    high-cycle fatigue behavior, 208–209
    initiation of cracks, 257
    LIGA technology, 41
    low-cycle fatigue behavior, 209–211
    mechanical properties, 332–340

metal matrix composites, 332–340
microstructure role, 256–261
modeling, 384–385
multilevel failure mechanisms, 118–119
nickel-base alloys, 484, 487–488
niobium alloys, 417–422
porous metallic materials, 116–119
response, aluminum and aluminum alloys,
    256–257
short crack growth, 374
silicon MEMS, 93
strain and stress amplitude, 332–338
temperature evolution, 212–219
thin films materials, 56–60
titanium alloys, 368–373
FCC (face-centered-cubic) form
    biomedical applications, 222
    cobalt-based alloys, 189–190
    corrosion-resistant alloys, 205
    high-cycle fatigue behavior, 208
    solid solution strengthening, 235
    wrought cobalt-based alloys, 201
Federation of European Producers and Abrasives
    (FEPA), 325
Feynman, Richard, 36
F foams. *See* As-fabricated (F) foams
FIB. *See* Focused ion-beam milling (FIB)
Fiberfrax, 284
Fiber optics applications, 91
Fiber reinforcement
    fundamentals, 279, 284–285
    titanium matrix composites, 390–391
Figaro Engineering Company, 89
Firstov studies, 172
Fishman, S., 468
Fixed-free cantilever beams, 57
Fleck and Hutchinson studies, 52–54
Fleck-Hutchinson deformation theory, 52
Fleck studies, 52
Fleischer studies, 236, 439
Florando studies, 46
Flow sensors, 88
Fluid shear flow, 29
Foams, metallic, 103–105, 110–111
Focused ion-beam milling (FIB), 76
FOD. *See* Foreign object damage (FOD)
Foil/fiber/foil method, 390
Force sensors, 85–86
Ford Motor Company, 7–8
Foreign object damage (FOD), 361
Forgings
    aluminum and aluminum alloys, 268
    secondary processing, discontinuous
       reinforcements, 302–303, 305
Four-strut unit cell model, 107–108

Fractures
microstructure role, 254–255
nickel-base alloys, 484, 487–488
niobium alloys, 412–417
resistance, 230
thin films materials, 55–56
titanium alloys, 368
toughness, mechanical properties, 318–338
Freels studies, 187–223
Free recovery, 148–149, 151
Frequency effects, 371
Fricke studies, 307
Future challenges and trends
nickel-base alloys, 490–491
NiTi-based high-temperature shape-memory
alloys, 177–180

G

Gamma-based titanium aluminides, 11
Gao, Nix and, studies, 18, 52–53
Garcia, Sehitoglu and, studies, 373
Garg, Anita, 180
Gas welding technique, modified, 300–301
General Electric, 432
General Motors, 7–8
Geometrically necessary dislocations (GNDs), 52
GFA. *See* Glass formation abilities (GFA),
Al-based alloys
Ghonem, Sansoz and, studies, 375, 380
Gibbs energies
CALPHAD models, 131, 134
heat of mixing, 137
multiple components effect, 141
Gibson and Ashby, Andrews, studies, 122
Gibson and Ashby studies, 106
Gibson-Ashby model, 105–107
Glass formation abilities (GFA),
Al-based alloys
atomic mobility, 136
CALPHAD models, 131–132
criterion, 127
crystallization, 127, 134–135
empirical rules, 127
enthalpy, 129–131
entropy estimation, 131
fundamentals, 125–126, 141
liquid heat, 136–138
metallic alloys, 126
mixing heat, 136–138
multiple components effect, 138–141
quasi-kinetics, primary crystallization, 134–135
short range ordering, 136–138
T_0-criterion, 132, 134
thermodynamic models, 131–141
thermodynamics, 128–131

Glow discharge, 77
GLV. *See* Grating light valve (GLV)
GNDs. *See* Geometrically necessary dislocations
(GNDs)
Goldberg studies, 164
Golf clubs, 8, 280
Gostelow, Evans and, studies, 374, 376
Goswami studies, 374
GP. *See* Guinier-Preston (GP) zones
Graphite fiber reinforcement, 341
Grating light valve (GLV), 91
Gregg and Cohen studies, 235–236
Greiff studies, 87
Growth, topical films, 72–74
Grummon studies, 161
Guinier-Preston (GP) zones
precipitation hardening, 236–242
solid solution strengthening, 235–236
tensile behavior, 254
Guinier studies, 239
Gyroscopes, 87–88

H

Hack and Leverant studies, 378, 380, 382
Hafnium additions, 12
Hall, Eylon and, studies, 373–374, 381
Hall, Kassner, Kosaka and, studies, 376, 381
Hall and Hammond studies, 373
Hall studies, 227
Hammond, Hall and, studies, 373
Han studies, 171
Hardening. *See* Precipitation hardening
Harper-Dorn creep, 426
Hartley, Craig, 432
Hassen studies, 332
Haynes, Elwood, 188, 194
HAYNES alloys
applications, 221
fundamentals, 207
high-cycle fatigue behavior, 208
high temperatures, 199
low-cycle fatigue, 209
Haynes International, Inc., 207
HDPCVD. *See* High density plasma chemical
vapor deposition (HDPCVD)
HDTV. *See* High definition TV (HDTV)
Heart-valve applications, 8
Heat, liquids, 136–138
Heat capacity, 307–308
Heat-treatable, systems, 250–251
Heat treatment effects, 323–325
Helmholtz free energy, 216–218
Hemker studies, 57
Henshall studies, 423
Heroult studies, 227

Hertzian contact mechanics, 19
Heteroepitaxy, 72
Heusler phase, 431
Hewlett Packard (HP), 38
High-cycle fatigue behavior, 208–209, 332
High definition TV (HDTV), 38
High density plasma chemical vapor deposition (HDPCVD), 73
High-temperature applications, 219, 221
High-temperature resistance, 198–204
High-velocity oxy-fuel spraying (HVOF), 295, 301–302
Hip implants, 8, 395
Historical developments
    aluminum and aluminum alloys, 228–230
    ternary alloys, 156–157
Honda Motor Company, 344
Honeywell, Inc., 90
Hooke's law, 213
Hosoda studies, 168
Hot-isostatically-pressed (HIPed) fibers, 390
Household appliance applications, 269
Howe, Clark and, studies, 87
Howmet Company, 10
Hsieh and Wu studies, 175
Hsieh studies, 174–176
Hughes, Jim, 96
Human prosthetic devices applications, 8
Hume-Rothery rule and phase, 130, 443
Hurd studies, 332
Hutchinson, Begley and, studies, 55
Hutchinson, Fleck and, studies, 52–53
HVOF. *See* High-velocity oxy-fuel spraying (HVOF)
Hydrogen embrittlement, 378
Hydrostatic extrusion, 305–306. *See also* Extrusions
Hydrostatic stress, 383–384
Hysitron, Inc., 50
Hysteresis
    fatigue, 212
    magnetic device application, 221
    (Ni,Pt)Ti alloys, 168
    NiTi alloy additions, 157–158
    Ni(Ti,Hf) alloys, 172
    Ni(Ti,Zr) alloys, 175

I

IACS. *See* International Annealed Copper Standard (IACS)
IADS. *See* International Alloy Designation System (IADS)
IMAQ Vision, 43
Implants, 8, 395
Impurities role, 232–234

INCONEL alloy, 221
Indentation testing, 48–51. *See also* Nanoindentation methods
Inertial sensors, 86–88
Inhomogeneous DRA structures, 330
Ink jet print heads, 38
Inoue studies, 126
IN718 polycrystals, 487
In situ composites
    fatigue crack growth resistance, 419–420, 422
    fracture resistance, 416–417
Intermetallic phase diagrams, 363, 365
Internal hydrogen, 378
International Alloy Designation System (IADS), 249
International Annealed Copper Standard (IACS), 243, 269
Ion beam milling, 73, 78
Ion etching, 78
Iron and iron-silicon alloys, 221
Irwin's formulation, 383
Isotropic etching
    microfabrication processes, 71, 76
    silicon dry etching, 78–79
Ito studies, 460, 463, 467

J

Jain studies, 461
Jha and Shankar, Ravichandran, studies, 381
Jiang studies, 187–223
JKR. *See* Johnson-Kendall-Roberts (JKR) model
Johnson, Rice and, studies, 328–329
Johnson-Kendall-Roberts (JKR) model, 19–21
Johnson-Mehl-Avrami's transformation, 134

K

Kamat studies, 328
Kassner, Kosaka and Hall studies, 376, 381
Kenina, Boriskina and, studies, 161
Kerner studies, 308
Khachin studies, 163
Kim studies, 26
Kinetics, 242
Klarstrom studies, 187–223
Kosaka and Hall, Kassner, studies, 376, 381
Koskinen studies, 43
Koss, Wojcik, Chan and, studies, 370, 372, 378, 382
Kumar studies, 437–468
Kwon studies, 107

L

Labview, 43
Lankford, Ritchie and, studies, 381
Laplace operator, 86

Laser Doppler velocimetry, 80–81
Laser traps and tweezers, 29
Laves phase
  Nb-Cr-X alloys, 404–405
  Nb-Si-Cr-Al-X alloys, 407
  niobium alloys, 402
  oxidation properties, 427–428
  in situ composites, 422, 425–427
  solid solution alloys and strengthening, 414, 416
  wear-resistant alloys, 197
  wrought cobalt-based alloys, 201–202
LCDF. *See* Low-cycle dwell fatigue (LCDF)
    behavior
LCF. *See* Low-cycle fatigue (LCF) behavior
Lederich and Sastry, Soboyejo, studies, 386
Leverant, Hack and, studies, 378, 380, 382
Lewandowski studies, 275–346, 409, 417
Liang studies, 174, 208
Liaw studies, 187–223
Life science applications, 91–92
Ligands, 24
LIGA technology
  fundamentals, 36–37, 39–41
  MEMS fabrication methods, 37–38
  microbeam testing, 47
  Ni thin film materials, 51–60
Lindquist and Wayman studies, 161, 167–168
Liquid heat, 136–138
Lithography, 75–76
Lockheed-Martin Corporation, 6
Long range ordering (LRO), 235–236
Lorentz force microactuators, 91
Lou studies, 35–61
Low-cycle dwell fatigue (LCDF) behavior
  crystallography, 379
  microstructural/microtextural effects, 375
  time dependent strain accumulation, 376, 378
  titanium alloys, 374
Low-cycle fatigue (LCF) behavior
  cobalt alloys and composites, 209–211
  crystallography, 379
  fatigue-creep synergism, 380
  microstructural/microtextural effects, 375
  stress and strain amplitude, 332
  time dependent strain accumulation, 376, 378
  titanium alloys, 374
Low pressure chemical vapor deposition (LPCVD)
  fundamentals, 73
  silicon MEMS, 94
  silicon oxide, 74
Low-pressure plasma deposition (LPPD), 295, 300
Low temperature properties, 263. *See also*
    Temperature
LPCVD. *See* Low pressure chemical vapor
    deposition (LPCVD)

LPPD. *See* Low-pressure plasma deposition (LPPD)
LRO. *See* Long range ordering (LRO)
Lucent Technologies, 39

M

MacDonald and Arora, Chu, studies, 379
Magnetic actuation, 83
Magnetic bead cytometry, 29
Magnetic device applications, 221
Magnetic sensors, 86
Martensite-austenite transformation
  fundamentals, 147–148
  (Ni,Pd)Ti alloys, 161, 165
  Ni(Ti,Hf) alloys, 173–174
  Ni(Ti,Zr) alloys, 174–175
Materials
  metal matrix composites, 285–286
  Mo-Si-B alloys, 439–440, 450–453
  silicon-based microelectromechanical systems,
    66–71
Mathematica software, 55
Mathews, A., 432
Matrix composites, 385–391
Matrix microstructure effects, 323–325
Matthey, Johnson, 180
Maugis-Dugdale (MD) model, 19–21
Maugis studies, 20
Mazza studies, 56
McBagonluri studies, 359–396
McDowell studies, 384
MD. *See* Maugis-Dugdale (MD) model
Mechanical characterization, MEMS thin films
    materials
  applications, 38–39
  fabrication methods, 37–38
  fatigue behavior, 56–60
  fracture behavior, 55–56
  fundamentals, 35–37, 61
  indentation testing, 48–51
  LIGA technologies, 39–41, 51–60
  mechanical testing methods, 42–51
  membrane testing, 44–46
  microbeam testing, 46–48
  microelectromechanical systems, 37–42
  reliability, 41–42
  size-dependent plasticity, 51–55
  strain gradient plasticity, 52
  uniaxial mechanical testing, 43–44
Mechanical properties
  cobalt alloys and composites, 192–218
  crack growth behavior, 338–340
  discontinuous reinforcements, 309–340
  ductility, 312–318
  elasticity modulus, 309

fatigue behavior, 332–340
fracture toughness, 318–338
heat treatment effects, 323–325
matrix microstructure effects, 323–325
mixed mode toughness, 327
modulus of elasticity, 309
Mo-Si-B alloys, 459–467
notch root radium change effects, 320
particle hardened materials, 328–329
properties, 307–340
reinforcements, 325–327
silicon, 67–68
single-phase alloys, 460–461
strain amplitude, 332–338
strength, 310–312
stress amplitude, 332–338
three-phase alloys, 465, 467
titanium alloys, 365–395
toughened MMCs, 329–332
two-phase alloys, 461–465
volume fraction change effects, 320–322
Mechanical properties, discontinuous
reinforcements
crack growth behavior, 338–340
fatigue behavior, 332–340
fracture toughness, 318–338
heat treatment effects, 323–325
matrix microstructure effects, 323–325
mechanical properties, 309–340
mixed mode toughness, 327
notch root radium change effects, 320
particle hardened materials, 328–329
properties, 307–340
reinforcements, 325–327
strain amplitude, 332–338
stress amplitude, 332–338
toughened MMCs, 329–332
volume fraction change effects, 320–322
Mechanical sensors, 84–88
Mechanical testing, 93–94
Mechanical testing methods
fundamentals, 42
indentation testing, 48–51
membrane testing, 44–46
microbeam testing, 46–48
uniaxial mechanical testing, 43–44
Mechanism-based strain gradient (MSG) plasticity
theory, 18, 53
Meisner and Sivokha studies, 168
Meisner studies, 174, 177
Membrane testing, 44–46
MEMS, thin films materials mechanical
characterization
applications, 38–39
fabrication methods, 37–38

fatigue behavior, 56–60
fracture behavior, 55–56
fundamentals, 35–37, 61
indentation testing, 48–51
LIGA technology, 39–41, 51–60
mechanical testing methods, 42–51
membrane testing, 44–46
microbeam testing, 46–48
microelectromechanical systems, 37–42
reliability, 41–42
size-dependent plasticity, 51–55
strain gradient plasticity, 52
uniaxial mechanical testing, 43–44
Mendiratta studies, 458
Meng studies, 171–174
Menon studies, 428
Mercer studies, 487
Merica studies, 237
Metallic materials, 3, 8, 11. *See also specific type*
Metallurgy
aluminum and aluminum alloys, 231–234
titanium alloys, 361–365
Metal matrix composites (MMCs)
applications, 340–346
atomization, spray, 295–298
automotive applications, 342–345
co-deposition processing, 296–298
commercial products, 346
conductivity, 307–308
crack growth behavior, 338–340
density, 307–308
deposition processing, 295–296, 298–300
discontinuously reinforcements, 289–305,
307–340
dispersion strengthened, 279–280
ductility, 312–318
elasticity modulus, 309
electronic packaging, 345–346
fatigue behavior, 332–340
fiber-reinforced, 279, 284–285
forging, 302–305
fracture toughness, 318–338
fundamentals, 276–278
heat capacity, 307–308
heat treatment effects, 323–325
high-velocity oxy-fuel spraying, 301–302
hydrostatic extrusion, 305–306
low pressure plasma deposition, 300
material, 285–286
matrix microstructure effects, 323–325
mechanical properties, 309–340
mixed mode toughness, 327
modified gas welding technique, 300–301
modulus of elasticity, 309
notch root radium change effects, 320

particle hardened materials, 328–329
particulate-reinforced, 279, 282
physical properties, 307–308
powder metallurgy, 290–294
premixed, 298–299
reinforcements, 279–285, 325–327
salient characteristics, 286–289
secondary processing, 302–306
solid phase processes, 290–294
space applications, 341–342
spray processing, 294–299
strain amplitude, 332–338
strength, 310–312
stress amplitude, 332–338
toughened MMCs, 329–332
types, 278–280
volume fraction change effects, 320–322
whisker-reinforced, 279, 281–282
Meyer's hardness definition, 16
MGW. See Modified gas welding (MGW)
Microbeam bending, 47
Microbeam testing, 46–48
Microelectromechanical systems, 37–42
Microfabricated post array detectors (mPAD), 29
Microfabrication processes
　anisotropic etching, 76–77
　chemical vapor deposition, 73–74
　dry etching, 77–79
　epitaxy, 72
　etching, 76–79
　evaporation, 73
　growth, topical films, 72–74
　isotropic etching, 76
　lithography, 75–76
　oxidation, 72
　polysilicon deposition, 73–74
　silicon-based microelectromechanical systems,
　　71–79
　silicon nitride, 74
　silicon oxide, 74
　spin-on method, 74
　sputtering, 72–73
　wet etching, 79
Micromachined gyroscopes, 87–88
Microphones, 85
Micropipette suction, 28
Microplate compression, 28
Microstructure effects and role
　cracks, 257–260
　creep, 261
　discontinuous reinforcements, 323–325
　dwell behavior, 374–376
　elevated temperature properties, 261–262
　fatigue behavior, 256–261, 368–370
　fractures, 254–255

fundamentals, 251
low temperature properties, 263
tensile behavior, 252–254
Microstructures
　aeroengines, 3
　niobium alloys, 402–408
Microtensile testing, 43, 55
Micro-texture effects
　dwell behavior, 374–376
　fatigue behavior, 368–370
Microvoid coalescence (MVC) process, 316
Microwave Integrated Multifunction Assemblies
　　(MIMAs), 345
Miedema model, 129–130, 136
Mikkola, Yang and, studies, 163
Mixed mode toughness, 327
Mixing heat, 136–138
MMCs. See Metal matrix composites (MMCs)
Models
　atomic mobility, 136
　CALPHAD, 131–132
　crystallization, quasi-kinetics, 134–135
　dwell fatigue, 384–385
　fatigue, 384–385
　glass formation abilities, 131–141
　heat, 136–138
　liquid heat, 136–138
　mixing heat, 136–138
　multiple components effect, 138–141
　short range ordering, 136–138
　T_0-criterion, 132, 134
　titanium alloys, 384–385
Modified gas welding (MGW), 295, 300–301
Modulus of elasticity. See Elasticity modulus
Moelwyn-Hughes model, 136
Mohr studies, 57–58
Monel 400/500 alloy, 477–478
Monotonic compression, 108–110
MOSFET devices, 90
Mo-Si-B alloys
　alloying behavior, 448–450
　binary systems, 441
　defect structures, 441–442
　diffusion behavior, 454–455
　fundamentals, 437–439, 468
　materials, 439–440, 450–453
　mechanical properties, 459–467
　oxidation behavior, 455–459
　phase equilibria, 441–442, 453–454
　powder processing, 453
　processing strategies, materials, 450–453
　quaternary Mo-RM-Si-B systems, 448–450
　refractory metals, 455–459
　refractory metal silicides/borosilicides, 442–450,
　　455–459

RM-B systems, 441
RM-Si-B systems, 441, 443–445
RM-Si systems, 441
selection of materials, 439–440
silicide phase structures, 446–448
single-phase alloys, 460–461
solidification processing, 450–452
stability, 441–445
ternary systems, 441
thermodynamic evaluation, 453–454
three-phase alloys, 465, 467
T_2 phase, 441–445
two-phase alloys, 461–465
Mott and Nabarro studies, 237
mPAD. *See* Microfabricated post array detectors (mPAD)
MP35N alloy, 221–222
MSG. *See* Mechanism-based strain gradient (MSG) plasticity theory
Mulder studies, 174–175
Multiaxial accelerometers, 87
Multicomponent rule, 129
Multilevel deformation
foam deformation, 110–111
fundamentals, 110
strut deformation, 111–113
struts, mechanical properties, 113–116
Multilevel failure mechanisms, 118–119
Multiple components effect, 138–141
MVC. *See* Microvoid coalescence (MVC) process

N

Nabarro, Mott and, studies, 237
Nanoindentation methods, 16–19, 50. *See also* Indentation testing
Nano-systems. *See* Small scale contact and adhesion
NASA, 278, 342
NASP. *See* National aerospace plane (NASP)
Nathal, Michael, 180
National aerospace plane (NASP), 4–8
"Negative heat of mixing" rule, 129
Nelson, 43
(Ni,Au)Ti alloy, 165–166
Nickel-base alloys
balanced properties engineering, 12
fatigue, 484, 487–488
fractures, 484, 487–488
fundamentals, 475–477, 491
future trends, 490–491
Ni-Cr alloys, 479
Ni-Cu alloys, 477–479
Ni-Fe-Cr alloys, 480–481
superalloys, 481, 483
Ni-Cr alloy, 479

Ni-Cu alloy, 477–479
Nieh studies, 467
Niesel studies, 307
Ni-Fe-Cr alloy, 480–481
Niobium alloy, Nb-Al-Ti-based intermetallics, 11–12
Niobium alloy, Nb-Al-X, 406
Niobium alloy, Nb-Cr-X, 404–405
Niobium alloy, Nb-Si-Cr-Al-X, 407–408
Niobium alloy, Nb-Si-X, 403–404
Niobium alloys
applications, 431–432
compositions, 402–408
creep, 422–426
fatigue crack growth resistance, 417–422
fracture resistance, 412–417
fundamentals, 401–402, 432
microstructures, 402–408
Nb-Al-X alloys, 406
Nb-Cr-X alloys, 404–405
Nb-Si-Cr-Al-X alloys, 407–408
Nb-Si-X alloys, 403–404
oxidation properties, 427–431
physical properties, 408–411
in situ composites, 416–417, 419–420, 422
solid solution alloys, 412–419
tensile properties, 408–411
types, 402–408
(Ni,Pd)Ti alloy, 161–165
(Ni,Pt)Ti alloy, 166–170
NiTi-based high-temperature shape-memory alloys
actuators, 150, 152–154
addition effects, 157–158
applications, 151–155
constrained recovery, 149–150, 152
damping, 155
free recovery, 148–149, 151
fundamentals, 146–147, 156–160, 177–180
future challenges, 177–180
historical background, 156–157
(Ni,Au)Ti alloys, 165–166
(Ni,Pd)Ti alloys, 161–165
(Ni,Pt)Ti alloys, 166–170
Ni(Ti,Hf) alloys, 170–174
Ni(Ti,Zr) alloys, 174–177
phase structures, 158–160
processing, 160–161
shape memory, 147–150
shape-memory effect, 148–150
superelastic behavior, 147–150, 154
superelasticity, 150
ternary alloys, 156–177
wear resistance, 155
work production, 150, 152–154

Ni(Ti,Hf) alloys
  ternary alloys, 170–174
Ni(Ti,Zr) alloys, 174–177
Nix, Doerner and, studies, 16
Nix, Saha and, studies, 17
Nix, Vlassak and, studies, 45
Nix and Gao studies, 18, 52–53
Nix studies, 45
Noburnium, 406, 431
Noebe, Nichaela, 180
Noebe, R.D., studies, 145–180
North Star cube corner tip, 53
North Star indenter, 50
Notch root radium change effects, 320

O

Occurrence, cobalt alloys and composites, 191
O foams. See Annealed (O) foams
OLEDs. See Organic light emitting devices (OLEDs)
Olier studies, 171–172
Oliver and Pharr studies, 16
Oliver-Pharr approach
  indentation testing, 48, 50
  nanoindentation methods, 18
Open cell foams, 105
Optical sensors, 86
Optical techniques, 80–81
Optical traps and tweezers, 29
OR. See Orientation relationship (OR)
Orbital Research, 161
Organic electronic devices, 26–28
Organic light emitting devices (OLEDs), 26–27
Orientation relationship (OR), 72
Orowan relationship, 237, 241
Osteolysis, 395
Otsuka and Ren studies, 158, 165
Otsuka studies, 156
Overaged (OA) aluminum alloys, 241
Owens Corning, 284
Oxidation
  balanced properties engineering, 12
  Mo-Si-B alloys, 455–459
  niobium alloys, 427–431
  topical films growth, 72
Oxy-fuel spraying. See High-velocity oxy-fuel
      spraying (HVOF)

P

Padhi, D., 347
Padula, S.A, II, studies, 145–180
Paris law regime
  fatigue crack growth behavior, 339
  niobium alloys, 419

two-phase alloys, 465
Paris studies, 329
Particle hardened materials, 328–329
Particulate reinforcement
  fundamentals, 279, 282
  titanium matrix composites, 385
  whisker comparison, 283
Pattering techniques, 71
PC projectors, 38
PDMS stamps, 26
PECVD. See Plasma enhanced chemical vapor
      deposition (PECVD)
Peierls-Nabarro (P-N) barrier energy
  niobium alloys, 409–411
  single-phase alloys, 461
  in situ composites, 426
  solid solution alloys and strengthening, 413–416
Peltier effect, 69
Perepezko studies, 437–468
Persistent slip bands (PSBs), 57
Pharr, Oliver and, studies, 16, 48
Phase equilibria
  cobalt alloys and composites, 188–191
  Mo-Si-B alloys, 441–442, 453–454
Phase Rule, 189
Phase structures, ternary alloys, 158–160
Photonics applications, 90–91
Physical metallurgy
  aluminum and aluminum alloys, 231–234
  titanium alloys, 361–365
Physical properties
  cobalt alloys and composites, 192–218
  discontinuous reinforcements, 307–308
  niobium alloys, 408–411
  silicon, 66–67
Pietrement and Troyon studies, 21
Piezoelectricity, 80, 82. See also Electrical
      properties
Piezoresistivity, 84. See also Electrical properties
Pitting, 261. See also Corrosion
Plasma enhanced chemical vapor deposition
      (PECVD), 73–74
Plasticity size effects, 53
Platinum cobalt alloy, 221
PM. See Powder metallurgy (PM) and alloys
PMCs. See Polymer matrix composites (PMCs)
PMMA. See Polymethylmethacrylate (PMMA)
      x-ray photoresists
Poiseuille flow approximation, 23–24
Poisson's ratios
  cobalt, 192
  contact mechanics, 19
  ductility, 316
  elasticity modulus, 309
  indentation testing, 49–50

nanoindentation methods, 17
sensing techniques, 79
silicon, 67
solid solution alloys and strengthening, 416
titanium alloys, 362
Polycrystalline silicon, 69
Polymer matrix composites (PMCs), 287–288
Polymethylmethacrylate (PMMA) x-ray
     photoresists, 39
Polysilicon deposition, 73–74
Polytech, 43
Poon studies, 126
Porous metallic materials
     creep behavior, 120–123
     cyclic strain accumulation, 116–118
     deformation, 110–116
     fatigue behavior, 116–119
     foams, 103–105, 110–111
     four-strut unit cell model, 107–108
     Gibson-Ashby dimensional model, 106–107
     monotonic compression, 108–110
     multilevel deformation, 110–116
     multilevel failure mechanisms, 118–119
     open cell foams, 105
     stress-strain behavior, 108–110
     structure unit models, 105–108
     strut deformation, 111–113
     struts, mechanical properties, 113–116
     tetrakaidecahedron unit cell model, 107
Potentiometric devices, 90
Powder metallurgy (PM) and alloys
     advantages and barriers, 247–248
     clustering, reinforcement, 326–327
     engineered products, 268
     fundamentals, 245–247
     metallic foam fabrication, 105
     primary processing, discontinuous
          reinforcements, 290–294
     titanium matrix composites, 10
     whisker reinforcement, 282, 386–387
Powder processing, 453
Pratt & Whitney
     metal matrix composites, 278
     Mo-Si-B alloys, 438
     national aerospace plane, 6
Precipitation hardening, 236–241
Precipitation kinetics, 242
Premixed MMCs, 298–299
Pressure sensors, 84–85
Preston studies, 239
Prestresses, 28–30
Primary processing, discontinuous reinforcements
     atomization, spray, 295–298
     co-deposition processing, 296–298
     deposition processing, 295–296, 298–300

high-velocity oxy-fuel spraying, 301–302
low pressure plasma deposition, 300
modified gas welding technique, 300–301
powder metallurgy, 290–294
premixed, 298–299
solid phase processes, 290–294
spray processing, 294–299
Processing
     cobalt alloys and composites, 191
     Mo-Si-B alloys, 450–453
     ternary alloys, 160–161
Production, aluminum and aluminum alloys,
     227–228
Properties
     aluminum and aluminum alloys, 242–243
     conductivity, 307–308
     crack growth behavior, 338–340
     density, 307–308
     ductility, 312–318
     elasticity modulus, 309
     fatigue behavior, 332–340
     fracture toughness, 318–338
     heat capacity, 307–308
     heat treatment effects, 323–325
     matrix microstructure effects, 323–325
     mechanical properties, 307–340
     mixed mode toughness, 327
     modulus of elasticity, 309
     notch root radium change effects, 320
     particle hardened materials, 328–329
     physical properties, 307–308
     reinforcements, 325–327
     strain amplitude, 332–338
     strength, 310–312
     stress amplitude, 332–338
     titanium alloys, 362
     toughened MMCs, 329–332
     volume fraction change effects, 320–322
Prosthetics. *See* Human prosthetic devices
     applications
PSBs. *See* Persistent slip bands (PSBs)
Pugh studies, 306
Pu studies, 174–177
PW1480 crystals, 488

Q

Quasi-kinetics, primary crystallization, 134–135
Quaternary Mo-RM-Si-B systems, 448–450

R

Radhakrishnan, Saxena and, studies, 369
Radio frequency (RF) applications
     silicon-based microelectromechanical systems, 92

silicon dry etching, 77–78
sputtering, 72
Rapid solidification (RS), 289
Rare-earth-metals, 448
Rare-earth-permanent magnets, 221
Ravichandran, Jha and Shankar studies, 381
Ravichandran studies, 381
Rayleigh-Maxwell equation, 307
Reactive ion etching (RIE), 38, 71
Reactive ion sputtering, 73
Read studies, 43
Receptor-ligand recognition, 24
Redlick-kister-Muggianu formulae, 131
Refractory alloys, 12
Refractory metals, 455–459
Refractory metal silicides/borosilicides, 442–450,
        455–459
Reinforcements, changes, 325–327. *See also*
        Discontinuous reinforcements
Reliability, 41–42, 93–95. *See also* Stability
Ren, Otsuka and, studies, 158, 165
Residual stresses, 93–94
Resistivity techniques, 79–80
Resonant beam tests, 47
Resonant pressure sensors, 85
Resonating sensors, 85–86
Resonating techniques, 81
Reynolds' number, 91–92
Rice and Johnson studies, 328–329
RIE. *See* Reactive ion etching (RIE)
Ritchie and Lankford studies, 381
Ritchie studies, 370
RM-B systems, 441
RM-Si-B systems, 441, 443–445
RM-Si systems, 441
Rogers studies, 26
Rolls-Royce Advanced Development Company, 432
ROM. *See* Rules-of-mixtures (ROM) calculations
Rosales and Schneibel studies, 461
R-ratio, 370–371
RS. *See* Rapid solidification (RS)
Rules-of-mixtures (ROM) calculations, 288

S

Saha and Nix studies, 17
Sakidja studies, 437–468
Salas, L., 432
Salient characteristics, 286–289
Sansoz and Ghonem studies, 375, 380
SAP. *See* Sintered aluminum products (SAP)
Sastry, Soboyejo, Lederich and, studies, 386
Sastry studies, 386
Saxena and Radhakrishnan studies, 369
Scanning electron microscopy (SEM)

fatigue behavior, LIGA technologies, 57
strut deformation, 111
Schneibel, Rosales and, studies, 461
Schneibel studies, 467
Scientific Advisory Board of the United States
        Air Force, 360
Scratch drive actuators (SDAs), 82
SCREAM process, 79
SDAs. *See* Scratch drive actuators (SDAs)
Secondary processing, discontinuous reinforce-
        ments, 302–303, 305–306
Seebeck effect, 69
Sehitoglu and Garcia studies, 373
Selection of materials, 439–440
SEM. *See* Scanning electron microscopy (SEM)
Sensing techniques, 79–81
Sensors and actuators, 79–84
Shademan and Soboyejo studies, 373
Shademan studies, 372–373
Shah, Anton and, studies, 405
Shankar, Ravichandran, Jha and, studies, 381
Shape memory alloys (SMAs)
    actuation techniques, 83
    actuators, 150, 152–154
    addition effects, 157–158
    applications, 151–155
    constrained recovery, 149–150, 152
    damping, 155
    effect, 148–150
    free recovery, 148–149, 151
    fundamentals, 146–148, 156–160, 177–180
    future challenges, 177–180
    historical background, 156–157
    (Ni,Au)Ti alloys, 165–166
    (Ni,Pd)Ti alloys, 161–165
    (Ni,Pt)Ti alloys, 166–170
    Ni(Ti,Hf) alloys, 170–174
    Ni(Ti,Zr) alloys, 174–177
    phase structures, 158–160
    processing, 160–161
    shape memory, 147–150
    shape-memory effect, 148–150
    superelastic behavior, 147–150, 154
    superelasticity, 150
    ternary alloys, 156–177
    wear resistance, 155
    work production, 150, 152–154
Shape memory effect (SME), 148–150
Sharpe studies, 43, 56
Shear assay experiments, 29
Shen, Soboyejo and Soboyejo studies, 370,
        374–375, 381
Shifler, D., 468
Shiflet studies, 125–141
Shimizu studies, 161, 163

Shockley partial dislocation, 190
Short range ordering (SRO), 136–138, 235
Shrotriya studies, 47, 61
Silicide phase structures, 446–448
Silicon
    anisotropic wet etching, 76–77
    dry etching, 77–79
    electrical properties, 68–69
    fundamentals, 66
    isotropic wet etching, 76
    mechanical properties, 67–68
    physical properties, 66–67
    sensors, 85
    thermal properties, 70
Silicon-based microelectromechanical systems
        (Si-MEMS)
    actuation techniques, 81–83
    anisotropic etching, 76–77
    applications, 79–92
    automotive applications, 90
    biochemical sensors, 88–90
    chemical vapor deposition, 73–74
    conductrimetric gas sensors, 89
    dry etching, 77–79
    electrical properties, 68–69
    epitaxy, 72
    etching, 76–79
    evaporation, 73
    flow sensors, 88
    force sensors, 85–86
    fundamentals, 66, 95
    growth, topical films, 72–74
    inertial sensors, 86–88
    isotropic etching, 76
    life science applications, 91–92
    lithography, 75–76
    materials, 66–71
    mechanical properties, 67–68
    mechanical sensors, 84–88
    mechanical testing, 93–94
    microfabrication processes, 71–79
    oxidation, 72
    photonics applications, 90–91
    polysilicon deposition, 73–74
    potentiometric devices, 90
    pressure sensors, 84–85
    reliability issues, 93–95
    RF applications, 92
    sensing techniques, 79–81
    sensors and actuators, 79–84
    silicon, 66–70
    silicon carbide, 70
    silicon nitride, 70, 74
    silicon oxide, 70, 74
    smart sensors, 84

    spin-on method, 74
    sputtering, 72–73
    torque sensors, 85–86
    wet etching, 79
Silicon carbide, 70
Silicon Light Machines, 91
Silicon nitride, 74
Silicon oxide, 74
Si-MEMS. *See* Silicon-based microelectro-
        mechanical systems (Si-MEMS)
Singh, P., 347
Single-phase alloys, 460–461
Sinha and Soboyejo studies, 373
Sinha studies, 374
Sintered aluminum products (SAP), 246
Sivokha, Meisner and, studies, 168
Size-dependent plasticity
    fatigue behavior, 56–60
    fracture behavior, 55–56
    fundamentals, 51–52
    LIGA technology, 53–60
    plasticity size effects, 53–55
    strain gradient plasticity, 52
Small scale contact and adhesion
    adhesion and biological adhesion, 21–26
    cell mechanical properties, 28–30
    contact mechanics, 19–21
    fundamentals, 15–16
    nanoindentation methods, 16–19
    organic electronic devices, 26–28
    prestresses, 28–30
    stamping methods, 26–28
Smart sensors, 84
SMAs. *See* Shape memory alloys (SMAs)
SME. *See* Shape memory effect (SME)
Soboyejo, Lederich and Sastry studies, 386
Soboyejo, Shademan and, studies, 373
Soboyejo, Shen, Soboyejo and, studies, 370,
        374–375, 381
Soboyejo, Sinha and, studies, 373
Soboyejo and Soboyejo, Shen, studies, 370,
        374–375, 381
Soboyejo studies
    advanced materials, 1–13
    nickel-base alloys, 475–491
    small scale contact and adhesion, 15–30
    thin film materials, 61
    titanium alloys, 359–396
SOG. *See* Spin-on glass (SOG)
Solidification processing, 450–452
Solid phase processes, 290–294
Solid solution alloys and strengthening
    deformation, 235–236
    fatigue crack growth resistance, 417–419
    fracture resistance, 412–416

niobium alloys, 412–416
Solid state actuators. *See* Actuators
Soriano, P., 432
Space applications, 341–342
Space Shuttle applications
  aluminum-lithium alloys, 265
  cobalt alloys, 221
  metal matrix composites, 278
Spence, Evans and Cope studies, 374, 376
Spin-on glass (SOG), 74
Spin-on method, 74
Sporting goods applications, 8
Spray processing
  co-deposition, 296–298
  deposition processing, 295–296
  fundamentals, 294–295
  premixed MMCs, 298–299
Sputtering, topical films growth, 72–73
Srivatsan studies, 225–270, 275–346
SRO. *See* Short range ordering (SRO)
SSD. *See* Statistically stored dislocations (SSD)
SSSS. *See* Super-saturated solid solution (SSSS)
Stability. *See also* Reliability
  Mo-Si-B alloys, 441–445
  oxidation behavior, 455
  silicide phase structures, 447
Stamping methods, 26–28
Static beam tests, 46–47
Statistically stored dislocations (SSD), 18, 52
Stefan-Boltzman constant, 216
STELLITE alloys
  biomedical applications, 222
  corrosion-resistant alloys, 205
  fundamentals, 194, 196–197
  high-temperature alloys, 198
Stiffness, titanium matrix composites, 10
Stölken and Evans studies, 47, 54
Strain amplitude, 332–338
Strain gradient plasticity, 52
Strain measurements, 43
Strength, 10–11, 310–312. *See also* Solid solution alloys and strengthening
Stress corrosion cracking (SCC), 229–230
Stresses
  fatigue behavior, 332–338
  polysilicon, 68
  ratio, dwell behavior, 380–381
  shielding, titanium alloys, 395
  silicon MEMS, 93–94
  single crystal wafers, 68
Stress-fatigue life data, 57–58, 60
Stress-strain behavior, 108–110
Stroh-type mechanism
  fatigue crack growth, 381–382
  short crack growth, 374

stress ratio effect, 380
Structure, cobalt alloys and composites, 188–191
Structure unit models
  four-strut unit cell model, 107–108
  fundamentals, 105–106
  Gibson-Ashby dimensional model, 106–107
  tetrakaidecahedron unit cell model, 107
Struts
  mechanical properties, multilevel deformation, 113–116
  multilevel deformation, 111–113
Subramanian studies, 412, 423
Subtractive cold welding, 26
Sulfur attack, 479
Superalloys, 481, 483
Superelastic behavior
  actuators, 150
  applications, 154
  constrained recovery, 149–150
  effect, 148–150
  free recovery, 148–149
  fundamentals, 147–148
  NiTi-based high-temperature shape-memory alloys, 147–150, 154
  superelasticity, 150
  work production, 150
Superelasticity, 150
Super-saturated solid solution (SSSS), 237
Surface micromachining, 37–38
Sutton studies, 281
Suzuki studies, 163

T

Tabor's parameter, 20
Tactile sensors, 86
Taguchi sensors, 89
Tan studies, 29
Taya and Arsenault studies, 312
Taylor relationship, 52
Taylor's series, 218
Tb-Dy-Fe material, 83
T_0-criterion, 132, 134
TEM. *See* Transmission electron microscopy (TEM)
Temperature
  aeroengines, 3
  alumina, 228
  balanced properties engineering, 11–12
  chemical vapor deposition, 73
  cobalt, 192
  creep, 121
  dwell behavior, 380
  evolution, fatigue, 212–219
  fatigue, 214

fatigue behavior, 212, 371
HAYNES 25 alloy, 199
low-cycle fatigue, 210–211
microstructural/microtextural effects, 375
Mo-Si-B alloys, 438
nickel-base alloys, 475–477, 479–480, 483
(Ni,Pd)Ti alloys, 163–164
(Ni,Pt)Ti alloys, 168
NiTi alloys, 146–147, 156–157
Ni(Ti,Hf) alloys, 170–171, 173
Ni(Ti,Zr) alloys, 174–176
oxidation behavior, 458
polysilicon, 73
silicon, 69
silicon MEMS, 94
silicon nitride, 74
silicon oxide, 74
superalloys, 490–491
three-phase alloys, 467
titanium matrix composites, 10
two-phase alloys, 463
ULTIMET alloy, 210–213, 218–219
wrought cobalt-based alloys, 201
Temper designation system, 249–251
Tensegrity cell models, 29
Tensile behavior, 252–254
Tensile properties, 41, 408–411
Tenteris, Anita, 180
Terfenol-D material, 83
Ternary alloys and systems
addition effects, 157–158
fundamentals, 156–160
historical background, 156–157
historical developments, 156–157
Mo-Si-B alloys, 441
(Ni,Au)Ti alloys, 165–166
(Ni,Pd)Ti alloys, 161–165
(Ni,Pt)Ti alloys, 166–170
Ni(Ti,Hf) alloys, 170–174
Ni(Ti,Zr) alloys, 174–177
phase structures, 158–160
processing, 160–161
Tetrakaidecahedron unit cell model, 105–107
Texas Instruments, 38, 90
T6 foams. See T6-strengthened (T6) foams
Thermal actuation, 82–83
Thermal properties, silicon, 70
Thermodynamics
evaluation, 453–454
glass formation abilities, 128–141
Thermoelectricity. See Electrical properties
Thermomechanical processing (TMP), 164, 257
Thin films materials mechanical characterization,
MEMS
applications, 38–39

fabrication methods, 37–38
fatigue behavior, 56–60
fracture behavior, 55–56
fundamentals, 35–37, 61
indentation testing, 48–51
LIGA technology, 39–41, 51–60
mechanical testing methods, 42–51
membrane testing, 44–46
microbeam testing, 46–48
microelectromechanical systems, 37–42
reliability, 41–42
size-dependent plasticity, 51–55
strain gradient plasticity, 52
uniaxial mechanical testing, 43–44
Thomson effect, 69
Three-phase alloys, 465, 467
TiB whiskers, 9–10. See also Whisker
reinforcement
Time dependent strain accumulation, 376, 378
Titanium alloy, IMI 685, 379
Titanium alloy, Ti-11, 379
Titanium alloy, Ti-6242, 373, 375–376
Titanium alloy, TiAl, 6, 8, 10
Titanium alloy, Ti-6Al-4V
aeroengines, 3
biomedical applications, 391
fatigue crack growth, 373–374
fundamentals, 361
Titanium alloy, Ti-8Al-1V-1Mo, 10
Titanium alloys
alloying elements, 363
applications, 362, 391–395
automotive applications, 9–10
biomedical applications, 391–395
bone detachment, 395
classifications, 363
cracks, 373–374, 381–384
creep, 366–368, 379–380
crystallography effects, 372–373, 378
cyclotoxicity, 393–395
dwell behavior, 374–381, 384–385
dwell fatigue, 381–384
dwell hold time, 381
engineering applications, 362
environmental effects, 370, 378
Evans and Bache studies, 382–383
Evans and Gostelow studies, 382
fatigue behavior, 368–373, 379–380,
384–385
fiber reinforcement, 390–391
fracture toughness, 368
frequency effects, 371
fundamentals, 360–361, 395–396
Hack and Leverant studies, 383–384
implants, 395

intermetallic phase diagrams, 363, 365
internal hydrogen, 378
matrix composites, 385–391
mechanical properties, 365–395
metallurgy, 361–365
microstructure effects, 368–370, 374–376
micro-texture effects, 368–370, 374–376
modeling, 384–385
Mo-Si-B alloys, 442–443
osteolysis, 395
particulate reinforcement, 385
physical metallurgy, 361–365
properties, 362
R-ratio, 370–371
stress ratio, 380–381
stress shielding, 395
temperature effects, 371, 380
TiB whiskers, 9–10
time dependent strain accumulation, 376, 378
whisker reinforcement, 385–388, 390
Wojcik, Chan and Koss studies, 382
Titanium matrix composites (TMC)
    automotive applications, 10
    fiber reinforcement, 390–391
    fundamentals, 385
    particulate reinforcement, 385
    whisker reinforcement, 385–388, 390
TMC. *See* Titanium matrix composites (TMC)
TMP. *See* Thermomechanical processing (TMP)
Topical films growth, 72–74
Torque sensors, 85–86
Toughened MMCs, 329–332
Toughness tree, 255
Toyota Motor Company
    automotive applications, 9–10
    composites, 343
    gamma titanium aluminides, 7–8
    titanium matrix composites, 10
    whisker reinforcement, 388, 390
T$_2$ phase, 441–445
Trace elements, 242
Transmission electron microscopy (TEM), 487
Trantina studies, 385
Triaxial tensile stresses, 317
TRIBALOY alloys, 197–198
TriboScope, 50
Troyon, Pietrement and, studies, 21
T6-strengthened (T6) foams
    deformation, 110–111
    stress-strain behavior, 108
    strut properties, 114–116
Turnbull's empirical equation, 134
Turner studies, 307
Two-phase alloys, 461–465

U

UHMWPE. *See* Ultra-high-molecular-weight-
        polyethylene (UHMWPE)
Ultimate tensile strength (UTS), 114–116
ULTIMET alloy
    corrosion-resistant alloys, 205, 207
    high-cycle fatigue behavior, 208
    low-cycle fatigue, 210
    temperature, 214, 218–219
Ultra-high-molecular-weight-polyethylene
        (UHMWPE), 395
Ultra-high temperature applications. *See* Mo-Si-B
        alloys
UMCO-50 alloy, 221
Uniaxial mechanical testing, 43–44
Unislide drive, 43
UTS. *See* Ultimate tensile strength (UTS)

V

van der Waals adhesion, 27–28
Van Vucht, Donkersloot and, studies, 165–166
Vasudevan, Satish, 225–270, 347
Velmex, 43
Vlassak, Xiang and, studies, 45
Vlassak and Nix studies, 45
Volume fraction change effects, 320–322

W

Walston studies, 430
Wang studies, 107, 126
Warren-Cowley parameters, 138
Waspalloy, 487
Watkins, DeLuca and, studies, 488
Wayman, Lindquist and, studies, 161, 167–168
Wayman, Wu and, studies, 166
WC-Co alloys, 222
Wear resistance
    applications, 155
    cobalt alloys and composites, 194, 196–198
    human prosthetic devices applications, 8
    titanium matrix composites, 10
Weldalite 049 alloy, 265
Wet etching, 71, 76–77, 79
Wetting theory, 22–23
Whisker reinforcement
    automotive applications, 9–10
    fundamentals, 279, 281–282
    particulate comparison, 283
    titanium matrix composites, 385–388, 390
Widmanstatten colonies, 368–370, 372, 378
Wiebe, Harold, 96
Wigner-Seiz cells, 129
Wilm, Alfred, 237

Wojcik, Chan and Koss studies, 370, 372, 378, 382
Wojcik studies, 431
Woodfield studies, 376
Work production, 150, 152–154
Wrought aluminum alloys, 243–244, 263
Wrought cobalt-based alloys, 201
Wrought non-heat-treatable aluminum alloys, 244–245
Wrought processing techniques, 10
Wu, Hsieh and, studies, 175
Wu and Wayman studies, 166

X

Xiang and Vlassak studies, 45
X-33 vehicle, 5–6

Y

Yang and Mikkola studies, 163
YF-22 airplane, 6
Yoder, Cooley and Crooker studies, 368
Yoder studies, 373
Yoshimi studies, 458, 460, 462
Young's equation, 23
Young's moduli. *See also* Elasticity modulus

biomedical titanium alloys, 395
cobalt, 192
contact mechanics, 19
fracture, LIGA technologies, 55–56
fracture toughness, 329
fundamentals, 309
Gibson-Ashby model, 106
indentation testing, 49–50
LIGA technology, 41
MEMS mechanical testing, 93
microbeam testing, 47
silicon, 67–68
solid solution alloys and strengthening, 416
stamping methods, 27
tetrakaidecahedron unit cell model, 107
thin films materials, 36
titanium alloys, 362

Z

Zhu studies, 125–141, 171
Zirconium
    cyclotoxicity, 394–395
    Mo-Si-B alloys, 442–443
    Ni(Ti,Zr) alloys, 174–177